CIVIL ENGINEERING AND ENGINEERING MECHANICS SERIES

N. M. NEWMARK AND W. J. HALL, EDITORS

PRENTICE-HALL INTERNATIONAL, INC., *London*
PRENTICE-HALL OF AUSTRALIA, PTY. LTD., *Sydney*
PRENTICE-HALL OF CANADA, LTD., *Toronto*
PRENTICE-HALL OF INDIA PRIVATE LIMITED, *New Delhi*
PRENTICE-HALL OF JAPAN, INC., *Tokyo*

ADVANCED STRENGTH OF MATERIALS

ADVANCED STRENGTH OF MATERIALS

Enrico Volterra
Professor of
Engineering Mechanics
The University of Texas
at Austin

J. H. Gaines
Associate Professor of
Aerospace Engineering
The University of Texas
at Arlington

PRENTICE-HALL, INC., Englewood Cliffs, N. J.

© 1971 by Prentice-Hall, Inc., Englewood Cliffs, N. J.
All rights reserved. No part of this book may be
reproduced in any form or by any means without permission
in writing from the publisher.

Current printing (last digit):

10 9 8 7 6 5 4

13-013854-1

Library of Congress Catalog Card Number: 70-132040

Printed in the United States of America

To

SIR GEOFFREY TAYLOR, O. M., F. R. S.

this book is dedicated

PREFACE

Under the general title *Advanced Strength of Materials* we discuss subjects which will be of interest to senior and graduate students of various engineering disciplines—including Aerospace Engineering, Civil Engineering, Engineering Mechanics, and Mechanical Engineering—as well as to researchers and practicing engineers. The first four chapters of the book cover subjects generally discussed in senior and graduate courses on the Mathematical Theory of Elasticity; while the remaining four chapters, also at the senior-graduate level, cover subjects in Strength of Materials.

Although it is our purpose to offer a balanced presentation of advanced subjects in the field of Strength of Materials, some special topics have been emphasized, as in Chapters Six and Seven where the bending of curved beams and the bending of curved beams resting on elastic foundations are discussed. The principal reason for this emphasis is that much of this material is not readily available; and the authors, who have made original contributions in these areas, have collected and brought together the results of several papers published over a number of years in various technical journals. Since most of the work in these areas provides answers to practical engineering problems (for example, those which arise in the case of foundations for water towers), it is believed that this part of the book will be of particular value to practicing engineers; therefore, tables are included. In the preparation of these numerical tables we gratefully acknowledge the kind assistance received from the "Istituto per le Applicazioni del Calcolo" of the Italian Consiglio Nazionale delle Ricerche and from the Computation Center of The University of Texas at Austin. We take this opportunity to thank the respective Directors, Professors Mauro Picone and Aldo Ghizzetti and Professor David M. Young.

We also acknowledge with thanks the help received in the preparation of the numerical results by Dr. T. C. Chang of the Department of Aerospace Engineering and Engineering Mechanics of The University of Texas at Austin and Dr. N. Al-Rashid of the Department of Civil Engineering of The University of Texas at Austin.

The mathematical derivations are presented in an elementary form without neglecting the necessary rigor. They do not necessitate greater mathematical knowledge than that which is generally required in engineering schools. All material dealing with the experimental aspects of the various topics discussed has been deliberately omitted, the principal object of the book being to provide students and practicing engineers with the fundamental bases on which the theories rest and to prepare them for more advanced study on these subjects. Various typical examples of the application of basic concepts to the solution of problems having practical importance are included.

Historical aspects receive consideration in the brief introduction to each chapter. The introduction also informs the reader of various aspects of the most important subjects discussed in the chapter. Numerous references to original sources are given at the end of each chapter.

Brief biographical sketches and footnotes are included concerning most of the eminent, deceased authors mentioned in the book. Reproductions of many of these authors have been added to enhance the interest in the biographical material.

The authors are grateful to the various persons and organizations that gave assistance in providing these reproductions: those of Professor A. E. H. Love and of Sir Charles Inglis were furnished by the Royal Society of London; that of Professor Morera by Professor Carlo Ferrari, President of the Accademia delle Scienze di Torino; that of Carlo Alberto Castigliano by Dottore Ing. Arturo Gaj, President of the Ordine degli Ingegneri of the Province of Asti where Castigliano was born; and that of Generale Menabrea by Professor Francesco Giacomo Tricomi of the University of Turin. All of the other reproductions are from the private collection of the late Professor Vito Volterra with the permission of the late Signora Virginia Volterra, wife of Vito Volterra.

The principal features of the present book are:

1. A general treatment of energy principles and their applications
2. A comprehensive yet introductory approach to two- and three-dimensional elasticity, with many numerical applications to problems of practical importance in civil and mechanical engineering
3. A thorough discussion of the theory of the bending of straight beams both in the elastic and in the plastic stages
4. An extensive treatment of the deflections of curved beams with circular axes

5. An ample treatment of beams on elastic foundations, including considerations of the cases of straight beams of infinite and finite lengths as well as circular beams
6. An extensive discussion of the bending of plates with many practical applications included

We are greatly indebted to the London Institution of Civil Engineers for their permission to reproduce graphs and drawings based on experiments with metallic beams bent beyond the elastic limit. These experiments were performed by Enrico Volterra thirty years ago in the Engineering Laboratory of Cambridge University at the suggestion of the late Sir Charles Inglis. The results of these early experiments were first published in the Journal of the London Institution of Civil Engineers.

We express our gratitude to friends and colleagues and particularly to Professor Bernard Budiansky of Harvard University and to Dr. Lloyd H. Donnell, for their advice during the preparation of the manuscript. We are very grateful to Mr. Francis T. Schmaus, M. A., M.A. in L.S., Engineering Librarian at the University of Texas at Austin, for his kind assistance in research of a bibliographical nature.

Finally, we wish to thank the following for their kind permission to reproduce material which appeared in their publications: the Royal Society of London, Accademia Nazionale dei Lincei, Accademia delle Scienze di Torino, the Italian Ministry of Public Works, London Institution of Civil Engineers, Ordine degli Ingegneri della Provincia di Asti, American Society of Mechanical Engineers, American Society of Civil Engineers, and the British Association for the Advancement of Science.

ENRICO VOLTERRA
J. H. GAINES

CONTENTS

1 STRESS AND STRAIN   1

1.0 Introduction, 1   **1.1** Stress, 6   **1.2** Strain and Strain Displacement Relations (Cauchy's Equations), 9   **1.3** Relationships between Stress and Strain (Generalized Hooke's Law), 12   **1.4** Equations of Equilibrium (Navier's Equations, Lamé's Equations), 19   **1.5** Compatibility Equations (Saint-Venant's Equations, Beltrami-Michell's Equations), 24   **1.6** Boundary Conditions, 27   **1.7** Thermal Stress Equations (Duhamel-Neumann's Equations), 29   **1.8** Saint-Venant's Principle, 31   **1.9** The Three Basic Problems of the Theory of Elasticity, 33   Bibliography, 33   Problems, 41

2 ENERGY PRINCIPLES AND GENERAL THEOREMS   44

2.0 Introduction, 44   **2.1** Strain Energy, 45   **2.2** Huber-Von Mises-Hencky Strength Theory, 49   **2.3** Other Strength Theories, 52   **2.4** Principle of Virtual Work, 54   **2.5** Uniqueness of Solution, 60   **2.6** Betti's and Maxwell's Reciprocity Theorems, 62   **2.7** Clapeyron's Theorem, 70   **2.8** Castigliano's First Theorem, 73   **2.9** Extension by Donati of Castigliano's Theorem, 80   **2.10** Engesser's Theorem on the Principle of Complementary Energy, 81   **2.11** Castigliano's Second Theorem or Menabrea's Theorem, 83   Bibliography, 87   Problems, 96

3 TWO-DIMENSIONAL ELASTICITY   100

3.0 Introduction, 100   **3.1** Plane Stress, 102   **3.2** Mohr's Circle for Stress, 104   **3.3** Plane Strain, 110   **3.4** Mohr's Circle for Strain and Strain Rosettes, 113   **3.5** Airy's Stress Function, 120   **3.6** Solution of Two-Dimensional Problems by the Use of Polynomials, 121   **3.7** Triangular and Rectangular Walls Subjected to Hydrostatic Pressures (Maurice Levy's Problems), 133   **3.8** Use of Polar Coordinates, 134

3.9 Bending of a Circular Bar (Golovin-Ribière Problem), 140   **3.10** Thick Tube Subjected to External and Internal Uniformly Distributed Pressures (Lamé's Problem), 145   **3.11** Shrink Fits, 150   **3.12** Rotating Disks and Cylinders, 153   **3.13** Stress Concentration due to a Circular Hole in a Stressed Plate (Kirsch's Problem), 159   **3.14** Concentrated Load Acting on the Vertex of a Wedge (Michell's Problem), 163   **3.15** Concentrated Load Acting on the Free Surface of a Plate (Flamant's Problem), 169   **3.16** Moment Acting on the Vertex of a Wedge (Inglis' Problem), 175   **3.17** Disk Subjected to Two Opposite Concentrated Forces (Hertz's Problem), 177   **3.18** Two-Dimensional Thermal Stresses, 186   **Appendix 3.A** Expression of the Laplacian Operator in Polar Coordinates, 191   **Appendix 3.B** Integration of Euler's Equation, 192   Bibliography, 194   Problems, 200

4 ELEMENTARY PROBLEMS IN THREE-DIMENSIONAL ELASTICITY   204

4.0 Introduction, 204   **4.1** Lamé's Solution of the Problem of the Thick Spherical Shell under Uniform Internal and External Pressures, 205   **4.2** Pure Bending of a Prismatic Bar, 212   **4.3** Coulomb's Theory of Torsion of a Circular Shaft, 217   **4.4** Navier's Theory of Torsion, 219   **4.5** Saint-Venant's Semi-Inverse Method for Solving Torsion Problems, 223   **4.6** Prandtl's Theory of Torsion, 235   **4.7** Prandtl's Membrane Analogy, 243   **4.8** Ritz's Method Applied to Torsion Problems, 248   Bibliography, 251   Problems, 256

5 BENDING OF STRAIGHT BEAMS   257

5.0 Introduction, 257   **5.1** Differential Equations of Equilibrium; Navier's Flexure Formula; Jourawsky's Shearing Stress Formula, 258   **5.2** Differential Equations for Deflections of Elastic Beams According to the Bernoulli-Euler Theory, 269   **5.3** Solution of Beam Deflection Problems by Direct Integration, 270   **5.4** Macaulay's Use of Singularity Functions for Studying Deflections of Beams, 271   **5.5** Use of Singularity Functions in the Case of Beams of Variable Cross Sections, 286   **5.6** Use of Taylor's and Maclaurin's Series for Studying Deflections of Beams, 291   **5.7** General Deflection Equation for Beams of Uniform Cross Section, 295   **5.8** Mohr's Conjugate Beam Method, 300   **5.9** Clapeyron's Three-Moment Equation for Solving Continuous Beams, 304   **5.10** Application of Trigonometric Series to the Study of Deflections of Beams, 311   **5.11** Elastic-Plastic Bending of Beams, 314   **5.12** Limit or Plastic Design of Beams, 324   **Appendix 5.A** Fourier Series Expansions, 327   Bibliography, 334   Problems, 341

6 BENDING OF A CURVED BEAM OUT OF ITS INITIAL PLANE   353

6.0 Introduction, 353   **6.1** Bending of a Curved Beam Out of Its Plane of Initial Curvature: Saint-Venant's Equations and Equations of

Equilibrium, 355 **6.2** *Bending of a Circular Beam Subjected to a Uniformly Distributed Load and Supported Symmetrically, 360* **6.3** *Bending of a Circular Arc Bow Girder Subjected to a Uniformly Distributed Load, 363* **6.4** *Bending of a Circular Arc Bow Girder Subjected to a Concentrated Load, 368 Bibliography, 376 Problems, 377*

7 BEAMS ON ELASTIC FOUNDATIONS 379

7.0 *Introduction, 379* **7.1** *The Equation of the Elastic Line for the Straight Bar According to the Winkler Hypothesis, 382* **7.2** *The Beam of Infinite Length, 384* **7.3** *The Beam of Finite Length, 390* **7.4** *The General Problem of the Straight Beam on an Elastic Foundation, 396* **7.5** *Bending of a Circular Beam Resting on an Elastic Foundation, 401* **7.6** *Bending of a Constrained Circular Beam Resting on an Elastic Foundation, 409* **7.7** *Deflections of Circular Beams Resting on Elastic Foundations Obtained by the Method of Harmonic Analysis, 417 Bibliography, 424 Problems, 426*

8 BENDING OF PLATES 429

8.0 *Introduction, 429* **8.1** *Derivation of the Lagrange Equilibrium Equation for a Thin Plate, 430* **8.2** *Deflection of a Circular Plate, 435* **8.3** *Bending of an Elliptic Plate Built-In at the Edge and Subjected to a Uniformly Distributed Load, 449* **8.4** *Navier's Solution for Simply Supported Rectangular Plates, 452* **8.5** *Maurice Levy's Solution for the Rectangular Plate, 460* **8.6** *Nadai's Solution for the Plate in the Form of an Isosceles Right Triangle with Simply Supported Edges, 464* **8.7** *Woinowsky-Krieger's Solution for the Plate in the Form of an Equilateral Triangle with Simply Supported Edges and Subjected to a Uniformly Distributed Load of Intensity p_0 466* **8.8** *Application of the Principle of Virtual Work to Obtain Rigorous and Approximate Solutions of the Problem of the Bending of Rectangular Plates, 467* **8.9** *Ritz's Method Applied to the Bending of Plates, 472* **8.10** *Application of Finite Difference Equations to the Bending of Plates, 476* **8.11** *Grashof's Method, 488* **8.12** *Bending of Plates on Elastic Foundations, 492 Bibliography, 499 Problems, 502*

AUTHOR INDEX 507

SUBJECT INDEX 515

LIST OF SYMBOLS

Roman Letter Symbols

	a	length, constant
	a_n	Fourier's coefficients of $\cos n\pi x/l$
	A	area, area of cross section of a beam, constant
	b	length, breadth of the web of a beam, constant
	b_n	Fourier's coefficients of $\sin n\pi x/l$
	B	breadth of a beam, constant
	c	curvature, constant
	c_{elastic}	elastic curvature
	c_{plastic}	plastic curvature
	c_{total}	$c_{\text{elastic}} + c_{\text{plastic}}$
	C	complementary energy, constant
	C_i	constants
	d	diameter, constant
$D = \dfrac{EH^3}{12(1-\nu^2)}$		flexural rigidity of a plate, constant, diameter
	e	volume expansion
	E	Young's modulus, constant
	E_c	Young's modulus for concrete
	E_s	Young's modulus for steel
	F	force in general
	G	center of gravity
$G = \dfrac{E}{2(1+\nu)}$		rigidity modulus
	h	height, height of the web of a beam, thickness of plate
	H	height of a beam
	I_x, I_y, I_z	moments of inertia of area about x, y, z axes
	I	moment of inertia of the cross section of a beam about a principal axis
	J	polar moment of inertia of an area

LIST OF SYMBOLS

	k	correction factor, foundation modulus
$k = \lambda + \dfrac{2G}{3}$		compressibility of a fluid
	l	length
	l, m, n	direction cosines
	M	bending moment
	M_A	couple
	M_{elastic}	maximum elastic bending moment
	M_{plastic}	maximum plastic bending moment
	M_T	torque
	n	an integer
	N	normal force
	p	lateral distributed force
	p_0	uniformly distributed load
	P	concentrated force
	Q	first moment of area about a principal axis
	r	radius of circle, radius of gyration
	r_2	external radius ⎫ of hollow disk
	r_1	internal radius ⎭
	R	radius of curvature
	R	second modulus of proportionality for linearly hardening materials
	R_x, R_y, R_z	reactive forces in the x, y, z directions
	s	intensity of a uniform stress in a bar
	u, v, w	components of the elastic displacement in the x, y, z directions
	t	thickness of a shell, of a hollow cylinder
	T	temperature
	U	strain energy
	V	transverse shear force, volume
	w	weight
$w = \dfrac{W_{\text{plastic}}}{W_{\text{elastic}}}$		coefficient of the plasticity of a cross section of a beam
	W	strain energy per unit volume, weight
	W_{elastic}	elastic section modulus
	W_{plastic}	plastic section modulus
	x, y, z	cartesian coordinates
	X, Y, Z	body forces in the x, y, z directions
	$\bar{X}, \bar{Y}, \bar{Z}$	surface forces in the x, y, z directions

Greek Letter Symbols

α	temperature coefficient of linear expansion, angle in general
α_{ij}	influence coefficient
β	angle in general
γ	angle in general
$\gamma_{xy}, \gamma_{yz}, \gamma_{zx}$	shear strains in cartesian coordinates
$\gamma_{r\theta}, \gamma_{z\theta}, \gamma_{zr}$	shear strains in cylindrical coordinates

LIST OF SYMBOLS

	δ	displacement
$\Delta = \left(\dfrac{\partial^2}{\partial x^2} + \dfrac{\partial^2}{\partial y^2} + \dfrac{\partial^2}{\partial z^2}\right)$		Laplacian operator
	$\epsilon_x, \epsilon_y, \epsilon_z$	normal strains in cartesian coordinates
	$\epsilon_r, \epsilon_\theta, \epsilon_z$	normal strains in cylindrical coordinates
	θ	angle in general, slope of the elastic curve of a bent beam
	Θ	$\sigma_x + \sigma_y + \sigma_z = \sigma_r + \sigma_\theta + \sigma_z$
$\lambda = \dfrac{\nu E}{(1+\nu)(1-2\nu)}$		Lamé's constant
	ν	Poisson's ratio
	ψ	Prandtl's stress function
	π	ratio of circumference of a circle to its diameter
	ρ	mass density, distance of a generic point of a circular cross section from the center
	$\sigma_x, \sigma_y, \sigma_z$	normal stresses in cartesian coordinates
	$\sigma_r, \sigma_\theta, \sigma_z$	normal stresses in cylindrical coordinates
	σ_m	ultimate strength
	σ_p	limit of proportionality
	σ_y	yield stress
	σ_y'	upper yield stress
	σ_y^*	lower yield stress
	σ_{yc}	yield stress in compression
	σ_{yt}	yield stress in tension
	$\tau_{xy}, \tau_{yz}, \tau_{zx}$	shear stresses in cartesian coordinates
	$\tau_{r\theta}, \tau_{\theta z}, \tau_{zr}$	shear stresses in cylindrical coordinates
	ϕ	angle in general
	Φ	Airy's stress function, warping function, angle in general
	χ	shear coefficient
	ω	angular velocity
	Ω	cross section area

The letter symbols conform in general with those approved by the American Standards Association.

CHAPTER

1

STRESS AND STRAIN

1.0 Introduction

In this first chapter the fundamentals of the theory of elasticity are discussed. The following assumptions are made in the analysis:

1. The deformations are assumed to be infinitesimal
2. The body is assumed to be elastic
3. The body is assumed to be homogeneous (i.e., the properties are the same at each point) and isotropic (i.e., the properties are the same in all directions).

Under assumptions (2) and (3) it can be shown that the elastic characteristics of the body can be specified by only two constants which are represented by the capital letter E called *Young's modulus* and the Greek letter v called *Poisson's ratio*. Figure 1.0.1 shows the so-called *tensile test diagram* for a bar of constant cross section. This diagram demonstrates the relationship between the tensile stress σ_x developed in the bar and its unit elongation ϵ_x. The diagram is similar to those obtained for ductile materials such as low carbon steel. The initial portion OA of this diagram is straight; therefore, the stress is proportional to the unit elongation, i.e.,

$$\sigma_x = E\epsilon_x \qquad (1.0.1)$$

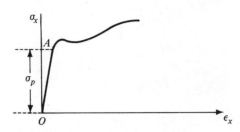

Figure 1.0.1

Point A of the diagram in Fig. 1.0.1 corresponding to the stress σ_p, called the *proportional limit*, terminates the straight part of the tensile test diagram where Eq. (1.0.1) is valid. If the stress is below the proportional limit then the longitudinal elongation of the bar ϵ_x is accompanied by equal lateral contractions ϵ_y and ϵ_z which are proportional to the elongation ϵ_x, i.e.,

$$\epsilon_y = \epsilon_z = -\nu\epsilon_x \qquad (1.0.2)$$

The factor of proportionality ν represents a constant for each material and is called *Poisson's ratio*. The reciprocal of Poisson's ratio $m = 1/\nu$ is called *Poisson's number*. Hooke's simple law is formulated by the two Eqs. (1.0.1) and (1.0.2).

The elastic body of which we are studying the deformations will in general be referred to a system of orthogonal cartesian coordinates $Oxyz$. The forces acting on the body will be distinguished as follows:

1. *External forces*
2. *Internal forces*

The external forces are:

(a) *Surface forces* which are forces distributed over the surface of the body like atmospheric pressure, hydraulic pressure or pressure exerted by one body on another. Their components on the x, y, and z axes of reference will be represented by the following notation: $\bar{X}, \bar{Y}, \bar{Z}$.

(b) *Body forces* which are forces distributed over the volume of the body like gravitational forces, magnetic forces, inertia forces, etc. Their components will be represented by the following notation: X, Y, and Z.

Young, Thomas. English physicist, physician, and Egyptologist (b. Milverton, Somerset 1773; d. London 1829). He was a physician in London, professor of physics at the Royal Institution, and Foreign Secretary of the Royal Society. He was the discoverer of the principle of interference of light, and introduced what is now known as "Young's modulus." He was among those who were first able to decipher Egyptian hieroglyphic inscriptions. [Reproduction from the Vito Volterra collection, Villa Volterra, Ariccia (Rome).]

Poisson, Simon-Denis. French mathematician (b. Pithiviers 1781; d. Paris 1840). He was professor of analysis and mechanics at the Ecole Polytechnique and at the Sorbonne. [Reproduction from the Vito Volterra collection, Villa Volterra, Ariccia (Rome).]

Hooke, Robert. English physicist (b. Freshwater, Isle of Wight 1635; d. London 1693). He was curator of experiments for the Royal Society, to which he was elected a Fellow in 1663, and was also its Secretary. He was professor of geometry at Gresham College. [Reproduction from the Vito Volterra collection, Villa Volterra, Ariccia (Rome).]

Cauchy, Augustin-Louis. French mathematician (b. Paris 1789; d. Sceaux, Seine, 1857). He was professor at the Ecole Polytechnique, the Sorbonne, and at the Collège de France. His fundamental contributions were in analysis, mechanics, mathematical theory of elasticity, and optics. His complete works, in twenty-eight volumes, were published by the French Academy of Sciences. [Reproduction from the Vito Volterra collection, Villa Volterra, Ariccia (Rome).]

Navier, Louis-Marie-Henri. French civil engineer and mechanicist (b. Dijon 1785; d. Paris 1836). He designed bridges in France and also in Rome, Italy, and was professor of applied mechanics at the Ecole des Ponts et Chaussées in Paris.

Lamé, Gabriel. French mathematician (b. Tours 1795; d. Paris 1870). He was professor of physics at the Ecole Polytechnique and as professor at the Sorbonne taught the theory of probability. His "Leçons sur la Théorie Mathématique de l'Elasticité des Corps Solides," the first book on the mathematical theory of elasticity, was published in 1852. [Reproduction from the Vito Volterra collection, Villa Volterra, Ariccia (Rome).]

Saint-Venant, Adhemar-Jean-Claude, Barré de. French civil engineer and mathematician (b. Filliers-en-Brie, Seine et Marne 1797; d. Saint-Ouen, Loire 1886). He was a civil engineer and professor at the Institut Agronomique at Versailles. He did work in hydrodynamics, but his most important contributions were on the mathematical theory of elasticity. His two best known works, "Sur la Torsion des Prismes" and "Mémoire sur la Flexion des Prismes," were published, respectively, in 1855 and 1856. [Reproduction from the Vito Volterra collection, Villa Volterra, Ariccia (Rome).]

Beltrami, Eugenio. Italian mathematician (b. Cremona 1835; d. Rome 1900). He was successively professor of mechanics and mathematical physics at the Universities of Pisa, Bologna, Pavia, and Rome. His important contributions were in the fields of geometry, hydrodynamics, potential theory, mathematical theory of elasticity, electricity and magnetism, and heat transfer. His complete works, in four volumes, were published during the years 1902–1920. [Reproduction from the Vito Volterra collection, Villa Volterra, Ariccia (Rome).]

Michell, John-Henry. Australian mathematician of English descent (b. Maldon, Victoria 1863; d. Melbourne 1943). After attending Melbourne University, he studied at Cambridge University, where in 1890 he became a Fellow of Trinity College. After his return to Australia, he held the professorship of mathematics at Melbourne University. His collected papers were published in 1964 together with those of his brother, Anthony-George-Maldon Michell (1870–1959) an eminent mechanical and hydraulic engineer. [Reproduction from the Vito Volterra collection, Villa Volterra, Ariccia (Rome).]

Duhamel, Jean-Marie-Constant. French mathematician (b. Saint-Malo 1797; d. Paris 1872). He was professor of analysis and mechanics at the Ecole Polytechnique, and at the University of Paris.

Neumann, Franz-Ernst. German mineralogist, physicist, and mathematician (b. Joachimstal 1798; d. Königsberg 1895). He was professor of mineralogy and physics at Königsberg.

The internal forces or reactive forces will be represented by the capital letter R and their components by the letters R_x, R_y, and R_z.

In the present chapter, stress is first defined. Afterwards strain is defined and then relationships between the components of elastic displacement and the components of strain are developed. These relationships were obtained by the French mathematician and mechanicist A. L. Cauchy. The relationship between stress and strain are then derived in generalized form (generalized Hooke's law). Next, the equations of equilibrium, Navier's equations and Lamés equations, are formulated. The compatibility equations are then derived in the forms given by Saint-Venant, Beltrami, and Michell. Afterwards, the boundary conditions are discussed. Following this, the thermal stress equations, derived independently by Duhamel and Neumann, are given and Saint-Venant's principle in the simple form as presented by Saint-Venant is shown.

The chapter ends with a discussion of the three basic problems in the theory of elasticity. Throughout the book, unless differently specified, the elastic bodies are assumed to be simply connected bodies.

1.1 Stress

Consider the body shown in Fig. 1.1.1 which is in equilibrium under the action of the system of external forces $P_1, P_2, \ldots, P_n$. Suppose now that the body is cut in two parts 1 and 2 by a plane. In order to keep the two separated parts of the body in equilibrium, one has to apply to parts 1 and 2 the forces $+R$ and $-R$, respectively, which are the resultant of the internal forces acting on the plane section of area A. Consider now an element of area ΔA of this section as shown in Fig. 1.1.2. The resultant internal force of intensity ΔR which acts on the area ΔA can be decomposed in two components: one ΔR_n acting in the direction of the normal to ΔA and the other ΔR_t acting in the plane of ΔA. Consider point P at the center of ΔA. The normal component of stress at point P is defined as

$$\sigma = \lim_{\Delta A \to 0} \frac{\Delta R_n}{\Delta A} \tag{1.1.1}$$

while the shear component of stress at the point is defined as

$$\tau = \lim_{\Delta A \to 0} \frac{\Delta R_t}{\Delta A} \tag{1.1.2}$$

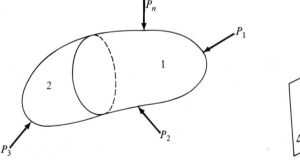

Figure 1.1.1

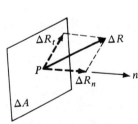

Figure 1.1.2

The normal stresses will always be denoted by the Greek letter σ and the shear stresses by the Greek letter τ. Normal stresses will always be considered positive if they are tension, negative if they are compression. The stress has dimensions force/length^2 and has a tensor character, being defined by its magnitude, its direction, and its position.

Consider now the small cube $OABCDEFG$ (see Fig. 1.1.3) with sides of lengths dx, dy, and dz. The outward normal to the face $ABEF$ is directed in the positive direction of the x-axis and face $ABEF$ is considered as positive. Also, faces $BCDE$ and $DEFG$ are considered as positive since the outward normals to these faces are directed, respectively, in the positive directions of the y- and z-axes. The other three remaining faces are considered as negative. On the face $ABEF$ are applied a normal stress σ_x (the letter σ indicates that it is a normal stress while subscript x indicates that this normal stress is applied on a face of the cube perpendicular to the x-axis) and the shear stresses τ_{xy} and τ_{xz} [letter τ indicates that the stress is a shear stress and the first subscript x indicates that the stress is applied on a face of the cube perpendicular to the x-axis while the second subscript y (or z) indicates that the shear stress component is directed in the y (or z) direction]. On the face $OCDG$, the directions of the stresses are as shown in Fig. 1.1.3. These three stresses σ_x, τ_{xy}, and τ_{xz} are positive (since the area $OCDG$ is negative and the forces $\sigma_x dydz$, $\tau_{xy} dydz$, and $\tau_{xz} dydz$ are directed in the negative directions of the x-, y-, and z-axes, the stresses

$$\sigma_x = \frac{-\sigma_x dydz}{-dydz} \qquad \tau_{xy} = \frac{-\tau_{xy} dydz}{-dydz} \quad \text{and} \quad \tau_{xz} = \frac{-\tau_{xz} dydz}{-dydz}$$

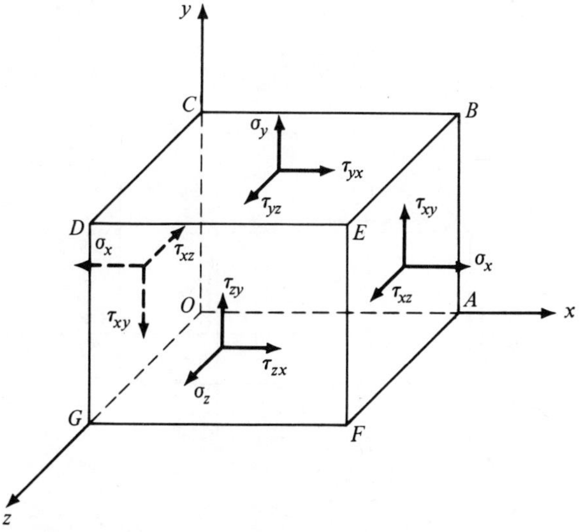

Figure 1.1.3

are positive quantities). This agrees in the case of normal stresses with our convention that stresses are positive if tension stresses, negative if compression. The stress tensor is therefore represented by the following nine quantities:

$$\begin{vmatrix} \sigma_x & \tau_{xy} & \tau_{xz} \\ \tau_{yx} & \sigma_y & \tau_{yz} \\ \tau_{zx} & \tau_{zy} & \sigma_z \end{vmatrix} \quad (1.1.3)$$

However, the following *laws of reciprocity for shearing stresses* are valid

$$\tau_{xy} = \tau_{yx}$$
$$\tau_{yz} = \tau_{zy} \quad (1.1.4)$$
$$\tau_{zx} = \tau_{xz}$$

and therefore the stress tensor is symmetrical and the nine quantities in Eq. (1.1.3) reduce to six independent quantities $\sigma_x, \sigma_y, \sigma_z, \tau_{xy}, \tau_{yz}$, and τ_{zx}. In order to prove Eqs. (1.1.4), consider the small cube of sides dx, dy, and dz as shown on Fig. 1.1.4. On its faces perpendicular to the x- and y-axes, the shear stresses τ_{xy} and τ_{yx} are acting. By expressing the condition of equilibrium, $\sum M_z = 0$, one obtains

$$(\tau_{xy}dydz)dx - (\tau_{yx}dxdz)dy = 0$$

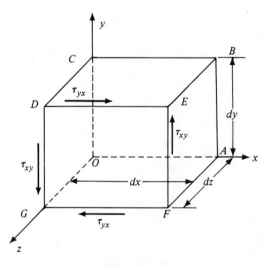

Figure 1.1.4

and it follows that

$$\tau_{xy} = \tau_{yx}$$

Similarly, the two other Eqs. (1.1.4) can be proved.

1.2 Strain and Strain Displacement Relations (Cauchy's Equations)

The small displacements of points in a deformed body will be resolved into components u, v, and w parallel to the x-, y-, and z-axes, respectively. Consider inside an unstrained elastic solid a point P of coordinates x, y, and z and the points P_1, P_2, and P_3 of coordinates $(x + dx, y, z)$, $(x, y + dy, z)$, and $(x, y, z + dz)$ (see Fig. 1.2.1). The coordinates of the four points before and after deformation are given in Table 1.2.1. The stretching of PP_1 in the x-direction is given by

$$\left(x + u + dx + \frac{\partial u}{\partial x}dx\right) - (x + u + dx) = \frac{\partial u}{\partial x}dx$$

and therefore the unit stretching in the x-direction is

$$\epsilon_x = \frac{\partial u}{\partial x}$$

In the same way one sees that the unit stretching in the y- and z-directions

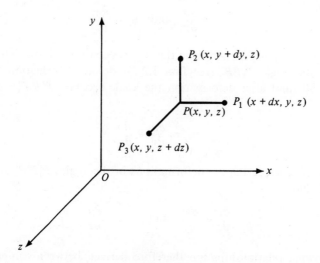

Figure 1.2.1

TABLE 1.2.1

Before Deformation		After Deformation	
Point	Coordinates	Point	Coordinates
P	x y z	P'	$x + u$ $y + v$ $z + w$
P_1	$x + dx$ y z	P_1'	$x + u + dx + \frac{\partial u}{\partial x}dx$ $y + v + \frac{\partial v}{\partial x}dx$ $z + w + \frac{\partial w}{\partial x}dx$
P_2	x $y + dy$ z	P_2'	$x + u + \frac{\partial u}{\partial y}dy$ $y + v + dy + \frac{\partial v}{\partial y}dy$ $z + w + \frac{\partial w}{\partial y}dy$
P_3	x y $z + dz$	P_3'	$x + u + \frac{\partial u}{\partial z}dz$ $y + v + \frac{\partial v}{\partial z}dz$ $z + w + dz + \frac{\partial w}{\partial z}dz$

is, respectively,

$$\epsilon_y = \frac{\partial v}{\partial y} \quad \text{and} \quad \epsilon_z = \frac{\partial w}{\partial z}$$

Let us consider now the changes in the angles due to the elastic deformation. Consider the angle P_1PP_2 (see Fig. 1.2.2). Before deformation the angle $P_1PP_2 = 90°$ and after deformation the angle becomes $P_1'P'P_2' = 90 - \gamma_{xy}$ with

$$\gamma_{xy} = \frac{\partial u}{\partial y} + \frac{\partial v}{\partial x}$$

Similarly, by considering the deformations of the angles $P_2'P'P_3'$ and $P_1'P'P_3'$ one finds that

$$\gamma_{yz} = \frac{\partial v}{\partial z} + \frac{\partial w}{\partial y} \quad \text{and} \quad \gamma_{xz} = \frac{\partial u}{\partial z} + \frac{\partial w}{\partial x}$$

The following relationships are therefore derived between components of displacement and strain (Cauchy's equations):

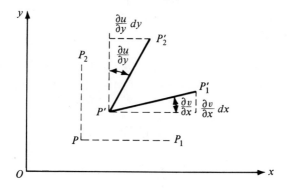

Figure 1.2.2

$$\begin{aligned} \epsilon_x &= \frac{\partial u}{\partial x} & \gamma_{xy} &= \frac{\partial u}{\partial y} + \frac{\partial v}{\partial x} \\ \epsilon_y &= \frac{\partial v}{\partial y} & \gamma_{yz} &= \frac{\partial v}{\partial z} + \frac{\partial w}{\partial y} \\ \epsilon_z &= \frac{\partial w}{\partial z} & \gamma_{zx} &= \frac{\partial w}{\partial x} + \frac{\partial u}{\partial z} \end{aligned} \quad (1.2.1)$$

and the strain tensor is given by

$$\begin{vmatrix} \epsilon_x & \gamma_{xy} & \gamma_{xz} \\ \gamma_{yx} & \epsilon_y & \gamma_{yz} \\ \gamma_{zx} & \gamma_{zy} & \epsilon_z \end{vmatrix}$$

which is a symmetric tensor since

$$\gamma_{xy} = \gamma_{yx}$$
$$\gamma_{yz} = \gamma_{zy}$$
$$\gamma_{zx} = \gamma_{xz}$$

The signs of the normal components of strain ϵ_x, ϵ_y, and ϵ_z are assumed to be positive if they correspond to an extension and negative if they correspond to a contraction. The signs of the shear components of strain γ_{xy}, γ_{yz}, and γ_{zx} are assumed to be positive if they correspond to a reduction of the angle during deformation and negative if the corresponding angle increases during deformation.

1.3 Relationships Between Stress and Strain (Generalized Hooke's Law)

Consider the small cube shown in Fig. 1.3.1 which is subjected to the normal stresses σ_x, σ_y, and σ_z. Consider separately the effects of σ_x, σ_y, and σ_z on the normal strain ϵ_x. Due to σ_x and in view of Eq. (1.0.1), one will have

$$\epsilon'_x = \frac{\sigma_x}{E} \tag{a}$$

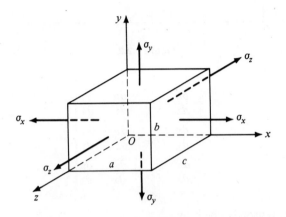

Figure 1.3.1

while due to σ_y and σ_z and in view of Eq. (1.0.2) one will have

$$\epsilon''_x = -\nu \frac{\sigma_y}{E} \tag{b}$$

$$\epsilon'''_x = -\nu \frac{\sigma_z}{E} \tag{c}$$

By superimposing Eqs. (a), (b), and (c) one finally obtains the unit strain ϵ_x. Strains ϵ_y and ϵ_z are found in the same manner. Thus,

$$\left. \begin{aligned} \epsilon_x &= \frac{1}{E}[\sigma_x - \nu(\sigma_y + \sigma_z)] \\ \epsilon_y &= \frac{1}{E}[\sigma_y - \nu(\sigma_z + \sigma_x)] \\ \epsilon_z &= \frac{1}{E}[\sigma_z - \nu(\sigma_x + \sigma_y)] \end{aligned} \right\} \tag{1.3.1}$$

By calling $e = \epsilon_x + \epsilon_y + \epsilon_z$ the volume expansion and letting

$$\Theta = \sigma_x + \sigma_y + \sigma_z$$

and adding Eqs. (1.3.1) together, one gets

$$\epsilon_x + \epsilon_y + \epsilon_z = e = \frac{1-2\nu}{E}(\sigma_x + \sigma_y + \sigma_z) = \frac{(1-2\nu)\Theta}{E} \quad (1.3.2)$$

From Eq. (1.3.2) it follows that

$$\sigma_y + \sigma_z = \frac{Ee}{1-2\nu} - \sigma_x \quad (1.3.3)$$

and by substituting Eq. (1.3.3) into the first of Eqs. (1.3.1) one obtains the stress σ_x as a function of the strain components. Expressions for σ_y and σ_z are obtained in the same manner. Thus,

$$\left.\begin{aligned}\sigma_x &= \frac{E}{1+\nu}\left(\epsilon_x + \frac{\nu e}{1-2\nu}\right) = \lambda e + 2G\epsilon_x \\ \sigma_y &= \lambda e + 2G\epsilon_y \\ \sigma_z &= \lambda e + 2G\epsilon_z \end{aligned}\right\} \quad (1.3.4)$$

where the constant

$$G = \frac{E}{2(1+\nu)}$$

is called the *shear modulus* or *modulus of rigidity* of the material, while

$$\lambda = \frac{\nu E}{(1+\nu)(1-2\nu)}$$

is called *Lamé's constant*.

Suppose that one has a hydrostatic pressure p acting on the cube of Fig. 1.3.1. In this case,

$$\sigma_x = \sigma_y = \sigma_z = -p$$

and Eq. (1.3.2) becomes

$$e = -\frac{3(1-2\nu)p}{E}$$

from which one finds that

$$-p = \frac{Ee}{3(1-2\nu)} = \left(\lambda + \frac{2G}{3}\right)e = ke$$

where

$$k = \left(\lambda + \frac{2G}{3}\right)$$

is called the *compressibility of the fluid*. In Table 1.3.1 the relationships between the elastic constants are given while in Table 1.3.2 the average physical properties of common metals are presented.

TABLE 1.3.1

RELATIONSHIPS BETWEEN ELASTIC CONSTANTS

	λ, G	k, G	G, ν	E, ν	E, G
$\lambda =$	λ	$k - \dfrac{2G}{3}$	$\dfrac{2G\nu}{1-2\nu}$	$\dfrac{\nu E}{(1+\nu)(1-2\nu)}$	$\dfrac{G(E-2G)}{(3G-E)}$
$G =$	G	G	G	$\dfrac{E}{2(1+\nu)}$	G
$k =$	$\lambda + \dfrac{2G}{3}$	k	$\dfrac{2G(1+\nu)}{3(1-2\nu)}$	$\dfrac{E}{3(1-2\nu)}$	$\dfrac{EG}{3(3G-E)}$
$E =$	$\dfrac{G(3\lambda + 2G)}{\lambda + G}$	$\dfrac{9kG}{3k+G}$	$2G(1+\nu)$	E	E
$\nu =$	$\dfrac{\lambda}{2(\lambda + G)}$	$\dfrac{3k-2G}{Gk+2G}$	ν	ν	$\dfrac{E}{2G} - 1$

Poisson's ratio is of the order 0.5 for rubber, 0.29 for steel, and 0.1 for concrete.

It can be shown that Poisson's ratio ν cannot be greater than 0.5. Consider a cube of sides a, b, and c in the x-, y-, and z-directions, respectively, and assume that tensile stresses of intensity σ are applied on the faces perpendicular to the x-axis. The initial volume of the cube is $V_0 = abc$. Since $\epsilon = \sigma/E$, the final volume is found to be

$$V' = a(1 + \epsilon)b(1 - \nu\epsilon)c(1 - \nu\epsilon) = V_0(1 + \epsilon)(1 - \nu\epsilon)^2$$

By neglecting powers of ϵ which are greater than 1, one obtains

$$V' = V_0 + V_0\epsilon(1 - 2\nu)$$

The increase in volume is

$$\Delta V = V' - V_0 = V_0\epsilon(1 - 2\nu)$$

and since $\Delta V \geq 0$ and $\epsilon > 0$ it follows that

$$(1 - 2\nu) \geq 0$$

or

$$\nu \leq \tfrac{1}{2}$$

According to Poisson's hypothesis, often used by seismologists in order to simplify computations, ν is assumed to be equal to 0.25, thus making

TABLE 1.3.2

AVERAGE PHYSICAL PROPERTIES OF COMMON METALS

Metal	Modulus of Elasticity (psi)		Proportional Limit (psi)		Ultimate Strength (psi)			Density (lb per cu in.)	Coefficient of Thermal Expansion (per °F)	Per Cent Elongation in/in.
	Tension E	Shear G	Tension	Shear	Tension	Comp.	Shear			
Aluminum cast (99% Al)	10×10^6	4×10^6	9,000		13,000		10,500	0.095	12.8×10^{-6}	20
Aluminum, hard-drawn	10×10^6	4×10^6	20,000		30,000			0.097	12.8×10^{-6}	4
Brass, cast (60% Cu, 40% Zn)	13×10^6	5×10^6	20,000		45,000			0.300	10.4×10^{-6}	20
Bronze, cast (90% Cu, 10% Sn)	12×10^6		20,000		33,000	56,000		0.295	10.0×10^{-6}	10
Common brass, rolled	14×10^6	5×10^6	25,000	15,000	60,000		50,000	0.310	10.4×10^{-6}	30
Copper, cast	13×10^6	6×10^6	8,000		30,000	45,000	27,000	0.322	9.3×10^{-6}	
Gray cast iron	15×10^6	6×10^6	6,000		20,000	80,000		0.260	6.0×10^{-6}	
Magnesium (extruded)	6.5×10^6	2.4×10^6	17,000		32,000		17,000	0.064	14.5×10^{-6}	7
Malleable cast iron	25×10^6	12.5×10^6	36,000	23,000	54,000		48,000	0.264	6.6×10^{-6}	18
Steel, cold rolled (0.2% carbon)	30×10^6	12×10^6	60,000	36,000	80,000		60,000	0.283		
Steel, hot rolled (0.2% carbon)	30×10^6	12×10^6	35,000	21,000	60,000	90,000	45,000	0.283	6.5×10^{-6}	30
Steel, hot rolled (0.8% carbon)	30×10^6	12×10^6	70,000	42,000	120,000		105,000	0.283	7.3×10^{-6}	10
Steel, oil quenched (3½% Ni, 0.4% C)	30×10^6	12×10^6	120,000	72,000	180,000		150,000	0.283		2
Nickel steel, oil quenched (3½% Ni, 0.4% C)	30×10^6	12×10^6	160,000	96,000	285,000					5
Wrought iron	27×10^6	10×10^6	30,000	18,000	50,000	60,000	40,000	0.278	6.7×10^{-6}	30

$$\lambda = G = \frac{2E}{5}$$

and

$$k = \frac{5G}{3} = \frac{2E}{3}$$

For incompressible solids $v = \frac{1}{2}$, thus making

$$\lambda = k = \infty \quad \text{and} \quad G = \frac{E}{3}$$

Example 1.3.1. A rod of steel 10 in. long having a square cross section of 1 in. by 1 in. is subjected to a tensile stress of 8 ton/in². in the direction of its length. Assuming that the modulus of elasticity is $E = 30 \times 10^6$ psi and $v = 0.30$, determine the change in volume of the rod.

If ϵ_x represents the strain in the direction of the axis of the rod, then

$$\epsilon_x = \frac{(8)(2000)}{30 \times 10^6} = \tfrac{16}{3}(10^{-4})$$

If ϵ_y and ϵ_z represent the strains parallel to the sides of the cross section, then

$$\epsilon_y = \epsilon_z = -v\epsilon_x = -0.3(\tfrac{16}{3})(10^{-4}) = -16 \times 10^{-5}$$

The increase in volume per unit volume is

$$\epsilon_x + \epsilon_y + \epsilon_z = (53 - 2 \times 16)(10^{-5}) = 21 \times 10^{-5}$$

The original volume was 10 in.³. The increase in volume is 0.0021 in.³.

The following relationships between shear stresses and shear strains (Hooke's law for shear)

$$\left.\begin{aligned} \tau_{xy} &= G\gamma_{xy} \\ \tau_{yz} &= G\gamma_{yz} \\ \tau_{zx} &= G\gamma_{zx} \end{aligned}\right\} \quad (1.3.5)$$

will now be derived. Consider the small cube of sides $dx = dy = 2a$ and unity in the z-direction as shown in Fig. 1.3.2. Suppose that on the faces perpendicular to the x- and y-axes the following normal stresses are applied: $-\sigma_x = \sigma_y = \sigma_0$. By writing the equations of equilibrium for the free body diagram in Fig. 1.3.3 it follows that

$$\tau\,ds = \sigma_0\,dy \cos 45° + \sigma_0\,dx \cos 45°$$

$$\tau = \sigma_0 \frac{dy}{ds} \cos 45° + \sigma_0 \frac{dx}{ds} \cos 45°$$

$$\tau = \sigma_0(\cos^2 45° + \cos^2 45°) = \sigma_0(\tfrac{1}{2} + \tfrac{1}{2}) = \sigma_0$$

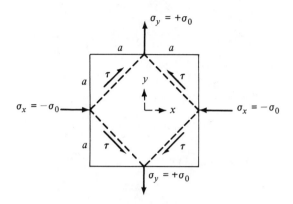

Figure 1.3.2

Therefore,

$$\tau = \sigma_0$$

The cube will be deformed as shown in Fig. 1.3.4. The following relationships are easily derived:

$$OA = OB = OC = OD = a$$
$$\sphericalangle ACB = 90°$$
$$\sphericalangle ACO = \frac{90°}{2} = 45°$$
$$\sphericalangle A'C'B' = 90° + \gamma$$
$$\sphericalangle A'C'O = 45° + \frac{\gamma}{2}$$

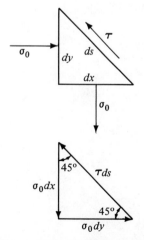

Figure 1.3.3

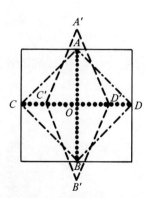

Figure 1.3.4

Therefore,

$$\tan \angle A'C'O = \tan\left(45° + \frac{\gamma}{2}\right) = \frac{OA'}{OC'}$$

But

$$OA' = a + \text{extension}$$
$$= a + a\epsilon_y = a + a\left[\frac{\sigma_0}{E} - \frac{v}{E}(-\sigma_0)\right]$$
$$= a + a\frac{\sigma_0}{E}(1 + v) = a\left[1 + \frac{\sigma_0(1 + v)}{E}\right]$$

Similarly,

$$OC' = a - \text{extension}$$
$$= a\left[1 - \frac{\sigma_0(1 + v)}{E}\right]$$

Thus,

$$\tan \angle A'C'O = \tan\left(45° + \frac{\gamma}{2}\right)$$
$$= \frac{1 + \frac{\sigma_0(1 + v)}{E}}{1 - \frac{\sigma_0(1 + v)}{E}}$$

But

$$\tan\left(45° + \frac{\gamma}{2}\right) = \frac{\left(\tan 45° + \tan\frac{\gamma}{2}\right)}{\left(1 - \tan 45° \tan\frac{\gamma}{2}\right)}$$
$$= \frac{1 + \tan\frac{\gamma}{2}}{1 - \tan\frac{\gamma}{2}}$$

Since $\gamma/2$ is very small it follows that

$$\tan\frac{\gamma}{2} \doteq \frac{\gamma}{2}$$

Thus,

$$\frac{1 + \frac{\gamma}{2}}{1 - \frac{\gamma}{2}} = \frac{1 + \frac{\sigma_0(1 + v)}{E}}{1 - \frac{\sigma_0(1 + v)}{E}}$$

from which one finds

$$\frac{\gamma}{2} = \sigma_0 \frac{(1+\nu)}{E}$$

However, since $\sigma_0 = \tau$, it follows that

$$\frac{\gamma}{2} = \tau \frac{(1+\nu)}{E}$$

From the above relation one obtains

$$\tau = \frac{E\gamma}{2(1+\nu)} = G\gamma$$

or, considering xy, then

$$\tau_{xy} = G\gamma_{xy}$$

Similarly, $\tau_{yz} = G\gamma_{yz}$ and $\tau_{zx} = G\gamma_{zx}$ as was to be shown.

1.4 Equations of Equilibrium (Navier's Equations, Lamé's Equations)

Consider the small cube of Fig. 1.4.1 and express the conditions of equilibrium of the forces (body forces and surface forces) acting on it. One has in the x-direction:

$$X dx dy dz = -\sigma_x dy dz + \left(\sigma_x + \frac{\partial \sigma_x}{\partial x} dx\right) dy dz - \tau_{yx} dz dx$$
$$+ \left(\tau_{yx} + \frac{\partial \tau_{yx}}{\partial x} dy\right) dx dz - \tau_{zx} dy dx + \left(\tau_{zx} + \frac{\partial \tau_{zx}}{\partial z} dz\right) dy dx = 0$$

which, after canceling $dxdydz$, becomes

Similarly,

$$\left.\begin{aligned}\frac{\partial \sigma_x}{\partial x} + \frac{\partial \tau_{xy}}{\partial y} + \frac{\partial \tau_{xz}}{\partial z} + X &= 0 \\ \frac{\partial \tau_{yx}}{\partial x} + \frac{\partial \sigma_y}{\partial y} + \frac{\partial \tau_{yz}}{\partial z} + Y &= 0 \\ \frac{\partial \tau_{zx}}{\partial x} + \frac{\partial \tau_{zy}}{\partial y} + \frac{\partial \sigma_z}{\partial z} + Z &= 0\end{aligned}\right\} \quad (1.4.1)$$

Equations (1.4.1) are Navier's equations of equilibrium for an elastic solid. If the elastic body is in motion one may apply D'Alembert's principle to account for the inertia forces. If ρ is the mass density of the elastic body then the inertia forces in the x-, y-, and z-directions are, respectively,

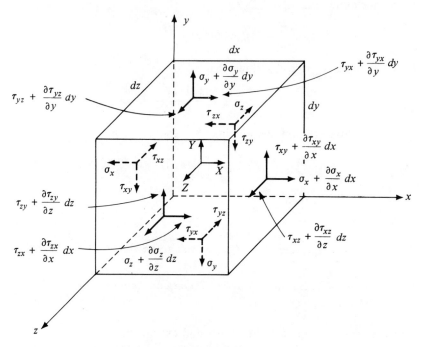

Figure 1.4.1

$$-\rho dx dy dz \frac{\partial^2 u}{\partial t^2}$$

$$-\rho dx dy dz \frac{\partial^2 v}{\partial t^2}$$

$$-\rho dx dy dz \frac{\partial^2 w}{\partial t^2}$$

By adding these forces to the body forces, Eqs. (1.4.1) may be written in the following forms (Navier's equations of motion):

$$\left.\begin{array}{l} \dfrac{\partial \sigma_x}{\partial x} + \dfrac{\partial \tau_{xy}}{\partial y} + \dfrac{\partial \tau_{xz}}{\partial z} + X = \rho \dfrac{\partial^2 u}{\partial t^2} \\[6pt] \dfrac{\partial \tau_{yx}}{\partial x} + \dfrac{\partial \sigma_y}{\partial y} + \dfrac{\partial \tau_{yz}}{\partial z} + Y = \rho \dfrac{\partial^2 v}{\partial t^2} \\[6pt] \dfrac{\partial \tau_{zx}}{\partial x} + \dfrac{\partial \tau_{zy}}{\partial y} + \dfrac{\partial \sigma_z}{\partial z} + Z = \rho \dfrac{\partial^2 w}{\partial t^2} \end{array}\right\} \quad (1.4.2)$$

By introducing Eqs. (1.3.4) and (1.3.5) into Eqs. (1.4.1) and using Eqs. (1.2.1) one obtains the following equations of equilibrium (Lamé's equations of equilibrium) expressed in terms of the components of elastic displacement:

$$\left.\begin{array}{l} G\Delta u + (\lambda + G)\dfrac{\partial e}{\partial x} + X = 0 \\[4pt] G\Delta v + (\lambda + G)\dfrac{\partial e}{\partial y} + Y = 0 \\[4pt] G\Delta w + (\lambda + G)\dfrac{\partial e}{\partial z} + Z = 0 \end{array}\right\} \qquad (1.4.3)$$

By introducing Eqs. (1.3.4) and (1.3.5) into Eqs. (1.4.2) and using Eqs. (1.2.1) one obtains the following equations of motion (Lamé's equations of motion) expressed in terms of the components of elastic displacement:

$$\left.\begin{array}{l} G\Delta u + (\lambda + G)\dfrac{\partial e}{\partial x} + X = \rho\dfrac{\partial^2 u}{\partial t^2} \\[4pt] G\Delta v + (\lambda + G)\dfrac{\partial e}{\partial y} + Y = \rho\dfrac{\partial^2 v}{\partial t^2} \\[4pt] G\Delta w + (\lambda + G)\dfrac{\partial e}{\partial z} + Z = \rho\dfrac{\partial^2 w}{\partial t^2} \end{array}\right\} \qquad (1.4.4)$$

In Eqs. (1.4.3) and (1.4.4)

$$\Delta = \dfrac{\partial^2}{\partial x^2} + \dfrac{\partial^2}{\partial y^2} + \dfrac{\partial^2}{\partial z^2}$$

represents the Laplacian operator.

D'Alembert, Jean Baptiste Le Rond. French mathematician, writer, and philosopher (b. Paris 1717; d. Paris 1783). When only twenty years old he became a member of the French Academy of Sciences and collaborated with Diderot on the "Dictionaire Encyclopédique" for which he wrote his famous "Discours Préliminaire." His scientific works are collected in eight volumes published during the years 1761–1780. His greatest scientific work is his "Traité de Dynamique" in which he applied his famous principle. [Reproduction from the Vito Volterra collection, Villa Volterra, Ariccia (Rome).]

Laplace, Pierre-Simon. French astronomer and mathematician (b. Beaumont-en-Auge 1749; d. Paris 1827). His most important contributions were in astronomy and in the mathematical theory of probability. His complete works, in thirteen volumes, were published during the years 1878–1904. [Reproduction from the Vito Volterra collection, Villa Volterra, Ariccia (Rome).]

Example 1.4.1. The stress components at point P are given by

$$\sigma_x = x^2 + y^2 \quad \sigma_y = y^2 + z^2$$
$$\tau_{xy} = xy \quad \tau_{yz} = yz$$
$$\tau_{xz} = xz \quad \sigma_z = z^2 + x^2$$

What must the body forces be in order to satisfy the conditions of equilibrium? It follows that

$$\frac{\partial \sigma_x}{\partial x} + \frac{\partial \tau_{xy}}{\partial y} + \frac{\partial \tau_{xz}}{\partial z} + X = 0 \quad \text{or} \quad 2x + x + x + X = 0$$

$$\frac{\partial \tau_{yx}}{\partial x} + \frac{\partial \sigma_y}{\partial y} + \frac{\partial \tau_{yz}}{\partial z} + Y = 0 \quad \text{or} \quad y + 2y + y + Y = 0$$

$$\frac{\partial \tau_{zx}}{\partial x} + \frac{\partial \tau_{zy}}{\partial y} + \frac{\partial \sigma_z}{\partial z} + Z = 0 \quad \text{or} \quad z + z + 2z + Z = 0$$

and finally that

$$X = -4x$$
$$Y = -4y$$
$$Z = -4z$$

Example 1.4.2. Given the following stress field:

$$\sigma_x = 50x^3 + 2y \text{ psi} \quad \tau_{xy} = 100z + 80y^2 \text{ psi}$$
$$\sigma_y = 40x^3 + 500 \text{ psi} \quad \tau_{yz} = 0 \text{ psi}$$
$$\sigma_z = 60y^2 + 30z^3 \text{ psi} \quad \tau_{zx} = xz^3 + 40x^2y \text{ psi}$$

Find the body force distribution required for equilibrium and the body force components at point (2, 2, 1).

By substituting the given stresses into the differential equations of equilibrium [Eqs. (1.4.1)] one finds that

$$(150x^2) + (160y) + (3xz^2) + X = 0$$
$$(0) + (0) + (0) + Y = 0$$
$$(z^3 + 80xy) + (0) + (90z^2) + Z = 0$$

From the above relations the body force distribution is found to be

$$X = -150x^2 - 160y - 3xz^2$$
$$Y = 0$$
$$Z = -80xy - z^3 - 90z^2$$

From these expressions the body force components at (2, 2, 1) are found to be

$$X = -(150)(2)^2 - (160)(2) - (3)(2)(1)^2$$
$$Y = 0$$
$$Z = -(80)(2)(2) - (1)^3 - (90)(1)^2$$

or finally

$$X = -926 \text{ lb/in.}^3$$
$$Y = 0$$
$$Z = -411 \text{ lb/in.}^3$$

Example 1.4.3. By assuming $E = 30 \times 10^6$ psi and $\nu = 0.25$, calculate the strains at point (2, 2, 1) of Example 1.4.2.

The stresses at point (2, 2, 1) are

$$\sigma_x = (50)(2)^3 + (2)(2) = 404 \text{ psi}$$
$$\sigma_y = (40)(2)^3 + 500 = 820 \text{ psi}$$
$$\sigma_z = (60)(2)^2 + (30)(1)^3 = 270 \text{ psi}$$
$$\tau_{xy} = (100)(1) + (80)(2)^2 = 420 \text{ psi}$$
$$\tau_{yz} = 0 \text{ psi}$$
$$\tau_{zx} = (2)(1)^3 + (40)(2)^2(2) = 322 \text{ psi}$$

By substituting the above values into Eqs. (1.3.1) and (1.3.5) with

$$G = \frac{E}{2(1+\nu)} = \frac{30 \times 10^6}{2(1 + 0.25)} = 12 \times 10^6$$

one obtains

$$\epsilon_x = \frac{404 - (0.25)(820 + 270)}{30 \times 10^6} = 4.383 \times 10^{-6} \text{ in./in.}$$

$$\epsilon_y = \frac{820 - (0.25)(404 + 270)}{30 \times 10^6} = 21.72 \times 10^{-6} \text{ in./in.}$$

$$\epsilon_z = \frac{270 - (0.25)(404 + 820)}{30 \times 10^6} = 1.200 \times 10^{-6} \text{ in./in.}$$

$$\gamma_{xy} = \frac{420}{12 \times 10^6} = 35.00 \times 10^{-6}$$

$$\gamma_{yz} = \frac{0}{12 \times 10^6} = 0$$

$$\gamma_{zx} = \frac{322}{12 \times 10^6} = 26.83 \times 10^{-6}$$

1.5 Compatibility Equations (Saint-Venant's Equations, Beltrami-Michell's Equations)

In paragraph 1.2 Cauchy's equations [Eqs. (1.2.1)] which express the components of strain in terms of the components of elastic displacement were presented in the following form:

$$\left. \begin{aligned} \epsilon_x &= \frac{\partial u}{\partial x} & \gamma_{xy} &= \frac{\partial u}{\partial y} + \frac{\partial v}{\partial x} \\ \epsilon_y &= \frac{\partial v}{\partial y} & \gamma_{yz} &= \frac{\partial v}{\partial z} + \frac{\partial w}{\partial y} \\ \epsilon_z &= \frac{\partial w}{\partial z} & \gamma_{zx} &= \frac{\partial w}{\partial x} + \frac{\partial u}{\partial z} \end{aligned} \right\} \quad (1.2.1)$$

As a consequence of Eqs. (1.2.1) it follows that the components of strain are not independent, i.e., they cannot be taken as arbitrary functions of the variables x, y, and z but are related by relationships called *compatibility equations*. They were first derived by Saint Venant and are valid if the body is simply connected. From Eqs. (1.2.1) it follows that

$$\frac{\partial^2 \epsilon_x}{\partial y^2} = \frac{\partial^3 u}{\partial y^2 \partial x} \quad \text{and} \quad \frac{\partial^2 \epsilon_y}{\partial x^2} = \frac{\partial^3 v}{\partial x^2 \partial y}$$

but since

$$\frac{\partial^2 \gamma_{xy}}{\partial x \partial y} = \frac{\partial^3 u}{\partial x \partial y^2} + \frac{\partial^3 v}{\partial x^2 \partial y}$$

It follows that

Similarly,
$$\left.\begin{aligned}\frac{\partial^2 \epsilon_x}{\partial y^2} + \frac{\partial^2 \epsilon_y}{\partial x^2} &= \frac{\partial^2 \gamma_{xy}}{\partial x \partial y} \\ \frac{\partial^2 \epsilon_y}{\partial z^2} + \frac{\partial^2 \epsilon_z}{\partial y^2} &= \frac{\partial^2 \gamma_{yz}}{\partial y \partial z} \\ \frac{\partial^2 \epsilon_z}{\partial x^2} + \frac{\partial^2 \epsilon_x}{\partial z^2} &= \frac{\partial^2 \gamma_{zx}}{\partial z \partial x}\end{aligned}\right\} \quad (1.5.1)$$

Also, from Eqs. (1.2.1) one derives that

$$\frac{\partial^2 \epsilon_x}{\partial y \partial z} = \frac{\partial^3 u}{\partial x \partial y \partial z} \quad \text{and} \quad \frac{\partial \gamma_{yz}}{\partial x} = \frac{\partial^2 v}{\partial x \partial z} + \frac{\partial^2 w}{\partial x \partial y}$$

$$\frac{\partial \gamma_{xz}}{\partial y} = \frac{\partial^2 u}{\partial y \partial z} + \frac{\partial^2 w}{\partial x \partial y} \quad \text{and} \quad \frac{\partial \gamma_{xy}}{\partial z} = \frac{\partial^2 u}{\partial y \partial z} + \frac{\partial^2 v}{\partial x \partial z}$$

from which it follows that

$$\frac{\partial^2 \gamma_{xy}}{\partial x \partial z} - \frac{\partial^2 \gamma_{yz}}{\partial x^2} + \frac{\partial^2 \gamma_{zx}}{\partial x \partial y} = 2\frac{\partial^3 u}{\partial x \partial y \partial z} = 2\frac{\partial^2 \epsilon_x}{\partial y \partial z}$$

or that

Similarly,
$$\left.\begin{aligned}\frac{\partial}{\partial x}\left[\frac{\partial \gamma_{xy}}{\partial z} - \frac{\partial \gamma_{yz}}{\partial x} + \frac{\partial \gamma_{zx}}{\partial y}\right] &= 2\frac{\partial^2 \epsilon_x}{\partial y \partial z} \\ \frac{\partial}{\partial y}\left[\frac{\partial \gamma_{yz}}{\partial x} - \frac{\partial \gamma_{zx}}{\partial y} + \frac{\partial \gamma_{xy}}{\partial z}\right] &= 2\frac{\partial^2 \epsilon_y}{\partial z \partial x} \\ \frac{\partial}{\partial z}\left[\frac{\partial \gamma_{zx}}{\partial y} - \frac{\partial \gamma_{xy}}{\partial z} + \frac{\partial \gamma_{yz}}{\partial x}\right] &= 2\frac{\partial^2 \epsilon_z}{\partial x \partial y}\end{aligned}\right\} \quad (1.5.2)$$

Equations (1.5.1) and (1.5.2) are not independent. If one substitutes into Eqs. (1.5.1) and (1.5.2), equations

$$\left.\begin{aligned}\epsilon_x &= \frac{1}{E}[\sigma_x - \nu(\sigma_y + \sigma_z)] \\ \epsilon_y &= \frac{1}{E}[\sigma_y - \nu(\sigma_z + \sigma_x)] \\ \epsilon_z &= \frac{1}{E}[\sigma_z - \nu(\sigma_x + \sigma_y)]\end{aligned}\right\} \quad (1.3.1) \quad \left.\begin{aligned}\gamma_{xy} &= \frac{\tau_{xy}}{G} \\ \gamma_{yz} &= \frac{\tau_{yz}}{G} \\ \gamma_{zx} &= \frac{\tau_{zx}}{G}\end{aligned}\right\} \quad (1.3.5)$$

and uses the equations of equilibrium [Eqs. (1.4.1)], one obtains the *Beltrami-Michell compatibility equations* in the following forms:

$$\Delta\sigma_x + \frac{1}{1+\nu}\frac{\partial^2\Theta}{\partial x^2} = -\frac{\nu}{1-\nu}\left(\frac{\partial X}{\partial x} + \frac{\partial Y}{\partial y} + \frac{\partial Z}{\partial z}\right) - 2\frac{\partial X}{\partial x}$$
$$\Delta\sigma_y + \frac{1}{1+\nu}\frac{\partial^2\Theta}{\partial y^2} = -\frac{\nu}{1-\nu}\left(\frac{\partial X}{\partial x} + \frac{\partial Y}{\partial y} + \frac{\partial Z}{\partial z}\right) - 2\frac{\partial Y}{\partial y} \quad (1.5.3)$$
$$\Delta\sigma_z + \frac{1}{1+\nu}\frac{\partial^2\Theta}{\partial z^2} = -\frac{\nu}{1-\nu}\left(\frac{\partial X}{\partial x} + \frac{\partial Y}{\partial y} + \frac{\partial Z}{\partial z}\right) - 2\frac{\partial Z}{\partial z}$$

and

$$\Delta\tau_{yz} + \frac{1}{1+\nu}\frac{\partial^2\Theta}{\partial y \partial z} = -\left(\frac{\partial Z}{\partial y} + \frac{\partial Y}{\partial z}\right)$$
$$\Delta\tau_{zx} + \frac{1}{1+\nu}\frac{\partial^2\Theta}{\partial z \partial x} = -\left(\frac{\partial X}{\partial z} + \frac{\partial Z}{\partial x}\right) \quad (1.5.4)$$
$$\Delta\tau_{xy} + \frac{1}{1+\nu}\frac{\partial^2\Theta}{\partial x \partial y} = -\left(\frac{\partial Y}{\partial x} + \frac{\partial X}{\partial y}\right)$$

By adding Eqs. (1.5.3) and letting $\Theta = \sigma_x + \sigma_y + \sigma_z$, one obtains the equation:

$$\Delta\Theta = -\frac{1+\nu}{1-\nu}\left(\frac{\partial X}{\partial x} + \frac{\partial Y}{\partial y} + \frac{\partial Z}{\partial z}\right) \quad (1.5.5)$$

If there are no body forces, or if body forces are constant, Eqs. (1.5.3) and (1.5.4) transform into the following equations:

$$\Delta\sigma_x + \frac{1}{1+\nu}\frac{\partial^2\Theta}{\partial x^2} = 0$$
$$\Delta\sigma_y + \frac{1}{1+\nu}\frac{\partial^2\Theta}{\partial y^2} = 0 \quad (1.5.6)$$
$$\Delta\sigma_z + \frac{1}{1+\nu}\frac{\partial^2\Theta}{\partial z^2} = 0$$

and

$$\Delta\tau_{yz} + \frac{1}{1+\nu}\frac{\partial^2\Theta}{\partial y \partial z} = 0$$
$$\Delta\tau_{zx} + \frac{1}{1+\nu}\frac{\partial^2\Theta}{\partial z \partial x} = 0 \quad (1.5.7)$$
$$\Delta\tau_{xy} + \frac{1}{1+\nu}\frac{\partial^2\Theta}{\partial x \partial y} = 0$$

Example 1.5.1. Does the distribution given by Example 1.4.2 satisfy the compatibility equations?

From Example 1.4.2 the stresses are

$$\sigma_x = 50x^3 + 2y \qquad \tau_{xy} = 100z + 80y^2$$
$$\sigma_y = 40x^3 + 500 \qquad \tau_{yz} = 0$$
$$\sigma_z = 60y^2 + 30z^3 \qquad \tau_{xz} = xz^3 + 40x^2y$$

and the body forces are

$$X = -150x^2 - 160y - 3xz^2$$
$$Y = 0$$
$$Z = -80xy - z^3 - 90z^2$$

Considering the first of Eqs. (1.5.3) one finds that

$$\Delta \sigma_x = 300x$$
$$\Theta = 90x^3 + 60y^2 + 2y + 30z^3 + 500$$
$$\frac{\partial^2 \Theta}{\partial x^2} = 540x$$
$$\frac{\partial X}{\partial x} + \frac{\partial Y}{\partial y} + \frac{\partial Z}{\partial z} = -300x - 6z^2 - 180z$$
$$\frac{\partial X}{\partial x} = -300x - 3z^2$$

By substituting the above quantities into the first of Eqs. (1.5.3) one finds that this compatibility equation is not satisfied.

1.6 Boundary Conditions

Consider the small tetrahedron $OABC$ shown in Fig. 1.6.1. If n is the normal to face ABC and $\widehat{nx}$, $\widehat{ny}$, and $\widehat{nz}$ are the angles between n and the x-, y-, and z-axes, respectively, and if A is the area of face ABC and A_x, A_y, and A_z are the areas of faces OBC, OAC, and OAB, respectively, the following relationships can be verified:

$$\frac{A_x}{A} = \cos \widehat{nx} \qquad \frac{A_y}{A} = \cos \widehat{ny} \qquad \frac{A_z}{A} = \cos \widehat{nz} \qquad (1.6.1)$$

If $\bar{X}$, $\bar{Y}$, and $\bar{Z}$ represent components of the surface forces applied per unit area of face ABC, the following *boundary conditions* can be derived:

$$\left.\begin{array}{l}\sigma_x \cos(\widehat{nx}) + \tau_{xy} \cos(\widehat{ny}) + \tau_{xz} \cos(\widehat{nz}) = \bar{X} \\ \tau_{yx} \cos(\widehat{nx}) + \sigma_y \cos(\widehat{ny}) + \tau_{yz} \cos(\widehat{nz}) = \bar{Y} \\ \tau_{zx} \cos(\widehat{nx}) + \tau_{zy} \cos(\widehat{ny}) + \sigma_z \cos(\widehat{nz}) = \bar{Z}\end{array}\right\} \qquad (1.6.2)$$

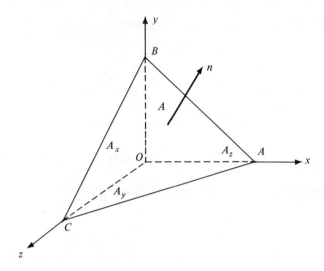

Figure 1.6.1

Equations (1.6.2) are obtained by expressing the conditions of equilibrium of the small tetrahedron subjected to the following forces:

$\bar{X}A, \bar{Y}A, \bar{Z}A$	on face ABC of area A
$\sigma_x A_x, \tau_{xy} A_x, \tau_{xz} A_x$	on face OBC of area A_x perpendicular to the x-axis
$\tau_{yx} A_y, \sigma_y A_y, \tau_{yz} A_y$	on face OAC of area A_y perpendicular to the y-axis
$\tau_{zx} A_z, \tau_{zy} A_z, \sigma_z A_z$	on face OAB of area A_z perpendicular to the z-axis

One observes that the body forces acting on the infinitesimal tetrahedron can be neglected due to the fact that in reducing the dimensions of the solid, the body forces acting on it diminish as the cube of the linear dimensions while the surface forces diminish as the square of the linear dimensions. In addition, since the tetrahedron is assumed to be small, variation of stresses over the sides can be neglected and the stresses can be assumed to be uniformly distributed over the faces of the tetrahedron. By expressing the condition of equilibrium of the components of the forces acting on the tetrahedron in the x-direction, one obtains

$$\bar{X}A - \sigma_x A_x - \tau_{xy} A_y - \tau_{xz} A_z = 0$$

and upon dividing by A and recalling Eqs. (1.6.1) the first of Eqs. (1.6.2) is obtained. Similarly, the other two boundary equations are derived.

Boundary equations [Eqs. (1.6.2)] can be expressed in terms of the components of elastic displacements. By introducing Eqs. (1.3.4), (1.3.5), and (1.2.1) into Eqs. (1.6.2) the following equations are obtained:

$$\begin{aligned}
\bar{X} &= \lambda e \cos(\widehat{nx}) + G\left[\frac{\partial u}{\partial x}\cos(\widehat{nx}) + \frac{\partial u}{\partial y}\cos(\widehat{ny}) + \frac{\partial u}{\partial z}\cos(\widehat{nz})\right] \\
&\quad + G\left[\frac{\partial u}{\partial x}\cos(\widehat{nx}) + \frac{\partial v}{\partial x}\cos(\widehat{ny}) + \frac{\partial w}{\partial x}\cos(\widehat{nz})\right] \\
\bar{Y} &= \lambda e \cos(\widehat{ny}) + G\left[\frac{\partial v}{\partial y}\cos(\widehat{ny}) + \frac{\partial v}{\partial z}\cos(\widehat{nz}) + \frac{\partial v}{\partial x}\cos(\widehat{nx})\right] \\
&\quad + G\left[\frac{\partial v}{\partial y}\cos(\widehat{ny}) + \frac{\partial w}{\partial y}\cos(\widehat{ny}) + \frac{\partial u}{\partial x}\cos(\widehat{nx})\right] \\
\bar{Z} &= \lambda e \cos(\widehat{nz}) + G\left[\frac{\partial w}{\partial z}\cos(\widehat{nz}) + \frac{\partial w}{\partial x}\cos(\widehat{nx}) + \frac{\partial w}{\partial y}\cos(\widehat{ny})\right] \\
&\quad + G\left[\frac{\partial w}{\partial z}\cos(\widehat{nz}) + \frac{\partial u}{\partial z}\cos(\widehat{nx}) + \frac{\partial v}{\partial z}\cos(\widehat{ny})\right]
\end{aligned} \quad (1.6.3)$$

1.7 Thermal Stress Equations (Duhamel-Neumann's Equations)

An elastic body will expand or contract when subjected to changes in temperature. If the body is free to expand or contract, it will not experience additional stresses when subjected to uniform temperature changes. However, if free expansion or contraction is prevented by external constraints, or if the temperature change is not uniform, then stresses known as *thermal stresses* will be produced.

Consider a bar of length l. Its change of length Δl due to a temperature change $T - T_0$ (T being the final temperature, T_0 the initial temperature) is

$$\Delta l = \alpha l(T - T_0) \quad (1.7.1)$$

where α (coefficient of thermal expansion) represents the change in length per unit length of the bar due to a change of temperature of 1° Fahrenheit. In Table 1.3.2, coefficients of thermal expansion for common metals are presented.

Since thermal expansion does not produce angular distortion in an isotropic material, Eqs. (1.3.5) will be unaffected by a change of temperature $\Delta T = T - T_0$, while Eqs. (1.3.1) will become

$$\begin{aligned}
\epsilon_x &= \frac{1}{E}[\sigma_x - \nu(\sigma_y + \sigma_z)] + \alpha \Delta T \\
\epsilon_y &= \frac{1}{E}[\sigma_y - \nu(\sigma_z + \sigma_x)] + \alpha \Delta T \\
\epsilon_z &= \frac{1}{E}[\sigma_z - \nu(\sigma_x + \sigma_y)] + \alpha \Delta T
\end{aligned} \quad (1.7.2)$$

Suppose now that the elastic body is completely restrained by a hydrostatic pressure

$$\sigma = \sigma_x = \sigma_y = \sigma_z \qquad (1.7.3)$$

when subjected to temperature change ΔT. By introducing Eqs. (1.7.3) and the conditions $\epsilon_x = \epsilon_y = \epsilon_z = 0$ into Eqs. (1.7.2) one obtains

while
$$\left. \begin{array}{l} \sigma_x = \sigma_y = \sigma_z = \sigma = -\dfrac{\alpha E \Delta T}{1 - 2\nu} \\[1em] \tau_{xy} = \tau_{yz} = \tau_{zx} = 0 \end{array} \right\} \qquad (1.7.4)$$

The components of stresses in Eqs. (1.7.4) must satisfy the equilibrium equations [Eqs. (1.4.1)] and the boundary conditions [Eqs. (1.6.2)]. By substituting Eqs. (1.7.4) into Eqs. (1.4.1) one obtains

$$\left. \begin{array}{l} X = -\dfrac{\alpha E}{1 - 2\nu} \dfrac{\partial \Delta T}{\partial x} \\[1em] Y = -\dfrac{\alpha E}{1 - 2\nu} \dfrac{\partial \Delta T}{\partial y} \\[1em] Z = -\dfrac{\alpha E}{1 - 2\nu} \dfrac{\partial \Delta T}{\partial z} \end{array} \right\} \qquad (1.7.5)$$

Upon substituting Eqs. (1.7.4) into Eqs. (1.6.2) one obtains

$$\left. \begin{array}{l} \bar{X} = -\dfrac{\alpha E \Delta T}{1 - 2\nu} \cos(\widehat{nx}) \\[1em] \bar{Y} = -\dfrac{\alpha E \Delta T}{1 - 2\nu} \cos(\widehat{ny}) \\[1em] \bar{Z} = -\dfrac{\alpha E \Delta T}{1 - 2\nu} \cos(\widehat{nz}) \end{array} \right\} \qquad (1.7.6)$$

If the elastic body is not restrained, thermal stresses inside the body will be obtained by superposition of the stresses given by Eq. (1.7.3) and stresses which are produced by body forces and surface forces whose components are expressed (with reversed sign) by Eqs. (1.7.5) and (1.7.6), respectively. Therefore, *thermal stresses* inside the body satisfy the following equations of equilibrium

$$\left. \begin{array}{l} \dfrac{\partial \sigma_x}{\partial x} + \dfrac{\partial \tau_{xy}}{\partial y} + \dfrac{\partial \tau_{xz}}{\partial z} = \dfrac{\alpha E}{1 - 2\nu} \dfrac{\partial \Delta T}{\partial x} \\[1em] \dfrac{\partial \tau_{yx}}{\partial x} + \dfrac{\partial \sigma_y}{\partial y} + \dfrac{\partial \tau_{yz}}{\partial z} = \dfrac{\alpha E}{1 - 2\nu} \dfrac{\partial \Delta T}{\partial y} \\[1em] \dfrac{\partial \tau_{zx}}{\partial x} + \dfrac{\partial \tau_{zy}}{\partial y} + \dfrac{\partial \sigma_z}{\partial z} = \dfrac{\alpha E}{1 - 2\nu} \dfrac{\partial \Delta T}{\partial z} \end{array} \right\} \qquad (1.7.7)$$

and the following boundary conditions

$$\left.\begin{array}{l}\sigma_x \cos(\widehat{nx}) + \tau_{xy}\cos(\widehat{ny}) + \tau_{xz}\cos(\widehat{nz}) = \dfrac{\alpha E \Delta T}{1-2\nu}\cos(\widehat{nx})\\[4pt] \tau_{yx}\cos(\widehat{nx}) + \sigma_y\cos(\widehat{ny}) + \tau_{yz}\cos(\widehat{nz}) = \dfrac{\alpha E \Delta T}{1-2\nu}\cos(\widehat{ny})\\[4pt] \tau_{zx}\cos(\widehat{nx}) + \tau_{zy}\cos(\widehat{ny}) + \sigma_z\cos(\widehat{nz}) = \dfrac{\alpha E \Delta T}{1-2\nu}\cos(\widehat{nz})\end{array}\right\} \quad (1.7.8)$$

1.8 Saint-Venant's Principle

In 1855, Saint-Venant enunciated the "principle of the elastic equivalence of statically equipollent systems of loads." According to this principle, if a system of forces acting on a portion of the boundary of an elastic body is replaced by a statically equivalent system of forces acting on the same por-

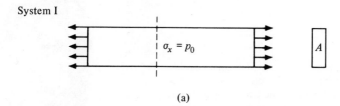

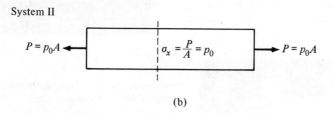

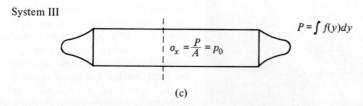

Figure 1.8.1

Kirchhoff, Gustav-Robert. German physicist (b. Königsberg 1824; d. Berlin 1887). He was successively professor of physics at the Universities of Breslau, Heidelberg, and Berlin. His fundamental contributions were in the fields of electricity, thermodynamics, and spectroscopic analysis. [Reproduction from the Vito Volterra collection, Villa Volterra, Ariccia (Rome).]

Morera, Giacinto. Italian mathematician (b. Novara 1856; d. Turin 1909). He was Professor of Mechanics at the University of Turin. His principal contributions were to the differential equations of dynamics, the equations of Pfaff, the functions of complex variables, and the mathematical theory of elasticity. (Reproduction kindly provided by Professor Carlo Ferrari, President of the Turin Academy of Sciences.)

Somigliana, Carlo. Italian mathematician (b. Como, 1860; d. Casanova Lanza, 1955). He was a descendant (on his mother's side) of Count Alessandro Volta, the famous Italian physicist (1745–1827). A pupil of Betti's at Pisa University, he was later successively Professor of Mathematical Physics at the Universities of Pavia and Turin. He was interested in many fields of applied science including electrostatics, geodesy, and geophysics, but his most celebrated and outstanding work was in the mathematical theory of elasticity (Somigliana's formulas, Somigliana's dislocation theory). [Reproduction from the Vito Volterra collection, Villa Volterra, Ariccia (Rome).]

tion of the boundary, then the stresses, strains, and elastic displacements in parts of the body sufficiently far removed from this portion of boundary remain approximately the same.

In order to explain Saint-Venant's principle, consider the three different systems of loading acting on the bar with rectangular cross section of area A as shown in Fig. 1.8.1. Although the three systems of external forces acting at the two ends of the bar are different, in the first case there being a uniformly distributed load of intensity p_0, in the second case a concentrated axial load of intensity P, and in the third case a variably distributed load of total value P, the longitudinal stress σ_x on the section at the center of the bar is the same in each case and is expressed by

$$\sigma_x = \frac{P}{A}$$

1.9 The Three Basic Problems of the Theory of Elasticity

Problems of elasticity present themselves in three forms:

(a) *First basic problem of the theory of elasticity*–when the loads on the surface as well as the body forces are prescribed.
(b) *Second basic problem*–when the components u, v, and w of the elastic displacement are prescribed on all points of the surface of the elastic body.
(c) *Mixed problem*–when loads are prescribed on part of the surface and displacements on the other.

In Chapter Two Kirchhoff's proof of the uniqueness of solutions of problems of the theory of elasticity will be given.

BIBLIOGRAPHY

Sections 1.1, 1.2, 1.3, 1.4, 1.5, 1.6. STRESS; STRAIN AND STRAIN DISPLACEMENT RELATIONS (CAUCHY'S EQUATIONS); RELATIONSHIPS BETWEEN STRESS AND STRAIN (GENERALIZED HOOKE'S LAW); EQUATIONS OF EQUILIBRIUM (NAVIER'S EQUATIONS, LAMÉ'S EQUATIONS); COMPATIBILITY EQUATIONS (SAINT-VENANT'S EQUATIONS, BELTRAMI-MICHELL'S EQUATIONS); BOUNDARY CONDITIONS

1. Almansi, E., "Introduzione alla Scienza delle Costruzioni," Clausen, Torino, Italy, 1901.
2. ——, "Sulle Equazioni dell'Elasticità," *Atti Accad. Nazl. Lincei, Rend.*, Serie V, Vol. XVI, 1st Sem. 1907, pp. 23–26.
3. Beltrami, E., "Sulle Equazioni Generali dell'Elasticità," *Ann. Mat.*, Serie II, T. X.,

1880–1882, pp. 188–211; *Nuovo Cimento, Serie III*, **T. XX,** 1886, pp. 186–192; **T. XXI,** 1887, pp. 25–36, 115–121.

4. ———, "Sull'Uso delle Coordinate Curvilinee nelle Teorie del Potenziale e dell'Elasticità," *Rend. Ist. Bologna*, 1884–85, pp. 76–78; *Mem. Ist. Bologna, Serie IV*, **T. VI,** 1884, pp. 401–448; *Nuovo Cimento, Serie III*, **T. XVIII,** 1885, pp. 190–192, 228–246; **T. XIX,** 1886, pp. 90–95, 97–121.

5. Betti, E., "Sopra le Equazioni di Equilibrio dei Corpi Solidi Elastici," *Ann. Mat. Serie II*, **T. VI,** 1873–1875, pp. 101–111.

6. Biezeno, C. B., and R. Grammel, "Engineering Dynamics," **Vol. II,** Blackie & Son, Ltd., Edinburgh, 1956.

7. Biot, M. A., "Mechanics of Incremental Deformations," John Wiley & Sons, Inc., New York, N.Y., 1965.

8. Cauchy, A. L., "Recherches sur l'Equilibre et le Mouvement Intérieur des Corps Solides ou Fluides, Elastiques ou non Elastiques," *Bull. Soc. Philomat.*, 1823.

9. ———, "De la Pression ou Tension dans un Corps Solide," *Excercices de Mathématiques*, 1827.

10. ———, "Sur la Condensation et la Dilatation des Corps Solides," *Exercices de Mathématiques*, 1827.

11. ———, "Sur les Relations qui existent dans l'Etat d'Equilibre d'un Corps Solide ou Fluide entre les Pressions ou Tensions et les Forces Accélératrices," *Exercices de Mathématiques*, 1827.

12. ———, "Sur les Equations qui expriment les Conditions d'Equilibre ou les Lois du Mouvement Intérieur d'un Corps Solide Elastique ou non Elastique," *Exercices de Mathématiques*, 1828.

13. ———, "Relations entre les Tensions et Déformations," *Exercices de Mathématiques*, 1829.

14. Cesáro, E., "Introduzione alla Teoria Matematica dell'Elasticità," Fratelli Bocca Editori, Torino, Italy, 1894.

15. Clebsh, A., "Théorie de l'Elasticité des Corps Solides," Leipzig, Germany, 1861. French translation by Barré de Saint-Venant and Flamant, Dunod, Paris, France, 1883.

16. Fung, Y. C., "Foundations of Solid Mechanics," Prentice-Hall, Inc., Englewood Cliffs, N.J., 1965.

17. Green, A. E., and W. Zerna, "Theoretical Elasticity," Oxford University Press, London, England, 1954.

18. Hooke, R., "Lectures de Potentia Restitutiva of Springs," London, England, 1678.

19. Karman, T. von, "Elastizität," Handworterbuch der Naturwissenschaften, **Vol. 3,** pp. 165–193, Verlag von Gustav Fischer, Jena, Germany, 1913. See also: Collected Works of Theodore von Karman, **Vol. I,** 1902–1913, pp. 420–468, Butterworth & Co. (Publishers) Ltd., London, England, 1956.

20. Lamé, G., "Leçons sur la Théorie Mathématique de l'Elasticité des Corps Solides," Paris, France, 1852.

21. Lamé, G., and B. P. E. Clapeyron, "Sur l'Equilibre Intérieur des Corps Solides Homogènes," *Acad. Sci.*, 1833.

22. Lauricella, G., "Sull'Equilibrio dei Corpi Elastici Isotropi," *Atti Accad. Nazl. Lincei, Rend.*, Serie V, **Vol. II**, 1st Sem. 1893, pp. 298–305.

23. ——, "Sull'Equilibrio dei Corpi Elastici Isotropi," *Nuovo Cimento, Serie III*, **T. XXXIV**, 1893, pp. 141–148.

24. ——, "Equilibrio dei Corpi Elastici Isotropi," *Ann. Scuola Normale Superiore Pisa*, **Vol. VIII**, 1895, pp. 1–120.

25. L'Hermite, R., "Résistance des Matériaux Théorique et Experimentale," **Tome I**, Dunod, Paris, France, 1959.

26. Lorenz, H., "Techniche Elastizitätslehre," Oldenbourg, Munich and Berlin, Germany, 1913.

27. Love, A. E. H., "A Treatise on the Mathematical Theory of Elasticity," 4th ed., Dover Publications, Inc., New York, N.Y., 1944.

28. Marcolongo, R., "Teoria Matematica dello Equilibrio dei Corpi Elastici," U. Hoepli, Milan, Italy, 1904.

29. Morera, G., "Sulle Equazioni Generali per l'Equilibrio dei Sistemi Continui a Tre Dimensioni," *Atti Accad. Torino*, **Vol. XX**, 1884–1885, pp. 43–53.

30. ——, "Intorno all'Equilibrio dei Corpi Elastici Isotropi," *Atti Accad. Torino*, **Vol. XLII**, 1906–1907, pp. 676–686.

31. Navier, L. M. H., "Recherches sur la Flexion des Plans Elastiques," *Bull. Soc. Philomat.*, 1825.

32. ——, "Mémoire sur les Lois de l'Equilibre et du Mouvement des Corps Solides Elastiques," *Acad. Sci.*, 1827.

33. ——, "Résumé des leçons données a l'Ecole des Ponts et Chausées sur l'application de la mécanique a l'établissement des constructions et des machines," Carilian-Goeury, Paris, France, 1833–1838.

34. Pérès, J., "Mécanique Générale," Masson, Paris, France, 1953.

35. Poincaré, H., "Leçons sur la Théorie de l'Elasticité," Dunod, Paris, France, 1882.

36. Poisson, S. D., "Sur l'Equilibre et le Mouvement des Corps Elastiques," *Acad. Sci.*, 1829.

37. ——, "Sur l'Equilibre et le Mouvement des Corps Cristalisés," *J. Ecole Polytech.*, 1831.

38. Prager, W., "Introduction to Mechanics of Continua," Ginn & Company, Boston, Mass., 1961.

39. Roy, M., "Mécanique des Milieux Continus et Déformables," Gauthier-Villars, Paris, 1950.

40. Sechler, E. E., "Elasticity in Engineering," John Wiley & Sons, Inc., New York, N.Y., 1952.

41. Sokolnikoff, I. S., "Mathematical Theory of Elasticity," McGraw-Hill Book Company, New York, N.Y., 1956.

42. Somigliana, C., "Sulle Equazioni dell'Elasticità," *Ann. Mat.*, Serie II, **T. XVII**, 1889–1890, pp. 37–64.

43. Sommerfeld, A., "Mechanics of Deformable Bodies," Lectures on Theoretical Physics, **Vol. II**, Academic Press, Inc., New York, N.Y., 1950.

44. Southwell, Sir Richard, "An Introduction to the Theory of Elasticity for Engineers and Physicists," Oxford University Press, London, England, 1941.
45. Tedone, O., "Sulle Equazioni dell'Elasticità in Coordinate Curvilinee," *Atti Accad. Torino,* **Vol. XXXIV,** 1898–1899, pp. 1054–1061.
46. Thomson, Sir William (Lord Kelvin), "Elasticity," *The Encyclopaedia Britannica,* 9th ed. (Reprinted in his *Mathematical and Physical Papers,* Cambridge, 1890.)
47. Thomson, W., and P. G. Tait, "Treatise on Natural Philosophy," 2nd ed., Cambridge University Press, London, England, 1879–1883.
48. Timoshenko, S. P., "History of Strength of Materials," McGraw-Hill Book Company, New York, N.Y., 1953.
49. Timoshenko, S. P., and J. N. Goodier, "Theory of Elasticity," McGraw-Hill Book Company, New York, N.Y., 1951.
50. Todhunter, I., and K. Pearson, "A History of the Theory of Elasticity and of the Strength of Materials," Cambridge University Press, **Vol. 1,** 1886; **Vol. 2,** 1893. Reprinted by Dover Publications, Inc., New York, N.Y., 1960.
51. Voigt, W., "Theoretische Studien uber die Elastizitäts Verhältnisse der Krystalle," Göttingen, Germany, 1887.
52. Wang, C. T., "Applied Elasticity," McGraw-Hill Book Company, New York, N.Y., 1953.
53. Westergaard, H. M., "Theory of Elasticity and Plasticity," Harvard University Press, Cambridge, Mass., 1952.

Section 1.7. THERMAL STRESS EQUATIONS (DUHAMEL-NEUMANN EQUATIONS)

1. Boley, B. A., and J. H. Weiner, "Theory of Thermal Stresses," John Wiley & Sons, Inc., New York, N.Y., 1960.
2. Borchardt, C. W., "Untersuchungen über die Elastizität Fester Isotroper Körper unter Berücksichtigung der Wärme," *Monatsber. Berliner Akad.,* January 1873, p. 9.
3. Carslaw, H. S., and J. C. Jaeger, "Conduction of Heat in Solids," 2nd ed., Oxford University Press, London, England, 1959.
4. Duhamel, J. M., "Mémoire sur le Calcul des Actions Moléculaires Developées par les Changements de Température," *Acad. Sci.,* 1838.
5. Fung, Y. C., "Foundations of Solid Mechanics," Prentice-Hall, Inc., Englewood Cliffs, N.J., 1965.
6. Gatewood, B. E., "Thermal Stresses," McGraw-Hill Book Company, New York, N.Y., 1957.
7. Hoff, N. J. (Ed.), "High Temperature Effects in Aircraft Structures," Pergamon, London, England, 1958.
8. L'Hermite, R., "Résistance des Matériaux Théorique et Experimentale," **Tome I,** Dunod, Paris, France, 1959.
9. Love, A. E. H., "A Treatise on the Mathematical Theory of Elasticity," 4th ed., Dover Publications, Inc., New York, N.Y., 1944.
10. Melan, E., and H. Parkus, "Wärmespannunge," Julius Springer, Vienna, Austria, 1953.

11. Neumann, F. E., "Die Gesetze der Doppelbrechung des Lichts in Comprimirten oder ungleichforming erwarmten unkrystallinischen Körpern," *Abhandl. Königl. Akad. Wiss. Berlin*, Zweiter Teil, 1841, pp. 1–254.
12. Nowacki, W., "Thermoelasticity," Addison-Wesley, Reading, Mass., 1962.
13. Parkus, H., "Thermal Stresses," Chapter 43, *Handbook of Engineering Mechanics*, W. Flugge (Ed.), McGraw-Hill Book Company, New York, N.Y., 1962.
14. Timoshenko, S. P., and J. N. Goodier, "Theory of Elasticity," McGraw-Hill Book Company, New York, N.Y., 1951.

Section 1.8. SAINT-VENANT'S PRINCIPLE

1. Boley, B. A., "Some Observations on Saint-Venant's Principle," *Proc. Third U.S. Natl. Congr. Appl. Mech.*, Providence, R.I., 1958, pp. 259–264.
2. Fung, Y. C., "Foundations of Solid Mechanics," Prentice-Hall, Inc., Englewood Cliffs, N.J., 1965.
3. Goodier, J. N., "A General Proof of Saint-Venant's Principle," *Phil. Mag.*, Series 7, **Vol. 23,** 1937, p. 607; **Vol. 24,** 1937, p. 325.
4. Hoff, N. J., "The Applicability of Saint-Venant's Principle to Airplane Structures," *J. Aeron. Sci.*, **Vol. 12,** 1945, pp. 455–460.
5. Horvay, G., "Some Aspects of Saint-Venant's Principle," *J. Mech. Phys. Solids*, **Vol. 5,** 1957, p. 77.
6. Locatelli, P., "Estensione del Principio di Saint-Venant a Corpi non perfettamente Elastici," *Atti Accad. Sci. Torino*, **Vol. 75,** 1940, p. 502; **Vol. 76,** 1941, p. 125.
7. Naghdi, P. M., "On Saint-Venant's Principle. Elastic Shells and Plates," *Univ. Calif. Inst. Eng. Res.*, **Tr. 1,** 1959.
8. Saint-Venant, B., "Mémoire sur la Torsion des Prismes," *Mémoire des Savants Etrangers*, Paris, France, 1855.
9. Southwell, Sir Richard, "On Castigliano's Theorem of Least Work and the Principle of Saint-Venant," *Phil. Mag.*, Series 6, **Vol. 45,** 1923, p. 193.
10. Sternberg, E., "On Saint-Venant's Principle," *Quart. Appl. Math.*, **Vol. 11,** 1954, pp. 393–402.
11. Von Mises, R., "On Saint-Venant's Principle," *Bull. Am. Math. Soc.*, **Vol. 51,** 1945, p. 555.
12. Zanaboni, O., "Dimostrazione Generale del Principio di Saint-Venant," *Atti Accad. Nazl. Lincei*, **Vol. 25,** 1937, p. 117.
13. ———, "Valutazione dell'Errore Massimo cui da luogo l'Applicazione del Principio del De Saint-Venant in un Solido Isotropo," *Atti Accad. Nazl. Lincei*, **Vol. 25,** 1937, p. 595.

Section 1.9. THE THREE BASIC PROBLEMS OF THE THEORY OF ELASTICITY

1. Betti, E., "Teoria dell'Elasticità," *Nuovo Cimento, Serie II*, **T. VII–VIII,** 1872, pp. 5–21, 69–97, 158–180; **T. IX,** 1873, pp. 34–43; **T. X,** pp. 58–84.
2. Biezeno, C. B., and R. Grammel, "Engineering Dynamics," **Vol. 11,** Blackie & Son, Ltd., Edinburgh, Scotland, 1956.

3. Block, V. I., "Stress-Functions in the Theory of Elasticity," *Prikl. Mat. i Mekh.*, **Vol. 14,** 1950, p. 415 (in Russian).

4. Boley, B. A., "A Method for the Construction of Fundamental Solutions in Elasticity Theory," *J. Math. Phys.*, **Vol. 3,** 1957, p. 261.

5. Boussinesq, M. J., "Application des Potentiels à l'Etude de l'Equilibre et du Mouvement des Solides Elastiques," Gauthier-Villars, Paris, France, 1885.

6. Cerruti, V., "Ricerche intorno all'Equilibrio dei Corpi Elastici Isotropi," *Atti Accad. Nazl. Lincei, Mem.*, *Serie III*, **Vol. XII,** 1881–1882, pp. 81–123.

7. ———, "Sulla Deformazione di un Corpo Elastico Isotropo per alcune Speciali Condizioni ai Limiti," *Atti Accad. Nazl. Lincei, Rend.*, *Serie IV.*, **Vol. IV,** 1st Sem. 1888, pp. 785–792; *Nuovo Cimento, Serie III*, **T. XXXIV,** 1893, pp. 115–124.

8. Clebsh, A., "Théorie de l'Elasticité des Corps Solides," Leipzig, 1861. French translation by Barré de Saint-Venant and Flamant, Dunod, Paris, France, 1883.

9. Cesáro, E., "Introduzione alla Teoria Matematica della Elasticità," Fratelli Bocca Editori, Torino, Italy, 1894.

10. Eubanks, R. A., and E. Sternberg, "On the Completeness of the Boussinesq-Papkovitch Stress Functions," *J. Rat. Mech. Anal.*, **Vol. 5,** 1956, p. 735.

11. Finzi, B., "Integrazione delle Equazioni Indefinite della Meccanica dei Sistemi Continui," *Atti Accad. Nazl. dei Lincei, Serie 6*, **Vol. 19,** 1934, p. 578.

12. Fredholm, I., "Sur les Equations de l'Equilibre d'un Corps Solide Elastique," *Acta Math.*, **Vol. 23,** 1900, p. 1.

13. Fung, Y. C., "Foundations of Solid Mechanics," Prentice-Hall, Inc., Englewood Cliffs, N.J., 1965.

14. Galerkin, B., "Contribution à la Solution Générale du Problème de la Théorie de l'Elasticité dans le Cas de Trois Dimensions," *Compt. Rend. Acad. Sci.*, Paris, **Vol. 190,** 1930, p. 1047.

15. Goodier, J. N., "A Survey of some Recent Researches in Theory of Elasticity," *Appl. Mech. Rev.*, **Vol. 4,** 1951, p. 330.

16. Green, A. E., and W. Zerna, "Theoretical Elasticity," Oxford University Press, London, England, 1954.

17. Krutkov, Y. A., "The Tensor of Stress-Functions and General Solutions in Elastostatics," Moscow, U.S.S.R., 1949 (in Russian).

18. Lamé, G., "Leçons sur la Théorie Mathématique de l'Elasticité des Corps Solides," Paris, France, 1852.

19. Lamé, G., and B. P. E. Clapeyron, "Sur l'Equilibre Intérieur des Corps Solides Homogènes," *Acad. Sci.*, 1833.

20. Lauricella, G., "Formole Generali Relative all'Integrazione delle Equazioni dell'Equilibrio dei Corpi Elastici. Applicazione al Caso di un Corpo Elastico Sferico," *Nuovo Cimento, Serie III*, **T. XXXVI,** 1894, pp. 314–321.

21. ———, "Studio degli Integrali del Somigliana relativi alla Elasticità," *Nuovo Cimento, Serie III*, **T. XXXVI,** 1894, pp. 225–235.

22. ———, "Sull'Integrazione delle Equazioni dell'Equilibrio dei Corpi Elastici," *Nuovo Cimento, Serie IV*, **T. I,** 1895, pp. 155–165.

23. ——, "Sull'Integrazione delle Equazioni dell'Equilibrio Elastico," *Ann. Mat., Serie II*, T. XXIII, 1895, pp. 288–308.

24. ——, "Sull'Integrazione delle Equazioni dell'Equilibrio dei Solidi Elastici Isotropi per dati Spostamenti in Superficie," *Nuovo Cimento, Serie IV*, T. IX, 1899, pp. 97–109; T. X, pp. 5–19.

25. ——, "Sull'Integrazione delle Equazioni dell'Equilibrio dei Corpi Elastici Isotropi," *Ann. Mat., Serie III*, T. XI, 1905, pp. 269–283.

26. ——, "Sull'Integrazione delle Equazioni dell'Equilibrio dei Corpi Elastici Isotropi," *Atti Accad. Nazl. Lincei, Rend., Serie V*, Vol. XV, 1st Sem., 1906, pp. 426–432.

27. ——, "Sulla Risoluzione del Problema di Dirichlet col Metodo di Fredholm e sull'Integrazione delle Equazioni dell'Equilibrio dei Solidi Elastici Indefiniti," *Atti Accad. Nazl. Lincei, Rend., Serie V*, Vol. XV, 1st Sem., 1906, pp. 611–619.

28. ——, "Sul Problema derivato di Dirichlet, sul Problema dell'Eletrostatica e sull'Integrazione delle Equazioni dell'Elasticita," *Atti Accad. Nazl. Lincei, Rend., Serie V*, Vol. XV, 2nd Sem., 1906, pp. 75–83.

29. L'Hermite, R., "Resistance des Matériaux Théorique et Experimentale," **Tome I**, Dunod, Paris, France, 1959.

30. Lorenz, H., "Techniche Elastizitätslehre," Oldenbourg, Munich and Berlin, Germany, 1913.

31. Love, A. E. H., "A Treatise on the Mathematical Theory of Elasticity," 4th ed., Dover Publications, Inc., New York, N.Y., 1944.

32. Marcolongo, R., "Teoria Matematica dello Equilibrio dei Corpi Elastici," U. Hoepli, Milano, Italy, 1904.

33. Mindlin, R. D., "Note on the Galerkin and Papkovitch Stress Functions," *Bull. Am. Math. Soc.*, Vol. 42, 1936, p. 373.

34. Morera, G., "Soluzione Generale delle Equazioni Indefinite dell'Equilibrio di un Corpo Continuo," *Atti Accad. Nazl. Lincei, Rend., Serie V;* Vol. I, 1st Sem., 1892, pp. 137–141.

35. ——, "Appendice alla Nota: Sulla Soluzione Generale delle Equazioni Indefinite, ecc.," *Atti Accad. Nazl. Lincei, Rend., Serie V*, Vol. 1, 1st Sem., 1892, pp. 233–234.

36. Orlando, L., "Sopra alcuni Problemi di Equilibrio Elastico," *Nuovo Cimento, Serie V;* T. VII, 1904, pp. 161–165.

37. Papkovitch, P. F., "Solution Generale des Equations Differéntielles Fondamentales d'Elasticité Exprimée par Trois Fonctions Harmoniques," *Compt. Rend. Acad. Sci.*, Paris, Vol. 195, 1932, p. 513.

38. ——, "Expressions Générales des Composantes des Tensions ne Renfermant comme Fonctions Arbitraires que des Fonctions Harmoniques," *Compt. Rend. Acad. Sci.*, Paris, Vol. 195, 1932, p. 754.

39. Poincaré, H., "Leçons sur la Théorie de l'Elasticité," Dunod, Paris, France, 1882.

40. Schaefer, H., "Die Spannungsfunktionen des Dreidimensionalen Kontinuums und des Elastischen Körpers," *Z. Angew. Math. Mech.*, Vol. 33, 1953, p. 356.

41. Slobodiansky, M. G., "General Forms of Solutions in Terms of Harmonic Functions to the Equations of Elasticity for Simply and Multiply Connected Regions," *Prikl. Mat. i Mekj.*, Vol. 18, 1954, p. 55 (in Russian).

42. Sokolnikoff, I. S., "Mathematical Theory of Elasticity," McGraw-Hill Book Company, New York, N.Y., 1956.

43. Somigliana, C., "Sopra l'Equilibrio di un Corpo Elastico Isotropo," *Nuovo Cimento, Serie III*, T. XVII, 1885, pp. 140–148, 272–276; T. XVIII, pp. 91–96, 161–166; T. XIX, 1886, pp. 84–90, 278–282; T. XX, pp. 181–185.

44. ———, "Sopra gli Integrali delle Equazioni della Isotropia Elastica," *Nuovo Cimento, Serie III*, T. XXXVI, 1894, pp. 28–39, 113–126.

45. ———, "Sul Potenziale Elastico," *Ann. Mat., Serie III*, T. VII, 1902, pp. 129–140.

46. ———, "Sull'Applicazione del Metodo delle Immagini alle Equazioni dell'Elasticità," *Atti Accad. Nazl. Lincei, Rend., Serie V;* **Vol. XIII**, 1st Sem., 1904, pp. 307–318.

47. ———, "Le Deformazioni Ausiliarie nei Problemi Alterni di Equilibrio Elastico," *Atti Accad. Nazl. Lincei, Rend., Serie V*, **Vol. XIII**, pp. 129–141.

48. Sommerfeld, A., "Mechanics of Deformable Bodies," Lectures on Theoretical Physics, **Vol. II**, Academic Press, Inc., New York, N.Y., 1950.

49. Southwell, Sir Richard, "An Introduction to the Theory of Elasticity for Engineers and Physicists," Oxford University Press, London, England, 1941.

50. Sternberg, E., "On some Recent Developments in the Linear Theory of Elasticity," *Structural Mech. Proc. First Symp. Naval Structural Mech.*, (Ed. J. N. Goodier and N. J. Hoff), Pergamon Press, London, England, 1960.

51. Tedone, O., "Sulla Integrazione delle Equazioni dei Corpi Elastici," *Atti Accad. Nazl. Lincei, Rend., Serie V*, **Vol. V**, 2nd Sem., 1896, pp. 460–467.

52. ———, "Sulle Formule che rappresentano lo Spostamento di un Punto di un Corpo Elastico in Equilibrio," *Nuovo Cimento, Serie IV*, T. XI, 1900, pp. 161–172.

53. ———, "Su alcuni Problemi di Equilibrio Elastico," *Atti Accad. Nazl. Lincei, Rend., Serie V*, **Vol. X**, 2nd Sem., 1901, pp. 251–258, 294–296.

54. ———, "Saggio di una Teoria Generale delle Equazioni dell'Equilibrio Elastico per un Corpo Isotropo," *Mem. I e Mem. II, Ann. Mat., Serie III*, T. VIII, 1902, pp. 129–180.

55. Thomson, Sir William (Lord Kelvin), "Elasticity," *The Encyclopaedia Britannica,* 9th ed. (Reprinted in his *Mathematical and Physical Papers,* Cambridge, 1890.)

56. Thomson W., and P. G. Tait, "Treatise on Natural Philosophy," 2nd ed., Cambridge University Press, London, England, 1879–1883.

57. Timoshenko, S. P., "History of Strength of Materials," McGraw-Hill Book Company, New York, N.Y., 1953.

58. Timoshenko, S. P., and J. N. Goodier, "Theory of Elasticity," McGraw-Hill Book Company, New York, N.Y., 1951.

59. Todhunter, I., and K. Pearson, "A History of the Theory of Elasticity and of the Strength of Materials," Cambridge University Press, **Vol. 1**, 1886; **Vol. 2**, 1893. Reprinted, Dover Publications, Inc., New York, N.Y., 1960.

60. Wang, C. T., "Applied Elasticity," McGraw-Hill Book Company, New York, N.Y., 1953.

61. Weber, C., "Spannungsfunktionen des Dreidimensionalen Kontinuums," *Z. Angew. Math. Mech.*, **Vol. 28**, 1948, p. 193.

62. Westergaard, H. M., "Theory of Elasticity and Plasticity," Harvard University Press, Cambridge, Mass., 1952.

PROBLEMS[1]

1-1. The strain in the direction of a thin straight rod is $\epsilon = 0.0004 + 0.00001x$ where x is the distance in inches from the end of the rod. Determine the change in length of the rod, if it is originally 50 in. long.

1-2. A square plate $ABCD$ has sides 2 in. in length. The x-axis is along side AB and the y-axis is along side AD. The plate is subjected to the shear strain $\gamma_{xy} = 0.004x + 0.002y$, where x and y are in inches. If side AB is fixed and lines parallel to AB remain parallel to it, determine the displacement of point D in the x-direction. Determine the change in the angle CDA.

1-3. The displacement components in a strained body are

$$u = 0.01x + 0.002y^2$$
$$v = 0.02x^2 + 0.02z^3$$
$$w = 0.001x + 0.005$$

What is the change in distance between two points which, before deformation, have coordinates $(3, 2, 0)$ and $(-1, 14, 5)$?

1-4. Given the displacements of Problem 1-3, determine the state of strain at point $(3, 1, 2)$.

1-5. The stress components at a point P are given by

$$\sigma_x = 2 + y^2 \quad \tau_{xy} = z$$
$$\sigma_y = 2x + z^3 \quad \tau_{yz} = x^3$$
$$\sigma_z = x^2 + y^2 \quad \tau_{xz} = y^3$$

Assuming that $E = 30 \times 10^6$ psi and $\nu = 0.29$, determine the state of strain at point $(2, 1, 3)$.

1-6. The stress components at a point P are given by

$$\sigma_x = y + 3z^2 \quad \tau_{xy} = z^3$$
$$\sigma_y = x + 2z \quad \tau_{yz} = x^2$$
$$\sigma_z = 2x + y \quad \tau_{xz} = y^2$$

Assuming that $E = 10 \times 10^6$ psi and $\nu = 0.30$, determine the state of strain at point $(3, 1, 2)$.

1-7. A steel bar with a 2 by 1 in. rectangular cross section is subjected to an axial pull of 15 tons. Determine the changes in the lengths of the sides of the cross section.

1-8. A solid cylinder 50 in. long and 2 in. in diameter is subjected to a tensile force of 20,000 lbs. One part of this cylinder of length L_1 is made of steel and the other part of length L_2 is made of aluminum.

[1] Unless specified differently, use the physical properties of common metals given by Table 1.3.2.

(a) Determine the lengths L_1 and L_2 so that the two parts elongate an equal amount.

(b) what is the total elongation of the cylinder?

1-9. A cube of steel the length of whose sides is 50 in. is subjected to a uniform pressure of 12 ton/in.² on two opposite faces. The other faces are prevented by lateral pressure from extending more than 0.015 in. Find the lateral pressure.

1-10. A bar weighing 200 lbs is held in a horizontal position by three vertical wires, one of steel attached to each end of the beam and one of brass attached to the middle of its length. The steel wires are 0.05 in. in diameter while the brass wire is 0.04 in. in diameter, all being of the same length. Find the stress in each wire.

1-11. Two vertical wires, one of brass and the other of copper, having the same lengths, are very close together and hold a weight of 100 lb. The brass wire has a diameter of 0.02 in. while the copper wire has a diameter of 0.03 in. Determine the force in each wire.

1-12. Two brass wires having equal diameters of 0.2 in. are attached at the same point and hold a weight of 100 lb. One of the wires has a length of 20 ft, while the other has a length of 20 ft and 0.015 in. Determine the stresses in the two wires.

1-13. Given the stress field of Problem 1-5, determine the body force distribution required for equilibrium.

1-14. Given the stress field of Problem 1-6, determine the body force distribution required for equilibrium.

1-15. Given the following stress field:

$$\sigma_x = 20x^3 + y^2 \text{ psi} \qquad \tau_{xy} = 100 + 80y^2 \text{ psi}$$
$$\sigma_y = 30x^3 + 100 \text{ psi} \qquad \tau_{yz} = 0 \text{ psi}$$
$$\sigma_z = 30y^2 + 30z^3 \text{ psi} \qquad \tau_{zx} = xz^3 + 30x^2y \text{ psi}$$

Find the body-force distribution that is required for equilibrium.

1-16. Does the stress field given in Problem 1-13 satisfy the compatibility equations?

1-17. Is the following state of strain

$$\epsilon_x = A(x^2 + y^2) \quad \epsilon_y = Ay^2 \quad \gamma_{xy} = 2Axy \quad \epsilon_z = \gamma_{xz} = \gamma_{yz} = 0$$

a possible one? A is a small constant.

1-18. The stress components at point P are given with respect to an xyz-coordinate system as

$$\sigma_x = 2 + y^2 \qquad \tau_{xy} = z$$
$$\sigma_y = 2x + z^3 \qquad \tau_{yz} = x^3$$
$$\sigma_z = x^2 + y^2 \qquad \tau_{zx} = y^3$$

In the absence of body forces, is this a physically possible stress distribution?

1-19. The stress components at point P are given with respect to an xyz-coordinate system as

$$\sigma_x = y + 3z^2 \quad \sigma_y = x + 2z$$
$$\tau_{xy} = z^3 \quad \tau_{zx} = x^2$$
$$\tau_{xz} = y^2 \quad \sigma_z = 2x + y$$

Determine if this is a physically possible stress distribution in the absence of body forces.

1-20. How much should a bar of brass with fixed ends be cooled in order to break it?

1-21. How much should a bar of cast iron with fixed ends be cooled in order to break it?

1-22. A bar of brass having a cross-sectional area of 0.20 in.^{2} is fastened to a bar of cast iron having a cross-sectional area of 0.40 in.^{2} when the temperature is 60° F. If the composite bar is fixed at the ends, find the stress in each part when the temperature is increased to 300° F. Assume that the brass and cast iron have the same length when the temperature is 60° F.

CHAPTER

2

ENERGY PRINCIPLES AND GENERAL THEOREMS

2.0 Introduction

In this chapter energy principles and general theorems are discussed. First, the strain energy of an elastic solid is defined. Then various strength theories are presented, among which are the Huber-Von Mises-Hencky theory, the Rankine theory, the Coulomb-Guest theory, the Saint-Venant theory, and finally the Beltrami-Haigh theory. The "Principle of Virtual Work" is then enunciated and is used for some important applications, among which is the derivation of the equations of equilibrium and the boundary conditions for an elastic solid. Kirchhoff's proof of the uniqueness of a solution for the elasticity problem of a simply connected solid is given. Betti's and Maxwell's reciprocity theorems are formulated and various applications of these theorems are shown. Castigliano's first theorem and Donati's extension of the above theorem to elastic continuous solids are discussed. Engesser's theorem on the principle of complementary energy follows and the chapter ends with the presentation of Castigliano's second theorem or Menabrea's theorem.

2.1 Strain Energy

Consider the small rectangular parallelepiped of sides dx, dy, and dz (see Fig. 2.1.1) inside an elastic body subjected to an external system of forces which vary slowly until they reach their final values.[1] The normal forces

$$\sigma_x dydz, \quad \sigma_y dzdx, \quad \sigma_z dxdy \qquad (2.1.1)$$

and the shear forces

$$\tau_{xy} dydz, \quad \tau_{xz} dydz, \quad \tau_{yz} dzdx, \quad \tau_{yx} dzdx, \quad \tau_{zx} dxdy, \quad \text{and} \quad \tau_{zy} dxdy \qquad (2.1.2)$$

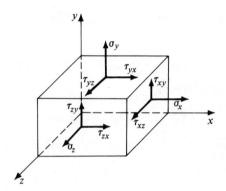

Figure 2.1.1

act on the six faces of the parallelepiped. Under the action of these forces, the faces of the parallelepiped will displace in the normal directions by the amounts $\epsilon_x dx$, $\epsilon_y dy$, and $\epsilon_z dz$ and distort by the amounts γ_{xy}, γ_{yz}, and γ_{zx}. The total work done by the forces in Eqs. (2.1.1) and (2.1.2) represents the mechanical energy stored in the element of volume $dxdydz$ and is called the *strain energy* of the element. The strain energy does not depend on the manner in which the forces in Eqs. (2.1.1) and (2.1.2) are applied but it depends only on their final values, as will be shown at the end of this section. The relation between the normal force $\sigma_x dydz$ and the extension $\epsilon_x dx$ as this force increases from zero to its final value is represented by the straight line OA in Fig. 2.1.2. The corresponding work is numerically equal to the area of the triangle OAB and is given by

$$\tfrac{1}{2}(\epsilon_x dx)(\sigma_x dydz) = \tfrac{1}{2}\sigma_x \epsilon_x dxdydz = \tfrac{1}{2}\sigma_x \epsilon_x dV$$

where $dV = dxdydz$ is the volume of the parallelepiped. For the other normal forces one gets

---

[1] The externally applied forces and therefore the internal systems of stresses and strains are assumed to vary slowly so that inertial effects can be neglected.

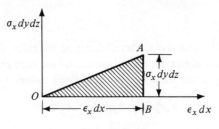

Figure 2.1.2

$$\tfrac{1}{2}\sigma_y\epsilon_y dV \quad \text{and} \quad \tfrac{1}{2}\sigma_z\epsilon_z dV$$

In the same way, for the shear force $\tau_{xy}dydz$ (see Fig. 2.1.3) the work is given by the area of triangle OAB, i.e.,

$$\tfrac{1}{2}(\gamma_{xy}dx)\tau_{xy}dydz = \tfrac{1}{2}\tau_{xy}\gamma_{xy}dxdydz = \tfrac{1}{2}\tau_{xy}\gamma_{xy}dV$$

and for the other shear forces by

$$\tfrac{1}{2}\tau_{yz}\gamma_{yz}dV \quad \text{and} \quad \tfrac{1}{2}\tau_{zx}\gamma_{zx}dV$$

The total strain energy stored in the parallelepiped is given by

$$dU = \tfrac{1}{2}[\sigma_x\epsilon_x + \sigma_y\epsilon_y + \sigma_z\epsilon_z + \tau_{xy}\gamma_{xy} + \tau_{yz}\gamma_{yz} + \tau_{zx}\gamma_{zx}]\,dV$$

and the total strain energy stored in the elastic body of volume V is given by

$$U = \frac{1}{2}\int_V [\sigma_x\epsilon_x + \sigma_y\epsilon_y + \sigma_z\epsilon_z + \tau_{xy}\gamma_{xy} + \tau_{yz}\gamma_{yz} + \tau_{zx}\gamma_{zx}]\,dV \quad (2.1.3)$$

In view of Eqs. (1.3.1) and (1.3.5), Eq. (2.1.3) can be written in the form

$$U = \frac{1}{4G}\int_V \left[\sigma_x^2 + \sigma_y^2 + \sigma_z^2 - \frac{\nu}{1+\nu}(\sigma_x + \sigma_y + \sigma_z)^2 + 2(\tau_{xy}^2 + \tau_{yz}^2 + \tau_{zx}^2)\right]dV \quad (2.1.4)$$

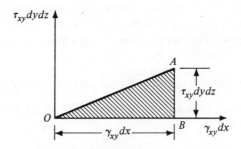

Figure 2.1.3

and in view of Eqs. (1.3.4) and (1.3.5), Eq. (2.1.3) can also be written in the form

$$U = G \int_V \left[\epsilon_x^2 + \epsilon_y^2 + \epsilon_z^2 + \frac{\nu e^2}{1 - 2\nu} + \frac{1}{2}(\gamma_{xy}^2 + \gamma_{yz}^2 + \gamma_{zx}^2) \right] dV \quad (2.1.5)$$

or

$$U = \int_V \left[\frac{\lambda}{2} e^2 + G(\epsilon_x^2 + \epsilon_y^2 + \epsilon_z^2) + \frac{G}{2}(\gamma_{xy}^2 + \gamma_{yz}^2 + \gamma_{zx}^2) \right] dV \quad (2.1.6)$$

where $e = \epsilon_x + \epsilon_y + \epsilon_z$. To prove that the strain energy of an elastic solid is independent of the manner in which the stresses are applied to the element and it depends only on their final values, suppose that for a parallelepiped of volume $dV = 1$, inside the elastic solid, stress σ_x increases first from zero to its final value, then normal stresses σ_y and σ_z increase from zero to their final values. Shear stresses are assumed to increase in any arbitrary manner. The work done by the normal forces will be:

during the increase of σ_x: $\quad \frac{1}{2}\left(\frac{\sigma_x}{E}\right)\sigma_x$

during the increase of σ_y: $\quad \frac{1}{2}\left(\frac{\sigma_y}{E}\right)\sigma_y - \nu\left(\frac{\sigma_y}{E}\right)\sigma_x$

during the increase of σ_z: $\quad \frac{1}{2}\left(\frac{\sigma_z}{E}\right)\sigma_z - \nu\left(\frac{\sigma_z}{E}\right)\sigma_x - \nu\left(\frac{\sigma_z}{E}\right)\sigma_y$

while the work due to shear will be

$$\frac{1}{2}\left(\frac{\tau_{xy}}{G}\right)\tau_{xy} + \frac{1}{2}\left(\frac{\tau_{yz}}{G}\right)\tau_{yz} + \frac{1}{2}\left(\frac{\tau_{zx}}{G}\right)\tau_{zx}$$

Adding all these works together, one obtains

$$\frac{1}{2E}[\sigma_x^2 + \sigma_y^2 + \sigma_z^2 - 2\nu(\sigma_x\sigma_y + \sigma_y\sigma_z + \sigma_z\sigma_x)] + \frac{1}{2G}[\tau_{xy}^2 + \tau_{yz}^2 + \tau_{zx}^2]$$

$$= \frac{1}{4G}\left[\sigma_x^2 + \sigma_y^2 + \sigma_z^2 - \frac{\nu}{(1+\nu)}(\sigma_x + \sigma_y + \sigma_z)^2 + 2(\tau_{xy}^2 + \tau_{yz}^2 + \tau_{zx}^2)\right]$$

which is identical to Eq. (2.1.4).

Example 2.1.1. Evaluate the strain energy stored in the circular bar of length l, subjected to the twisting moment M_T applied at its free end as shown in Fig. 2.1.4. One uses the Coulomb formula for torsion of circular shafts;[2] thus

$$\tau_\rho = \frac{M_T \rho}{J}$$

[2] See Eq. (4.3.1).

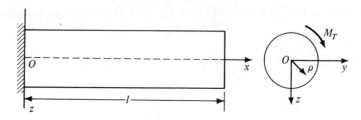

Figure 2.1.4

where $J = \pi D^4/32$ is the polar moment of inertia of the cross-sectional area of the shaft of diameter D, ρ is the distance of a generic point of the cross section from the axis of the bar. In this case

$$\tau_\rho^2 = \tau_{xy}^2 + \tau_{xz}^2$$

and

$$U = \frac{1}{2G} \iiint_V (\tau_{xy}^2 + \tau_{xz}^2)\, dxdydz = \frac{1}{2G} \iiint_V \tau_\rho^2 dxdydz$$
$$= \frac{1}{2GJ^2} \int_l (M_T)^2 dx \iint_\Omega \rho^2 dydz = \frac{1}{2G} \int_l \frac{M_T^2}{J} dx$$

Example 2.1.2. To evaluate the strain energy stored in the prismatic bar of length l (shown in Fig. 2.1.5) which is subjected to external forces acting in the plane of symmetry xz of the beam, one considers a generic section of the beam. The forces

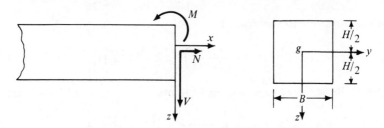

Figure 2.1.5

acting on the section of cross-sectional area Ω are a moment M, a shear force V, and a normal force N. The corresponding stresses are:

$$\sigma_x = \frac{N}{\Omega} + \frac{Mz}{I} \qquad \sigma_y = \sigma_z = 0$$
$$\tau_{xy} = \tau_{yz} = 0 \qquad \tau_{xz} = \frac{3}{2}\frac{V}{\Omega}\left[1 - 4\left(\frac{z}{H}\right)^2\right] \qquad (2.1.7)$$

By substituting Eqs. (2.1.7) into Eq. (2.1.4) one obtains, since

$$\iint_\Omega dydz = \Omega \quad \iint_\Omega z^2 dydz = I \quad \iint_\Omega z\, dydz = 0$$

that

$$U = \frac{1}{4G}\iiint_V \left(1 - \frac{v}{1+v}\right)\sigma_x^2\, dxdydz + \frac{1}{2G}\iiint_V \tau_{xz}^2\, dxdydz$$

$$= \frac{1}{2E}\iiint_V \left(\frac{N}{\Omega} + \frac{Mz}{I}\right)^2 dxdydz + \frac{9}{8G\Omega^2}\iiint_V V^2\left[1 - 4\left(\frac{z}{H}\right)^2\right]^2 dxdydz$$

$$= \frac{1}{2E\Omega^2}\int_l N^2 dx \iint_\Omega dydz + \frac{1}{2EI^2}\int_l M^2 dx \iint_\Omega z^2 dydz$$

$$+ \frac{1}{E\Omega I}\int_l NM dx \iint_\Omega z dydz + \frac{9}{8G\Omega^2}\int_l V^2 dx \int_{-B/2}^{B/2} dy \int_{-H/2}^{H/2}\left[1 - 4\left(\frac{z}{H}\right)^2\right]^2 dz$$

$$= \frac{1}{2}\int_l \frac{N^2}{E\Omega}\, dx + \frac{1}{2}\int_l \frac{M^2}{EI}\, dx + \frac{3}{5}\int_l \frac{V^2}{G\Omega}\, dx$$

or

$$U = \frac{1}{2}\int_l \frac{N^2}{E\Omega}\, dx + \frac{1}{2}\int_l \frac{M^2}{EI}\, dx + \frac{1}{2}\int_l \frac{\chi V^2}{G\Omega}\, dx \tag{2.1.8}$$

where the shear coefficient $\chi = \frac{6}{5} = 1.2$

In the case in which dimensions H and B of the cross section of the bar are small as compared with its length l, terms depending on the shear force V and normal force N are usually small as compared with the one depending on the bending moment M and can be neglected. Equation (2.1.8) reduces in this case to

$$U = \frac{1}{2}\int_l \frac{M^2}{EI}\, dx \tag{2.1.9}$$

The strain energy in the case of pin jointed trusses is given by

$$U = \sum_{i=1}^n \frac{S_i^2 l_i}{2E\Omega_i} \tag{2.1.10}$$

where S_i, l_i, and Ω_i are, respectively, the force, length, and cross-sectional area of ith member of the truss and n is the number of members in the truss.

2.2 Huber-Von Mises-Hencky Strength Theory

From Eq. (2.1.4) one sees that the strain energy per unit volume of an elastic solid can be written as

$$U_1 = \frac{1}{4G}\left[\sigma_x^2 + \sigma_y^2 + \sigma_z^2 - \frac{v}{1+v}(\sigma_x + \sigma_y + \sigma_z)^2 + 2(\tau_{xy}^2 + \tau_{yz}^2 + \tau_{zx}^2)\right] \tag{2.2.1}$$

and can be decomposed into two parts

$$U_1 = (U_1)_v + (U_1)_d \tag{2.2.2}$$

where

$$(U_1)_v = \frac{1-2\nu}{6E}(\sigma_x + \sigma_y + \sigma_z)^2 \qquad (2.2.3)$$

represents strain energy corresponding to a change in volume and

$$(U_1)_d = \frac{1}{12G}[(\sigma_x - \sigma_y)^2 + (\sigma_y - \sigma_z)^2 + (\sigma_z - \sigma_x)^2 \\ + 6(\tau_{xy}^2 + \tau_{yz}^2 + \tau_{zx}^2)] \qquad (2.2.4)$$

represents strain energy corresponding to a distortion of the elastic body. The decomposition in Eq. (2.2.2) is obtained from Eq. (2.2.1) as follows:

$$\begin{aligned}
U_1 &= \frac{1}{4G}\left[\sigma_x^2 + \sigma_y^2 + \sigma_z^2 - \frac{\nu}{1+\nu}(\sigma_x + \sigma_y + \sigma_z)^2 + 2(\tau_{xy}^2 + \tau_{yz}^2 + \tau_{zx}^2)\right] \\
&= \frac{1}{4G}\left\{\frac{1}{3}[(\sigma_x + \sigma_y + \sigma_z)^2 + (\sigma_x - \sigma_y)^2 + (\sigma_y - \sigma_z)^2 + (\sigma_z - \sigma_x)^2]\right. \\
&\quad \left. - \frac{\nu}{1+\nu}(\sigma_x + \sigma_y + \sigma_z)^2 + 2(\tau_{xy}^2 + \tau_{yz}^2 + \tau_{zx}^2)\right\} \\
&= \frac{1}{4G}\left\{\frac{1-2\nu}{3(1+\nu)}(\sigma_x + \sigma_y + \sigma_z)^2 + \frac{1}{3}(\sigma_x - \sigma_y)^2 + (\sigma_y - \sigma_z)^2\right. \\
&\quad \left. + (\sigma_z - \sigma_x)^2] + 2(\tau_{xy}^2 + \tau_{yz}^2 + \tau_{zx}^2)\right\} = (U_1)_v + (U_1)_d
\end{aligned}$$

Consider now the strain energy $(U_1)_v$ given by Eq. (2.2.3) which corresponds to a state of stress caused by the uniform hydrostatic pressure

$$p = \frac{\sigma_x + \sigma_y + \sigma_z}{3}$$

and which produces the volume expansion

$$e = \frac{1-2\nu}{E}(\sigma_x + \sigma_y + \sigma_z) = \frac{3(1-2\nu)}{E}p \neq 0$$

In this first decomposition, the element is subjected to a change of volume, but not to a change of form since the specific deformation is independent of direction. The strain energy $(U_1)_d$ given by Eq. (2.2.4) corresponds to a complementary state of stress in which the stress components are:

$$\sigma_x' = \sigma_x - p \quad \sigma_y' = \sigma_y - p \quad \sigma_z' = \sigma_z - p$$
$$\tau_{xy} \qquad\qquad \tau_{yz} \qquad\qquad \tau_{zx}$$

In this decomposition the volume expansion is zero since

$$e = \frac{1-2\nu}{E}[\sigma'_x + \sigma'_y + \sigma'_z] = \frac{1-2\nu}{E}[\sigma_x + \sigma_y + \sigma_z - 3p] = 0$$

The element which does not change in volume but generally changes its form, has a distortion. The strain energy of distortion per unit volume $(U_1)_d$ has been chosen as a criterion of failure of a material in the failure theory proposed by Huber-Von Mises-Hencky. According to this theory, yielding begins when the distortion energy reaches the value of the distortion energy at the yield point in a simple tension test.

Consider a specimen of steel subjected to a tension test. If in this case, $\sigma_x \neq 0$ then $\sigma_y = \sigma_z = \tau_{xy} = \tau_{yz} = \tau_{xz} = 0$ and in view of Eq. (2.2.4) one has

$$(U_1)_d^{(1)} = \frac{\sigma_x^2}{6G} = \frac{(1+\nu)}{3E}(\sigma_x)^2$$

Consider now a specimen of the same steel subjected to pure shear. In this case $\tau_{xy} \neq 0$ and $\sigma_x = \sigma_y = \sigma_z = \tau_{yz} = \tau_{zx} = 0$ and in view of Eq. (2.2.4) one has in this case:

$$(U_1)_d^{(2)} = \frac{\tau_{xy}^2}{2G} = \frac{(1+\nu)}{E}(\tau_{xy})^2$$

At failure, in accordance with Huber-Von Mises-Hencky theory of failure, one must have that

$$(U_1)_d^{(1)} = (U_1)_d^{(2)}$$

Thus, if σ_y and τ_y represent the yield stress in tension and the yield stress in shear, respectively, one has

$$\frac{(1+\nu)}{3E}(\sigma_y)^2 = \frac{(1+\nu)}{E}(\tau_y)^2$$

or

$$\tau_y = \frac{1}{\sqrt{3}}\sigma_y = 0.557\sigma_y \qquad (2.2.5)$$

Experiments done on steel show that the ratio between the yield stress in tension and the yield stress in shear are in good agreement with those given by Eq. (2.2.5). Moreover, Bridgman's[3] experiments showed that in case of

[3] **Bridgman, Percy Williams.** American physicist (b. Cambridge, Mass. 1882; d. Randolph, N. H. 1961). He was Higgins Professor of physics at Harvard University and Nobel laureate in physics (1946). He is very well known for his research work on materials at high pressures and on their thermodynamic behavior. See Bibliography, Section 2.2, Nos. 3 and 4, at the end of this chapter.

ductile materials, the material did not become inelastic under a triaxial state of stress produced by a very high hydrostatic pressure.

2.3 Other Strength Theories

The Huber-Von Mises-Hencky strength theory is not the only such theory which has been proposed. Some of the principal strength theories will be discussed in this section and their results compared. The main object of these theories is to furnish a way, once the behavior of a material under a simple uniaxial compression or tension test is known, to predict when failure will occur under any kind of combined stress. Among the various strength theories proposed are:

(a) *The maximum stress theory or Rankine's theory.* According to this theory, yielding in the material begins only when the maximum (or minimum) principal stress has reached a value equal to the tension (or compression) elastic limit of the material as found in a simple tension (or compression) test. It follows that

$$\sigma_x = \sigma_{yp} = \frac{P}{\Omega}$$

Although this theory has been contradicted by many experiments, it has been found to be in good agreement with results of tests on brittle materials.

(b) *The maximum shear stress theory or Coulomb-Guest's theory.* According to this theory inelastic action in the material starts when the maximum shearing stress becomes equal to the maximum shearing stress at the yield point in a simple tension test. Since the maximum shearing stress is equal to half the difference between the maximum and minimum principal stresses[4] then according to this theory

$$\tfrac{1}{2}(\sigma_{max} - \sigma_{min}) = \tau_{yp}$$

This theory gives good experimental results when applied to ductile materials.

(c) *The maximum strain theory or Saint Venant's theory.* According to this theory, inelastic action in the material starts when the maximum strain (in elongation or in compression) becomes equal to the strain at the yield point (in elongation or in compression, respectively) in

---

[4] See Chapter 3, Section 2.

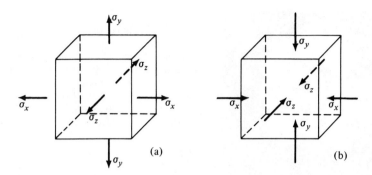

Figure 2.3.1

a simple uniaxial state of stress. It follows then that failure occurs when [see Figs. 2.3.1(a) and (b)]

$$\frac{\sigma_x}{E} - \frac{\nu}{E}(\sigma_y + \sigma_z) = \frac{(\sigma_{yt})}{E} \quad \text{(for case a)}$$

$$\frac{\sigma_x}{E} - \frac{\nu}{E}(\sigma_y + \sigma_z) = \frac{(\sigma_{yc})}{E} \quad \text{(for case b)}$$

where σ_{yt} and σ_{yc} are, respectively, the tensile and compressive yield stresses. Experiments with materials under hydrostatic pressure seem to contradict this theory.

(d) *The total energy or Beltrami-Haigh theory.* According to this theory, inelastic action in the material starts when the energy per unit volume absorbed by the material is equal to the energy per unit volume absorbed by the material upon reaching the yield point under an uniaxial state of stress. The Huber-Von Mises-Hencky theory, which was

Rankine, William John Macquorn. Scottish engineer and physicist (b. Edinburgh 1820; d. Glasgow 1872). He was professor of civil engineering at the University of Glasgow and is considered one of the founders of thermodynamics on which he wrote one of the first formal treatises. [Reproduction from the Vito Volterra collection, Villa Volterra, Ariccia (Rome).]

Levi-Civita, Tullio. Italian mathematician (b. Padova 1873; d. Rome 1941). He was professor of rational mechanics at the University of Padova when only 24 years old and was professor of advanced analysis and of mechanics at the University of Rome from 1919 until 1938 when he was dismissed as a consequence of the fascist racial laws. He is considered one of the greatest mathematicians of the twentieth century who left outstanding contributions in relativistic mechanics, in the three-body problem, and in celestial mechanics. His complete works in four volumes were published by the Accademia Nazionale dei Lincei. [Reproduction from the Vito Volterra collection, Villa Volterra, Ariccia (Rome).]

discussed earlier, is a particular case of the Beltrami-Haigh theory since it considers that part of the total strain energy stored in the strained body which produces changes of shape.

In a paper published in 1901, T. Levi-Civita showed how Beltrami's theory could be extended from the static to the dynamic case.

2.4 Principle of Virtual Work

The principle of virtual work for an elastic solid states that in every virtual deformation of an elastic solid the sum of the work done by the external forces is equal to the variation of the strain energy of the solid, i.e.,

$$\sum_{i=1}^{n} \vec{F}_i \cdot \vec{\delta r}_i = \delta U$$

By letting

$$\delta L = \sum_{i=1}^{n} \vec{F}_i \cdot \vec{\delta r}_i$$

represent the total work done by the external forces in the virtual displacement of the solid, the above relation may be written as follows:

$$\delta(L - U) = 0 \tag{2.4.1}$$

Example 2.4.1. The equations of equilibrium for an elastic solid can be derived by the use of Eq. (2.4.1). The work L done by the external forces (body forces per unit mass ρX, ρY, and ρZ and surface forces $\bar{X}$, $\bar{Y}$, and $\bar{Z}$ on the free surface S) on the body is

Sec. 2.4 ENERGY PRINCIPLES AND GENERAL THEOREMS

$$L = \iiint_V \rho(Xu + Yv + Zw)dxdydz + \iint_S (\bar{X}u + \bar{Y}v + \bar{Z}w)dS \quad (2.4.2)$$

The strain energy is given by Eq. (2.1.6), or

$$U = \iiint_V W dxdydz$$

where W, the strain energy per unit volume, is

$$W = \frac{\lambda}{2} e^2 + G(\epsilon_x^2 + \epsilon_y^2 + \epsilon_z^2) + \frac{G}{2}(\gamma_{xy}^2 + \gamma_{yz}^2 + \gamma_{zx}^2) \quad (2.4.3)$$

From the above relations, one finds that

$$\begin{aligned}
\delta U &= \iiint_V \delta W dxdydz \\
&= \iiint_V \left\{ \left[\frac{\partial}{\partial x}\left(\frac{\partial W}{\partial \epsilon_x}\right) + \frac{\partial}{\partial y}\left(\frac{\partial W}{\partial \gamma_{xy}}\right) + \frac{\partial}{\partial z}\left(\frac{\partial W}{\partial \gamma_{xz}}\right) \right] \delta u \right. \\
&\quad + \left[\frac{\partial}{\partial x}\left(\frac{\partial W}{\partial \gamma_{yx}}\right) + \frac{\partial}{\partial y}\left(\frac{\partial W}{\partial \epsilon_y}\right) + \frac{\partial}{\partial z}\left(\frac{\partial W}{\partial \gamma_{yz}}\right) \right] \delta v + \ldots \Big\} dxdydz \\
&\quad + \iint_S \left\{ \left[\frac{\partial W}{\partial \epsilon_x} \alpha + \frac{\partial W}{\sigma \gamma_{xy}} \beta + \frac{\partial W}{\partial \gamma_{xz}} \gamma \right] \delta u \right. \\
&\quad + \left[\frac{\partial W}{\partial \gamma_{yx}} \alpha + \frac{\partial W}{\partial \epsilon_y} \beta + \frac{\partial W}{\partial \gamma_{yz}} \gamma \right] \delta v \\
&\quad + \left[\frac{\partial W}{\partial \gamma_{zx}} \alpha + \frac{\partial W}{\partial \gamma_{zy}} \beta + \frac{\partial W}{\partial \epsilon_z} \gamma \right] \delta w \Big\} dS
\end{aligned} \quad (2.4.4)$$

where α, β, and γ are the cosines of the normal to the surface S directed positively outward.

Equation (2.4.4) is derived in the following way:
Since

$$\epsilon_x = \frac{\partial u}{\partial x}, \ldots$$

it follows that

$$\frac{\partial W}{\partial \epsilon_x} \delta \epsilon_x = \frac{\partial W}{\partial \epsilon_x} \frac{\partial \delta u}{\partial x} = \frac{\partial}{\partial x}\left(\frac{\partial W}{\partial \epsilon_x} \delta u\right) - \frac{\partial}{\partial x}\left(\frac{\partial W}{\partial \epsilon_x}\right) \delta u$$

and Gauss's formula gives

$$\iiint_V \frac{\partial}{\partial x}\left(\frac{\partial W}{\partial \epsilon_x} \delta u\right) dV = \iint_S \frac{\partial W}{\partial \epsilon_x} \delta u \alpha \, dS$$

From Eqs. (2.4.2) one finds that

Gauss, Karl Friedrich. German mathematician, physicist, and astronomer (b. Brunswick 1777; d. Göttingen 1855). From 1807 until his death, he was professor at Göttingen University and director of Göttingen Observatory. His many outstanding contributions in the fields of differential geometry, analysis, theory of numbers, method of least squares, mathematical physics, and astronomy, collected in seven volumes, were published from 1863 to 1871 by the Royal Society of Göttingen. [Reproduction from the Vito Volterra collection, Villa Volterra, Ariccia (Rome).]

$$\delta L = \iiint_V \rho(X\delta u + Y\delta v + Z\delta w)dV \\ + \iint_S (\bar{X}\delta u + \bar{Y}\delta v + \bar{Z}\delta w)dS \qquad (2.4.5)$$

In view of Eqs. (2.4.4) and (2.4.5) Eq. (2.4.1) becomes:

$$\iiint_V \left\{ \left[\frac{\partial}{\partial x}\left(\frac{\partial W}{\partial \epsilon_x}\right) + \frac{\partial}{\partial y}\left(\frac{\partial W}{\partial \gamma_{xy}}\right) + \frac{\partial}{\partial z}\left(\frac{\partial W}{\partial \gamma_{xz}}\right) + \rho X \right]\delta u \right. \\ + \left[\frac{\partial}{\partial x}\left(\frac{\partial W}{\partial \gamma_{yx}}\right) + \frac{\partial}{\partial y}\left(\frac{\partial W}{\partial \epsilon_y}\right) + \frac{\partial}{\partial z}\left(\frac{\partial W}{\partial \gamma_{yz}}\right) + \rho Y \right]\delta v \\ + \left. \left[\frac{\partial}{\partial x}\left(\frac{\partial W}{\partial \gamma_{zx}}\right) + \frac{\partial}{\partial y}\left(\frac{\partial W}{\partial \gamma_{zy}}\right) + \frac{\partial}{\partial z}\left(\frac{\partial W}{\partial \epsilon_z}\right) + \rho Z \right]\delta w \right\} dxdydz \\ + \iint_S \left\{ \left[\bar{X} - \frac{\partial W}{\partial \epsilon_x}\alpha - \frac{\partial W}{\partial \gamma_{xy}}\beta - \frac{\partial W}{\partial \gamma_{xz}}\gamma \right]\delta u \right. \\ + \left[\bar{Y} - \frac{\partial W}{\partial \gamma_{yx}}\alpha - \frac{\partial W}{\partial \epsilon_y}\beta - \frac{\partial W}{\partial \gamma_{yz}}\gamma \right]\delta v \\ + \left. \left[\bar{Z} - \frac{\partial W}{\partial \gamma_{zx}}\alpha - \frac{\partial W}{\partial \gamma_{zy}}\beta - \frac{\partial W}{\partial \epsilon_z}\gamma \right]\delta w \right\} dS$$

Since δu, δv, and δw are arbitrary, one obtains the following equations of equilibrium

$$\left.\begin{array}{l} \frac{\partial}{\partial x}\left(\frac{\partial W}{\partial \epsilon_x}\right) + \frac{\partial}{\partial y}\left(\frac{\partial W}{\partial \gamma_{xy}}\right) + \frac{\partial}{\partial z}\left(\frac{\partial W}{\partial \gamma_{xz}}\right) + \rho X = 0 \\ \frac{\partial}{\partial x}\left(\frac{\partial W}{\partial \gamma_{yx}}\right) + \frac{\partial}{\partial y}\left(\frac{\partial W}{\partial \epsilon_y}\right) + \frac{\partial}{\partial z}\left(\frac{\partial W}{\partial \gamma_{yz}}\right) + \rho Y = 0 \\ \frac{\partial}{\partial x}\left(\frac{\partial W}{\partial \gamma_{zx}}\right) + \frac{\partial}{\partial y}\left(\frac{\partial W}{\partial \gamma_{zy}}\right) + \frac{\partial}{\partial z}\left(\frac{\partial W}{\partial \epsilon_z}\right) + \rho Z = 0 \end{array}\right\} \qquad (2.4.6)$$

and the following boundary conditions on S

$$\left.\begin{aligned} \bar{X} &= \frac{\partial W}{\partial \epsilon_x}\alpha + \frac{\partial W}{\partial \gamma_{xy}}\beta + \frac{\partial W}{\partial \gamma_{xz}}\gamma \\ \bar{Y} &= \frac{\partial W}{\partial \gamma_{yx}}\alpha + \frac{\partial W}{\partial \epsilon_y}\beta + \frac{\partial W}{\partial \gamma_{yz}}\gamma \\ \bar{Z} &= \frac{\partial W}{\partial \gamma_{zx}}\alpha + \frac{\partial W}{\partial \gamma_{zy}}\beta + \frac{\partial W}{\partial \epsilon_z}\gamma \end{aligned}\right\} \quad (2.4.7)$$

In view of Eqs. (2.4.3) and (1.3.4), Eqs. (2.4.6) can be written in the more familiar form

$$\left.\begin{aligned} \frac{\partial \sigma_x}{\partial x} + \frac{\partial \tau_{xy}}{\partial y} + \frac{\partial \tau_{xz}}{\partial z} + \rho X &= 0 \\ \frac{\partial \tau_{yx}}{\partial x} + \frac{\partial \sigma_y}{\partial y} + \frac{\partial \tau_{yz}}{\partial z} + \rho Y &= 0 \\ \frac{\partial \tau_{zx}}{\partial x} + \frac{\partial \tau_{zy}}{\partial y} + \frac{\partial \sigma_z}{\partial z} + \rho Z &= 0 \end{aligned}\right\} \quad (2.4.8)$$

Example 2.4.2. The deflection line of a simply supported beam subjected to a vertical force P applied at distance a from the left support as shown in Fig. 2.4.1 will now be determined.

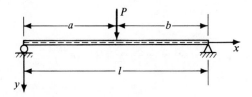

Figure 2.4.1

Assume that the beam has a vertical plane of symmetry xy which contains force P. The elastic line v will also be situated in this plane. Neglecting the shear force the strain energy will be, in accordance with Eq. (2.1.9),

$$U = \frac{1}{2} \int_0^l \frac{M^2}{EI} dx \quad (2.1.9)$$

From the strength of materials it is known that

$$\frac{1}{R} = \frac{d^2 v}{dx^2} = \frac{M}{EI}$$

or that

$$M = EI \frac{d^2 v}{dx^2}$$

It follows from Eq. (2.1.9) that

$$U = \frac{EI}{2} \int_0^l \left(\frac{d^2v}{dx^2}\right)^2 dx \tag{2.4.9}$$

Suppose that the function $v(x)$ be developed in a sine series; then

$$v(x) = C_1 \sin \frac{\pi x}{l} + C_2 \sin \frac{2\pi x}{l} + C_3 \sin \frac{3\pi x}{l} + \ldots$$
$$= \sum_{i=1}^{\infty} C_i \sin \frac{i\pi x}{l} \tag{2.4.10}$$

where the $C_i (i = 1, 2, 3, \ldots, \infty)$ are constants. It can be easily seen that function $v(x)$ expressed by Eq. (2.4.10) satisfies the boundary conditions of simple support for the beam, i.e.,

$$v(0) = EI \frac{d^2v(0)}{dx^2} = v(l) = EI \frac{d^2v(l)}{dx^2} = 0$$

In order to define a virtual displacement of the beam, an infinitely small variation δC_j is given only to coefficient C_j. The corresponding variation of the strain energy (Eq. 2.4.9) is

$$\delta U = EI \int_0^l \frac{d^2v(x)}{dx^2} \delta\left(\frac{d^2v(x)}{dx^2}\right) dx \tag{2.4.11}$$

but,

$$\frac{d^2v(x)}{dx^2} = -\frac{\pi^2}{l^2} \sum_{i=1}^{\infty} i^2 C_i \sin \frac{i\pi x}{l} \tag{2.4.12}$$

and, therefore,

$$\delta\left[\frac{d^2v(x)}{dx^2}\right] = -\frac{\pi^2}{l^2} j^2 \sin \frac{j\pi x}{l} \delta C_j \tag{2.4.13}$$

By introducing Eqs. (2.4.12) and (2.4.13) into Eq. (2.4.11), one obtains

$$\delta U = \frac{\pi^4 EI}{l^4} \int_0^l \left[C_1 \sin \frac{\pi x}{l} + (2)^2 C_2 \sin \frac{2\pi x}{l} + (3)^2 C_3 \sin \frac{3\pi x}{l} + \ldots \right.$$
$$\left. + (i)^2 C_i \sin \frac{i\pi x}{l} + \ldots \right] j^2 \sin \frac{j\pi x}{l} \delta C_j dx \tag{2.4.14}$$

Consideration of the orthogonality properties of the circular functions shows that

$$\int_0^l \sin \frac{i\pi x}{l} \sin \frac{j\pi x}{l} dx = \begin{cases} = 0 \text{ for } i \neq j \\ = \frac{l}{2} \text{ for } i = j \end{cases}$$

and Eq. (2.4.14) reduces to

$$\delta U = \frac{\pi^4 EI}{2l^3} i^4 C_i \delta C_i$$

The virtual work of the external forces reduces in this case to the work of force P which is

$$\sum_{i=1}^{n} \vec{F_i} \cdot \vec{\delta r_i} = P(\delta v)_{x=a} = P \sin \frac{i\pi a}{l} \delta C_i$$

By introducing the above expressions into Eq. (2.4.1) one obtains

$$P \sin \frac{i\pi a}{l} \delta C_i - \frac{\pi^4 EI}{2l^3} i^4 C_i \delta C_i = 0$$

from which it follows that

$$C_i = \frac{2Pl^3}{\pi^4 EI} \frac{1}{(i)^4} \sin \frac{i\pi a}{l} \qquad (2.4.15)$$

By substituting Eq. (2.4.15) into Eq. (2.4.10) one obtains for the deflection curve the expression

$$v(x) = \frac{2Pl^3}{\pi^4 EI} \sum_{i=1}^{\infty} \frac{1}{i^4} \sin \frac{i\pi a}{l} \sin \frac{i\pi x}{l} \qquad (2.4.16)$$

In the particular case in which load P is at the center of the bar ($a = l/2$) Eq. (2.4.16) becomes

$$v(x) = \frac{2Pl^3}{\pi^4 EI} \left[\sin \frac{\pi x}{l} - \frac{1}{(3)^4} \sin \frac{3\pi x}{l} + \frac{1}{(5)^4} \sin \frac{5\pi x}{l} \cdots \right]$$

or

$$v(x) = \frac{2Pl^3}{\pi^4 EI} \sum_{i=1}^{\infty} \left[\frac{(-1)^{i+1}}{(2i-1)^4} \sin \frac{(2i-1)\pi x}{l} \right] \qquad (2.4.17)$$

It follows that the deflection at the center of the beam is

$$v\left(\frac{l}{2}\right) = \frac{2Pl^3}{\pi^4 EI} \left[1 + \frac{1}{3^4} + \frac{1}{5^4} + \cdots \right] = \frac{2Pl^3}{\pi^4 EI} \sum_{i=1}^{\infty} \left[\frac{1}{(2i-1)^4} \right] \qquad (2.4.18)$$

The above series converges very rapidly. Taking only the first term of the series one has

$$v\left(\frac{l}{2}\right) = \frac{Pl^3}{48.7 EI}$$

which is 1.46% below the value

$$v\left(\frac{l}{2}\right) = \frac{Pl^3}{48EI}$$

given by the strength of materials. If one takes the first two terms of the series in Eq. (2.4.18) one has

$$v\left(\frac{l}{2}\right) = \frac{Pl^3}{48.11EI}$$

and the error is reduced to 0.23 per cent.

2.5 Uniqueness of Solution

It was proved by G. Kirchhoff that for given surface and body forces there exists a unique state of stress inside a simply connected elastic solid.

Suppose that at each point inside the elastic solid there exist the following components of body force:

$$X, Y, \text{and } Z$$

and at each point on the surface there exist the following components of surface force:

$$\bar{X}, \bar{Y}, \text{and } \bar{Z}$$

In order to prove Kirchhoff's theorem, a proof by absurdum will be given, i.e., that if two solutions exist they must be identical. Suppose then that two such solutions exist. To the first will correspond the following stress tensor:

$$\begin{vmatrix} \sigma'_x & \tau'_{xy} & \tau'_{xz} \\ \tau'_{yx} & \sigma'_y & \tau'_{yz} \\ \tau'_{zx} & \tau'_{zy} & \sigma'_z \end{vmatrix} \tag{2.5.1}$$

To the second will correspond the following stress tensor:

$$\begin{vmatrix} \sigma''_x & \tau''_{xy} & \tau''_{xz} \\ \tau''_{yx} & \sigma''_y & \tau''_{yz} \\ \tau''_{zx} & \tau''_{zy} & \sigma''_z \end{vmatrix} \tag{2.5.2}$$

The first components of stress will satisfy the following equilibrium equations:

ENERGY PRINCIPLES AND GENERAL THEOREMS

$$\left.\frac{\partial \sigma'_x}{\partial x} + \frac{\partial \tau'_{xy}}{\partial y} + \frac{\partial \tau'_{xz}}{\partial z} + X = 0 \atop \cdots \atop \cdots \right\} \quad (2.5.3)$$

and the following boundary conditions:

$$\left.\bar{X} = \sigma'_x l + \tau'_{xy} m + \tau'_{xz} n \atop \cdots \atop \cdots \right\} \quad (2.5.4)$$

where $l = \cos(\widehat{nx})$, $m = \cos(\widehat{ny})$, and $n = \cos(\widehat{nz})$. The second components of stress will satisfy the following equilibrium equations:

$$\left.\frac{\partial \sigma''_x}{\partial x} + \frac{\partial \tau''_{xy}}{\partial y} + \frac{\partial \tau''_{xz}}{\partial z} + X = 0 \atop \cdots \atop \cdots \right\} \quad (2.5.5)$$

and the following boundary conditions:

$$\left.\bar{X} = \sigma''_x l + \tau''_{xy} m + \tau''_{xz} n \atop \cdots \atop \cdots \right\} \quad (2.5.6)$$

Consider now the system:

$$\begin{vmatrix} \sigma_x = \sigma'_x - \sigma''_x & \tau_{xy} = \tau'_{xy} - \tau''_{xy} & \tau_{xz} = \tau'_{xz} - \tau''_{xz} \\ \tau_{yx} = \tau'_{yx} - \tau''_{yx} & \sigma_y = \sigma'_y - \sigma''_y & \tau_{yz} = \tau'_{yz} - \tau''_{yz} \\ \tau_{zx} = \tau'_{zx} - \tau''_{zx} & \tau_{zy} = \tau'_{zy} - \tau''_{zy} & \sigma_z = \sigma'_z - \sigma''_z \end{vmatrix} \quad (2.5.7)$$

System (2.5.7) will satisfy the following equilibrium equations:

$$\left.\frac{\partial \sigma_x}{\partial x} + \frac{\partial \tau_{xy}}{\partial y} + \frac{\partial \tau_{xz}}{\partial z} = 0 \atop \cdots \atop \cdots \right\} \quad (2.5.8)$$

and the following boundary conditions:

$$\left.\sigma_x l + \tau_{xy} m + \tau_{xz} n = 0 \atop \cdots \atop \cdots \right\} \quad (2.5.9)$$

Equations (2.5.8) and (2.5.9) are obtained by subtracting Eqs. (2.5.5) from Eqs. (2.5.3) and Eqs. (2.5.6) from Eqs. (2.5.4).

The state of stress corresponding to the stress tensor given in Eqs. (2.5.7) has no body forces and no surface forces. Its total strain energy must therefore be equal to zero. The expression for the strain energy of an elastic body is given by the integral

$$U = \iiint_V \left[\tfrac{1}{2}\lambda e^2 + G(\epsilon_x^2 + \epsilon_y^2 + \epsilon_z^2) + \frac{G}{2}(\gamma_{xy}^2 + \gamma_{yz}^2 + \gamma_{zx}^2) \right] dV \quad (2.5.10)$$

In order that the integral of Eq. (2.5.10) be zero, every term of it must be zero, i.e.,

$$e = \epsilon_x = \epsilon_y = \epsilon_z = \gamma_{xy} = \gamma_{yz} = \gamma_{zx} = 0 \quad (2.5.11)$$

But as a consequence of Eqs. (2.5.11) it follows that

$$\sigma_x = \sigma_y = \sigma_z = \tau_{xy} = \tau_{yz} = \tau_{zx} = 0$$

and this implies that

$$\sigma'_x = \sigma''_x \quad \sigma'_y = \sigma''_y \quad \sigma'_z = \sigma''_z$$
$$\tau'_{xy} = \tau''_{xy} \quad \tau'_{yz} = \tau''_{yz} \quad \tau'_{zx} = \tau''_{zx}$$

as we wanted to prove.

2.6 Betti's and Maxwell's Reciprocity Theorems

(a) Introduction. The principle of superposition is not valid for the strain energy since the strain energy of an elastic solid is a quadratic function of the forces involved. It follows that in the case in which more than one external force is applied to an elastic solid, the total work due to external forces is not equal to the sum of the works which would be obtained by applying the single forces separately.

Suppose now that on an elastic solid there are acting two external forces, P_1 and P_2. Since the order of application of the forces is arbitrary, let us first apply force P_1 which will produce work L_1. Then, let us apply force P_2 which will produce work L_2. This is the same work which would be produced by force P_2, if it were acting alone on the body. During application of force P_2, force P_1, still acting, produces additional work L_{12}, because its point of application is displaced due to the deformation produced by P_2. The total work L is

$$L = L_1 + L_2 + L_{12}$$

Work L_{12} done by force P_1 due to the application of force P_2 is called the *mutual* or *indirect work* of the two forces. It is zero, if and only if, the displacement of the point of application of force P_1 produced by force P_2 is zero or is perpendicular to the direction of P_1. (This happens for instance in a prismatic straight beam for forces N, M, V, and M_T whose mutual works are in fact zero.) The mutual work can be positive or negative. It therefore follows that work due to several forces can be greater or smaller than the sum of single works.

Example 2.6.1. Consider the axial force $P = P_1$ acting on the column of length l shown in Fig. 2.6.1. The work done by force P_1 is

$$L_1 = \frac{1}{2}(P_1)(P_1 l/E\Omega) = \frac{P_1^2 l}{2E\Omega}$$

Now suppose that force P_1 doubles, i.e., $P_2 = 2P_1$. It follows that

$$L_2 = \frac{2P_1^2 l}{E\Omega} = 4L_1$$

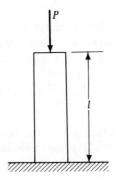

Figure 2.6.1

(*b*) *Betti's reciprocity theorem.* Consider an elastic solid with fixed or elastic supports subjected to two systems of external forces A and B (see Fig. 2.6.2). Assume force system A is applied first. It will produce work L_A. Then force system B is applied. It will produce work L_B while forces A produce the mutual work L_{AB}. The total work will be:

$$L = L_A + L_B + L_{AB} \tag{2.6.1}$$

Now, assume that the order of application of the external forces is reversed. Forces B which produce work L_B will first be applied; then forces A, which will produce work L_A while forces B produce the mutual work L_{BA} are applied. It follows that the total work L is

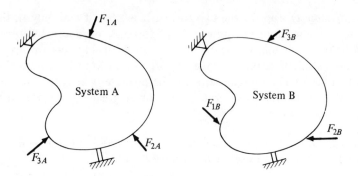

Figure 2.6.2

$$L = L_B + L_A + L_{BA} \tag{2.6.2}$$

Work L must be the same in Eqs. (2.6.1) and (2.6.2) because at the end, the elastic solid is loaded by the same system of external forces; it follows that

$$L_{AB} = L_{BA} \tag{2.6.3}$$

Equation (2.6.3) expresses Betti's reciprocity theorem: The indirect or mutual work done by a system of external forces A acting on an elastic solid during the application of a new system of external forces B is equal to the indirect or mutual work which would be done by the system of external forces B, if they were already acting during the application of a new system of external forces A.

Example 2.6.2. Consider the simply supported beam AB (see Fig. 2.6.3) subjected to the action of a concentrated load P at the center and to the couple M_A at support A. The deflection of the center of the beam will be:

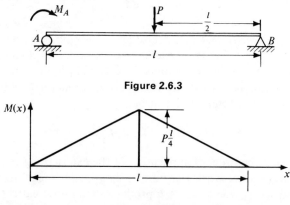

Figure 2.6.3

Figure 2.6.4

(a) Due to the effect of the concentrated load P (obtained by applying the conjugate beam method[5] and using the bending moment diagram of Fig. 2.6.4)

$$v_1 = \frac{Pl^2}{16EI}\frac{l}{2} - \frac{Pl^2}{16EI}\frac{l}{6} = \frac{Pl^3}{48EI}$$

(b) Due to the effect of the couple M_A (obtained by applying the conjugate beam method and using the bending moment diagram of Fig. 2.6.5)

$$v_2 = \frac{M_A l}{6EI}\frac{l}{2} - \left(\frac{M_A x}{EI}\frac{x}{2}\frac{x}{3}\right)_{x=l/2} = \frac{M_A l^2}{16EI}$$

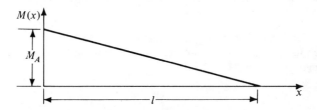

Figure 2.6.5

The rotation of the end A of the beam will be:

(a) Due to the effect of the concentrated load P (obtained by applying the conjugate beam method and using the bending moment diagram of Fig. 2.6.4)

$$\phi_{A1} = \frac{Pl^2}{16EI}$$

(b) Due to the effect of the couple M_A (obtained by applying the conjugate beam method and using the bending moment diagram of Fig. 2.6.5)

$$\phi_{A2} = \frac{Ml}{3EI}$$

Suppose that force P (Force A) is first applied. The work will be:

$$L_A = \frac{1}{2}Pv_1 = \frac{P^2 l^3}{96EI}$$

Then the couple M_A (Force B) is applied. The work will be:

$$L_B = \frac{1}{2}M_A\phi_{A2} = \frac{M_A^2 l}{6EI}$$

$$L_{AB} = Pv_2 = \frac{PM_A l^2}{16EI}$$

---

[5] See Chapter 5, Sec. 8.

Now suppose that the couple M_A (Force B) is first applied. The work will be:

$$L_B = \frac{1}{2} M_A \phi_{A2} = \frac{M_A^2 l}{6EI}$$

Then force P (Force A) is applied. The work will be:

$$L_A = \frac{1}{2} P v_1 = \frac{P^2 l^3}{96EI}$$

$$L_{BA} = M_A \phi_{A1} = \frac{M_A P l^2}{16EI}$$

From the above relations it is found that $L_{AB} = L_{BA}$, which is in agreement with Betti's theorem.

Example 2.6.3. Consider the simply supported beam of length l subjected to the concentrated load P at the center and to the uniformly distributed load p_0 acting over the length of the bar (Fig. 2.6.6).

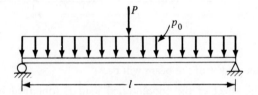

Figure 2.6.6

The deflection equation for the concentrated load P (Force A) is

$$v_A(x) = \frac{P}{48EI}(3l^2 x - 4x^3)$$

The deflection equation for the distributed load p_0 (Force B) is

$$v_B(x) = \frac{p_0}{24EI}(l^3 x - 2l^3 + x^4)$$

It follows that

$$v_B\left(\frac{l}{2}\right) = \frac{5p_0 l^4}{384EI}$$

and that

$$L_{AB} = P v_B\left(\frac{l}{2}\right) = \frac{5Pp_0 l^4}{384EI}$$

One finds that

$$L_{BA} = 2 \int_0^{l/2} p_0 \frac{P}{48EI}(3l^2 x - 4x^3) dx = \frac{5Pp_0 l^4}{384EI}$$

and, therefore, $L_{AB} = L_{BA}$ which is in agreement with Betti's theorem.

Betti, Enrico. Italian mathematician (b. near Pistoia 1823; d. Soiana near Pisa 1892). From 1857 until his death he was professor of mathematical physics at the University of Pisa and was also for a while a member of the Italian Parliament and afterwards Senator of the Italian Kingdom. He is considered one of the most eminent mathematicians of the nineteenth century. He left outstanding contributions in algebra (where he extended the unfinished work of E. Galois), in topology, in elliptic functions, in potential theory, and in elasticity. As director of the "Scuola Normale" of Pisa from 1865, he contributed to making this school and the University of Pisa famous centers of mathematical research. Among his pupils were most of the best known Italian mathematicians. [Reproduction from the Vito Volterra collection, Villa Volterra, Ariccia (Rome).]

Maxwell, James Clerk. Scotch physicist and mathematician (b. Edinburgh 1831; d. Cambridge 1879). After graduating in 1854 from Cambridge University as second wrangler (the senior wrangler being that year E. J. Routh) he was elected Fellow of Trinity College and was, in succession, professor of natural philosophy at Marischal College at Aberdeen, professor of physics and astronomy at King's College in London, and finally first occupant of the newly founded chair of experimental physics and director of the newly founded Cavendish Laboratory at Cambridge University. His principal works were in color sensation, the dynamic theory of gases, and in electricity and magnetism. He translated Faraday's ideas into mathematical notation and also made important contributions to the mathematical theory of elasticity. His collected works were published in 1890. [Reproduction from the Vito Volterra collection, Villa Volterra, Ariccia (Rome).]

(c) *Maxwell's reciprocity theorem.* Betti's theorem was given in 1872[6] while Maxwell's theorem was given earlier in 1864.[7] Maxwell's theorem is a particular and important case of Betti's theorem and can be derived from the latter. Consider an elastic solid subjected to a single force $P_A = 1$ acting at a point A and in a direction α [see Fig. 2.6.7(a)] and a second force $P_B = 1$

---

[6] See Bibliography, Sec. 2.6 No. 5 at the end of this chapter.
[7] See Bibliography, Sec. 2.6 No. 13 at the end of this chapter.

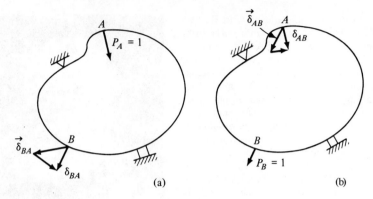

Figure 2.6.7

acting at a point B in a direction β [see Fig. 2.6.7(b)]. Denoting by δ_{AB} the displacement of A in the α-direction produced by unit force P_B and by δ_{BA} the displacement of B in the β-direction produced by unit force P_A it follows from Betti's theorem that

$$P_A \cdot \delta_{AB} = P_B \cdot \delta_{BA}$$

from which

$$\delta_{AB} = \delta_{BA}$$

Equation (2.6.4) expresses Maxwell's reciprocity theorem: The displacement of point A in the α-direction produced by unit force acting at point B in the β-direction is equal to the displacement of point B in the β-direction produced by unit force acting at point A in the α-direction.

Example 2.6.4. The cantilever beam AB shown in Fig. 2.6.8 is subjected to a concentrated force P at its end A. The deflection Δ of point C at a distance a from A is to be computed.

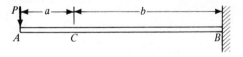

Figure 2.6.8

The solution will be obtained by applying a load P at C (see Fig. 2.6.9) and computing the deflection at A. Then, in view of Maxwell's theorem, the deflection at A due to load P applied at C is equal to the deflection at C due to load P applied at A. Hence, calling δ_A and δ_C the deflections of the beam at A and C, respectively, and α_C the slope of the beam at point C, one has from Fig. 2.6.9 the following relations:

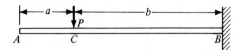

Figure 2.6.9

$$\delta_A = \delta_C + \alpha_C a$$

where

$$\delta_C = (Pb^3/3EI) \qquad \alpha_C = (Pb^2/2EI)$$

If follows that

$$\Delta = \delta_A = (Pb^3/3EI) + (Pb^2/2EI)a = (Pb^2/6EI)(2b + 3a)$$
$$= (Pb^2/6EI)(3l - b)$$
$$= (P/6EI)(2l^3 - 3l^2a - a^3)$$

Example 2.6.5. A load $P_R = 1200$ lb acting at a point R of a beam (see Fig. 2.6.10) produces the following vertical displacements at three points A, B, and C of the beam:

$$\delta_A = 0.3 \text{ in.} \qquad \delta_B = 0.8 \text{ in.} \qquad \delta_C = 0.5 \text{ in.}$$

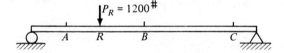

Figure 2.6.10

Find the deflection of point R produced by the following loads:

$$P_A = 1500 \text{ lb} \qquad P_B = 700 \text{ lb} \qquad P_C = 1000 \text{ lb}$$

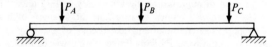

Figure 2.6.11

acting at points A, B, and C, respectively (see Fig. 2.6.11).

A load of 1 lb acting at point R will produce at points A, B, and C the following displacements in inches

$$v_{A_R} = \frac{0.30}{1200} \qquad v_{B_R} = \frac{0.8}{1200} \qquad v_{C_R} = \frac{0.5}{1200}$$

which, in view of Maxwell's theorem, are the deflections of point R due to a 1 lb force acting at A, at B, and at C. It follows that the deflection at point R due to forces P_A, P_B, and P_C is

$$v_R = \frac{1500(0.30) + 700(0.8) + 1000(0.5)}{1200} = 1.26 \text{ in.}$$

(d) Other theorems of reciprocity. Betti's and Maxwell's theorems are not the only theorems of reciprocity of the mathematical theory of elasticity. In 1887, R. Land enunciated another theorem of reciprocity between the state of elastic constraint and the elastic deformation due to the action of external forces acting on an elastic body.[8] In 1912, G. Colonnetti[9] gave a general demonstration of this reciprocity theorem which he called the second reciprocity theorem.[10] This theorem in current literature is referred to as the Land-Colonnetti reciprocity theorem.

In 1905, Vito Volterra, in his theory of elastic dislocations in multiconnected elastic bodies, formulated a reciprocity theorem between two elastic dislocations (Volterra's reciprocity theorem).[11] Both Land-Colonnetti's and Volterra's theorems of reciprocity find important applications in the mechanics of elastic bodies.[12]

In 1938, Vito Volterra enunciated a most general theorem of reciprocity.[13] The theorem of reciprocity of Betti, the theorem of reciprocity of Land-Colonnetti, and the theorem of reciprocity of Volterra are particular cases of this most general theorem of reciprocity of Vito Volterra.[14]

2.7 Clapeyron's Theorem

Clapeyron's[15] theorem states that the strain energy U of an elastic body subjected to the action of a system of n statically applied forces P_1, P_2, P_3,

---

[8] See Bibliography, Sec. 2.6 No. 11 at the end of this chapter.

[9] **Colonnetti, Gustavo.** Italian mathematician and civil engineer (b. Turin 1886; d. Turin 1968). Professor of strength of materials at the University of Pisa and at the Polytechnic Institute of Turin. He wrote well-known treatises on the strength of materials and on mechanics, and many papers on the mathematical theory of elasticity and on plasticity. He suggested an ingenious plan for stabilizing the leaning tower of Pisa.

[10] See Bibliography, Sec. 2.6 Nos. 6 and 7 at the end of this chapter.

[11] See Bibliography, Sec. 2.6 No. 15 at the end of this chapter.

[12] See Bibliography, Sec. 2.6 No. 17 at the end of this chapter.

[13] See Bibliography, Sec. 2.6 No. 17 at the end of this chapter.

[14] See Bibliography, Sec. 2.6 Nos. 16 and 17 at the end of this chapter.

[15] **Clapeyron, Benoit Paul Emile.** French mathematician and civil engineer (b. Paris 1799; d. Paris 1864). After graduating from the École Polytechnique, he went to Russia with Lamé where both taught pure and applied mathematics at the Institute of Engineering of Ways of Communication in St. Petersburg. Clapeyron helped with the design of various important structural works including suspension bridges in St. Petersburg. Upon his return to France, he was interested mainly in the construction of French railroads and

... P_n[15] is equal to one half the sum of the products of the intensities of the forces and the components of the displacements of their points of application in the directions of the forces. Let L_e be the work done by external forces. For conservative forces it follows that

$$U = L_e$$

Since the work done by external forces is independent of the way in which the forces are applied and depends only on their final values, one can assume that the forces vary from their initial values (zero) to their final values P_k ($k = 1, 2, 3, \ldots, n$) through the intermediate values αP_k where α is a numerical coefficient common to each force. The value of α is assumed to vary slowly from zero to its final value 1.

Let s_k be the final value of the displacement of point k in the direction of force P_k which is applied at point k. Since the body is assumed to be elastic, when the generic force applied at point k has reached the intermediate value αP_k the corresponding displacement of point k in the direction of the force will have reached the intermediate value αs_k. When the force increases from αP_k to the value $(\alpha + d\alpha)P_k$, its point of application will be displaced by the amount $ds_k = s_k d\alpha$ in the direction of force P_k and the corresponding work done by force P_k will be (neglecting infinitesimal quantities of the second order)

$$\alpha P_k ds_k = \alpha P_k s_k d\alpha$$

and the infinitesimal work done by n forces will be

$$dL_e = \sum_{k=1}^{n} P_k s_k \alpha \, d\alpha$$

The total work done by n forces in the passage from the initial to the final configuration will be

$$L_e = U = \sum_{k=1}^{n} P_k s_k \int_0^1 \alpha \, d\alpha = \tfrac{1}{2} \sum_{k=1}^{n} P_k s_k \qquad (2.7.1)$$

The strain energy U of an elastic system under the action of concentrated forces P_k ($k = 1, 2, 3, \ldots, n$) can be expressed as a quadratic homogeneous function of the forces by introducing influence coefficients. The influence coefficient α_{ij} is defined as the displacement of the point of application i

---

in the application of thermodynamics to locomotive design. His classical paper, "Sur la puissance motrice de la chaleur," in which Carnot's ideas are translated into the language of mathematics, and on which modern thermodynamics is based, was published in 1834 in the Journal de l'Ecole Polytechnique.

[15] These forces are external forces and reactions. They are assumed to be concentrated forces but the theorem can be extended to the case of distributed forces.

of force P_i in the direction of force P_i produced by a unit force applied at point j and in the direction of force P_j.[16] Under the action of forces $P_k(k = 1, 2, 3, \ldots, n)$, the total displacement of point i will be

$$s_i = P_1\alpha_{1i} + P_2\alpha_{2i} + P_3\alpha_{3i} + \ldots + P_n\alpha_{ni}$$

and Eq. (2.7.1) becomes

$$\begin{aligned} U = L_e = \tfrac{1}{2}[&P_1(P_1\alpha_{11} + P_2\alpha_{12} + P_3\alpha_{13} + \ldots + P_n\alpha_{1n}) \\ +\, &P_2(P_1\alpha_{21} + P_2\alpha_{22} + P_3\alpha_{23} + \ldots + P_n\alpha_{2n}) \\ +\, &P_3(P_1\alpha_{31} + P_2\alpha_{32} + P_3\alpha_{33} + \ldots + P_n\alpha_{3n}) + \ldots \\ +\, &P_n(P_1\alpha_{n1} + P_2\alpha_{n2} + P_3\alpha_{n3} + \ldots + P_n\alpha_{nn})] \end{aligned}$$

or, finally

$$= \tfrac{1}{2} \sum_{i=1}^{n} \sum_{j=1}^{n} \alpha_{ij} P_i P_j \qquad (2.7.2)$$

Example 2.7.1. As an application of Clapeyron's theorem the deflection of a simply supported bar of length l under the action of a concentrated force of intensity P will be determined.

The force is applied at the section of the bar which is at distances a and b from the two supports (see Fig. 2.4.1). The bending moment will be

$$M_1 = \frac{Pbx}{l} \quad \text{for } 0 \leq x \leq a$$

$$M_2 = \frac{Pax'}{l} \quad \text{for } 0 \leq x' \leq b$$

The strain energy will be

$$U = \frac{P^2 b^2}{2EIl^2} \int_0^a x^2 dx + \frac{P^2 a^2}{2EIl^2} \int_0^b x'^2 dx' = \frac{P^2 a^2 b^2}{6EIl}$$

It δ is the deflection under load P, then in view of Clapeyron's theorem it follows that

$$L_e = \frac{1}{2} P\delta = U = \frac{P^2 a^2 b^2}{6EIl}$$

From the above relation the deflection is found to be

$$\delta = \frac{Pa^2 b^2}{3EIl}$$

---

[16] From Maxwell's theorem of reciprocity:

$$\alpha_{ij} = \alpha_{ji}$$

In the particular case in which the load is acting in the middle of the beam $a = b = \frac{l}{2}$ and it follows that

$$\delta = \frac{Pl^3}{48EI}$$

Example 2.7.2. By use of Clapeyron's theorem, the elastic energy of a vertical conical bar supported at its circular base and subjected to the action of its own weight will be evaluated.

Calling P the total weight of the bar, h its height, Ω_0 the area of the base, and $\Omega(x)$ the area of a generic section at distance x from the vertex, one finds that the cross-sectional area $\Omega(x) = \Omega_0(x^2/h^2)$ and that the load $P(x) = P(x^3/h^3)$. It follows that

$$dU = \frac{P^2\left(\frac{x^3}{h^3}\right)^2 dx}{2E\Omega_0 \frac{x^2}{h^2}} = \frac{P^2}{2E\Omega_0}\frac{x^4}{h^4}dx$$

and finally

$$U = \frac{P^2}{2E\Omega_0 h^4}\int_0^h x^4 dx = \frac{P^2 h}{10E\Omega_0}$$

2.8 Castigliano's First Theorem

Before discussing Castigliano's theorem it will be necessary to explain the difference between strain energy and complementary energy. Consider the elastic spring shown in Fig. 2.8.1 subjected to a gradually increasing force F. If the load deflection relationship is not linear, force F will be related to the deflection y of the spring by curve OB of Fig. 2.8.2. Area OBD below and to the right of the curve represents the work done upon the spring by the force and it represents the strain energy U stored in the spring. It is expressed by

$$U = \int_0^y F dy \qquad (2.8.1)$$

By differentiating Eq. (2.8.1) with respect to y one obtains

$$\frac{dU}{dy} = F \qquad (2.8.2)$$

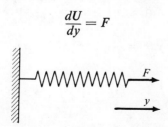

Figure 2.8.1

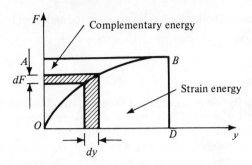

Figure 2.8.2

i.e., the force applied to the spring when the deflection of the spring is y. Consider now area OBA above and to the left of the curve shown in Fig. 2.8.2. It is represented by

$$C = \int_0^F y\,dF \tag{2.8.3}$$

Engesser called this quantity "Ergänzungsarbeit." This German word was translated in 1939 by A. W. Adkins of the Massachusetts Institute of Technology into the English language as "Complementary Energy."

Differentiation of the complementary energy with respect to force F gives

$$\frac{dC}{dF} = y \tag{2.8.4}$$

Complementary energy is a mathematical quantity and has in general no physical status; there are however, cases in which strain and complementary energies are interchangeable. Suppose for instance that force F and the deflection y of the spring of Fig. 2.8.1 are related by the equation

$$F = ay^n \tag{2.8.5}$$

where a and n are constants. In the case in which $n = 1$, one has the linear system. By substituting Eq. (2.8.5) into Eqs. (2.8.1) and (2.8.3) one obtains

$$U = \int_0^y F\,dy = a\int_0^y y^n\,dy = \frac{1}{na^{1/n}}\int_0^F F^{1/n}\,dF$$

$$C = \int_0^F y\,dF = \frac{1}{a^{1/n}}\int_0^F F^{1/n}\,dF = an\int_0^y y^n\,dy$$

It follows that

$$\left.\begin{aligned}\frac{dU}{dy} &= F & \frac{dU}{dF} &= \frac{1}{n}\left(\frac{F}{a}\right)^{1/n} = \frac{1}{n}y \\ \frac{dC}{dF} &= y & \frac{dC}{dy} &= any^n = nF\end{aligned}\right\} \tag{2.8.6}$$

When $n = 1$ Eqs. (2.8.6) become

$$\frac{dU}{dF} = y \qquad (2.8.7)$$

$$\frac{dC}{dy} = F \qquad (2.8.8)$$

Thus when $n = 1$ the strain and complementary energies are interchangeable (see Fig. 2.8.3), but when $n \neq 1$, factor $1/n$ or n must be introduced. Equation (2.8.7) contains what is usually referred to as Castigliano's first theorem of strain energy for the case when $n = 1$; the principle of least work (or Menabrea's theorem or Castigliano's second theorem) can be derived from the first theorem. Castigliano was an engineer and as a consequence was mainly concerned with materials used for construction purposes which have linear force-deflection characteristics. It is therefore not surprising that Castigliano, having discovered the possibility of using strain energy to derive

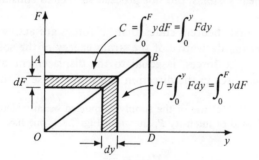

Figure 2.8.3

Castigliano, Carlo Alberto. Italian mathematician and railroad engineer (b. Asti 1847; d. Milan 1884). His celebrated work on the mathematical theory of elasticity was published in the book entitled "Théorie de l'équilibre des systèmes élastiques et ses applications" (Turin, 1881). (Reproduction kindly provided by Dr. Ing. Arturo Gaj, president of the "Ordine degli Ingegneri" of the province of Asti, Italy.)

Menabrea, Count Luigi Federico. Italian mathematician, statesman, soldier, and diplomat (b. Chambery, Savoy 1809; d. Saint Cassin, near Chambery 1896). From 1846 to 1860 he was professor of strength of materials at the University of Turin. He was many times minister and also prime minister for King Victor Emanuel II, general in the Italian army, and Italian ambassador to London and Paris. (Reproduction kindly provided by Professor Francesco Giacomo Tricomi of the University of Turin.)

deflections of linear systems, did not proceed further to consider the general problem.

Castigliano's first theorem states that "if forces are acting on an elastic solid then the partial derivative of the strain energy of the solid taken with respect to one of the forces, is equal to the displacement of the point of application of the force in the direction and in the sense of the force."

Example 2.8.1. In the case of the simply supported beam of length l subjected to a concentrated load of intensity P, shown in Fig. 2.8.4, one has

$$M(x) = \frac{Px}{2}$$

for $0 \leq x \leq l/2$; thus,

$$U = \frac{1}{2} \int_0^l \frac{M^2(x)}{EI} dx = \frac{P^2}{4EI} \int_0^{l/2} x^2 dx = \frac{P^2 l^3}{96 EI}$$

and

$$\frac{\partial U}{\partial P} = \frac{Pl^3}{48EI} = v\left(\frac{l}{2}\right)$$

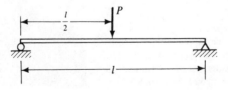

Figure 2.8.4

Example 2.8.2. For the cantilever beam of length l subjected to a couple M at the free end, shown in Fig. 2.8.5, one has

$$U = \frac{1}{2} \int_0^l \frac{M^2}{EI} dx = \frac{M^2 l}{2EI}$$

and

$$\frac{\partial U}{\partial M} = \frac{Ml}{EI} = v'(0)$$

Example 2.8.3. For the cantilever beam of length l subjected to a concentrated force P at the free end, shown in Fig. 2.8.6, one has

$$U = \frac{1}{2} \int_0^l \frac{M^2 dx}{EI} = \frac{P^2}{2EI} \int_0^l x^2 dx = \frac{P^2 l^3}{6EI}$$

and

$$\frac{\partial U}{\partial P} = \frac{Pl^3}{3EI} = v(0)$$

Castigliano's first theorem can be easily demonstrated, in the case of concentrated forces by using Clapeyron's theorm

$$U = \frac{1}{2} \sum_{i=1}^n \sum_{j=1}^n \alpha_{ij} P_i P_j \qquad (2.7.2)$$

The terms containing P_k in Eq. (2.7.2) are

$$\tfrac{1}{2}[P_1 P_k \alpha_{1k} + P_2 P_k \alpha_{2k} + P_3 P_k \alpha_{3k} + \ldots$$
$$+ P_k(P_1 \alpha_{k1} + P_2 \alpha_{k2} + \ldots + P_k \alpha_{kn} + \ldots P_n \alpha_{kn}) + \ldots$$
$$+ P_n P_k \alpha_{nk}]$$

or, by Maxwell's reciprocity theorem

$$P_k P_1 \alpha_{k1} + P_k P_2 \alpha_{k2} + \ldots + \tfrac{1}{2} P_k^2 \alpha_{kk} + \ldots + P_k P_n \alpha_{kn}$$

By differentiating with respect to P_k one finds that

$$\frac{\partial U}{\partial P_k} = P_1 \alpha_{k1} + P_2 \alpha_{k2} + \ldots + P_k \alpha_{kk} + \ldots + P_n \alpha_{kn} = s_k$$

as was to be demonstrated.

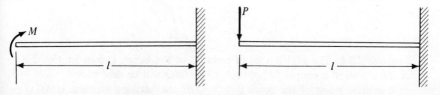

Figure 2.8.5   Figure 2.8.6

Example 2.8.4. The vertical displacement of point A of the system shown in Fig. 2.8.7 will be determined for the case in which the two bars AB and AC are composed of the same material and the cross-sectional area Ω_2 of AC is equal to 3Ω, Ω being the cross-sectional area of AB.

The force in bar AB is $S_{AB} = \sqrt{3}\,P$ while the force in bar AC is $S_{AC} = -2P$. It follows that the strain energy is

$$U = \frac{(\sqrt{3}\,P)^2 l}{2E\Omega} + \frac{(-2P)^2 2l/\sqrt{3}}{2E(3\Omega)} = \frac{27 + 8\sqrt{3}}{18}\frac{P^2 l}{E\Omega}$$

and the displacement of point A is found to be

$$S_a = \frac{\partial U}{\partial P} = \frac{27 + 8\sqrt{3}}{9}\frac{Pl}{E\Omega}$$

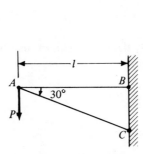

Figure 2.8.7   Figure 2.8.8

Example 2.8.5. The displacement of point A of the system shown in Fig. 2.8.8 will be determined for the case in which the five bars have the same cross-sectional area Ω and are composed of the same material.

The forces in the bars are

$$S_1 = S_2 = S_4 = S_5 = \frac{P}{\sqrt{3}} \qquad S_3 = -\frac{P}{\sqrt{3}}$$

It follows that the strain energy is

$$U = \frac{5\frac{P^2}{3}l}{2E\Omega} = \frac{5}{6}\frac{P^2 l}{E\Omega}$$

and the displacement of point A is

$$S_a = \frac{\partial U}{\partial P} = \frac{5}{3}\frac{Pl}{E\Omega}$$

Example 2.8.6. The rotation ϕ_c of point C of the beam, shown in Fig. 2.8.9, under the action of a couple M applied at its center will be determined.

At a generic point of the beam $0 \leq x \leq l/2$ the bending moment is $M(x) = -M(x/L)$. It follows for the strain energy that

$$U = 2\frac{1}{2EI}\int_0^{l/2}\left(-\frac{Mx}{l}\right)^2 dx = \frac{M^2 l}{24EI}$$

and for the rotation ϕ_c that

$$\phi_c = \frac{\partial U}{\partial M} = \frac{Ml}{12EI}$$

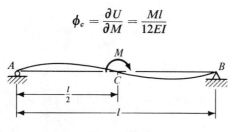

Figure 2.8.9

In most cases it is not necessary, in applying Castigliano's theorem, to compute the strain energy U and afterwards to differentiate it with respect to P_k [if Eq. (2.1.9) is used for the strain energy] or with respect to S_k [if Eq. (2.1.10) is used for the strain energy]. Instead it is easier to first differentiate Eqs. (2.1.9) and (2.1.10) from which it follows that

$$s_k = \int_0^l \frac{M}{EI}\frac{\partial M}{\partial P_k} dx \tag{2.8.9}$$

$$s_k = \sum_{i=1}^n S_i \frac{\partial S_i}{\partial P_k}\frac{P_i}{E\Omega_i} \tag{2.8.10}$$

This procedure will be illustrated in the following two examples.

Example 2.8.7. The relative displacement s_{AB} of points A and B of the frame, shown in Fig. 2.8.10, will be determined.

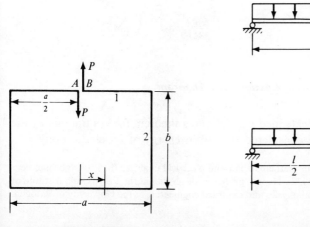

Figure 2.8.10

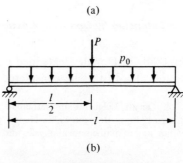

Figure 2.8.11

In bar 1, one has

$$M = Px \qquad \frac{\partial M}{\partial P} = x$$

while in bar 2

$$M = \frac{Pa}{2} \qquad \frac{\partial M}{\partial P} = \frac{a}{2}$$

It follows that

$$s_{AB} = \frac{4}{EI}\int_0^{a/2}(Px)(x)dx + \frac{2}{EI}\int_0^b\left(\frac{Pa}{2}\right)\left(\frac{a}{2}\right)dx = \frac{Pa^3}{6EI} + \frac{Pa^2b}{2EI}$$

Example 2.8.8. To compute the displacement at the center of the simply supported horizontal bar [see Fig. 2.8.11(a)] which is subjected to a uniformly distributed load of intensity p_0, we add a fictitious concentrated load of intensity P at the center of the beam [see Fig. 2.8.11(b)]. It follows that for $0 \leq x \leq l/2$

$$M(x) = \left(\frac{p_0 l}{2} + \frac{P}{2}\right)x - \frac{p_0 x^2}{2}$$

and

$$\frac{\partial M(x)}{\partial P} = \frac{x}{2}$$

Thus

$$v\left(\frac{l}{2}\right) = \frac{2}{EI}\int_0^{l/2} M(x)\frac{\partial M(x)}{\partial P}dx = \frac{2}{EI}\int_0^{l/2}\left(p_0\frac{lx}{2} + \frac{Px}{2} - \frac{p_0 x^2}{2}\right)\frac{x}{2}dx$$

$$= \frac{5p_0 l^4}{384EI} + \frac{Pl^3}{48EI}$$

By letting $P = 0$ one has the result that

$$v\left(\frac{l}{2}\right) = \frac{5p_0 l^4}{384EI}$$

2.9 Extension by Donati of Castigliano's Theorem

In 1888, L. Donati[17] extended Castigliano's theorem for continuous elastic bodies. To do this he considered the strain energy of an elastic body as a

---

[17] **Donati, Luigi.** Italian mathematician and electrical engineer (b. Fossombrone, near Pesaro 1846; d. Bologna 1932). He was professor of physics at the Polytechnic Institute of Milan, and of mathematical physics and electrical engineering at the University of Bologna.

functional of the components u, v, and w of the displacements of the different points of the body. Components of the displacements are functions of the points of the body, but, at the same time can be considered as functions of components X, Y, and Z of the intensity of the field of force at any point of the body. Since the functional derivatives of the strain energy with respect to functions X, Y, and Z are, respectively, u, v, and w, Donati's theorem follows. In the same way, the strain energy of an elastic body can be considered as a functional of components X, Y, and Z of the field of force at any point, and these components can be considered as functions of the components of the displacements. It follows that the functional derivatives of the strain energy with respect to the functions u, v, and w are X, Y, and Z, respectively.

2.10 Engesser's Theorem on the Principle of Complementary Energy

F. Engesser generalized Castigliano's theorem to the case in which displacements are non-linear functions of external forces by introducing the notion of complementary energy. He showed that derivatives of the complementary energy with respect to independent forces give displacements.[18] Engesser's work received little attention until 1942 when H. M. Westergaard drew attention to Engesser's contribution.[19] Engesser's principle has been used in the solution of problems of plasticity.

Variational principles in the theory of plasticity have been presented also by G. Colonnetti,[20] P. Hodge and W. Prager,[21] and H. J. Greenberg.[22] Engesser's principle will be illustrated by the following problem.[23]

Example 2.10.1. Two identical horizontal bars of original lengths l are hinged together at one end and hinged to fixed supports A and B at the other ends as shown in Fig. 2.10.1. Determine the distance δ that point C will displace vertically under the action of vertical force P.

The following relationship is valid for small displacements

$$\sin \alpha = \tan \alpha = \frac{\delta}{l}$$

Calling S the forces in the bar, Ω the cross-sectional areas of the bars, and ϵ the

[18] See Bibliography, Sec. 2.10 No. 4 at the end of this chapter.
[19] See Bibliography, Sec. 2.10 No. 15 at the end of this chapter.
[20] See Bibliography, Sec. 2.10 No. 2 at the end of this chapter.
[21] See Bibliography, Sec. 2.10 No. 8 at the end of this chapter.
[22] See Bibliography, Sec. 2.10 No. 7 at the end of this chapter.
[23] This problem is taken from S. P. Timoshenko's "History of Strength of Materials," pp. 292–293.

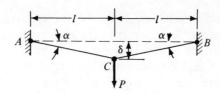

Figure 2.10.1

unit elongation of each bar it follows that

$$S = \frac{P}{2 \sin \alpha} \doteq \frac{Pl}{2\delta}$$

and

$$\epsilon = \frac{\sqrt{l^2 + \delta^2} - l}{l} \doteq \frac{1}{2}\frac{\delta^2}{l^2}$$

Also

$$\epsilon = \frac{S}{\Omega E} = \frac{Pl}{2\delta \Omega E}$$

From the last two relations it follows that

$$\frac{Pl}{2\delta \Omega E} = \frac{1}{2}\frac{\delta^2}{l^2}$$

and finally

$$\delta = l\sqrt[3]{\frac{P}{E\Omega}} \tag{2.10.1}$$

Equation (2.10.1) shows that deflection δ is not a linear function of the load, although the material is assumed to obey Hooke's law. The strain energy is given by

$$U = \int_0^\delta P d\delta = \frac{lP^{4/3}}{4\sqrt[3]{\Omega E}}$$

and is not a function of the second degree of force P and $\partial U/\partial P$ is not equal to the displacement δ. The complementary energy is

$$C = \int_0^P \delta dP = \frac{l}{\sqrt[3]{\Omega E}} \int_0^P P^{1/3} dP = \frac{3lP^{4/3}}{4\sqrt[3]{\Omega E}} \tag{2.10.2}$$

From Eq. (2.10.2) it follows that

$$\frac{dC}{dP} = l\sqrt[3]{\frac{P}{E\Omega}} = \delta$$

2.11 Castigliano's Second Theorem or Menabrea's Theorem

Castigliano's theorem is very useful in the analysis of statically indeterminate structures. A structure is called statically indeterminate if the number n_r of external reactions is greater than the number of independent equations n_s which can be written by applying the principles of statics. The difference $(n_r - n_s)$ is called the degree of indeterminateness of the structure. In the case of the continuous beam resting on three supports as shown in Fig. 2.11.1(a), the free body diagram of Fig. 2.11.1(b) shows that the number n_r of external reactions is 3 (R_A, R_B, and R_C). The number n_s of independent equations furnished by statics is 2 ($\sum M = 0$, $\sum Y = 0$). It follows that the degree of indetermination of the system $(n_r - n_s)$ is one in this case.

In the case of the beam with one end fixed and the other simply supported as shown in Fig. 2.11.2(a), the free body diagram of Fig. 2.11.2(b) shows that the number n_r of external reactions is 3 (R_A, R_B, and M). The number n_s of independent equations furnished by statics is 2 ($\sum M = 0$, $\sum Y = 0$). It follows that the degree of indetermination of the system $(n_r - n_s)$ is one in this case.

A statically indeterminate system can be replaced by a statically determinate one called the fundamental system by removing $(n_r - n_s)$ conditions of support and replacing them by convenient forces (or couples) called statically indeterminate quantities. These statically indeterminate quantities may be computed by the use of Castigliano's theorem. To do this, consider a system k times statically indeterminate. On this system are applied n external forces (or couples) $F_1, F_2, F_3, \ldots, F_n$, the k statically indeterminate quantities which will be indicated by the letters

$$X_1, X_2, X_3, \ldots, X_k$$

as well as the reactions at the supports. These reactions can always be ex-

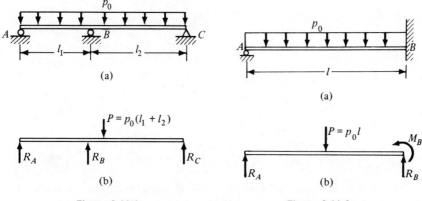

Figure 2.11.1　　　　　Figure 2.11.2

pressed, through the conditions of equilibrium, as functions of the quantities $F_i(i = 1, 2, 3, \ldots, n)$ and $X_i(i = 1, 2, 3, \ldots, k)$. The strain energy of the fundamental system can be expressed as a function of the above quantities, i.e.,

$$U = U(F_1, F_2, F_3, \ldots, F_n; X_1, X_2, X_3, \ldots, X_k)$$

The displacements (or rotations) of the point of application of the forces (or couples) X_i in the direction of these forces must be zero. This implies, in view of Castigliano's theorem, the k conditions

$$\frac{\partial U}{\partial X_i} = 0 \quad (i = 1, 2, 3, \ldots, k) \tag{2.11.1}$$

The number of Eqs. (2.11.1) is equal to the degree of indetermination of the system. By the use of Eqs. (2.11.1), the statically indeterminate quantities $X_i(i = 1, 2, 3, \ldots, k)$ can be determined.

Equations (2.11.1) constitute the necessary conditions for the strain energy of the system, considered as a function of the statically indeterminate quantities X_i, to have a stationary value. However, in the case of stable equilibrium it can be shown that the values of $X_i(i = 1, 2, 3, \ldots, k)$ satisfying Eqs. (2.11.1) correspond to a minimum of the strain energy. This important principle, expressed by Eqs. (2.11.1), is called the principle of least work. It was first stated by F. Menabrea.[24] The complete proof of the principle was given by Castigliano who made use of this principle to solve statically indeterminate systems.[25] This principle is often referred to as Castigliano's second theorem.

Example 2.11.1. By the use of Castigliano's second theorem the reaction at the support B of the beam, shown in Fig. 2.11.1, will be determined for the case in which $l_1 = l_2 = l$.

Letting $R_B = X$, it follows that for $0 \leq x \leq l$

$$M(x) = \left(p_0 l - \frac{X}{2}\right)x - \frac{p_0 x^2}{2}$$

and

$$\frac{\partial M(x)}{\partial X} = -\frac{x}{2}$$

Hence, it follows that

$$\frac{\partial U}{\partial X} = \frac{2}{EI}\int_0^l M(x)\frac{\partial M(x)}{\partial X}dx = \frac{1}{EI}\int_0^l \left[-\left(p_0 l - \frac{X}{2}\right)\frac{x^2}{2} + \frac{p_0 x^3}{4}\right]dx = 0$$

---

[24] See Bibliography, Sec. 2.11 Nos. 18, 19, 20, 21, and 22 at the end of this chapter.
[25] See Bibliography, Sec. 2.11 Nos. 5 and 6 at the end of this chapter.

or

$$-\frac{p_0 l^4}{6} + \frac{X l^3}{12} + \frac{p_0 l^4}{16} = 0$$

from which it is found that

$$X = \frac{5 p_0 l}{4}$$

Example 2.11.2. By the use of Castigliano's second theorem the reaction $R_A = X$ at the left support of the statically indeterminate system, shown in Fig. 2.11.2, will be determined.

The following relationships are immediately derived for $0 \leq x \leq l$

$$M(x) = Xx - \frac{p_0 x^2}{2} \qquad \frac{\partial M(x)}{\partial X} = x$$

and thus

$$\frac{\partial U}{\partial X} = \frac{1}{EI} \int_0^l \left[Xx - \frac{p_0 x^2}{2} \right] x \, dx = 0$$

from which

$$\frac{X l^3}{3} - \frac{p_0 l^4}{8} = 0$$

and finally

$$X = \tfrac{3}{8} p_0 l$$

Example 2.11.3. By use of Castigliano's second theorem, the reaction of the central support C of the structure, shown in Fig. 2.11.3, will be determined.

If X is the unknown reaction of the support CD, then for a generic section of the beam AB at a distance $x \leq l$ from A one will have

$$M(x) = p_0 l x - \frac{X}{2} x - \frac{p_0 x^2}{2}$$

The strain energy of the whole system will be

$$U = 2 \int_0^l \frac{M^2 dx}{2EI} + \frac{X^2 h}{2 E_1 \Omega_1}$$

From the condition that $\partial U / \partial X = 0$ it follows that

$$X = \frac{\dfrac{5}{24} \dfrac{p_0 l^4}{EI}}{\dfrac{l^3}{6EI} + \dfrac{h}{E_1 \Omega_1}}$$

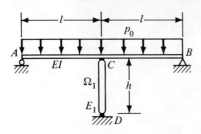
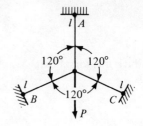

Figure 2.11.3   Figure 2.11.4

In the case in which the intermediate support is rigid ($E_1 = \infty$), the above formula reduces to $X = \frac{5}{4} p_0 l$ as in Example 2.11.1.

Example 2.11.4. By the use of Castigliano's second theorem, the forces acting in the three bars of the elastic system, shown in Fig. 2.11.4, will be determined.

Calling X the force acting on the vertical bar OA, the other two bars will each be acted upon by forces $(P - X)$. The strain energy of the whole system will be:

$$U = \frac{X^2 l}{2E\Omega} + \frac{2(P - X)^2 l}{2E\Omega}$$

From the condition that

$$\frac{\partial U}{\partial X} = \frac{Xl}{E\Omega} - \frac{2(P - X)l}{E\Omega} = 0$$

it follows that

$$X = \tfrac{2}{3} P$$

Example 2.11.5. The double symmetrical frame of Fig. 2.11.5(a) will now be studied by applying Castigliano's second theorem.

Owing to the symmetry of the structure and of the load, reactions at B will be the same as those at A while at C there will be only a vertical reaction. The system is therefore only twice statically indeterminate. The two unknown components of the reactions at A, V_A, and H_A will be assumed as statically indeterminate quantities. In a generic section S_1 of AD and in a generic section S_2 of DF one will have [see Fig. 2.11.5(b)]

$$M = -H_A x \qquad \frac{\partial M}{\partial V_A} = 0 \qquad \frac{\partial M}{\partial H_A} = -x \quad \text{for } S_1$$

$$M = V_A x - H_A h - \frac{p_0 x^2}{2} \qquad \frac{\partial M}{\partial V_A} = x \qquad \frac{\partial M}{\partial H_A} = -h \quad \text{for } S_2$$

It follows that

$$\frac{\partial U}{\partial V_A} = \frac{1}{EI} \int_0^l \left[V_A x - H_A h - \frac{p_0 x^2}{2} \right] x \, dx = 0$$

$$\frac{\partial U}{\partial H_A} = \frac{1}{E_1 I_1} \int_0^h (-H_A x)(-x) \, dx + \frac{1}{EI} \int_0^l \left(V_A x - H_A h - \frac{p_0 x^2}{2} \right)(-h) \, dx = 0$$

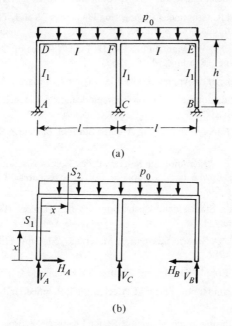

Figure 2.11.5

from which the two equations

$$8V_A - 12H_Ah = 3p_0l^2$$
$$3\alpha V_A l - 2(3\alpha + 1)H_Ah = \alpha p_0 l^2$$

are derived. Finally

$$V_A = \frac{3(\alpha + 1)}{2(3\alpha + 4)}p_0l$$

$$H_A = \frac{\alpha}{4(3\alpha + 4)}\frac{p_0l^2}{h}$$

where

$$\alpha = \frac{E_1I_1l}{EIh}$$

In the above example the effects of shear as well as the effects of normal forces have been neglected.

BIBLIOGRAPHY

Sections 2.1, 2.4, 2.5, 2.7. STRAIN ENERGY; PRINCIPLE OF VIRTUAL WORK; UNIQUENESS OF SOLUTION; CLAPEYRON'S THEOREM

1. Argyris, J. H., and S. Kelsey, "Energy Theorems and Structural Analysis," Butterworth & Co. (Publishers) Ltd., London, England, 1960.

2. Betti, E., "Teoria della Elasticità," *Nuovo Cimento, Serie 2*, T. VII and VIII, 1872.
3. Biezeno, C. B., and R. Grammel, "Engineering Dynamics," **Vol. I**, Theory of Elasticity, Blackie & Son Ltd., Glasgow, Scotland, 1955.
4. Bisplinghoff, R. L., J. W. Mar, and T. H. H. Pian, "Statics of Deformable Solids," Addison-Wesley, Inc., Reading, Mass., 1965.
5. Ceradini, C., "Meccanica Applicata alle Costruzioni," Vallardi, Milan, Italy, 1921.
6. Cesáro, E., "Introduzione alla Teoria Matematica della Elasticità," Fratelli Bocca Editori, Turin, Italy, 1894.
7. Charlton, T. M., "Energy Principles in Applied Statics," Blackie & Son Ltd., Glasgow, Scotland, 1959.
8. Clapeyron, E. B. P., "Mémoire sur le Travail des Forces Elastiques dans un Corps Solide Elastique Déformé par l'action des Forces Extérieures," *Comp. Rend., Paris,* **Vol. XLVI**, 1858, p. 208.
9. Colonnetti, G., "La Statica delle Costruzioni," **Vol. I**, Unione Tipografica Editrice Torinese, Turin, Italy, 1928.
10. Den Hartog, J. P., "Advanced Strength of Materials," McGraw-Hill Book Company, New York, N.Y., 1952.
11. Foppl, A., and L. Foppl, "Drang und Zwang," Oldenbourg, Munich, Germany, 1924.
12. Fung, Y. C., "Foundations of Solid Mechanics," Prentice-Hall, Inc., Englewood Cliffs, N. J., 1965.
13. Girtler, R., "Einführung in die Mechanik Fester Elastischer Körper," Julius Springer, Berlin, Germany, 1931.
14. Graffi, D., "Elementi di Meccanica Razionale," Casa Editrice Patron, Bologna, Italy, 1961.
15. Green, A. E., and W. Zerna, "Theoretical Elasticity," Oxford University Press, London, England, 1954.
16. Guidi, C., "Lezioni sulla Scienza delle Costruzioni," Vincenzo Bona, Turin, Italy, 1905.
17. Hoff, N. J., "The Analysis of Structures," John Wiley & Sons, Inc., New York, N.Y., 1956.
18. Kirchhoff, G., "Vorlesungen über Matematische Physik," **Bd. I**: Mechanik, B. G. Teubner, Leipzig, Germany, 1897.
19. Lamb, E. H., "The Principle of Virtual Velocities and Its Application to the Theory of Elastic Structures," *Inst. Civil Engrs.*, "Selected Engineering Papers," **No. 10**, 1923.
20. Langhaar, H. L., "Energy Methods in Applied Mechanics," John Wiley & Sons, Inc., New York, N.Y., 1962.
21. Levi-Civita, T., and U. Amaldi, "Lezioni di Meccanica Razionale," **Vol. I**, Zanichelli, Bologna, Italy, 1950.
22. L'Hermite, R., "Resistance des Matériaux Théorique et Experimentale," **Tome I**, Dunod, Paris, France, 1959.
23. Lorenz, H., "Techniche Elastizitätslehre," Oldenbourg, Munich and Berlin, Germany, 1913.
24. Love, A. E. H., "A Treatise on the Mathematical Theory of Elasticity," 4th ed., Dover Publications, Inc., New York, N.Y., 1944.

25. Muller-Breslau, H., "Die Neueren Methoden der Festigkeitslehre," Kroner, Leipzig, Germany, 1886.

26. Nadai, A., "Theory of Flow and Fracture of Solids," **Vol. 2**, McGraw-Hill Book Company, New York, N.Y., 1963.

27. Neal, B. G., "Structural Theorems and Their Applications," The Macmillan Company, New York, N.Y., 1964.

28. Oravas, G. A. E., and L. McLean, "Historical Development of Energetical Principles in Elastomechanics," Part I and Part II, *Appl. Mech. Rev.*, **Vol. 19**, Nos. 8 and 11, August and November, 1966.

29. Pippard, A. J. S., and J. F. Baker, "The Analysis of Engineering Structures," Edward Arnold (Publishers) Ltd., London, England, 1945.

30. Pirlet, J., "Statik der Baukonstruktionen," Julius Springer, Berlin, Germany, 1921–1923.

31. Shanley, F. R., "Strength of Materials," McGraw-Hill Book Company, New York, N.Y., 1957.

32. Sokolnikoff, I. S., "Mathematical Theory of Elasticity," McGraw-Hill Book Company, New York, N.Y., 1956.

33. Sommerfeld, A., "Mechanics of Deformable Bodies," Lectures on Theoretical Physics, **Vol. II**, Academic Press, Inc., New York, N.Y., 1950.

34. Southwell, Sir Richard, "An Introduction to the Theory of Elasticity for Engineers and Physicists," Oxford University Press, London, England, 1941.

35. Timoshenko, S. P., "History of Strength of Materials," McGraw-Hill Book Company, New York, 1953.

36. Timoshenko, S. P., and J. N. Goodier, "Theory of Elasticity," McGraw-Hill Book Company, New York, N.Y., 1951.

37. Timoshenko, S. P., and D. H. Young, "Engineering Mechanics," McGraw-Hill Book Company, New York, N.Y., 1937.

38. ———, "Theory of Structures," 2nd ed., McGraw-Hill Book Company, New York, N.Y., 1964.

39. Todhunter, I., and K. Pearson, "A History of the Theory of Elasticity and of the Strength of Materials," Cambridge University Press, **Vol. 1**, 1886; **Vol. 2**, 1893, reprinted by Dover Publications, Inc., New York, N.Y., 1960.

40. Wang, C. T., "Applied Elasticity," McGraw-Hill Book Company, New York, N.Y., 1953.

41. Weingarten, J. L., "Ueber das Clapeyronsche Theorem in der Technichen Elastizitätstheorie," *Z. Architek. Ing.*, **Vol. 55**, Col. 515, 1909.

42. Westergaard, H. M., "Theory of Elasticity and Plasticity," Harvard University Press, Cambridge, Mass., 1952.

Sections 2.2, 2.3. HUBER-VON MISES-HENCKY STRENGTH THEORY; OTHER STRENGTH THEORIES

1. Beltrami, E., "Sulle Condizioni di Resistenza dei Corpi Elastici," *Rend. Ist. Lombardo*, Serie II, **Vol. XVIII**, 1885, pp. 704–714; *Nuovo Cimento*, Serie III, **T. XVIII**, pp. 145–155.

2. Bresler, B., and K. Pister, "Failure of Plain Concrete under Combined Stresses," *Trans. Am. Soc. Civil Engrs.*, **Vol. 122**, 1957, pp. 1049–1059.

3. Bridgman, P. W., "Studies in Large Plastic Flow and Fracture with Special Emphasis on the Effects of Hydrostatic Pressure," 1st ed., McGraw-Hill Book Company, New York, N.Y., 1952.

4. Bridgman, P. W., "The Physics of High Pressure," 2nd ed., The Macmillan Company, New York, N. Y., 1950.

5. Case, J., "The Strength of Materials," Chapter VI, Edward Arnold (Publishers) Ltd., London, England, 1943.

6. Cook, G., "The Elastic Limit of Metals Exposed to Triaxial Stress," *Proc. Roy. Soc. London, Serie A*, **Vol. 137**, 1932, p. 55.

7. Griffith, A. A., "The Phenomena of Rupture and Flow of Solids," *Phil. Trans. Roy. Soc. London, Serie A*, **Vol. 221**, 1920, pp. 163–198.

8. Guest, J. J., "Yield Surface in Combined Stress," *Phil. Mag.*, **Vol. 150**, 1920, p. 261; **Vol. 50**, 1900, p. 69.

9. Haigh, B. P., *Engineering*, **Vol. 190**, 1920, p. 158.

10. Hencky, H., "Zur Theorie Plastischer Deformationen," *Z. Angew. Math. Mech.*, **Vol. 4**, 1924, p. 323.

11. ——, "Über das Wesen der Plastischen Verformung," *Z. Ver. Deut. Ing.*, **Vol. 69**, 1925, p. 695.

12. Irwin, G. R., "Fracture Mechanics," *Proc. First Symp. Naval Struct. Mech.*, Pergamon Press, London, England, 1958.

13. Karman, T. von, "Festigkeitsversuche unter Allseitigem Druck," *Z. Ver. Deut. Ing.*, **Vol. 55**, 1911, pp. 1749–1757.

14. Lessells, J. M., and C. W. McGregor, "Combined Stress Experiments on a Nickel-Chrome—Molybdenum Steel," *J. Franklin Inst.*, **Vol. 230**, 1940, p. 163.

15. Levi-Civita, T., "Sul Massimo Cimento Dinamico nei Sistemi Elastici," *Nuovo Cimento, Serie V*, **T. II**, 1901, pp. 188–196; Opere Matematiche, Publicate a cura dell'Accademia Nazionale dei Lincei, **Vol. II**, Bologna, Italy, 1956, pp. 145–152.

16. Lode, W., "Versuche über den Einfluss der mittleren Hauptspannung auf das Fliessen der Metalle Eisen, Kupfer und Nickel," *Z. Physik*, **Bd. 36**, 1926, pp. 913–939.

17. ——, "Der Einfluss der mittleren Hauptspannung auf das Fliessen der Metalle," Forschungsarb. *Ver. Deut. Ing.*, Berlin, **Heft 303**, 1928.

18. Ludvik, P., "Bruchgefahr und Materialprüfung," Schweiz. Verband Materialprüfung., Tech. Ber., **Vol. 13**, Zurich, Switzerland, 1928.

19. ——, "Elemente der Technologischen Mechanik," Julius Springer, Berlin, Germany, 1909.

20. Mises, R. von, "Mechanik der Festen Körper in Plastisch-Deformablen Zustand," *Nachr. Akad. Wiss. Göttingen, Math-physik, Kl.*, 1913.

21. Mohr, O., "Abhandlungen aus dem Gebiete der Technischen Mechanik," 2nd ed., W. Ernst und Sohn, Berlin, Germany, 1914.

22. Murphy, G., "Advanced Mechanics of Materials," McGraw-Hill Book Company, New York, N.Y., 1964.

23. Nadai, A., "Zur Mechanik der bildsamen Formänderungen," *Stahl Eisen*, 1925.

24. ——, "Versuche über die Fliessgrenze des Eisens," *Proc. 2nd Intern. Congr. Appl. Mech.*, Zurich, Switzerland, Sept., 1926.

25. ——, "Theories of Strength," *J. Appl. Mech.*, **Vol. 1, No. 3,** 1933, pp. 111–129.
26. ——, "Theory of Flow and Fracture of Solids," McGraw-Hill Book Company, New York, N.Y., 1950.
27. Peterson, R. E., "Strength Theories Applied to Fatigue of Ductile Materials," *Proc. Soc. Exptl. Stress Anal.*, **Vol. 1, No. 1,** 1943, pp. 124–127.
28. Popov, E. P., "Introduction to Mechanics of Solids," Chapter IX, Part C, Prentice-Hall, Inc., Englewood Cliffs, N.J., 1968.
29. Roŝ, M., and A. Eichinger, "Versuche zur Klärung der Frage der Bruchgefahr," Eidgenossenschaft Materialprufungsanstalt, E.T.H., Zurich, Switzerland.
30. Seely, F. B., and J. O. Smith, "Advanced Mechanics of Materials," John Wiley & Sons, Inc., New York, N.Y., 1952.
31. Siebel, E., and A. Maier, "Einfluss mehrachsiger Spannungszustaende auf das Formaenderungs-Vermoegen Metallischer Werkstoffe," *Z. Ver. Deut. Ing.*, **Vol. 77,** 1933, p. 1345.
32. Soderberg, C. R., "Factor of Safety and Working Stress," *Trans. Am. Soc. Mech. Engrs.*, **Vol. 52, No. 11,** 1930.
33. Taylor, Sir Geoffrey, "The Distortion of an Aluminum Crystal during a Tensile Test," (Bakerian Lecture to the Royal Society, delivered 22 February 1923.) *Proc. Roy. Soc., London, Series A*, **Vol. CII,** 1923, pp. 643–667.
34. Taylor, Sir Geoffrey, and H. Quinney, "The Plastic Distortion of Metals," *Phil. Trans. Roy. Soc., London, Series A*, **Vol. 230,** 1931, pp. 323–362.
35. Tresca, H., *Compt. Rend. Savants Etrangers*, Paris, 1865, 1868, and 1870.
36. Westergaard, H. M., "On the Resistance of Ductile Materials to Combined Stresses," *J. Franklin Inst.*, 1920, p. 627.

Section 2.6. BETTI'S AND MAXWELL'S RECIPROCITY THEOREMS

1. Albenga, G., "Sul Teorema di Reciprocità di Land," *Reale Accad. Sci. Torino*, 1915.
2. Beggs, G. E., "Solution Précise, au Moyen de Modèles en Papier, de Problèmes Statiquement Indéterminés," *Compt. Rend.*, **Vol. 176,** 1923, pp. 885–886.
3. ——, "The Use of Models in the Solution of Indetermined Structures," *Proc. Second Intern. Congr. Appl. Mech.*, Zurich, Switzerland, Sept., 1926, pp. 301–304.
4. ——, "The Use of Models in the Solution of Indeterminate Structures," *J. Franklin Inst.*, 1927, pp. 375–386.
5. Betti, E., "Teoria dell'Elasticità," *Nuovo Cimento, Serie II*, **T. VII** and **VIII,** 1872.
6. Colonnetti, G., "Sul Principio di Reciprocità," *Atti Accad. Nazl. Lincei, Rend.*, Serie 5, **Vol. 21,** 1st Sem., 1912, pp. 393–398.
7. ——, "Sul Secondo Principio di Reciprocità," *Atti Accad. Sci. Torino*, **Vol. 50,** 1915.
8. ——, "Su di una Reciprocità tra Deformazioni e Distorsioni," *Atti Accad. Nazl. Lincei, Rend.*, Serie 5, **Vol. 24,** 1st Sem., 1915, pp. 404–408.
9. ——, "La Statica delle Costruzioni," Vol. I, Unione Tipografica Editrice Torinese, Turin, Italy, 1928.
10. ——, "Il Secondo Principio di Reciprocità e le sue Applicazioni al Calcolo delle Deformazioni Permanenti" (three papers), *Atti Accad. Nazl. Lincei, Rend.*, **Vol. 27,** 1st Sem., 1938.

11. Land, R., "Die Gegenseitigkeit elastischer Formänderungen, etc." *Wochbl. Baukunde*, 1887, p. 25.

12. ——, "Ueber die Ermittlung und die Gegenseitigen Beziehungen der Einflusslinien für Träger," *Z. Bauwesen*, 1890, p. 165.

13. Maxwell, J. C., "On the Calculation of the Equilibrium and Stiffness of Frames," *Phil. Mag.*, Vol. **27**, 1864, pp. 294.

14. Ricci, C. L., "Meccanica Applicata alle Costruzioni," Edizione Politecnica, Naples, Italy, 1942.

15. Volterra, Vito, "Sur l'Equilibre des Corps Elastiques Multiplement Connexes," *Ann. Sci. Ecole Normale Sup.*, 3 Serie, Vol. **24**, 1907, pp. 401–518.

16. ——, "Opere Matematiche," Vol. **3**, Accademia Nazionale dei Lincei, Rome, Italy, 1957.

17. Volterra, Vito, and Enrico Volterra, "Sur les Distorsions des Corps Elastiques, Théorie et Applications," Mémorial des Sciences Mathématiques, Fascicule **CXLVII**, Gauthier-Villars, Paris, France, 1960.

Section 2.8. CASTIGLIANO'S FIRST THEOREM

1. Andrews, E. S., "Elastic Stresses in Structures," Scott Greenwood and Son, London, England, 1919.

2. Castigliano, A. C., "Intorno ai Sistemi Elastici," Dissertazione di Laurea presentata alla Commissione Esaminatrice della Reale Scuola degli Ingegneri di Torino, Vincenzo Bona, Turin, Italy, 1873.

3. ——, "Intorno all'Equilibrio dei Sistemi Elastici," *Atti Accad. Sci. Torino*, Serie 2, Vol. **X**, 1875, p. 10.

4. ——, "Nuova Teoria intorno all'Equilibrio dei Sistemi Elastici," *Atti Accad. Sci. Torino*, Vol. **XI**, 1875, p. 127.

5. ——, "Théorie de l'Equilibre des Systèmes Elastiques et ses Applications," A. F. Negro, Turin, Italy, 1879.

6. ——, "Intorno ad una Proprietà dei Sistemi Elastici," *Atti Accad. Sci. Torino*, Vol. **XVII**, 1882, p. 705.

7. Cotterill, J. H., "Applied Mechanics," Macmillan & Co., Ltd., London, England, 1884.

8. Crotti, F., "Esposizione del Teorema di Castigliano e suo Raccordo colla Teoria dell'Elasticità," *Atti Collegio Ingegneri ed Architetti Milan*, Vol. **11**, 1878 (also published in *Politecnico*, Vol. **27**, 1879, p. 45).

9. ——, "Commemorazione di Alberto Castigliano," *Politecnico*, Vol. **32**, Nos. **11** and **12**, 1884.

10. ——, "La Teoria dell'Elasticità nei Suoi Principi Fondamentali e nelle sue Applicazioni Pratiche alle Costruzioni," U. Hoeple, Milan, Italy, 1888.

11. Matheson, J. A. L., "Castigliano's Theorem of Compatibility," *Engineering*, Vol. **180**, 1955, p. 828.

12. ——, "Hyperstatic Structures," Vol. **I**, Butterworth & Co. (Publishers) Ltd., London, England, 1959.

Section 2.9. EXTENSION BY DONATI OF CASTIGLIANO'S THEOREM

1. Donati, L., "Sul Lavoro di Deformazione dei Sistemi Elastici," *Mem. Accad. Sci. Bologna, Serie IV*, T. **IX**, 1888, p. 345.

2. ——, "Illustrazione al Teorema del Menabrea," *Mem. Accad. Sci. Bologna, Serie IV*, T. **X**, 1889, p. 267.

3. ——, "Ulteriori Osservazioni intorno al Teorema del Menabrea," *Mem. Accad. Sci. Bologna, Serie V*, T. **IV**, 1894, p. 449.

4. ——, "Introduzione Teorica al Corso di Fisica Tecnica nella Reale Scuola d'Applicazione per gli Ingegneri di Bologna," Litografia Sauer e Barigazza, 1900–1901.

5. Volterra, Vito, "Theory of Functionals and of Integral and Integro-Differential Equations," Chapter VI, Section III, No. 3 (The Strain Energy Functional in the Theory of Elasticity), Dover Publications, Inc., New York, N.Y., 1959.

Section 2.10. ENGESSER'S THEOREM ON THE PRINCIPLE OF COMPLEMENTARY ENERGY

1. Charlton, T. M., "Analysis of Statically Indeterminate Structures by the Complementary Energy Method," *Engineering*, Vol. **174**, 1952, p. 389.

2. Colonnetti, G., "De L'équilibre des systèmes élastiques dans lesquels se produisent des déformations plastiques," *J. Math. Pures Appl.*, Vol. **9**, 17, 1938, pp. 233–235.

3. Den Hartog, J. P., "Advanced Strength of Materials," McGraw-Hill Book Company, New York, N.Y., 1952.

4. Engesser, F., "Über Statisch Unbestimmte Träger bei Beliebigen Formänderungs Gesetze Ergänzungsarbeit," *Z. Architekten Ingenieur-Vereins zu Hannover*, Vol. **35**, 1889, pp. 733–744.

5. ——, "Über die Berechung Statisch Unbestimmter Systeme," *Zentr. Bauerwaltung*, 1907, p. 606.

6. Fung, Y. C., "Foundations of Solid Mechanics," Prentice-Hall, Inc., Englewood Cliffs, N.J., 1965.

7. Greenberg, H. J., "On the Variational Principles of Plasticity," O.N.R. Report, Graduate Div. Appl. Math., Brown University, Providence, R.I., 1949.

8. Hodge, P., and W. Prager, "A Variational Principle for Plastic Materials with Strain-Hardening," *J. Math. Phys.*, Vol. **27**, 1948, pp. 1–10.

9. Reissner, E., "On a Variational Theorem in Elasticity," *J. Math. Phys.*, Vol. **29**, 1950, pp. 90–95.

10. ——, "On a Variational Theorem for Finite Elastic Deformations," *J. Math. Phys.*, Vol. **32**, 1953, pp. 129–135.

11. ——, "On Variational Principles in Elasticity," *Proc. Symp. Appl. Math.*, Vol. **8**, 1958, pp. 1–6.

12. Sokolnikoff, I. S., "Mathematical Theory of Elasticity," McGraw-Hill Book Company, New York, N.Y., 1956.

13. Southwell, Sir Richard, "An Introduction to the Theory of Elasticity for Engineers and Physicists," Oxford University Press, London, England, 1941.

14. Wang, C. T., "Applied Elasticity," McGraw-Hill Book Company, New York, N.Y., 1953.

15. Westergaard, H. M., "On the Method of Complementary Energy and Its Application to Structures Stressed Beyond the Proportional Limit, to Buckling and Vibrations, and to Suspension Bridges," *Trans. Am. Soc. Civil. Engrs.*, Vol. 112, No. 107, 1942, pp. 765–803.

16. ——, "Theory of Elasticity and Plasticity," Harvard University Press, Cambridge, Mass., 1952.

Sections 2.11. CASTIGLIANO'S SECOND THEOREM OR MENABREA'S THEOREM

1. Argyris, J. H., and S. Kelsey, "Energy Theorems and Structural Analysis," Butterworth & Co. (Publishers), Ltd., London, England, 1960.

2. Bellavitis, G., "Sopra alcuni Studi dei Signori Dorna e Menabrea intorno alle Pressioni sopra piu di Tre Appoggi Considerazione," *Riv. Lavori Ist. Accad. Sci., Lettere ed Arti Padova*, Vol. 9, 1860, p. 33.

3. Biezeno, C. B., and R. Grammel, "Engineering Dynamics," Vol. II, Blackie & Son, Ltd., Glasgow, Scotland, 1956.

4. Brown, E. H., "The Energy Theorems of Structural Analysis," *Engineering*, Vol. 179, 1955, p. 305.

5. Castigliano, A., "Nuova Teoria intorno all'Equilibrio dei Sistemi Elastici," *Reale Accad. Sci., Torino*, 1875.

6. ——, "Théorie de l'Equilibre des Systèmes Elastiques et ses Applications," A. F. Negro, Turin, Italy, 1879.

7. Cerruti, V., "Sopra un Teorema del Signor Menabrea-Nota," *Atti Accad. Lincei, Rend.*, Serie 2, 1875, p. 570.

8. Colonnetti, G., "L'equilibrio elastico dal punto di vista energetico," *Reale Accad. Sci. Torino*, 1912, p. 479.

9. ——, "La Statica delle Costruzioni," Vol. I, Unione Tipografics Editrice Torinese, Turin, Italy, 1928.

10. Cotterill, J. H., "Applied Mechanics," Macmillan & Co., Ltd., London, England, 1884.

11. Domke, O., "Über Variationsprinzipien in der Elastizitätslehre," *Z. Math. Phys.*, Vol. 63, 1914.

12. Donati, L., "Illustrazione al teorema di Menabrea," *Mem. Accad. Sci. Bologna*, 1889, p. 267.

13. ——, "Ulteriori osservazioni intorno al teorema di Menabrea," *Mem. Accad. Sci. Bologna*, 1894, p. 449.

14. Dorna, A., "Memoria sulle Pressioni Sopportate dai Punti di Appoggio di un Sistema Equilibrato ed in Istato prossimo al Moto," *Mem. Accad. Sci. Torino, Serie II*, T. XVIII, 1857.

15. Hoff, N. J., "The Analysis of Structures," John Wiley & Sons, Inc., New York, N. Y. 1956.

16. L'Hermite, R., "Résistance des Matériaux Théorique et Expérimentale," Vol. I, Dunod, Paris, France, 1959.

17. Lorenz, H., "Techniche Elastizitätslehre," Oldenbourg, Munich and Berlin, Germany, 1913.

18. Menabrea, L. F., "Principio Generale per determinare le Tensioni e le Pressioni in un Sistema Elastico," A seminar presented to the Reale Accademia delle Scienze di Torino, Turin, Italy, 1857.

19. ——, "Nouveau Principe sur la Distribution des Tensions dans les Systèmes Elastiques," Compt. Rend. Paris, Vol. XLVI, 1858, p. 1056.

20. ——, "Etude de Statique Physique. Principe Général pour déterminer les Pressions et les Tensions dans un Système Elastique," Mem. Accad. Sci. Torino, Serie II, T. XXV, 1868, p. 141.

21. ——, "Principe d'élasticité ou principe du moindre travail," Reale Accad. Sci. Torino, 1871.

22. ——, "Sulla Determinazione delle Tensioni e delle Pressioni nei Sistemi Elastici," Atti Accad. Nazl. Lincei, Rend., Serie II, T. 2, 1875, p. 201.

23. Menabrea, L. F., and J. L. F. Bertrand, "An Abstract of Bertrand's Letter to General Menabrea," published by Menabrea in Atti Accad. Sci. Torino, T. V, 1869–1870, p. 702.

24. Neal, B. G., "Structural Theorems and Their Applications," The Macmillan Company, New York, N.Y., 1964.

25. Pippard, A. J. S., "Strain Energy Methods of Stress Analysis," Longmans, Green & Company, Ltd., London, England, 1928.

26. Poschl, T., "Elementare Festigkeitslehre," Julius Springer, Berlin, Germany, 1936.

27. Ricci, C. L., "Generalizzazione del secondo teorema di Castigliano," Atti XIX Riunione Soc. Ital. Progr. Sci., Bolzano-Trento, Italy, 1930.

28. ——, "Meccanica Applicata alle Costruzioni," Edizione Politecnica, Naples, Italy, 1942.

29. Southwell, Sir Richard, "An Introduction to the Theory of Elasticity for Engineers and Physicists," Oxford University Press, London, England, 1941.

30. Thomson, Sir William, "On the General Theory of the Equilibrium of an Elastic Solid," Phil. Trans. Roy. Soc., London, Vol. 163, p. 583.

31. Timoshenko, S. P., "History of Strength of Materials," McGraw-Hill Book Company, New York, N.Y., 1953.

32. Timoshenko, S. P., and J. N. Goodier, "Theory of Elasticity," McGraw-Hill Book Company, New York, N.Y., 1951.

33. Timoshenko, S. P., and D. H. Young, "Theory of Structures," 2nd ed., McGraw-Hill Book Company, New York, N.Y., 1964.

34. Wang, C. T., "Applied Elasticity," McGraw-Hill Book Company, New York, N.Y., 1953.

35. Westergaard, H. M., "Theory of Elasticity and Plasticity," Harvard University Press, Cambridge, Mass., 1952.

36. Williams, D., "The Relations between the Energy Theorems Applicable in Structural Theory," Phil. Mag., Vol. 26, 1938, p. 617.

37. ——, "The Use of the Principle of Minimum Potential Energy in Problems of Static Equilibrium," Aer. Res. Ctte. R and M., No. 1827, 1938.

PROBLEMS

2-1. In what ratio are the amounts of strain energy stored in two identical beams, one simply supported, the other built-in at its ends, and both subjected to equal uniformly distributed loads over the entire lengths?

2-2. Solve Problem 2-1 for the case in which the two beams are subjected to equal concentrated loads applied at the centers.

2-3. Express the strain energy of a two-dimensional elastic solid for the case of plane stress.

2-4. Express the strain energy of a two-dimensional elastic solid for the case of plane strain.

2-5. In a thin-walled closed cylinder subjected to internal pressure, the principal stresses are $\sigma_y = 2\sigma_x = \sigma$. If the yield stress in simple tension of the material is σ_{yp} and Poisson's ratio $\nu = 0.30$, determine the value σ at which yielding starts in terms of σ_{yp} using the various failure theories and considering the least favorable conditions.

2-6. A thin-walled closed hollow cylinder is subjected simultaneously to an internal pressure and to an axial pressure. If the tangential stress $\sigma_y = 0.75\sigma_x$, what are the values of σ_x in conjunction with σ_y that will stress the material to $\sigma_{yp} = 40,000$ psi according to the various strength theories discussed?

2-7. What internal pressure and additional axial tensile stress will cause yielding according to the various strength theories discussed, if the diameter of the cylinder of Problem 2.6 is 40 in., the wall thickness $\frac{1}{4}$ in., $\sigma_y = 0.75\sigma_x$, and $\sigma_{yp} = 40,000$ psi?

2-8. For the beam shown in Fig. P.2.1. assume that the deflection is given by the series

$$v(x) = \sum_{n=2,4,\ldots}^{\infty} C_n\left(1 - \cos\frac{n\pi x}{l}\right)$$

and verify that the boundary conditions are satisfied. Determine the expressions for the constants C_n. Use one, then two, and then three terms from the series and compute the midpoint deflection. Compare these results with the results from elementary beam theory.

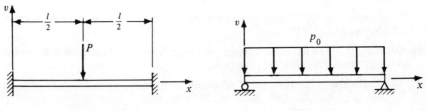

Figure P.2.1   Figure P.2.2

ENERGY PRINCIPLES AND GENERAL THEOREMS

2-9. By use of a sine series representation, determine an equation for the elastic curve of a simply supported beam subjected to a uniformly distributed load (see Fig. P.2.2). Use one and then two terms from the series and compute the midspan deflection. Compare these results with the results from elementary beam theory.

2-10. For the horizontal beam shown in Fig. P.2.3 a vertical displacement of support B of 0.25 in. causes a reaction $A_B = 2$ tons at A. Determine the reaction B_A at B due to a vertical displacement of 0.35 in. at support A. (Hint: apply Betti's theorem.)

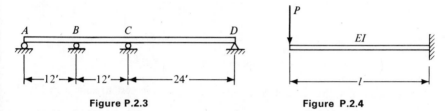

Figure P.2.3   Figure P.2.4

2-11. By use of Clapeyron's theorem, determine the deflection at the end of a cantilever beam subjected to a concentrated end load P (see Fig. P.2.4).

2-12. By use of Clapeyron's theorem, determine the rotation at the end of a cantilever beam subjected to a concentrated end moment M (see Fig. P.2.5).

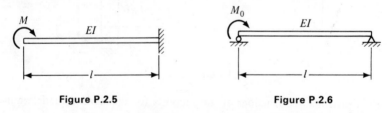

Figure P.2.5   Figure P.2.6

2-13. By use of Clapeyron's theorem, determine the rotation at the left-hand support of the beam shown in Fig. P.2.6.

2-14. Determine the axial force N, the shear force Q, and the bending moment M in a generic section of the ring shown in Fig. P.2.7 due to the action of two equal and opposite forces P applied at the extremities of a diameter.

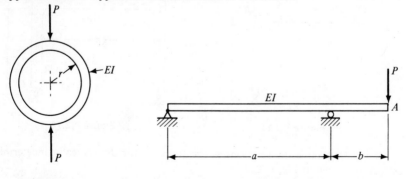

Figure P.2.7   Figure P.2.8

2-15. Determine the displacement of point A in the elastic system shown in Fig. P.2.8.

2-16. Determine the displacement of point A in the elastic system shown in Fig P.2.9.

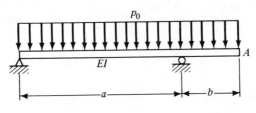

Figure P.2.9

2-17. Determine the rotation at point A in the elastic system shown in Fig. P.2.10.

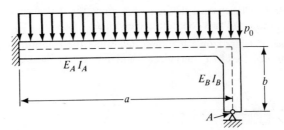

Figure P.2.10

2-18. Determine the reactions at the supports for the statically indeterminate structure shown in Fig. P.2.11.

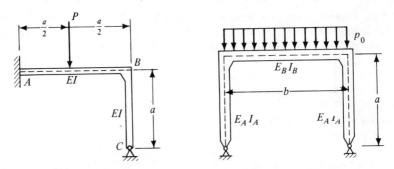

Figure P.2.11   Figure P.2.12

2-19. Determine the reactions at the supports for the statically indeterminate structure shown in Fig. P.2.12.

2-20. Determine the reactions at the supports for the statically indeterminate structure shown in Fig. P.2.13.

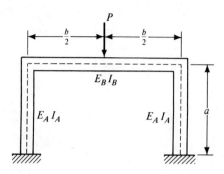

Figure P.2.13

CHAPTER

3

TWO-DIMENSIONAL ELASTICITY

3.0 Introduction

In this chapter some problems of two-dimensional elasticity will be discussed. By considering two-dimensional instead of three-dimensional elasticity, a considerable mathematical simplification is introduced. This simplification allows closed form solutions to be obtained for many problems having important practical applications.

The state of *plane stress* will first be defined and the graphical representation of stresses at a point, as suggested by Otto Mohr,[1] will be explained. Then the state of *plane strain* will be defined. The stress function, first introduced by Airy, will be presented and it will be shown that many important problems in two-dimensional elasticity can be solved by using polynomials to represent the stress function. For instance, the problems of the triangular

[1] **Mohr, Christian Otto.** German engineer and mechanicist (b. Wesselburen, Holstein, 1835; d. Dresden 1918). He was a railroad engineer and in 1887 he became professor at the Polytechnic School in Stuttgart.

Airy, Sir George Biddel. English astronomer (b. Alnwick, Northumberland 1801; d. Greenwich 1892). He was Lucasian Professor of Mathematics and then Plumian Professor of astronomy at Cambridge University and the first Director of the Cambridge Observatory. He became Astronomer Royal and director of Greenwich Observatory in 1835 and was author of various treatises on celestial mechanics, optics, and magnetism. The paper in which Airy's function was first used was published in 1862 in the Report of the British Association for the Advancement of Science. [Reproduction from the Vito Volterra collection, Villa Volterra, Ariccia (Rome).]

Descartes, du Perron, René. French philosopher, physicist, and mathematician (b. La Haye in Touraine 1596; d. Stockholm 1650). After a period in the army as a soldier under Maurice of Orange in Holland, Germany, and Hungary, he returned to France and later moved to Holland where he remained until 1649, when he went to Stockholm. He dedicated himself to science and philosophy. His famous "Discours de la Méthode" was published in Leiden in 1637. He was the founder of Analytical Geometry and was also author of important philosophical works. [Reproduction from the Vito Volterra collection, Villa Volterra, Ariccia (Rome).]

Inglis, Sir Charles Edward. British mechanicist (b. Worcester 1875; d. Suffolk 1952). He was professor of mechanical sciences and head of the department of engineering at Cambridge University from 1919 to 1943. Although his principal work was in bridge vibrations, he left important contributions in the mathematical theory of elasticity including a paper written in 1913 on the concentration of stresses in a plate due to the presence of elliptical holes. He was president of the London Institution of Civil Engineers and a Fellow of the Royal Society. (Photograph reproduced with permission of the Royal Society of London.)

Hertz, Heinrich Rudolph. German physicist (b. Hamburg 1857; d. Bonn 1894). He was a pupil of H. von Helmholtz in Berlin and was successively professor at the Karlsruhe Polytechnic School and at the University of Bonn. At the latter, he succeeded R. J. E. Clausius in his chair of physics. His most famous work is found in physics where, in 1887, he gave experimental proof of the existence of electromagnetic waves which were foreseen sixteen years earlier by J. C. Maxwell. He also left outstanding contributions in mechanics and in the mathematical theory of elasticity. His book, *Die Principien der Mechanik*, was published after his death by H. von Helmoltz, while his complete works in three volumes were published in 1894–95. [Reproduction from the Vito Volterra collection, Villa Volterra, Ariccia (Rome).]

and rectangular walls under hydrostatic pressure (Maurice Levy's[2] problems) can be solved.

Until now cartesian coordinates have been used. However, in the solution of many two-dimensional elasticity problems it is more convenient to use polar coordinates. By the use of polar coordinates, the following problems will be discussed in succession: bending of a circular bar (Golovin-Ribière's problem), a thick tube subjected to external and internal uniformly distributed pressures (Lamé's problem), stress concentration due to a circular hole in a stressed plate (Kirsch's problem), concentrated load acting on the vertex of a wedge (Michell's problem), concentrated load acting on the free surface of a plate (Flamant's[3] problem), moment acting on the vertex of a wedge (Inglis' problem), and a disk subjected to two opposite, concentrated forces (Hertz's problem). Two-dimensional thermal stresses are then discussed.

3.1 Plane Stress

One has a state of plane stress when the stresses satisfy the following conditions:

$$\sigma_z = \tau_{xz} = \tau_{yz} = 0 \qquad (3.1.1)$$

[2] Levy, Maurice. French engineer and mathematician (b. Ribeauville 1838; d. Paris 1910). He was professor of applied mechanics at the Ecole Centrale des Arts et Manufactures and professor of celestial and analytical mechanics at the Collège de France, where he succeeded Serret in 1885.

[3] Flamant, Alfred Aimé. French engineer (b. Noyal 1839; d. Paris?). As an engineer he was responsible for the design of many important hydraulic works in France. He was a pupil and collaborator of Barré de Saint-Venant and translated with him the great work

The stress tensor then reduces to

$$\begin{vmatrix} \sigma_x & \tau_{xy} \\ \tau_{yx} & \sigma_y \end{vmatrix}$$

The conditions expressed by Eqs. (3.1.1) represent, with only a very small error, the state of stress in a thin plate which is subjected to forces applied at the boundary, parallel to the plane of the plate, and uniformly distributed over the thickness (see Fig. 3.1.1).

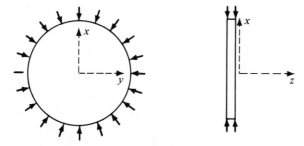

Figure 3.1.1

In the case of plane stress, the equilibrium equations [Eqs. (1.4.1)] become

$$\left. \begin{aligned} \frac{\partial \sigma_x}{\partial x} + \frac{\partial \tau_{xy}}{\partial y} + X = 0 \\ \frac{\partial \tau_{yx}}{\partial x} + \frac{\partial \sigma_y}{\partial y} + Y = 0 \end{aligned} \right\} \quad (3.1.2)$$

The compatibility equations [Eqs. (1.5.1)] reduce to

$$\frac{\partial^2 \epsilon_x}{\partial y^2} + \frac{\partial^2 \epsilon_x}{\partial x^2} = \frac{\partial^2 \gamma_{xy}}{\partial x \partial y} \quad (3.1.3)$$

while Eqs. (1.3.1) become

$$\left. \begin{aligned} \epsilon_x &= \frac{1}{E} (\sigma_x - \nu \sigma_y) \\ \epsilon_y &= \frac{1}{E} (\sigma_y - \nu \sigma_x) \\ \epsilon_z &= -\frac{\nu}{E} (\sigma_x + \sigma_y) \end{aligned} \right\} \quad (3.1.4)$$

---

of Clebsch entitled "Théorie de l'Elasticité des Corps Solides." He authored a book entitled "Mécanique Générale et Hydraulique" and was professor at the Ecole des Ponts et Chaussées and at the Ecole Centrale des Arts et Manufactures.

and Eqs. (1.3.5) become

$$\left.\begin{array}{c}\gamma_{xy} = \dfrac{\tau_{xy}}{G} \\ \gamma_{yz} = \gamma_{zx} = 0\end{array}\right\} \quad (3.1.5)$$

By substituting the first two of Eqs. (3.1.4) and the first of Eqs. (3.1.5) into Eq. (3.1.3) one finds that

$$\frac{\partial^2}{\partial x^2}(\sigma_y - \nu\sigma_x) + \frac{\partial^2}{\partial y^2}(\sigma_x - \nu\sigma_y) = 2(1+\nu)\frac{\partial^2 \tau_{xy}}{\partial x \partial y} \quad (3.1.6)$$

By differentiating the first of Eqs. (3.1.2) with respect to x, the second with respect to y, and adding them together one obtains

$$\frac{\partial^2 \tau_{xy}}{\partial x \partial y} = -\frac{1}{2}\left[\frac{\partial^2 \sigma_x}{\partial x^2} + \frac{\partial^2 \sigma_y}{\partial y^2} + \frac{\partial X}{\partial x} + \frac{\partial Y}{\partial y}\right] \quad (3.1.7)$$

By substituting Eq. (3.1.7) into Eq. (3.1.6), the equation

$$\left(\frac{\partial^2}{\partial x^2} + \frac{\partial^2}{\partial y^2}\right)(\sigma_x + \sigma_y) = -(1+\nu)\left(\frac{\partial X}{\partial x} + \frac{\partial Y}{\partial y}\right) \quad (3.1.8)$$

follows.

In the case of plane stress, the boundary conditions [Eqs. (1.6.2)] reduce to two equations

$$\left.\begin{array}{c}\bar{X} = \sigma_x \cos(\widehat{nx}) + \tau_{xy} \cos(\widehat{ny}) \\ \bar{Y} = \tau_{yx} \cos(\widehat{nx}) + \sigma_y \cos(\widehat{ny})\end{array}\right\} \quad (3.1.9)$$

It should be pointed out that in the case of plane stress while $\sigma_z = 0$, $\epsilon_z \neq 0$, but is given by the third of Eqs. (3.1.4).

3.2 Mohr's Circle for Stress

Suppose that at a generic point O inside an elastic body in equilibrium one has a state of plane stress defined by the stress components σ_x, σ_y, and τ_{xy}. This state of stress is represented in Fig. 3.2.1 by the stresses σ_x, σ_y, and τ_{xy} acting on the cube of sides dx and dy and unit thickness in the z-direction, i.e., perpendicular to the xy plane. The normal stresses are assumed positive since they correspond to a tension, while the shear stresses τ_{xy} are assumed as positive if they are clockwise, negative if they are counterclockwise. The problem which has to be solved is to find how the intensity of the normal

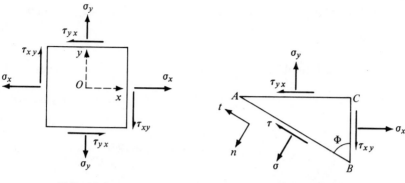

Figure 3.2.1 **Figure 3.2.2**

stress σ and the shear stress τ, acting on a plane whose normal makes an angle Φ with the x-axis, vary as a function of the angle Φ.

By expressing the conditions of equilibrium of the forces acting on the small prism of sides $AC(=dx)$, $BC(=dy)$, $AB(=ds)$, and unit thickness, this question can be immediately answered (see Fig. 3.2.2). In fact, by projecting the forces in the direction of the normal to face AB and noting that $\sum F_n = 0$, we find that

$$\sigma\, ds = \sigma_x\, dy \cos \Phi + \sigma_y\, dx \sin \Phi - \tau_{xy}\, dy \sin \Phi - \tau_{yx}\, dx \cos \Phi$$

But since

$$\frac{dx}{ds} = \sin \Phi, \qquad \frac{dy}{ds} = \cos \Phi, \qquad \tau_{xy} = \tau_{yx}$$

the above equation becomes

$$\sigma = \sigma_x \cos^2 \Phi + \sigma_y \sin^2 \Phi - 2\tau_{xy} \sin \Phi \cos \Phi \qquad (3.2.1)$$

From trigonometry, we have the following double angle relations:

$$\cos^2 \Phi = \tfrac{1}{2}(1 + \cos 2\Phi)$$
$$\sin^2 \Phi = \tfrac{1}{2}(1 - \cos 2\Phi)$$
$$\sin 2\Phi = 2 \sin \Phi \cos \Phi$$

and Eq. (3.2.1) can be written as

$$\sigma = \frac{\sigma_x + \sigma_y}{2} + \frac{\sigma_x - \sigma_y}{2} \cos 2\Phi - \tau_{xy} \sin 2\Phi \qquad (3.2.2)$$

By projecting the forces acting on prism ABC in the direction of tangent t and noting that $\sum F_t = 0$, we find that

$$\tau \, ds = \sigma_x \, dy \sin \Phi - \sigma_y \, dx \cos \Phi + \tau_{xy} \, dy \cos \Phi - \tau_{yx} \, dx \sin \Phi$$

from which

$$\tau = \sigma_x \cos \Phi \sin \Phi - \sigma_y \sin \Phi \cos \Phi + \tau_{xy} \cos^2 \Phi - \tau_{xy} \sin^2 \Phi \quad (3.2.3)$$

By use of the double angle formulas, the preceding equation may be written as

$$\tau = \frac{\sigma_x - \sigma_y}{2} \sin 2\Phi + \tau_{xy} \cos 2\Phi \quad (3.2.4)$$

Equations (3.2.2) and (3.2.4) express the relationships between the stresses σ and τ and the angle Φ. From these equations, it can be seen that the locus of points with σ and τ as coordinates is a circle. This can be demonstrated in the following way: first, rewrite Eq. (3.2.2) in the form

$$\sigma - \frac{\sigma_x + \sigma_y}{2} = \frac{\sigma_x - \sigma_y}{2} \cos 2\Phi - \tau_{xy} \sin 2\Phi \quad (3.2.5)$$

then square both sides of Eqs. (3.2.4) and (3.2.5) and add to obtain

$$\left[\sigma - \frac{\sigma_x + \sigma_y}{2}\right]^2 + \tau^2 = \left[\frac{\sigma_x - \sigma_y}{2}\right]^2 + \tau_{xy}^2 \quad (3.2.6)$$

Equation (3.2.6) is the equation of a circle in the $\sigma\tau$-plane[4] (see Fig. 3.2.3). The radius of the circle is

$$r = \sqrt{\left(\frac{\sigma_x - \sigma_y}{2}\right)^2 + \tau_{xy}^2} \quad (3.2.7)$$

and its center is on the σ axis at a distance a from the origin where

$$a = \frac{\sigma_x + \sigma_y}{2} \quad (3.2.8)$$

Each point on the circle, shown in Fig. 3.2.3, represents the state of stress on

[4] To prove this, compare the equation of the circle, shown in Fig. 3.2.4, i.e.,

$$(x - a)^2 + y^2 = r^2 \quad (a)$$

with Eq. (3.2.6). Equations (3.2.6) and (a) are mathematically the same, if

$$x = \sigma \quad y = \tau$$

$$a = \frac{\sigma_x + \sigma_y}{2} \quad r = \sqrt{\left(\frac{\sigma_x - \sigma_y}{2}\right)^2 + \tau_{xy}^2}$$

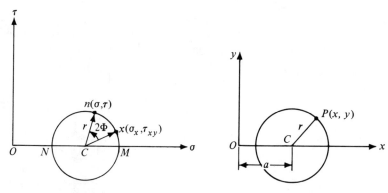

Figure 3.2.3   Figure 3.2.4

a plane passing through point O. Thus, point x represents the state of stress on the vertical plane passing through point O and point n represents the state of stress on the n-plane passing through point O (the vertical and n-planes are shown in Fig. 3.2.2).

From Fig. 3.2.3 we find that the maximum and minimum normal stresses occur on planes where the shear stress is zero. These planes are called principal planes. The maximum normal stress is

$$\sigma_{\max} = OM = OC + CM$$

or

$$\sigma_{\max} = \frac{\sigma_x + \sigma_y}{2} + \sqrt{\left(\frac{\sigma_x - \sigma_y}{2}\right)^2 + \tau_{xy}^2} \qquad (3.2.9)$$

Similarly, we find that the minimum normal stress is

$$\sigma_{\min} = \frac{\sigma_x + \sigma_y}{2} - \sqrt{\left(\frac{\sigma_x - \sigma_y}{2}\right)^2 + \tau_{xy}^2} \qquad (3.2.10)$$

and the extreme values of shear stress are

$$\tau_{\substack{\max \\ \min}} = \pm \sqrt{\left(\frac{\sigma_x - \sigma_y}{2}\right)^2 + \tau_{xy}^2} \qquad (3.2.11)$$

It should also be noted that an angular measurement Φ on the element (Fig. 3.2.2) corresponds to angular measurement 2Φ on the circle (Fig. 3.2.3) and that angles are measured in the same sense on the element and on the circle.

The graphical representation of the state of biaxial-stress was introduced by Otto Mohr and is commonly known as Mohr's circle.

Example 3.2.1. The stresses at a point in a body are $\sigma_x = 1000$ psi, $\sigma_y = -500$ psi, and $\tau_{xy} = 600$ psi. This state of stress is shown in Fig. 3.2.5 (a). By use of Mohr's circle find the magnitudes of the normal and shearing stresses acting on:

1. The principal planes
2. The planes of maximum shearing stresses

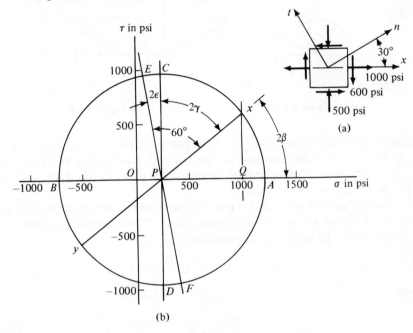
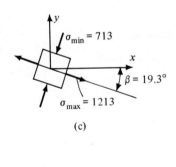
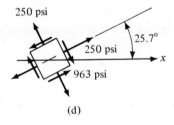
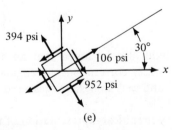

Figure 3.2.5

3. The plane whose normal n makes an angle of 30° with the x-axis
4. The plane at right angles to that in part 3.

Since stresses σ_x, σ_y, and τ_{xy} are given, the state of stress at the point is known and the stresses on any plane (which is perpendicular to the xy-plane) through the point can be determined. According to Eq. (3.2.7), the radius of the circle is

$$r = \sqrt{\left[\frac{1000 - (-500)}{2}\right]^2 + (600)^2} = 963 \text{ psi}$$

and according to Eq. (3.2.8), the center of the circle is at distance

$$a = \frac{1000 + (-500)}{2} = 250 \text{ psi}$$

from the origin and along the σ-axis. The Mohr circle construction, based on the above values, is shown in Fig. 3.2.5 (b). Points x and y represent, respectively, stress conditions on vertical and horizontal planes passing through the point.

1. The stresses on the planes of maximum and minimum normal stress, represented, respectively, by points A and B on Mohr's circle, are

$$\sigma_{max} = a + R = 250 + 963 = 1213 \text{ psi} \qquad \tau = 0$$
$$\sigma_{min} = a - R = 250 - 963 = -713 \text{ psi} \qquad \tau = 0$$

and the angle 2β is

$$2\beta = \text{arc tan } Qx/PQ = \text{arc tan } \tfrac{600}{750} = 38.6°$$

Angle β ($= 19.3°$) is the angle between the x-plane and the plane of maximum normal stress. Points A and B are at opposite ends of a diameter and are separated by 180° on Mohr's circle; therefore, the planes of maximum and minimum normal stresses are perpendicular. These planes are shown in Fig. 3.2.5 (c).

2. From Fig. 3.2.5 (b) it is readily seen that points C and D are the points of maximum and minimum shear stress, respectively. Thus, we have

$$\tau_{max} = R = 963 \text{ psi} \qquad \sigma = a = 250 \text{ psi}$$
$$\tau_{min} = -R = -963 \text{ psi} \qquad \sigma = a = 250 \text{ psi}$$

and the angle 2γ is

$$2\gamma = \text{arc tan } PQ/Qx = \text{arc tan } \tfrac{750}{600} = 51.4°$$

Angle γ ($= 25.7°$) is the angle between the x-plane and the plane of maximum shear stress. The planes of maximum and minimum shear stress and the corresponding stresses are shown in Fig. 3.2.5 (d). It should be noted that points C and A, representing stress conditions on the planes of maximum shear stress and maximum normal stress, respectively, are separated by 90°

on Mohr's circle; therefore, the planes of maximum shear stress and maximum normal stress are separated by 45°.

3, 4. The stress condition on a plane whose normal n makes an angle of 30° with the x-axis is represented by point E in Fig. 3.2.5 (b). The stress condition on the plane perpendicular to the n-plane, i.e., the t-plane, is represented by point F in Fig. 3.2.5 (b). The angle 2ϵ is

$$2\epsilon = 60° - 2\gamma = 8.6°$$

Thus, stresses on the n-plane are:

$$\sigma_n = a - R \sin 2\epsilon = 250 - (963)(0.1495) = 106 \text{ psi}$$
$$\tau_{nt} = R \cos 2\epsilon = (963)(0.9888) = 952 \text{ psi}$$

and stress on the t-plane are:

$$\sigma_t = a + R \sin 2\epsilon = 250 + (963)(0.1495) = 394 \text{ psi}$$
$$\tau_{nt} = -R \cos 2\epsilon = -(963)(0.9888) = -952 \text{ psi}$$

The n- and t-planes and the corresponding stresses are shown in Fig. 3.2.5 (e).

3.3 Plane Strain

One has a state of *plane strain* when the following conditions are satisfied by the components of the elastic diaplacement:

$$\left.\begin{array}{l} u = u(x, y) \\ v = v(x, y) \\ w = 0 \end{array}\right\} \qquad (3.3.1)$$

From Eqs. (3.3.1) it follows for the components of strain that

$$\left.\begin{array}{l} \gamma_{xz} = \dfrac{\partial u}{\partial z} + \dfrac{\partial w}{\partial x} = 0 \\ \gamma_{yz} = \dfrac{\partial v}{\partial z} + \dfrac{\partial w}{\partial y} = 0 \\ \epsilon_z = \dfrac{\partial w}{\partial z} = 0 \end{array}\right\} \qquad (3.3.2)$$

and the strain tensor is reduced to

$$\begin{vmatrix} \epsilon_x & \gamma_{xy} \\ \gamma_{yx} & \epsilon_y \end{vmatrix}$$

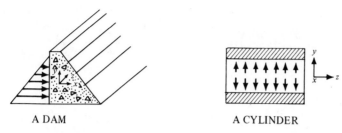

A DAM   A CYLINDER

Figure 3.3.1   Figure 3.3.2

Conditions expressed by Eqs. (3.3.1) and (3.3.2) are very closely approximated in the case of a long dam restrained at both ends (see Fig. 3.3.1), in the case of a long cylinder subjected to internal pressure (see Fig. 3.3.2), and in the case of a tunnel (see Fig. 3.3.3).

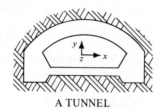

A TUNNEL

Figure 3.3.3

In the case of plane strain, the equilibrium equations [Eqs. (1.4.1)] become

$$\left.\begin{array}{l}\dfrac{\partial \sigma_x}{\partial x} + \dfrac{\partial \tau_{xy}}{\partial y} + X = 0 \\ \dfrac{\partial \tau_{yx}}{\partial x} + \dfrac{\partial \sigma_y}{\partial y} + Y = 0 \end{array}\right\} \qquad (3.3.3)$$

which are the same as Eqs. (3.1.2) for the plane stress case. The stress strain relations [Eqs. (1.3.1)] for this case become

$$\left.\begin{array}{l}\epsilon_z = \dfrac{1}{E}[\sigma_z - v(\sigma_x + \sigma_y)] = 0 \quad \text{or} \quad \sigma_z = v(\sigma_x + \sigma_y) \\ \epsilon_x = \dfrac{1}{E}[\sigma_x - v(\sigma_y + \sigma_z)] = \dfrac{1+v}{E}[\sigma_x(1-v) - v\sigma_y] \\ \epsilon_y = \dfrac{1}{E}[\sigma_y - v(\sigma_z + \sigma_x)] = \dfrac{1+v}{E}[\sigma_y(1-v) - v\sigma_x] \end{array}\right\} \qquad (3.3.4)$$

and Eq. (1.5.5), i.e.,

$$\Delta\Theta = \left(\dfrac{1+v}{1-v}\right)\left[\dfrac{\partial X}{\partial x} + \dfrac{\partial Y}{\partial y} + \dfrac{\partial Z}{\partial z}\right] \qquad (1.5.5)$$

becomes

$$\left(\frac{\partial^2}{\partial x^2} + \frac{\partial^2}{\partial y^2}\right)(\sigma_x + \sigma_y) = -\frac{1}{1-\nu}\left(\frac{\partial X}{\partial x} + \frac{\partial Y}{\partial y}\right) \quad (3.3.5)$$

since

$$\Delta = \frac{\partial^2}{\partial x^2} + \frac{\partial^2}{\partial y^2}$$

$$\Theta = \sigma_x + \sigma_y + \sigma_z = (1+\nu)(\sigma_x + \sigma_y)$$

$$\frac{\partial Z}{\partial z} = 0$$

The boundary conditions [Eqs. (1.6.2)] for the state of plane strain are

$$\begin{aligned} \bar{X} &= \sigma_x \cos(\widehat{nx}) + \tau_{xy} \cos(\widehat{ny}) \\ \bar{Y} &= \tau_{yx} \cos(\widehat{nx}) + \sigma_y \cos(\widehat{ny}) \end{aligned} \quad (3.3.6)$$

which are the same as Eqs. (3.1.9) for the case of plane stress.[5]

If one compares the equilibrium equations, the compatibility equations,

---

[5] Although the conditions expressed by Eqs. (3.3.1) and (3.3.2) represent cases of *plane strain*, the general case of plane strain is expressed by less restrictive conditions. In fact, one considers an elastic body to be *in a state of plane strain* when the following three conditions are satisfied:

(a) Each section, originally plane and perpendicular to the z-axis, remains, during deformation, plane and perpendicular to the z-axis. This condition implies that

$$\frac{\partial w}{\partial x} = \frac{\partial w}{\partial y} = 0$$

(b) The deformation which occurs in the plane of a generic section perpendicular to the z-axis is independent of the position of the section. This condition implies that

$$\frac{\partial u}{\partial z} = \frac{\partial v}{\partial z} = 0$$

(c) The translation of the plane of a generic section, perpendicular to the z-axis, in the z-direction is a linear function of variable z. This last condition implies that

$$w = Az + B$$

where A and B are two constants.

In view of the above conditions Eqs. (3.3.3), (3.3.5), and (3.3.6) remain the same while σ_z becomes

$$\sigma_z = EA + \nu(\sigma_x + \sigma_y)$$

since $\epsilon_z = A$. However, in all two-dimensional strain cases which will be considered, Eqs. (3.3.1) and (3.3.2) will be satisfied and the constant A will be zero.

and the boundary conditions for the cases of plane stress and plane strain, one sees that in the second member of Eq. (3.1.8) there is the factor $(1 + v)$, while in Eq. (3.3.5) there is the factor $(1/1 - v)$. Factor $(1 + v)$ varies between 1 and 1.5, while factor $(1/1 - v)$ varies between 1 and 2. Since

$$\frac{1}{1-v} = 1 + v + v^2 + \ldots$$

the difference between $(1 + v)$ and $(1/1 - v)$ is of the order of v^2. The main difference between the two cases of two-dimensional elasticity is that in the case of plane stress $\sigma_z = 0$ and $\epsilon_z = -v/E(\sigma_x + \sigma_y)$, while in the case of plane strain $\epsilon_z = 0$ and $\sigma_z = v(\sigma_x + \sigma_y)$. If body forces are absent or constant, Eqs. (3.1.8) and (3.3.5) become the same equation, i.e.,

$$\left(\frac{\partial^2}{\partial x^2} + \frac{\partial^2}{\partial y^2}\right)(\sigma_x + \sigma_y) = 0 \qquad (3.3.7)$$

In this case, the elastic constants E and v do not appear in the equations of the problem and the solution of the elasticity problem is independent of the elastic characteristics of the material. In the cases which will be considered in the next paragraphs, only these conditions will be considered.

3.4 Mohr's Circle for Strain and Strain Rosettes

Suppose that at a generic point O inside an elastic body in equilibrium, one has a state of plane strain defined by the strain components ϵ_x, ϵ_y, and γ_{xy}. When the body deforms, axes Oxy become $O^*x'y'$. It is required to find the extensional strain ϵ and the shear strain γ with respect to a new system of orthogonal axes nt, the n-axis making the angle θ with the x-axis (see Figs. 3.4.1 and 3.4.2). From Fig. 3.4.1 (b) the following relationships are derived:

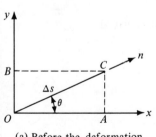
(a) Before the deformation

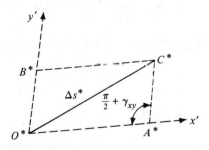
(b) After the deformation

Figure 3.4.1

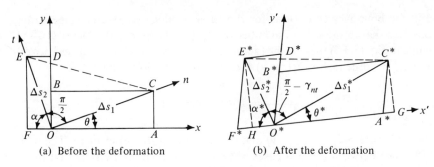

(a) Before the deformation (b) After the deformation

Figure 3.4.2

$$\angle O^*A^*C^* = \frac{\pi}{2} + \gamma_{xy}$$

$$O^*A^* = (1 + \epsilon_x)OA = (1 + \epsilon_x) \cos \theta \, \Delta s$$

$$O^*B^* = (1 + \epsilon_y)OB = (1 + \epsilon_y) \sin \theta \, \Delta s$$

$$(\Delta s^*)^2 = (O^*C^*)^2 = (O^*A^*)^2 + (O^*B^*)^2$$
$$\quad - 2(O^*A^*)(O^*B^*) \cos O^*A^*C^*$$
$$= (1 + \epsilon_x)^2 \, \Delta s^2 \cos^2 \theta + (1 + \epsilon_y)^2 \, \Delta s^2 \sin^2 \theta$$
$$\quad - 2(1 + \epsilon_x)(1 + \epsilon_y) \, \Delta s^2 \sin \theta \cos \theta \cos\left(\frac{\pi}{2} + \gamma_{xy}\right)$$
$$= \Delta s^2 [1 + 2\epsilon_x \cos^2 \theta + 2\epsilon_y \sin^2 \theta + 2\gamma_{xy} \sin \theta \cos \theta]$$

since strains ϵ_x, ϵ_y, and γ_{xy} are considered infinitesimal quantities. It follows that

$$\Delta s^* = \Delta s[1 + \epsilon_x \cos^2 \theta + \epsilon_y \sin^2 \theta + \gamma_{xy} \sin \theta \cos \theta]$$

and for the extensional strain in the n-direction

$$\epsilon_n = \lim_{\Delta s \to 0} \frac{\Delta s^* - \Delta s}{\Delta s} = \epsilon_x \cos^2 \theta + \epsilon_y \sin^2 \theta + \gamma_{xy} \sin \theta \cos \theta \quad (3.4.1)$$

From Fig. 3.4.2 (b) the following relationships are derived:

$$\alpha^* + \frac{\pi}{2} - \gamma_{nt} + \theta^* = \pi$$

from which

$$\gamma_{nt} = (\alpha^* + \theta^*) - \frac{\pi}{2}$$

Since γ_{nt} is a small quantity:

$$\gamma_{nt} = \sin \gamma_{nt} = \sin\left[(\alpha^* + \theta^*) - \frac{\pi}{2}\right] = -\cos(\alpha^* + \theta^*)$$
$$= -\cos\alpha^* \cos\theta^* + \sin\alpha^* \sin\theta^* \tag{a}$$

$$\sin\theta^* = \frac{GC^*}{O^*C^*} = \frac{A^*C^* \sin\left(\frac{\pi}{2} - \gamma_{xy}\right)}{O^*C^*} = \frac{(1+\epsilon_y)\Delta s \sin\theta \cos\gamma_{xy}}{(1+\epsilon_n)\Delta s}$$
$$= (\sin\theta)(1 + \epsilon_y - \epsilon_n) \tag{b}$$

$$\sin\alpha^* = \frac{O^*E^*}{O^*E^*} = \frac{F^*E^* \sin\left(\frac{\pi}{2} - \gamma_{xy}\right)}{O^*E^*} = \frac{(1+\epsilon_y)\Delta s \cos\theta \cos\gamma_{xy}}{(1+\epsilon_t)\Delta s}$$
$$= (\cos\theta)(1 + \epsilon_y - \epsilon_t) \tag{c}$$

$$\cos\theta^* = \frac{O^*A^* + A^*G}{O^*C^*} = \frac{O^*A^* + A^*C^* \cos\left(\frac{\pi}{2} - \gamma_{xy}\right)}{O^*C^*}$$
$$= \frac{(1+\epsilon_x)\Delta s \cos\theta + (1+\epsilon_y)\Delta s \sin\theta \sin\gamma_{xy}}{(1+\epsilon_n)\Delta s}$$
$$= (\cos\theta)(1 + \epsilon_x - \epsilon_n) + \gamma_{xy} \sin\theta \tag{d}$$

$$\cos\alpha^* = \frac{O^*F^* - HF^*}{O^*E^*} = \frac{O^*F^* - F^*E^* \cos\left(\frac{\pi}{2} - \gamma_{xy}\right)}{O^*E^*}$$
$$= \frac{(1+\epsilon_x)\Delta s \sin\theta - (1+\epsilon_y)\Delta s \cos\theta \sin\gamma_{xy}}{(1+\epsilon_t)\Delta s}$$
$$= (\sin\theta)(1 + \epsilon_x - \epsilon_t) - \gamma_{xy} \cos\theta \tag{e}$$

By substituting Eqs. (b), (c), (d), and (e) into Eq. (a) and neglecting higher order terms in the strains, one finds that

$$\gamma_{nt} = [-(\sin\theta)(1 + \epsilon_x - \epsilon_t) + \gamma_{xy}\cos\theta][(\cos\theta)(1 + \epsilon_x - \epsilon_n) + \gamma_{xy}\sin\theta]$$
$$+ [(\cos\theta)(1 + \epsilon_y - \epsilon_t)][(\sin\theta)(1 + \epsilon_y - \epsilon_n)]$$
$$= 2(\epsilon_y - \epsilon_x)\sin\theta\cos\theta + \gamma_{xy}(\cos^2\theta - \sin^2\theta) \tag{3.4.2}$$

Eqs. (3.4.1) and (3.4.2) can be written in the following forms:

$$\epsilon_n = \frac{1}{2}(\epsilon_x + \epsilon_y) + \frac{1}{2}(\epsilon_x - \epsilon_y)\cos 2\theta + \frac{1}{2}\gamma_{xy}\sin 2\theta \tag{3.4.3}$$

$$\frac{\gamma_{nt}}{2} = -\frac{1}{2}(\epsilon_x - \epsilon_y)\sin 2\theta + \frac{1}{2}\gamma_{xy}\cos 2\theta \tag{3.4.4}$$

Equations (3.4.3) and (3.4.4) express the relationships between strains ϵ_n and γ_{nt} and the angle θ. From Eqs. (3.4.3) and (3.4.4) it can be shown that the locus of points with ϵ_n and $-\gamma_{nt}/2$ as coordinates, and for the strain condition specified by ϵ_x, ϵ_y, and γ_{xy}, is a circle with radius

$$r = \sqrt{\left(\frac{\epsilon_x - \epsilon_y}{2}\right)^2 + \left(\frac{1}{2}\gamma_{xy}\right)^2}$$

and center C on the ϵ-axis at a distance

$$a = \frac{\epsilon_x + \epsilon_y}{2}$$

from the origin (see Fig. 3.4.3). From Fig. 3.4.3 one also obtains

$$\epsilon_{\max} = \frac{\epsilon_x + \epsilon_y}{2} + \sqrt{\left(\frac{\epsilon_x - \epsilon_y}{2}\right)^2 + \left(\frac{1}{2}\gamma_{xy}\right)^2} \quad (3.4.5)$$

$$\epsilon_{\min} = \frac{\epsilon_x + \epsilon_y}{2} - \sqrt{\left(\frac{\epsilon_x - \epsilon_y}{2}\right)^2 + \left(\frac{1}{2}\gamma_{xy}\right)^2} \quad (3.4.6)$$

$$\left(\frac{\gamma_{nt}}{2}\right)_{\substack{\max \\ \min}} = \pm\sqrt{\left(\frac{\epsilon_x - \epsilon_y}{2}\right)^2 + \left(\frac{1}{2}\gamma_{xy}\right)^2} \quad (3.4.7)$$

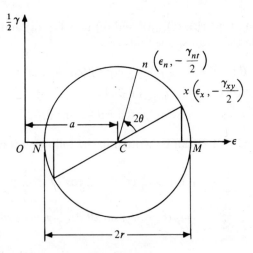

Figure 3.4.3

Example 3.4.1. At a point in an elastic body subjected to plane strain there are the following strains:[6]

$$\epsilon_x = 400\mu \quad \epsilon_y = 50\mu \quad \gamma_{xy} = 400 \times 10^{-6} \text{ radians}$$

Determine by use of Mohr's circle:

(a) The principal strains (i.e., the maximum and minimum normal strains) and the principal directions.
(b) Strains ϵ_n, ϵ_t, and γ_{nt} where n and t are perpendicular axes with the angle between the x- and n-axes being 30° [see Fig. 3.4.4 (a)].

The Mohr circle for strain may be obtained by plotting points $x(\epsilon_x, -\gamma_{xy}/2)$ and $y(\epsilon_y, \gamma_{xy}/2)$ in the ϵ-$\gamma/2$ plane as shown in Fig. 3.4.4(b). Line xy, which is a diameter of the circle, is then drawn and its point of intersection C with the ϵ-axis is the center of the circle. The distance a from the origin of coordinates to C is

$$a = \frac{\epsilon_x + \epsilon_y}{2} = \frac{400 + 50}{2} = 225\,\mu$$

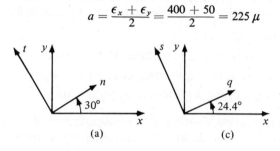

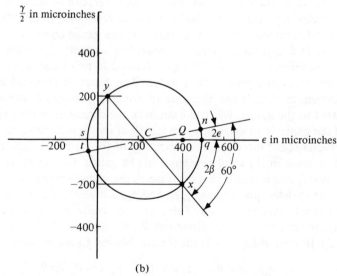

(b)

Figure 3.4.4

[6] The symbol μ stands for micro-, i.e., 10^{-6}. Thus a strain $\epsilon = 400\,\mu$ is actually $\epsilon = 400 \times 10^{-6}$ in./in.

and the radius is

$$r = \sqrt{(\overline{CQ})^2 + (\overline{Qx})^2} = \sqrt{(400-225)^2 + (200)^2} = 266\mu$$

(a) The principal strains and directions can be read directly from Fig. 3.4.4 (b). Alternately, these quantities can be computed as follows:

$$\epsilon_{max} = a + r = 225 + 266 = 491\mu$$
$$\epsilon_{min} = a - r = 225 - 266 = -41\mu$$
$$2\beta = \text{arc tan}\frac{Qx}{CQ} = \text{arc tan}\frac{200}{175} = 48.8°$$

The direction of the maximum normal strain is $\beta = 24.4°$ counterclockwise from the x-direction. The principal strain directions are the q- and s-directions shown in Fig. 3.4.4 (c).

(b) Points n and t [Fig. 3.4.4 (b)] represent strain conditions associated with the n- and t-directions, respectively. These strains are computed as follows:

$$2\epsilon = 60 - 2\beta = 11.2°$$
$$\epsilon_n = a + r \cos 2\epsilon = 225 + 266(\cos 11.2°) = 486\mu$$
$$\epsilon_t = a - r \cos 2\epsilon = 225 - 266(\cos 11.2°) = -36\mu$$
$$\gamma_{nt} = -2r \sin 2\epsilon = -(2)(266)(\sin 11.2°) = -105\mu$$

The problem of measuring shearing strains is an extremely difficult problem and simple techniques for accurately measuring small angle changes have not been developed. Normal strains, however, can be accurately measured by means of extensometers or by means of resistance strain gauges.

The most common form of a resistance strain gauge consists of a length of fine wire laid to form a system of close, but separated, parallel strands, connected in series at the ends. The grid formed in this fashion is cemented or bonded between layers of thin paper. The gauge is then cemented to the specimen so that, when the area of contact is strained, the strain is transmitted to the gauge. When the strain in the direction of the axis of the wires of the gauge is an extension, the length of the strands increases and the diameter decreases; therefore, the resistivity of the wire increases, resulting in a net increase in the total resistance of the gauge. On the other hand, a compressional strain results in a decrease in the gauge's resistance.

In order to determine strains ϵ_x, ϵ_y, and γ_{xy} at a point of a strained elastic body the *rosette method* is used. Three normal strains ϵ_1, ϵ_2, and ϵ_3 corresponding, respectively, to the directions θ_1, θ_2, and θ_3 are determined (see Fig. 3.4.5). In view of Eq. (3.4.1) the three following equations are derived:

$$\begin{aligned}\epsilon_1 &= \epsilon_x \cos^2\theta_1 + \epsilon_y \sin^2\theta_1 + \gamma_{xy} \cos\theta_1 \sin\theta_1 \\ \epsilon_2 &= \epsilon_x \cos^2\theta_2 + \epsilon_y \sin^2\theta_2 + \gamma_{xy} \cos\theta_2 \sin\theta_2 \\ \epsilon_3 &= \epsilon_x \cos^2\theta_3 + \epsilon_y \sin^2\theta_3 + \gamma_{xy} \cos\theta_3 \sin\theta_3\end{aligned} \quad (3.4.8)$$

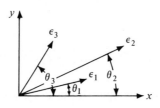

Figure 3.4.5

By solving this system of linear equations strains ϵ_x, ϵ_y, and γ_{xy} are determined. Two types of strain rosettes are commonly used:

1. The rectangular rosette or 45° strain rosette
2. The equiangular rosette or 60° strain rosette

The 45° strain rosette. In this case $\theta_1 = 0$, $\theta_2 = 45°$, $\theta_3 = 90°$ and Eqs. (3.4.8) become

$$\epsilon_1 = \epsilon_x$$

$$\epsilon_2 = \frac{\epsilon_x + \epsilon_y}{2} + \frac{\gamma_{xy}}{2}$$

$$\epsilon_3 = \epsilon_y$$

from which

$$\epsilon_x = \epsilon_1$$

$$\epsilon_y = \epsilon_3$$

$$\gamma_{xy} = 2\epsilon_2 - \epsilon_1 - \epsilon_3$$

It follows that the maximum and minimum strains are

$$\epsilon_{\substack{max \\ min}} = \tfrac{1}{2}[\epsilon_1 + \epsilon_3 \pm \sqrt{2}\sqrt{(\epsilon_1 - \epsilon_2)^2 + (\epsilon_2 - \epsilon_3)^2}]$$

and the directions of the principal strains are given by

$$\tan 2\theta = \frac{2\epsilon_2 - (\epsilon_1 + \epsilon_3)}{\epsilon_1 - \epsilon_3}$$

The 60° strain rosette. In this case $\theta_1 = 0$, $\theta_2 = 60°$, $\theta_3 = 120°$ and Eqs. (3.4.8) become

$$\epsilon_x = \epsilon_1$$

$$\epsilon_y = \frac{1}{3}(2\epsilon_2 + 2\epsilon_3 - \epsilon_1)$$

$$\gamma_{xy} = \frac{2}{\sqrt{3}}(\epsilon_2 - \epsilon_3)$$

The principal strains are

$$\epsilon_{\substack{max\\min}} = \frac{1}{3}(\epsilon_1 + \epsilon_2 + \epsilon_3) \pm \frac{\sqrt{2}}{3}\sqrt{(\epsilon_1 - \epsilon_2)^2 + (\epsilon_1 - \epsilon_3)^2 + (\epsilon_2 - \epsilon_3)^2}$$

while the directions of the principal strains are given by

$$\tan 2\theta = \frac{\sqrt{3}\,(\epsilon_2 - \epsilon_3)}{2\epsilon_1 - \epsilon_2 - \epsilon_3}$$

3.5 Airy's Stress Function

If it is assumed that the body forces are constant and equal to Y_0[7] then the equilibrium equations for the case of plane stress [Eqs. (3.1.2)] and for the case of plane strain [Eqs. (3.3.3)] reduce to the same equations; thus,

$$\left.\begin{array}{l} \dfrac{\partial \sigma_x}{\partial x} + \dfrac{\partial \tau_{xy}}{\partial y} = 0 \\[6pt] \dfrac{\partial \tau_{yx}}{\partial x} + \dfrac{\partial \sigma_y}{\partial y} + Y_0 = 0 \end{array}\right\} \quad \text{equations of equilibrium} \quad (3.5.1)$$

Similarly, the compatibility equations [Eqs. (3.1.8) and (3.3.5)] for the two cases have the same form, which is

$$\left(\frac{\partial^2}{\partial x^2} + \frac{\partial^2}{\partial y^2}\right)(\sigma_x + \sigma_y) = 0 \quad \text{equation of compatibility} \quad (3.5.2)$$

In order to solve Eqs. (3.5.1) and (3.5.2) G. B. Airy introduced a stress function $\Phi = \Phi(x, y)$ which is related to the components of stress by the following equations:

$$\left.\begin{array}{l} \sigma_x = \dfrac{\partial^2 \Phi}{\partial y^2} \\[6pt] \sigma_y = \dfrac{\partial^2 \Phi}{\partial x^2} \\[6pt] \tau_{xy} = -\dfrac{\partial^2 \Phi}{\partial x \partial y} - Y_0 x \end{array}\right\} \quad (3.5.3)$$

By substituting Eqs. (3.5.3) into Eqs. (3.5.1) we find that the equilibrium

---

[7] In the case of gravitational force, by assuming the x-axis horizontal and the y-axis vertical and directed positively down and calling ρ the density of the elastic body and g the acceleration due to gravity, one has

$$X = 0, \quad Y = Y_0 = \rho g$$

equations are identically satisfied. By substituting Eqs. (3.5.3) into Eq. (3.5.2) the equation

$$\left(\frac{\partial^2}{\partial x^2} + \frac{\partial}{\partial y^2}\right)\left(\frac{\partial^2 \Phi}{\partial x^2} + \frac{\partial^2 \Phi}{\partial y^2}\right) = \frac{\partial^4 \Phi}{\partial x^4} + 2\frac{\partial^4 \Phi}{\partial x^2 \partial y^2} + \frac{\partial^4 \Phi}{\partial y^4} = 0 \quad (3.5.4)$$

or

$$\Delta\Delta\Phi = 0 \quad (3.5.4')$$

is obtained. It follows that in a two-dimensional elasticity problem (plane stress or plane strain), when body forces are constant or absent, the first boundary value problem is solved when the function $\Phi = \Phi(x, y)$ is found which satisfies the biharmonic equation [Eq. (3.5.4)] and the boundary conditions expressed by the equations

$$\left.\begin{array}{l}\sigma_x \cos{(\widehat{nx})} + \tau_{xy} \cos{(\widehat{ny})} = \bar{X} \\ \tau_{yx} \cos{(\widehat{nx})} + \sigma_y \cos{(\widehat{ny})} = \bar{Y}\end{array}\right\} \quad (3.5.5)$$

Once, the components of stress σ_x, σ_y, and τ_{xy} are obtained, the problem of finding the components of the elastic displacement can be reduced in the case of plane stress to the integration of the equations

$$\left.\begin{array}{l}\dfrac{\partial u}{\partial x} = \epsilon_x = \dfrac{1}{E}(\sigma_x - \nu\sigma_y) \\[4pt] \dfrac{\partial v}{\partial y} = \epsilon_y = \dfrac{1}{E}(\sigma_y - \nu\sigma_x) \\[4pt] \dfrac{\partial u}{\partial y} + \dfrac{\partial v}{\partial x} = \gamma_{xy} = \dfrac{2(1+\nu)}{E}\tau_{xy}\end{array}\right\} \quad (3.5.6)$$

and in the case of plane strain to the integration of the equations

$$\left.\begin{array}{l}\dfrac{\partial u}{\partial x} = \epsilon_x = \dfrac{1}{E}[(1-\nu^2)\sigma_x - \nu(1+\nu)\sigma_y] \\[4pt] \dfrac{\partial v}{\partial y} = \epsilon_y = \dfrac{1}{E}[(1-\nu^2)\sigma_y - \nu(1+\nu)\sigma_x] \\[4pt] \dfrac{\partial u}{\partial y} + \dfrac{\partial v}{\partial x} = \gamma_{xy} = \dfrac{2(1+\nu)}{E}\tau_{xy}\end{array}\right\} \quad (3.5.7)$$

3.6 Solution of Two-Dimensional Problems by the Use of Polynomials[8]

In this section, Eq. (3.5.4) will be satisfied by expressing Airy's function $\Phi(x, y)$ in the form of homogeneous polynomials.

[8] Unless differently specified in this and in the following sections, it will be assumed that body forces are absent.

(a) Polynomial of the second degree.

$$\Phi_2(x, y) = \frac{a_2}{2} x^2 + b_2 xy + \frac{c_2}{2} y^2 \qquad (3.6.1)$$

Equation (3.6.3) satisfies Eq. (3.5.4). The corresponding stresses are given by

$$\left. \begin{array}{l} (\sigma_x)_2 = \dfrac{\partial^2 \Phi_2}{\partial y^2} = c_2 \\[4pt] (\sigma_y)_2 = \dfrac{\partial^2 \Phi_2}{\partial x^2} = a_2 \\[4pt] (\tau_{xy})_2 = -\dfrac{\partial^2 \Phi_2}{\partial x \partial y} = -b_2 \end{array} \right\} \qquad (3.6.2)$$

The state of stress corresponding to Eqs. (3.6.2) is shown in Fig. 3.6.1.

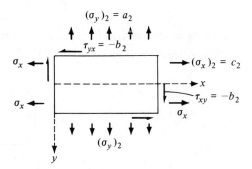

Figure 3.6.1

(b) Polynomial of the third degree.

$$\Phi_3(x, y) = \frac{a_3}{2 \cdot 3} x^3 + \frac{b_3}{1 \cdot 2} x^2 y + \frac{c_3}{1 \cdot 2} xy^2 + \frac{d_3}{2 \cdot 3} y^3 \qquad (3.6.3)$$

Equation (3.6.3) satisfies Eq. (3.5.4). The corresponding stresses are given by

$$\left. \begin{array}{l} (\sigma_x)_3 = \dfrac{\partial^2 \Phi_3}{\partial x^2} = c_3 x + d_3 y \\[4pt] (\sigma_y)_3 = \dfrac{\partial^2 \Phi_3}{\partial y^2} = a_3 x + b_3 y \\[4pt] (\tau_{xy})_3 = -\dfrac{\partial^2 \Phi_3}{\partial x \partial y} = -b_3 x - c_3 y \end{array} \right\} \qquad (3.6.4)$$

In the case in which $a_3 = b_3 = c_3 = 0$ Eqs. (3.6.4) reduce to

$$\left.\begin{array}{l}(\sigma_x)_3 = d_3 y \\ (\sigma_y)_3 = (\tau_{xy})_3 = 0\end{array}\right\} \quad (3.6.4a)$$

Equations (3.6.4a) correspond to pure bending on the face perpendicular to the x-axis, as shown in Fig. 3.6.2.

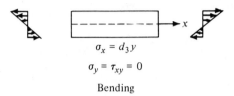

$\sigma_x = d_3 y$
$\sigma_y = \tau_{xy} = 0$

Bending

Figure 3.6.2

(c) **Polynomial of the fourth degree.**

$$\Phi_4(x, y) = \frac{a_4}{3 \cdot 4} x^4 + \frac{b_4}{2 \cdot 3} x^3 y + \frac{c_4}{1 \cdot 2} x^2 y^2 + \frac{d_4}{2 \cdot 3} xy^3 + \frac{e_4}{3 \cdot 4} y^4 \quad (3.6.5)$$

Equation (3.6.5) satisfies Eq. (3.5.4), if

$$e_4 = -(2c_4 + a_4)$$

The corresponding stresses are given by

$$\left.\begin{array}{l}(\sigma_x)_4 = \dfrac{\partial^2 \Phi_4}{\partial y^2} = c_4 x^2 + d_4 xy - (2c_4 + a_4) y^2 \\ (\sigma_y)_4 = \dfrac{\partial^2 \Phi_4}{\partial x^2} = a_4 x^2 + b_4 xy + c_4 y^2 \\ (\tau_{xy})_4 = -\dfrac{\partial^2 \Phi_4}{\partial x \partial y} = -\dfrac{b_4}{2} x^2 - 2c_4 xy - \dfrac{d_4}{2} y^2\end{array}\right\} \quad (3.6.6)$$

(d) **Polynomial of the fifth degree.**

$$\Phi_5(x, y) = \frac{a_5}{4 \cdot 5} x^5 + \frac{b_5}{3 \cdot 4} x^4 y + \frac{c_5}{2 \cdot 3} x^3 y^2 + \frac{d_5}{2 \cdot 3} x^2 y^3 + \frac{e_5}{3 \cdot 4} xy^4 + \frac{f_5}{4 \cdot 5} y^5$$

$$(3.6.7)$$

Equation (3.6.7) satisfies Eq. (3.5.4), if

$$(3a_5 + 2c_5 + e_5)x + (b_5 + 2d_5 + 3f_5)y = 0$$

or, if the two equations

$$3a_5 + 2c_5 + e_5 = 0 \quad \text{or} \quad e_5 = -3a_5 - 2c_5$$
$$b_5 + 2d_5 + 3f_5 = 0 \quad \text{or} \quad b_5 = -2d_5 - 3f_5$$

are satisfied. The corresponding stresses are given by

$$\left.\begin{aligned}
(\sigma_x)_5 &= \frac{\partial^2 \Phi_5}{\partial y^2} = c_5 \frac{x^3}{3} + d_5 x^2 y - (3a_5 + 2c_5)xy^2 + f_5 y^3 \\
(\sigma_y)_5 &= \frac{\partial^2 \Phi_5}{\partial x^2} = a_5 x^3 - (3f_5 + 2d_5)x^2 y + c_5 xy^2 + \frac{d_5}{3} y^3 \\
(\tau_{xy})_5 &= -\frac{\partial^2 \Phi_5}{\partial x \partial y} = (3f_5 + 2d_5)\frac{x^3}{3} - c_5 x^2 y - d_5 xy^2 + (3a_5 + 2c_5)\frac{y^3}{3}
\end{aligned}\right\} \quad (3.6.8)$$

(e) Polynomial of the sixth degree.

$$\left.\begin{aligned}
\Phi_6(x, y) &= \frac{a_6}{5 \cdot 6} x^6 + \frac{b_6}{4 \cdot 5} x^5 y + \frac{c_6}{3 \cdot 4} x^4 y^2 + \frac{d_6}{2 \cdot 3} x^3 y^3 + \frac{e_6}{3 \cdot 4} x^2 y^4 \\
&\quad + \frac{f_6}{4 \cdot 5} xy^5 + \frac{g_6}{5 \cdot 6} y^6
\end{aligned}\right\} \quad (3.6.9)$$

Equation (3.6.9) satisfies Eq. (3.5.4), if

$$2x^2[6a_6 + 2c_6 + e_6] + 6xy[b_6 + f_6 + d_6] + 2y^2[c_6 + 6g_6 + e_6] = 0$$

or, if the following three equations are satisfied:

$$6a_6 + e_6 + 2c_6 = 0 \quad \text{or} \quad e_6 = -6a_6 - 2c_6$$
$$b_6 + f_6 + 2d_6 = 0 \quad \text{or} \quad b_6 = -f_6 - 2d_6$$
$$c_6 + 6g_6 + 2e_6 = 0 \quad \text{or} \quad g_6 = 2a_6 + \frac{c_6}{2}$$

The corresponding stresses are given by

$$\left.\begin{aligned}
(\sigma_x)_6 &= \frac{\partial^2 \Phi_6}{\partial y^2} = \frac{c_6}{6} x^4 + d_6 x^3 y - (6a_6 + 2c_6)x^2 y^2 + f_6 xy^3 + (4a_6 + c_6)\frac{y^4}{2} \\
(\sigma_y)_6 &= \frac{\partial^2 \Phi_6}{\partial x^2} = a_6 x^4 - (f_6 + 2d_6)x^3 y + c_6 x^2 y^2 + d_6 xy^3 - (3a_6 + c_6)\frac{y^4}{3} \\
(\tau_{xy})_6 &= -\frac{\partial^2 \Phi_6}{\partial x \partial y} = (f_6 + 2d_6)\frac{x^4}{4} - \frac{2}{3} c_6 x^3 y - \frac{3}{2} d_6 x^2 y^2 \\
&\quad + \frac{4}{3}(3a_6 + c_6)xy^3 - \frac{f_6}{4} y^4 - (4a_6 + c_6)\frac{y^5}{10}
\end{aligned}\right\}$$

$$(3.6.10)$$

Example 3.6.1. By the use of polynomials obtain the solution of the cantilever bar shown in Fig. 3.6.3. The bar has length l, rectangular cross section of height $2c$, and unit breadth and is subjected to a concentrated load of intensity P applied at the free end.

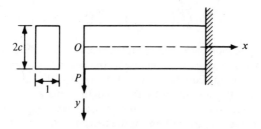

Figure 3.6.3

We shall seek the solution by superimposing the solutions given by the polynomial of the second degree [Eq. (3.6.1)] with $a_2 = c_2 = 0$, i.e.,

$$\Phi_2 = b_2 xy$$

and that given by the polynomial of the fourth degree [Eq. (3.6.5)] with $a_4 = b_4 = c_4 = e_4 = 0$, i.e.,

$$\Phi_4 = \frac{d_4}{2 \cdot 3} xy^3$$

The corresponding components of stress obtained from Eqs. (3.6.2) and (3.6.6), are

$$\left. \begin{array}{l} \sigma_x = d_4 xy \\ \sigma_y = 0 \\ \tau_{xy} = -b_2 - \dfrac{d_4}{2} y^2 \end{array} \right\} \quad (3.6.11)$$

The following boundary conditions must be satisfied:

$$\left. \begin{array}{ll} (\tau_{xy})_{y=\pm c} = 0 & \text{from which} \quad d_4 = -\dfrac{2b_2}{c^2} \\ \displaystyle\int_{-c}^{c} \tau_{xy}(1\,dy) = -P & \text{from which}^9 \quad b_2 = \dfrac{3}{4}\dfrac{P}{c} \end{array} \right\} \quad (3.6.12)$$

[9] Quantity b_2 is derived by first substituting the expression for d_4 into the last of Eqs. (3.6.11) to obtain

$$\tau_{xy} = -b_2 \left(1 - \frac{y^2}{c^2}\right)$$

and then substituting this expression for shear stress into the boundary condition

$$\int_{-c}^{c} \tau_{xy}(1)\,dy = -P$$

The components of stress may now be written as

$$\left.\begin{aligned}\sigma_x &= -\frac{3}{2}\frac{P}{c^3}xy = \frac{My}{I} \\ \sigma_y &= 0 \\ \tau_{xy} &= \frac{3P}{4c}\left(1 - \frac{y^2}{c^2}\right) = \frac{P}{I}\left(\frac{c^2 - y^2}{2}\right)\end{aligned}\right\} \quad (3.6.13)$$

where $I = \frac{2}{3}c^3$ is the moment of inertia of the cross section of the beam with respect to the neutral axis, and $M = -Px$ represents the bending moment acting at a generic section of the beam. Equations (3.6.13) show that the results obtained for this problem coincide with those given by the strength of materials.

In order to determine the components of the elastic displacement Eqs. (3.5.6) are used. These equations are written for the present case in the following form:

$$\left.\begin{aligned}\frac{\partial u}{\partial x} &= \epsilon_x = \frac{\sigma_x}{E} = -\frac{Pxy}{EI} \\ \frac{\partial v}{\partial y} &= \epsilon_y = -\frac{\nu\sigma_x}{E} = \frac{\nu Pxy}{EI} \\ \frac{\partial u}{\partial y} + \frac{\partial v}{\partial x} &= \gamma_{xy} = \frac{\tau_{xy}}{G} = \frac{P}{2IG}(c^2 - y^2)\end{aligned}\right\} \quad (3.6.14)$$

By integrating the first two of Eqs. (3.6.14) with respect to x and y, respectively, one obtains

$$\left.\begin{aligned}u &= -\frac{Px^2 y}{2EI} + f(y) \\ v &= \frac{\nu Pxy^2}{2EI} + f_1(x)\end{aligned}\right\} \quad (3.6.15)$$

From Eqs. (3.6.15) one finds that

$$\left.\begin{aligned}\frac{\partial u}{\partial y} &= -\frac{Px^2}{2EI} + \frac{df(y)}{dy} \\ \frac{\partial v}{\partial x} &= \frac{\nu Py^2}{2EI} + \frac{df_1(x)}{dx}\end{aligned}\right\} \quad (3.6.16)$$

Combining the third of Eqs. (3.6.14) with Eqs. (3.6.16) it follows that

$$\gamma_{xy} = \frac{P}{2GI}(c^2 - y^2) = \frac{\partial u}{\partial y} + \frac{\partial v}{\partial x} = -\frac{Px^2}{2EI} + \frac{df(y)}{dy} + \frac{\nu Py^2}{2EI} + \frac{df_1(x)}{dx} \quad (3.6.17)$$

Now let

$$F(x) = -\frac{Px^2}{2EI} + \frac{df_1(x)}{dx} \quad (3.6.18)$$

$$G(y) = \frac{df(y)}{dy} + \frac{\nu Py^2}{2EI} + \frac{Py^2}{2GI} \quad (3.6.19)$$

$$K = \frac{Pc^2}{2GI} = \text{constant}$$

and Eq. (3.6.17) can be written as

$$F(x) + G(y) = K \tag{3.6.20}$$

Equation (3.6.20) means that $F(x)$ and $G(y)$ are both constant. These constants are denoted by A and B, i.e.,

$$F(x) = A$$
$$G(y) = B$$

Now, in view of Eqs. (3.6.18) and (3.6.19) it follows that

$$\frac{df_1(x)}{dx} = \frac{Px^2}{2EI} + A$$

$$\frac{df(y)}{dy} = -\nu\frac{Py^2}{2EI} + \frac{Py^2}{2GI} + B$$

By integration of these equations one obtains

$$\left.\begin{aligned} f(y) &= -\nu\frac{Py^3}{6EI} + \frac{Py^3}{6GI} + By + C \\ f_1(x) &= \frac{Px^3}{6EI} + Ax + D \end{aligned}\right\} \tag{3.6.21}$$

where C and D are constants. By substituting Eqs. (3.6.21) into Eqs. (3.6.15) one gets

$$\left.\begin{aligned} u &= -\frac{Px^2 y}{2EI} - \nu\frac{Py^3}{6EI} + \frac{Py^3}{6GI} + By + C \\ v &= \frac{\nu Pxy^2}{2EI} + \frac{Px^3}{6EI} + Ax + D \end{aligned}\right\} \tag{3.6.22}$$

Components of elastic displacement u and v given by Eqs. (3.6.22) have to satisfy the following boundary conditions:

$$\left.\begin{aligned} u = v &= 0 \\ \frac{\partial v}{\partial x} &= 0 \end{aligned}\right\} \quad \text{for } x = l \text{ and } y = 0 \tag{3.6.23}$$

It follows that

$$\left.\begin{aligned} C = 0 \quad & A = -\frac{Pl^2}{2EI} \\ D = -\frac{Pl^3}{6EI} - Al &= \frac{Pl^3}{3} \end{aligned}\right\} \tag{3.6.24}$$

Example 3.6.2. By the use of polynomials obtain the solution of the simply supported beam subjected to a uniformly distributed load of intensity p_0 as shown in Fig. 3.6.4. The bar has length $2l$ and rectangular cross section of height $2c$, and unit breadth.

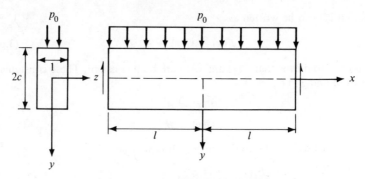

Figure 3.6.4

We shall seek the solution by superimposing the solutions given by the polynomial of the second degree [Eq. (3.6.1)] with $c_2 = b_2 = 0$, i.e.,

$$\Phi_2 = \frac{a_2}{2} x^2$$

and that given by the polynomial of the third degree [Eq. (3.6.3)] with $a_3 = c_3 = 0$, i.e.,

$$\Phi_3 = \frac{b_3}{2} x^2 y + \frac{d_3}{2 \cdot 3} y^3$$

and that given by the polynomial of the fifth degree [Eq. (3.6.7)] with $a_5 = b_5 = c_5 = e_5 = 0$, i.e.,

$$\Phi_5 = \frac{d_5}{2 \cdot 3} x^2 y^3 - \frac{2 d_5}{3 \cdot 4 \cdot 5} y^5$$

The corresponding components of stress are, from Eqs. (3.6.2), (3.6.4), and (3.6.8)

$$\left. \begin{aligned} \sigma_x &= d_3 y + d_5 \left(x^2 y - \frac{2}{3} y^3 \right) \\ \sigma_y &= a_2 + b_3 y + \frac{d_5}{3} y^3 \\ \tau_{xy} &= -b_3 x - d_5 x y^2 \end{aligned} \right\} \quad (3.6.25)$$

The following boundary conditions must be satisfied:

$$\left. \begin{aligned} (\tau_{xy})_{y=\pm c} &= 0 \\ (\sigma_y)_{y=+c} &= 0 \\ (\sigma_y)_{y=-c} &= -p_o \end{aligned} \right\} \quad (3.6.26)$$

and

$$\left. \begin{aligned} \int_{-c}^{c} (\sigma_x)_{x=\pm l} (1 dy) &= 0 \\ \int_{-c}^{c} (\tau_{xy})_{x=\pm l} (1 dy) &= \mp p_o l \\ \int_{-c}^{c} (\sigma_x)_{x=\pm l} (y dy) &= 0 \end{aligned} \right\} \quad (3.6.27)$$

From Eqs. (3.6.26) one obtains

$$\left.\begin{array}{r}-b_3 - d_5 c^2 = 0 \\ a_2 + b_3 c + \dfrac{d_5}{3} c^3 = 0 \\ a_2 - b_3 - \dfrac{d_5}{3} c^3 = -p_o\end{array}\right\}$$

from which

$$\left.\begin{array}{r}a_2 = -\dfrac{p_o}{2} \\ b_3 = \dfrac{3p_o}{4c} \\ d_5 = -\dfrac{3p_o}{4c^3}\end{array}\right\}$$

To evaluate the last constant d_3 one uses the third of Eqs. (3.6.27) and obtains

$$\int_{-c}^{c} (\sigma_x)_{x=\pm l}(y\,dy) = \int_{-c}^{c} [d_3 y + d_5(l^2 y - \tfrac{2}{3} y^3)]\,y\,dy = 0$$

and finally

$$d_3 = -d_5(l^2 - c^2) = \frac{3p_o}{4c}\left(\frac{l^2}{c^2} - \frac{2}{5}\right)$$

The following values are obtained for the components of stress:

$$\left.\begin{array}{l}\sigma_x = \dfrac{3}{4}\dfrac{p_o}{c}\left(\dfrac{l^2}{c^2} - \dfrac{2}{5}\right)y - \dfrac{3p_o}{4c^3}\left(x^2 y - \dfrac{2}{3}y^3\right) \\ \sigma_y = -\dfrac{p_o}{2} + \dfrac{3p_o}{4c} y - \dfrac{p_o}{4c^3} y^3 \\ \tau_{xy} = -\dfrac{3p_o}{4c} x + \dfrac{3p_o}{4c^3} xy^2\end{array}\right\}$$

By introducing the moment of inertia of the cross-sectional area of the beam, $I = \tfrac{2}{3} c^3$, the above equations become

$$\left.\begin{array}{l}\sigma_x = \dfrac{p_o}{2I}(l^2 - x^2)y + \dfrac{p_o}{I}\left(\dfrac{y^3}{3} - \dfrac{c^2 y}{5}\right) \\ \sigma_y = -\dfrac{p_o}{2I}\left(\dfrac{y^3}{3} - c^2 y + \dfrac{2}{3} c^3\right) \\ \tau_{xy} = -\dfrac{p_o}{2I} x(c^2 - y^2)\end{array}\right\} \quad (3.6.28)$$

The results given by the strength of materials in this case are:

$$\left.\begin{array}{l}\sigma_x = \dfrac{My}{I} = \dfrac{p_o}{2I}(l^2 - x^2)y \\ \sigma_y = 0 \\ \tau_{xy} = \dfrac{VQ}{I} = -\dfrac{p_o}{2I} x(c^2 - y^2)\end{array}\right\} \quad (3.6.29)$$

where

$$M = \frac{p_o}{2}(l^2 - x^2) \quad \text{bending moment}$$

$$V = -p_o x \quad \text{shear force}$$

$$Q = \tfrac{1}{2}(c^2 - y^2) \quad \text{static moment with respect to the neutral axis of the shaded area shown in Fig. 3.6.5}$$

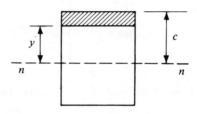

Figure 3.6.5

Comparison between the values of σ_x given by Eqs. (3.6.28) and (3.6.29) are shown in Fig. 3.6.6. Figure 3.6.7 shows the diagram of σ_y in accordance with the second of Eqs. (3.6.28). The strength of materials gives $\sigma_y = 0$. The values for τ_{xy} agree in both theories.

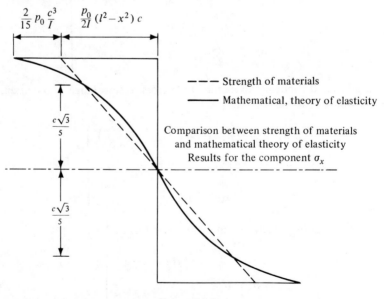

Figure 3.6.6

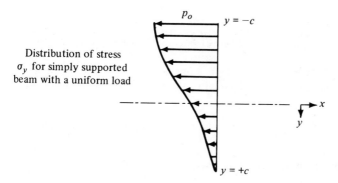

Figure 3.6.7

Distribution of stress σ_y for simply supported beam with a uniform load

In order to derive the components of the elastic displacement Eqs. (3.5.6) are used. These equations are written for the present case in the following forms:

$$\left.\begin{aligned}\frac{\partial u}{\partial x}=\epsilon_x &=\frac{1}{E}(\sigma_x-\nu\sigma_y)=\frac{p_o}{2EI}\bigg[(l^2-x^2)y+\frac{2}{3}y^3-\frac{2}{5}c^2 y \\ &\quad+\nu\Big(\frac{1}{3}y^3-c^2 y+\frac{2}{3}c^3\Big)\bigg] \\ \frac{\partial v}{\partial y}=\epsilon_y &=\frac{1}{E}(\sigma_y-\nu\sigma_x)=-\frac{p_o}{2EI}\bigg[\Big(\frac{1}{3}y^3-c^2 y+\frac{2}{3}c^3\Big) \\ &\quad+\nu(l^2-x^2)y+\nu\Big(\frac{2}{3}y^3-\frac{2}{5}c^2 y\Big)\bigg] \\ \frac{\partial u}{\partial y}+\frac{\partial v}{\partial x}=\gamma_{xy} &=\frac{2(1+\nu)}{E}\tau_{xy}=-\frac{p_o}{2GI}x(c^2-y^2)\end{aligned}\right\} \quad (3.6.30)$$

By integrating the first of Eqs. (3.6.30) with respect to x one gets

$$u=\frac{p_o}{2EI}\bigg[\Big(l^2 x-\frac{x^3}{3}\Big)y+x\Big(\frac{2}{3}y^3-\frac{2}{5}c^2 y\Big)+\nu x\Big(\frac{y^3}{3}-c^2 y+\frac{2}{3}c^3\Big)\bigg]+f(y) \quad (3.6.31)$$

By integrating the second of Eqs. (3.6.30) with respect to y one gets

$$v=-\frac{p_o}{2EI}\bigg[\Big(\frac{1}{12}y^4-\frac{c^2 y^2}{2}+\frac{2}{3}c^3 y\Big)+\nu(l^2-x^2)\frac{y^2}{2}+\nu\Big(\frac{y^4}{6}-\frac{1}{5}c^2 y^2\Big)\bigg]+F(x) \quad (3.6.32)$$

By differentiating u with respect to y and v with respect to x and by using the third of Eqs. (3.6.30) one gets

$$\frac{\partial u}{\partial y}+\frac{\partial v}{\partial x}=\frac{p_o}{2EI}\bigg[l^2 x-\frac{x^3}{3}+x\Big(2y^2-\frac{2}{5}c^2\Big)+\nu x(y^2-c^2)\bigg]$$
$$+\frac{df(y)}{dy}+\frac{p_o}{2EI}\nu xy^2+\frac{dF(x)}{dx}=\gamma_{xy}=-\frac{p_o}{2GI}x(c^2-y^2)$$

or

$$\frac{df(y)}{dy} = -\frac{dF(x)}{dx} + \frac{p_o}{6EI}x^3 - \frac{p_o}{2EI}\left[l^2 + \left(\frac{5}{8}+v\right)c^2\right]x \qquad (3.6.33)$$

Since the left side of Eq. (3.6.33) is a function of y only and the right side is a function of x only, they must be equal to the same constant A. From Eq. (3.6.33) the two equations

$$\frac{df(y)}{dy} = A \qquad \frac{dF(x)}{dx} = \frac{p_o}{6EI}x^3 - \frac{p_o}{2EI}\left[l^2 + \left(\frac{8}{5}+v\right)c^2\right]x - A$$

follow. By integration one obtains

$$f(y) = Ay + B$$

$$F(x) = \frac{p_o}{24EI}x^4 - \frac{p_o}{4EI}\left[l^2 + \left(\frac{8}{5}+v\right)c^2\right]x^2 - Ax + C$$

where A, B, and C are constants to be determined from the boundary conditions. By substituting the above expressions for $f(y)$ and $F(x)$ into Eqs. (3.6.31) and (3.6.32), one obtains

$$u = \frac{p_o}{2EI}\left[\left(l^2 x - \frac{x^3}{3}\right)y + x\left(\frac{2}{3}y^3 - \frac{2}{5}c^2 y\right) + vx\left(\frac{y^3}{3} - c^2 y + \frac{2}{3}c^3\right)\right] + Ay + B$$

$$v = -\frac{p_o}{2EI}\left[\frac{y^4}{12} - \frac{c^2 y^2}{2} + \frac{2}{3}c^3 y + v\left\{(l^2 - x^2)\frac{y^2}{2} + \frac{y^4}{6} - \frac{c^2}{5}y^2\right\}\right.$$
$$\left. - \frac{x^4}{12} + \left\{\frac{l^2}{2} + \left(\frac{4}{5} + \frac{v}{2}\right)c^2\right\}x^2\right] - Ax + C$$

From the boundary condition that $u = 0$ for $x = 0$ and for any value of y, one obtains $A = B = 0$ and from the boundary condition that $v = 0$ for $x = \pm l$, $y = 0$ one obtains

$$C = \frac{5}{24}\frac{p_o}{E}\frac{l^4}{I}\left[1 + \frac{12}{5}\left(\frac{4}{5} + \frac{v}{2}\right)\frac{c^2}{l^2}\right]$$

The following equations are now obtained for u and v:

$$u = \frac{p_o}{2EI}\left[\left(l^2 x - \frac{x^3}{3}\right)y + x\left(\frac{2}{3}y^3 - \frac{2}{5}c^2 y\right) + vx\left(\frac{y^3}{3} - c^2 y + \frac{2}{3}c^3\right)\right]$$

$$v = -\frac{p_o}{2EI}\left[\frac{y^4}{12} - \frac{c^2 y^2}{2} + \frac{2}{3}c^3 y + v\left\{(l^2 - x^2)\frac{y^2}{2} + \frac{y^4}{6} - \frac{c^2}{5}y^2\right\}\right.$$
$$\left. - \frac{x^4}{12} + \left\{\frac{l^2}{2} + \left(\frac{4}{5} + \frac{v}{2}\right)c^2\right\}x^2\right] + \frac{5}{24}\frac{p_o}{E}\frac{l^4}{I}\left[1 + \frac{12}{5}\left(\frac{4}{5} + \frac{v}{2}\right)\frac{c^2}{l^2}\right]$$

From the second of these equations one obtains for $y = 0$, the deflection of the neutral axis; hence,

$$(v)_{y=0} = -\frac{p_o}{2EI}\left[-\frac{x^4}{12} + \left\{\frac{l^2}{2} + \left(\frac{4}{5} + \frac{v}{2}\right)c^2\right\}x^2\right] + \frac{5}{24}\frac{p_o l^4}{EI}\left[1 + \frac{12}{5}\left(\frac{4}{5} + \frac{v}{2}\right)\frac{c^2}{l^2}\right]$$

(3.6.34)

The maximum deflection is obtained from Eq. (3.6.34) with $x = 0$; thus

$$(v)_{\substack{x=0 \\ y=0}} = v_{max} = \frac{5}{24}\frac{p_0 l^4}{EI}\left[1 + \frac{12}{5}\left(\frac{4}{5} + \frac{v}{2}\right)\frac{c^2}{l^2}\right] \tag{3.6.35}$$

The corresponding value given by the strength of materials is

$$v_{max} = \frac{5}{24}\frac{p_0 l^4}{EI} \tag{3.6.36}$$

which is less than the one given by Eq. (3.6.35), the difference being of the order $(c/l)^2$.

3.7 Triangular and Rectangular Walls Subjected to Hydrostatic Pressures (Maurice Levy's Problems)

By selecting a stress function $\Phi(x, y)$ in the form of a polynomial, Maurice Levy was able to solve the problem of a triangular retaining wall having weight ρg per unit volume of the wall and subjected to an hydrostatic pressure

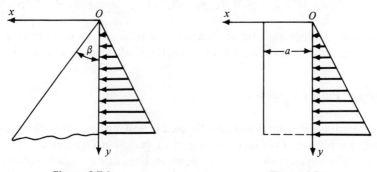

Figure 3.7.1   Figure 3.7.2

γy (see Fig. 3.7.1). He found that the following system of stresses, derived from a fifth degree polynomial, satisfy the boundary conditions:

$$\left.\begin{array}{l} \sigma_x = -\gamma y \\ \sigma_y = \left(\dfrac{\rho g}{\tan \beta} - \dfrac{2\gamma}{\tan^3 \beta}\right)x + \left(\dfrac{\gamma}{\tan^2 \beta} - \rho g\right)y \\ \tau_{xy} = -\dfrac{\gamma}{\tan^2 \beta}x \end{array}\right\} \tag{3.7.1}$$

In the case of a rectangular wall of thickness a subjected to an hydrostatic pressure (see Fig. 3.7.2), Maurice Levy, by selecting polynomials up to the sixth degree, found the solution in the following form:

$$\sigma_x = -\gamma y \left(1 - \frac{x}{a}\right)^2 \left(1 + 2\frac{x}{a}\right)$$

$$\sigma_y = -\rho g y - \frac{\gamma y^3}{a^2}\left(2\frac{x}{a} - 1\right) + \gamma y \left(4\frac{x^3}{a^3} - 6\frac{x^2}{a^2} + \frac{12}{5}\frac{x}{a} - \frac{1}{5}\right) \quad (3.7.2)$$

$$\tau_{xy} = -\frac{\gamma x}{a}\left(1 - \frac{x}{a}\right)\left[3\frac{y^2}{a} - \frac{a}{5} + x\left(1 - \frac{x}{a}\right)\right]$$

It can be shown that the system of stresses given by Eqs. (3.7.2) satisfy the boundary conditions:

$$\text{For } x = 0 \quad \sigma_x = -\gamma y \quad \tau_{xy} = 0$$
$$\text{For } x = a \quad \sigma_x = 0 \quad \tau_{xy} = 0$$

while the condition $\tau_{xy} = 0$ for $y = 0$ is not satisfied; instead, for $y = 0$

$$\tau_{xy} = \frac{\gamma x}{a}\left(1 - \frac{x}{a}\right)\left[x\left(1 - \frac{x}{a}\right)\frac{a}{5}\right]$$

However, the resultant of these stresses is zero, i.e.,

$$\int_0^a (\tau_{xy})_{y=0}\, dx = 0$$

and therefore, since the stresses reduce to a balanced system of forces, in accordance with Saint-Venant's principle, their influence will be small.

3.8 Use of Polar Coordinates

Polar coordinates r and θ will now be used in solving some two-dimensional elasticity problems. Consider the system of cartesian coordinates Oxy and the orthogonal system Ort, the r-axis being inclined at the angle θ with respect to the x-axis (see Fig. 3.8.1). Consider also the two faces which are perpendicular to the r- and t-axes. On these faces, the normal stresses σ_r and σ_θ and the shearing stresses $\tau_{r\theta}$ and $\tau_{\theta r}$ will be acting. For equilibrium

$$\tau_{r\theta} = \tau_{\theta r}$$

The Airy stress function $\Phi(x, y)$ was introduced in Sec. 3.5 and it was shown that this function must satisfy Eq. (3.5.4′), i.e.,

$$\Delta\Delta\Phi = 0 \qquad (3.5.4')$$

Since, in polar coordinates (see Appendix 3.A), the Laplacian operator is

$$\Delta = \frac{\partial^2}{\partial r^2} + \frac{1}{r}\frac{\partial}{\partial r} + \frac{1}{r^2}\frac{\partial^2}{\partial \theta^2}$$

Sec. 3.8 TWO-DIMENSIONAL ELASTICITY

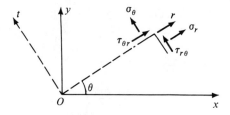

Figure 3.8.1

then Eq. (3.5.4') becomes

$$\left(\frac{\partial^2}{\partial r^2} + \frac{1}{r}\frac{\partial}{\partial r} + \frac{1}{r^2}\frac{\partial^2}{\partial \theta^2}\right)\left(\frac{\partial^2 \Phi}{\partial r^2} + \frac{1}{r}\frac{\partial \Phi}{\partial r} + \frac{1}{r^2}\frac{\partial^2 \Phi}{\partial \theta^2}\right) = 0 \quad (3.8.1)$$

Also, the following equations can be derived (see Appendix 3.A) which relate derivatives in rectangular and polar coordinates:

$$\left.\begin{aligned}
\frac{\partial^2 \Phi}{\partial x^2} &= \frac{\partial^2 \Phi}{\partial r^2}\cos^2\theta - \frac{\partial^2 \Phi}{\partial r \partial \theta}\frac{\sin(2\theta)}{r} + \frac{\partial^2 \Phi}{\partial \theta^2}\frac{\sin^2\theta}{r^2} + \frac{\partial \Phi}{\partial r}\frac{\sin^2\theta}{r} + \frac{\partial \Phi}{\partial \theta}\frac{\sin(2\theta)}{r^2} \\
\frac{\partial^2 \Phi}{\partial y^2} &= \frac{\partial^2 \Phi}{\partial r^2}\sin^2\theta + \frac{\partial^2 \Phi}{\partial r \partial \theta}\frac{\sin(2\theta)}{r} + \frac{\partial^2 \Phi}{\partial \theta^2}\frac{\cos^2\theta}{r^2} + \frac{\partial \Phi}{\partial r}\frac{\cos^2\theta}{r} - \frac{\partial \Phi}{\partial \theta}\frac{\sin(2\theta)}{r^2} \\
\frac{\partial^2 \Phi}{\partial x \partial y} &= \frac{\partial^2 \Phi}{\partial r^2}\frac{\sin(2\theta)}{2} + \frac{\partial^2 \Phi}{\partial r \partial \theta}\frac{\cos 2\theta}{r} - \frac{\partial^2 \Phi}{\partial \theta^2}\frac{\sin(2\theta)}{2r^2} - \frac{\partial \Phi}{\partial r}\frac{\sin(2\theta)}{2r} - \frac{\partial \Phi}{\partial \theta}\frac{\cos(2\theta)}{r^2}
\end{aligned}\right\}$$

$$(3.8.2)$$

From Fig. 3.8.1 one sees that

$$\sigma_r = (\sigma_x)_{\theta=0}$$
$$\sigma_\theta = (\sigma_y)_{\theta=0}$$
$$\tau_{r\theta} = (\tau_{xy})_{\theta=0}$$

It follows from Eqs. (3.8.2) that

$$\left.\begin{aligned}
\sigma_r &= \left(\frac{\partial^2 \Phi}{\partial y^2}\right)_{\theta=0} = \frac{1}{r^2}\frac{\partial^2 \Phi}{\partial \theta^2} + \frac{1}{r}\frac{\partial \Phi}{\partial r} \\
\sigma_\theta &= \left(\frac{\partial^2 \Phi}{\partial x^2}\right)_{\theta=0} = \frac{\partial^2 \Phi}{\partial r^2} \\
\tau_{r\theta} &= -\left(\frac{\partial^2 \Phi}{\partial x \partial y}\right)_{\theta=0} = -\frac{1}{r}\frac{\partial^2 \Phi}{\partial r \partial \theta} + \frac{1}{r^2}\frac{\partial \Phi}{\partial \theta} = -\frac{\partial}{\partial r}\left(\frac{1}{r}\frac{\partial \Phi}{\partial \theta}\right)
\end{aligned}\right\} \quad (3.8.3)$$

If Φ and the components of stress are independent of the variable θ, then Eqs. (3.8.1) and (3.8.3) reduce to the following (Euler's equation):

$$\left(\frac{d^2}{dr^2} + \frac{1}{r}\frac{d}{dr}\right)\left(\frac{d^2\Phi}{dr^2} + \frac{1}{r}\frac{d\Phi}{dr}\right) = \frac{d^4\Phi}{dr^4} + \frac{2}{r}\frac{d^3\Phi}{dr^3} - \frac{1}{r^2}\frac{d^2\Phi}{dr^2} + \frac{1}{r^3}\frac{d\Phi}{dr} = 0$$

(3.8.4)

and

$$\left.\begin{aligned} \sigma_r &= \frac{1}{r}\frac{d\Phi}{dr} \\ \sigma_\theta &= \frac{d^2\Phi}{dr^2} \\ \tau_{r\theta} &= 0 \end{aligned}\right\} \quad (3.8.5)$$

The general solution of Eq. (3.8.4) is (see Appendix 3.B)

$$\Phi(r) = A \log r + Br^2 \log r + Cr^2 + D \quad (3.8.6)$$

where A, B, C, and D are constants. By substituting Eq. (3.8.6) into Eqs. (3.8.5) one finds that

$$\left.\begin{aligned} \sigma_r &= \frac{A}{r^2} + B(1 + 2\log r) + 2C \\ \sigma_\theta &= -\frac{A}{r^2} + B(3 + 2\log r) + 2C \\ \tau_{r\theta} &= 0 \end{aligned}\right\} \quad (3.8.7)$$

The polar components of strain ϵ_r, ϵ_θ, and $\gamma_{r\theta}$ are expressed in terms of the polar components of displacement u and v (see Fig. 3.8.2) by the following relationships:

$$\left.\begin{aligned} \epsilon_r &= \frac{\partial u}{\partial r} \\ \epsilon_\theta &= \frac{u}{r} + \frac{1}{r}\frac{\partial v}{\partial \theta} \\ \gamma_{r\theta} &= \frac{1}{r}\frac{\partial u}{\partial \theta} + \frac{\partial v}{\partial r} - \frac{v}{r} \end{aligned}\right\} \quad (3.8.8)$$

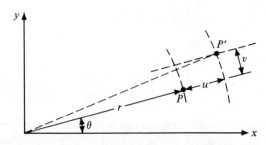

Figure 3.8.2

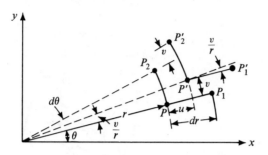

Figure 3.8.3

In order to derive Eqs. (3.8.8), suppose that the total deformation which displaces generic point P to its final position P' is composed of two distinct deformations:

1. A deformation in which components u and v have everywhere the same value as they have at point P.
2. A deformation in which only the variation of components u and v are considered.

Considering deformation 1 first (see Fig. 3.8.3), the following relationships are obtained:

$$P'P'_1 = PP_1$$
$$PP_2 = r\,d\theta$$
$$P'P'_2 = (r+u)d\theta$$

Thus,

$$\left.\begin{array}{l} (\epsilon_\theta)_1 = \dfrac{P'P'_2 - PP_2}{PP_2} = \dfrac{(r+u)d\theta - r\,d\theta}{r\,d\theta} = \dfrac{u}{r} \\[2mm] (\epsilon_r)_1 = 0 \qquad (\gamma_{r\theta})_1 = -\dfrac{v}{r} \end{array}\right\} \qquad (3.8.9)$$

In deformation 2 (see Fig. 3.8.4), the following relationships are derived:

$$(\epsilon_r)_2 = \frac{P'P''_1 - P'P'_1}{PP_1} = \frac{\dfrac{\partial u}{\partial r}dr}{dr} = \frac{\partial u}{\partial r}$$

$$(\epsilon_\theta)_2 = \frac{P'P''_2 - P'P'_2}{PP_2} = \frac{\dfrac{\partial v}{\partial \theta}d\theta}{r\,d\theta} = \frac{1}{r}\frac{\partial v}{\partial \theta} \qquad (3.8.10)$$

$$(\gamma_{r\theta})_2 = \frac{1}{r}\frac{\partial u}{\partial \theta} + \frac{\partial v}{\partial r}$$

By adding Eqs. (3.8.9) and (3.8.10) Eqs. (3.8.8) are obtained.

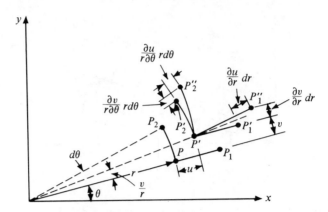

Figure 3.8.4

In order to derive expressions for the components of the elastic displacement u and v when the components of stress σ_r and σ_θ ($\tau_{r\theta} = 0$) are independent of θ one uses Eqs. (3.8.7) and (3.8.8) and the stress-strain relations, i.e.,

$$\left. \begin{array}{l} \epsilon_r = \dfrac{\partial u}{\partial r} = \dfrac{1}{E}(\sigma_r - v\sigma_\theta) \\[6pt] \epsilon_\theta = \dfrac{u}{r} + \dfrac{1}{r}\dfrac{\partial v}{\partial \theta} = \dfrac{1}{E}(\sigma_\theta - v\sigma_r) \\[6pt] \gamma_{r\theta} = \dfrac{1}{r}\dfrac{\partial u}{\partial \theta} + \dfrac{\partial v}{\partial r} - \dfrac{v}{r} = \dfrac{\tau_{r\theta}}{G} = 0 \end{array} \right\} \qquad (3.8.11)$$

The first of Eqs. (3.8.11) becomes

$$\frac{\partial u}{\partial r} = \frac{1}{E}\left[\frac{(1+v)}{r^2}A + 2(1-v)B\log r + (1-3v)B + 2(1-v)C\right]$$

and after integration, the displacement component u is obtained; thus,

$$u = \frac{1}{E}\left[-\frac{(1+v)}{r}A + 2(1-v)Br\log r - (1+v)Br + 2(1-v)Cr\right] + f(\theta) \qquad (3.8.12)$$

The second of Eqs. (3.8.11) can be written as

$$\frac{\partial v}{\partial \theta} = \frac{r}{E}(\sigma_\theta - v\sigma_r) - u = \frac{4Br}{E} - f(\theta)$$

and after integration, the displacement component v is obtained; thus,

$$v = \frac{4Br\theta}{E} - \int_0^\theta f(\theta)d\theta + F(r) \qquad (3.8.13)$$

By substituting Eqs. (3.8.12) and (3.8.13) into the third of Eqs. (3.8.11) one obtains

$$\frac{1}{r}\frac{df(\theta)}{d\theta} + \frac{dF(r)}{dr} + \frac{1}{r}\int_0^\theta f(\theta)d\theta - \frac{F(r)}{r} = 0$$

which can be written in the form

$$\frac{df(\theta)}{d\theta} + \int_0^\theta f(\theta)\,d\theta = F(r) - r\frac{dF(r)}{dr} = D = \text{constant}$$

which yields the two equations

$$\left.\begin{array}{c} \dfrac{df(\theta)}{d\theta} + \displaystyle\int_0^\theta f(\theta)\,d\theta = D \\[1em] F(r) - r\dfrac{dF(r)}{dr} = D \end{array}\right\} \qquad (3.8.14)$$

By differentiating the first of Eqs. (3.8.14) with respect to θ, one gets

$$\frac{d^2 f(\theta)}{d\theta^2} + f(\theta) = 0$$

whose general integral is

$$f(\theta) = J\cos\theta + K\sin\theta \qquad (3.8.15)$$

where J and K are constants. By substituting Eq. (3.8.15) into the first of Eqs. (3.8.14), one finds that

$$-J\sin\theta + K\cos\theta + J\sin\theta - K\cos\theta + K = D$$

or

$$K = D$$

and Eq. (3.8.15) can be written as

$$f(\theta) = J\cos\theta + D\sin\theta \qquad (3.8.16)$$

By differentiating the second of Eqs. (3.8.14) with respect to r, one obtains

$$\frac{dF(r)}{dr} - r\frac{d^2 F(r)}{dr^2} - \frac{dF(r)}{dr} = 0$$

or
$$\frac{d^2F(r)}{dr^2} = 0$$

whose integral is

$$F(r) = Hr + G \tag{3.8.17}$$

where H and G are constants. By substituting Eq. (3.8.17) into the second of Eqs. (3.8.14) one finds that

$$G = D$$

and Eq. (3.8.17) can be written as

$$F(r) = Hr + D \tag{3.8.18}$$

By substituting Eqs. (3.8.16) and (3.8.18) into Eqs. (3.8.12) and (3.8.13) one finally obtains the following expressions for the components of elastic displacement u and v in polar coordinates:

$$u = \frac{1}{E}\left[-\frac{(1+v)}{r}A + 2(1-v)B\log r - (1+v)Br + 2(1-v)Cr\right]$$
$$+ J\cos\theta + D\sin\theta \tag{3.8.19}$$
$$v = \frac{4Br\theta}{E} - J\sin\theta + D\cos\theta + Hr$$

where A, B, C, D, H, and J are constants. By the use of Eqs. (3.8.7) and (3.8.19) various problems in two-dimensional elasticity will be solved in the next sections.

3.9 Bending of a Circular Bar (Golovin-Ribière Problem)[10]

Consider the circular bar shown in Fig. 3.9.1. It is limited by two concentric circles whose radii a and b are of the same order of magnitude while its unit thickness is small as compared with its height $b - a$. Two equal and opposite couples M are applied at the ends of the bar.

The following boundary conditions must be satisfied:

[10] According to S. Timoshenko, the solution of this problem was first obtained in 1881 by the Russian mechanicist H. Golovin. It was independently solved eight years later by the French mechanicist M. C. Ribière (See Bibliography Sec. 3.9 at the end of this chapter, Nos. 9 and 18).

TWO-DIMENSIONAL ELASTICITY

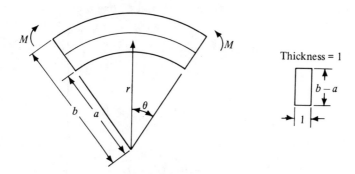

Figure 3.9.1

$$\text{for } \begin{cases} r = a \\ r = b \end{cases} \sigma_r = 0 \tag{3.9.1}$$

$$\int_a^b \sigma_\theta(r)(1)\,dr = M \tag{3.9.2}[11]$$

By substituting the first of Eqs. (3.8.7) into Eqs. (3.9.1) the two equations

$$\left. \begin{array}{l} \dfrac{A}{a^2} + B(1 + 2\log a) + 2C = 0 \\[4pt] \dfrac{A}{b^2} + B(1 + 2\log b) + 2C = 0 \end{array} \right\} \tag{3.9.3}$$

are obtained. By substituting the second of Eqs. (3.8.5) into Eq. (3.9.2) one obtains

$$\int_a^b \sigma_\theta(r)(1)\,dr = \int_a^b \frac{d^2\Phi}{dr^2} r\,dr = \int_a^b r\,d\left(\frac{d\Phi}{dr}\right) = r\frac{d\Phi}{dr}\bigg]_a^b - \int_a^b \frac{d\Phi}{dr}\,dr$$

$$= r\frac{d\Phi}{dr}\bigg]_a^b - \Phi\bigg]_a^b = -\Phi\bigg]_a^b = -M$$

since $r\dfrac{d\Phi}{dr}\bigg]_a^b = 0$ [this follows from Eqs. (3.9.3)]. It follows that

$$A \log\left(\frac{b}{a}\right) + B(b^2 \log b - a^2 \log a) + c(b^2 - a^2) = M \tag{3.9.4}$$

---

[11] The condition $\int_a^b \sigma_\theta(r)(1)dr = 0$ is satisfied once Eqs. (3.9.1) are satisfied. In fact, by substituting the second of Eqs. (3.8.5) into this equation, one obtains

$$\int_a^b \sigma_\theta 1\,dr = \int_a^b \frac{d^2\Phi}{dr^2}\,dr = \frac{d\Phi}{dr}\bigg]_a^b = \left[\frac{A}{b} + Bb(1 + 2\log b) + 2Cb\right]$$

$$- \left[\frac{A}{a} + Ba(1 + 2\log a) + 2Ca\right] = 0$$

in view of Eqs. (3.9.3).

Solving Eqs. (3.9.3) and (3.9.4) for A, B, and C, one obtains

$$A = -\frac{4M}{K} a^2 b^2 \log \frac{b}{a}$$
$$B = -\frac{2M}{K} (b^2 - a^2)$$
$$C = \frac{M}{K} [b^2 - a^2 + 2(b^2 \log b - a^2 \log a)]$$ (3.9.5)

where

$$K = (b^2 - a^2) - 4a^2 b^2 \left(\log \frac{b}{a}\right)^2$$

By substituting Eqs. (3.9.5) into Eqs. (3.8.7), one has for the components of stress the following expressions:

$$\sigma_r = -\frac{4M}{K} \left[\frac{a^2 b^2}{r^2} \log \frac{b}{a} + b^2 \log \frac{r}{b} + a^2 \log \frac{a}{r}\right]$$
$$\sigma_\theta = -\frac{4M}{K} \left[-\frac{a^2 b^2}{r^2} \log \frac{b}{a} + b^2 \log \frac{r}{b} + a^2 \log \frac{a}{r} + b^2 - a^2\right]$$ (3.9.6)
$$\tau_{r\theta} = 0$$

It is of interest to note that stresses σ_θ are not linearly distributed across the section. Figure 3.9.2 shows the diagrams of σ_r and σ_θ for the case in which

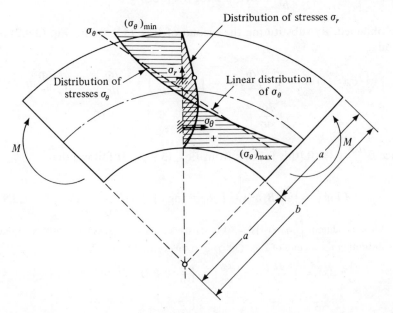

Figure 3.9.2

$a = b/2$. One notices that, due to the curvature of the bar, σ_θ increases near the concave side and decreases near the convex side. Letting

$$(\sigma_\theta)_{\substack{\max \\ \min}} = \alpha \frac{M}{a^2}$$

one finds

$$\alpha = \frac{(\sigma_\theta)_{\substack{\max \\ \min}} a^2}{M} \qquad (3.9.7)$$

and the coefficient α will depend only on the ratio b/a. In Table 3.9.1, the values of α given by Eq. (3.9.7) in accordance with the mathematical theory of elasticity are compared, for three different values of the ratio b/a, with those given by the strength of materials in the case of the linear stress distribution (Bernoulli-Navier theory)[12] and in the case of the hyperbolic stress distribution (Winkler theory).[13]

TABLE 3.9.1*

$\dfrac{b}{a}$	According to Strength of Materials				According to the Mathematical Theory of Elasticity	
	Navier-Bernoulli		Winkler			
	α for $(\sigma_\theta)_{\max}$	α for $(\sigma_\theta)_{\min}$	α for $(\sigma_\theta)_{\max}$	α for $(\sigma_\theta)_{\min}$	α for $(\sigma_\theta)_{\max}$	α for $(\sigma_\theta)_{\min}$
1.3	+66.67	−66.67	+72.98	−61.27	+73.05	−61.35
2	+ 6.00	− 6.00	+ 7.725	− 4.863	+ 7.755	− 4.917
3	+ 1.50	− 1.50	+ 2.285	− 1.095	+ 2.292	− 1.130

* These results were obtained by V. Billevicz. See Bibliography, Sec. 3.9, No. 2, at the end of this chapter.

In order to compute the components of the elastic displacement from Eqs. (3.8.19), i.e., to compute the values of the three constants D, H, and J since the constants A, B, and C of the problem are given by Eqs. (3.9.5), it will be

---

[12] **Bernoulli, Daniel.** Swiss mathematician (b. Groningen 1700; d. Basel 1782). He was the son of Johann Bernoulli. After studying medicine he spent seven years as a mathematician at the Russian Academy, then was successively Professor of botany, anatomy, and natural philosophy at Basel University. His "Hydrodynamica" dealing with the statics and dynamics of fluids was published in 1738. In addition to his outstanding contribution in the field of mechanics he did important work also in probability and its application to statistics.

[13] **Winkler, Emil.** German engineer (b. Torgau, Saxony, 1835; d. Berlin 1888). He was professor of bridge and railroad engineering successively at the Polytechnic Institutes of Prague, Vienna, and Berlin.

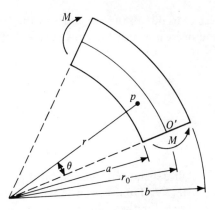

Figure 3.9.3

assumed that point O' (see Fig. 3.9.3) is fixed, and that the tangent to the curved axis of the bar $[r = r_0 = (a + b)/2]$ has at point O' a fixed direction. This implies that

$$u = v = \frac{dv}{dr} = 0 \quad \text{for } r = r_0 = \frac{a+b}{2} \text{ and } \theta = 0 \quad (3.9.8)$$

Conditions (3.9.8) expressed with the aid of Eqs. (3.8.19) give the following values for the constants:

$$\left.\begin{array}{l} D = H = 0 \\ J = \dfrac{1}{E}\left[\dfrac{(1 + v)A}{r_0} - 2(1 - v)Br_0 \log r_0 + (1 + v)Br_0 - 2(1 - v)Cr_0\right] \end{array}\right\} \quad (3.9.9)$$

Component v of the elastic displacement is then found to be

$$v = \frac{4Br\theta}{E} - J \sin \theta \quad (3.9.10)$$

where constant J is given by Eq. (3.9.9) while constant B is given by the second of Eqs. (3.9.5).

Equation (3.9.10) shows that at a given section of the curved beam ($\theta =$ constant) one has

$$v = \beta r + \gamma \quad (3.9.11)$$

where β and γ are constants given, respectively, by

$$\beta = \frac{4B\theta}{E}$$

and
$$\gamma = -J \sin \theta$$

Equation (3.9.11) proves that plane cross sections of the curved bar remain plane for the case of pure bending. Thus the Bernoulli–Navier hypothesis is valid for this case.

3.10 Thick Tube Subjected to External and Internal Uniformly Distributed Pressures (Lamé's Problem)

Consider the thick cylinder of length l, internal radius r_1, and external radius r_2 which is subjected to the internal and external uniformly distributed pressures p_1 and p_2 as shown in Fig. 3.10.1. The bases of the cylinder at $z = 0$ and $z = l$ are assumed to be completely restrained; therefore, $w = 0$ (see Sec. 3.3) and it follows that the problem is a case of plane strain. Moreover, because of symmetry, stresses σ_r and σ_θ will be independent of θ and stress $\tau_{r\theta} = 0$. The stresses will be expressed by Eqs. (3.8.7), i.e.,

$$\left. \begin{array}{l} \sigma_r = \dfrac{A}{r^2} + B(1 + 2\log r) + 2C \\[6pt] \sigma_\theta = -\dfrac{A}{r^2} + B(3 + 2\log r) + 2C \end{array} \right\} \quad (3.10.1)$$

Also for symmetry, component u of the elastic displacement will be independent of θ and component v will be zero. Equations (3.8.8) will in this case become

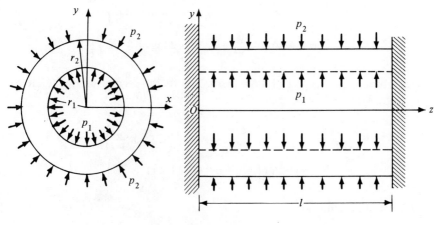

Figure 3.10.1

$$\left.\begin{array}{l}\epsilon_r = \dfrac{du}{dr} \\[4pt] \epsilon_\theta = \dfrac{u}{r} \\[4pt] \gamma_{r\theta} = 0\end{array}\right\} \qquad (3.10.2)$$

Hooke's law is given by the following equations:

$$\left.\begin{array}{l}\epsilon_r = \dfrac{1}{E}[\sigma_r - \nu(\sigma_\theta + \sigma_z)] \\[4pt] \epsilon_\theta = \dfrac{1}{E}[\sigma_\theta - \nu(\sigma_z + \sigma_r)] \\[4pt] \epsilon_z = \dfrac{1}{E}[\sigma_z - \nu(\sigma_r + \sigma_\theta)]\end{array}\right\} \qquad (3.10.3)$$

However, since $\epsilon_z = 0$ it follows from the third of Eqs. (3.10.3) that

$$\sigma_z = \nu(\sigma_r + \sigma_\theta) \qquad (3.10.4)$$

By substituting Eq. (3.10.4) into the first two of Eqs. (3.10.3), one obtains

$$\left.\begin{array}{l}\epsilon_r = \dfrac{1}{E}[(1 - \nu^2)\sigma_r - \nu(1 + \nu)\sigma_\theta] \\[4pt] \epsilon_\theta = \dfrac{1}{E}[(1 - \nu^2)\sigma_\theta - \nu(1 + \nu)\sigma_r]\end{array}\right\} \qquad (3.10.5)$$

By use of Eqs. (3.10.1), the first of Eqs. (3.10.2), and the first of Eqs. (3.10.5), one obtains

$$\left.\begin{array}{l}\dfrac{du}{dr} = \dfrac{1}{E}\left[(1 - \nu^2)\left\{\dfrac{A}{r^2} + B(1 + 2\log r) + 2C\right\}\right. \\[4pt] \left. \qquad - \nu(1 + \nu)\left\{-\dfrac{A}{r^2} + B(3 + 2\log r) + 2C\right\}\right]\end{array}\right\} \qquad (3.10.6)$$

By combining the second of Eqs. (3.10.2) with the second of Eqs. (3.10.3) and using Eqs. (3.10.1) one obtains

$$u = \dfrac{r}{E}\left[(1 - \nu^2)\left\{-\dfrac{A}{r^2} + B(3 + 2\log r) + 2C\right\} \right.$$
$$\left. - \nu(1 + \nu)\left\{\dfrac{A}{r^2} + B(1 + 2\log r) + 2C\right\}\right]$$

and upon differentiation

$$\frac{du}{dr} = \frac{1}{E}\left[(1-v^2)\left\{-\frac{A}{r^2} + B(3+2\log r) + 2C\right\}\right.$$
$$\left. - v(1+v)\left\{\frac{A}{r^2} + B(1+2\log r) + 2C\right\}\right] \qquad (3.10.7)$$
$$+ \frac{r}{E}\left[(1-v^2)\left\{\frac{2A}{r^3} + \frac{2B}{r}\right\} - v(1+v)\left\{-\frac{2A}{r^3} + \frac{2B}{r}\right\}\right]$$

By equating Eqs. (3.10.6) and (3.10.7) one gets

$$4(1-v)B = 0 \quad \text{or} \quad B = 0$$

With this value for constant B, Eqs. (3.10.1) become

$$\left.\begin{array}{l} \sigma_r = \dfrac{A}{r^2} + 2C \\[2mm] \sigma_\theta = -\dfrac{A}{r^2} + 2C \end{array}\right\} \qquad (3.10.8)$$

The two boundary conditions

$$(\sigma_r)_{r=r_1} = \frac{A}{r_1^2} + 2C = -p_1$$

$$(\sigma_r)_{r=r_2} = \frac{A}{r_2^2} + 2C = -p_2$$

give the following values for constants A and C:

$$A = \frac{r_1^2 r_2^2 (p_2 - p_1)}{r_2^2 - r_1^2}$$

$$2C = \frac{r_1^2 p_1 - r_2^2 p_2}{r_2^2 - r_1^2}$$

and Eqs. (3.10.8) become

$$\left.\begin{array}{l} \sigma_r = \dfrac{r_1^2 r_2^2 (p_2 - p_1)}{r_2^2 - r_1^2} \cdot \dfrac{1}{r^2} + \dfrac{r_1^2 p_1 - r_2^2 p_2}{r_2^2 - r_1^2} \\[3mm] \sigma_\theta = -\dfrac{r_1^2 r_2^2 (p_2 - p_1)}{r_2^2 - r_1^2} \cdot \dfrac{1}{r^2} + \dfrac{r_1^2 p_1 - r_2^2 p_2}{r_2^2 - r_1^2} \end{array}\right\} \qquad (3.10.9)$$

By substituting Eqs. (3.10.9) into Eq. (3.10.4) one finds that

$$\sigma_z = \frac{2v(r_1^2 p_1 - r_2^2 p_2)}{r_2^2 - r_1^2} = \text{constant} \qquad (3.10.10)$$

If the ends of the cylinder (at $z = 0$ and $z = l$) are free, one must superimpose upon the normal stress σ_z, given by Eq. (3.10.10), an equal and opposite stress $-\sigma_z$. This will cause a longitudinal strain given by

$$\epsilon_z = -\frac{\sigma_z}{E} = -\frac{2\nu(r_1^2 p_1 - r_2^2 p_2)}{E(r_2^2 - r_1^2)} \qquad (3.10.11)$$

Stresses σ_r and σ_θ will remain the same as those given by Eq. (3.10.9).
If $p_1 \neq 0$ and $p_2 = 0$, Eqs. (3.10.9) become

$$\left. \begin{aligned} \sigma_r &= \frac{r_1^2 p_1}{r_2^2 - r_1^2}\left(1 - \frac{r_2^2}{r^2}\right) \\ \sigma_\theta &= \frac{r_1^2 p_1}{r_2^2 - r_1^2}\left(1 + \frac{r_2^2}{r^2}\right) \end{aligned} \right\} \qquad (3.10.12)$$

which shows that $\sigma_r \leq 0$ (compression) since $r_2 \geq r$, while $\sigma_\theta > 0$ (tension). The extreme values are for $r = r_1$ where

$$\left. \begin{aligned} (\sigma_r)_{\min} &= (\sigma_r)_{r=r_1} = -p_1 \\ (\sigma_\theta)_{\max} &= (\sigma_\theta)_{r=r_1} = \frac{r_1^2 p_1}{r_2^2 - r_1^2}\left(1 + \frac{r_2^2}{r_1^2}\right) \end{aligned} \right\} \qquad (3.10.13)$$

When $r_2 = \infty$ (see Fig. 3.10.2) one has the case of a circular hole inside an infinite elastic medium and Eqs. (3.10.12) become

$$\left. \begin{aligned} \sigma_r &= -\frac{r_1^2}{r^2} p_1 \\ \sigma_\theta &= +\frac{r_1^2}{r^2} p_1 \end{aligned} \right\} \qquad (3.10.14)$$

In this particular case $\epsilon_z = \sigma_z = 0$, and one has at the same time a case of plane stress and of plane strain. The deformation of the tube is in this case given by

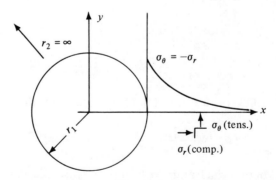

Figure 3.10.2

$$u = \frac{r}{E}[(1-v^2)\sigma_\theta - v(1+v)\sigma_r] = \frac{(1+v)r_1^2 p_1}{Er}$$

and

$$u_{max} = (u)_{r=r_1} = \frac{(1+v)r_1 p_1}{E}$$

while $u = 0$ for $r \to \infty$.

From the second of Eqs. (3.10.12), one finds that when the thickness $t = r_2 - r_1$ of the cylinder becomes very small in comparison with the radius $r_0 = (r_2 + r_1)/2$ the following relation is obtained

$$\sigma_\theta = \frac{r_1^2 p}{r_2^2 - r_1^2}\left(1 + \frac{r_2^2}{r^2}\right) = \frac{r_0^2 p}{(r_2+r_1)(r_2-r_1)}\left(1 + \frac{r_2^2}{r^2}\right)$$
$$= \frac{r_0^2 p(1+1)}{(2r_0)t} = p\frac{r_0}{t} \qquad (3.10.15)^{14}$$

Example 3.10.1. Determine the stresses in a steel tube of internal radius 8 in. and outside radius 16 in., if the internal pressure is $p = 15{,}000$ psi.

At the inside of the tube

$$\sigma_\theta = 15{,}000\frac{(16)^2 + (8)^2}{(16)^2 - (8)^2} = 25{,}000 \text{ psi}$$

From Strength of Materials

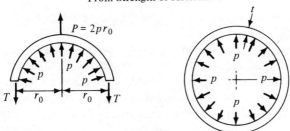

Figure 3.10.3

[14] Equation (3.10.15) is the formula derived in the strength of materials for hoop stresses in cylindrical containers subjected to pressure p. This formula is then derived directly in the following way (see Fig. 3.10.3):

$$2T = 2pr_0$$

or

$$2\sigma_\theta \cdot t \cdot 1 = 2pr_0$$

and finally

$$\sigma_\theta = \frac{pr_0}{t}$$

At the outside of the tube

$$\sigma_\theta = 2(15{,}000)\frac{(8)^2}{(16)^2 - (8)^2} = 10{,}000 \text{ psi}$$

The stresses at the middle of the tube ($r = 12$ in.) are

$$\sigma_r = 15{,}000\frac{(8)^2}{(16)^2 - (8)^2}\left(1 - \frac{(16)^2}{(12)^2}\right) = -4084 \text{ psi}$$

$$\sigma_\theta = 15{,}000\frac{(8)^2}{(16)^2 - (8)^2}\left(1 + \frac{(16)^2}{(12)^2}\right) = 13{,}894 \text{ psi}$$

By using Eq. (3.10.15) one would have obtained

$$\sigma_\theta = 15{,}000 \times \tfrac{12}{8} = 22{,}500 \text{ psi}$$

with an error of -10 per cent.

3.11 Shrink Fits

The formulas derived in Sec. 3.10 show that stresses in thick tubes due to internal pressure are very inconveniently distributed. The second of Eqs. (3.10.13) shows that

$$\sigma_\theta > p_1$$

and that σ_θ becomes equal to p_1 only in the case in which the external radius of the cylinder becomes infinite.

In order to reduce stresses inside a thick tube subjected to a high internal pressure, as in the case of guns, two or more tubes are assembled together by "shrink fitting." In the case of two tubes, the outside tube has an internal diameter slightly less than the external diameter of the inside tube. The outside tube is expanded by heating and the two tubes assembled together. After the tubes attain the same temperature the outside tube causes circumferential compressive stresses in the inner tube and the inside tube causes circumferential tensile stresses in the outside tube. In this way, by building a "prestressed composite tube" it is possible to reduce stresses inside the tube when an internal pressure is acting.

Substantially the same process of "shrink fitting" is used in the building of locomotive wheels. Since wheels of locomotives are fitted with specially hardened rims which can be replaced in case of wear, the rim or tire has an internal diameter slightly smaller than the external diameter of the wheel. The tire is expanded by heating and applied to the wheel. After attaining the same temperature tire and wheel exert on one another a pressure sufficient to prevent any relative movement.

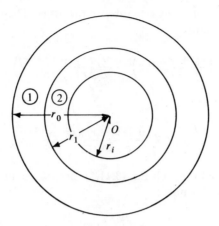

Figure 3.11.1

In discussing the problem of "shrink fit" in guns, let us consider the case of two cylinders, 1 and 2 (see Fig. 3.11.1). Combining the second of Eqs. (3.10.5) and the second of Eqs. (3.10.2) one obtains

$$u = r\epsilon_\theta = \frac{r}{E}[(1-v^2)\sigma_\theta - v(1+v)\sigma_r] \tag{3.11.1}$$

By setting $r_1 = r_i$ and $r_2 = r_0$ in the equations for σ_r and σ_θ [Eqs. (3.10.9)] and substituting the resulting expressions into Eq. (3.11.1), one obtains

$$u = \frac{1-v}{E}\frac{p_2 r_0^2 - p_1 r_i^2}{r_0^2 - r_i^2}r + \frac{1+v}{E}(p_2 - p_1)\frac{r_0^2 r_i^2}{r_0^2 - r_i^2}\frac{1}{r} \tag{3.11.2}$$

If, in Eq. (3.11.2), $p_1 \neq 0$ and $p_2 = 0$, the deformation u_i at the inside of the tube (for $r = r_i$) will be

$$u_i = -\frac{p_1 r_i}{E}\left[\frac{r_0^2 + r_i^2}{r_0^2 - r_i^2} + v\right] \tag{3.11.3}$$

If, in Eq. (3.11.2), $p_1 = 0$ and $p_2 \neq 0$, the deformation u_0 at the outside of the tube (for $r = r_0$) will be

$$u_0 = \frac{p_2 r_0}{E}\left[\frac{r_0^2 + r_i^2}{r_0^2 - r_i^2} - v\right] \tag{3.11.4}$$

Now, in order to study the case of the two tubes shrunk together, consider the two tubes, 1 and 2, of Fig. 3.11.1. Actually the external radius of the inside tube is larger than the internal radius of the outside tube. Call δ the difference and p the pressure exerted by one tube on the other. Then from Eqs. (3.11.3) and (3.11.4) one obtains

$$\frac{pr_1}{E}\left(\frac{r_0^2 + r_1^2}{r_0^2 - r_1^2} + \nu\right) + \frac{pr_1}{E}\left(\frac{r_1^2 + r_i^2}{r_1^2 - r_i^2} - \nu\right) = \delta$$

from which it follows that

$$p = E\frac{\delta}{r_1}\frac{(r_1^2 - r_i^2)(r_0^2 - r_1^2)}{2r_1^2(r_0^2 - r_i^2)} \tag{3.11.5}$$

Once p is determined through Eq. (3.11.5), prestresses in the inside tube are computed from the formulas

$$\left.\begin{array}{l} \sigma_r = \dfrac{p_1 r_1^2}{r_1^2 - r_i^2}\left(1 - \dfrac{r_i^2}{r^2}\right) \\[2mm] \sigma_\theta = \dfrac{pr_1^2}{r_1^2 - r_i^2}\left(1 + \dfrac{r_i^2}{r^2}\right) \end{array}\right\} \tag{3.11.6}$$

while prestresses in the outside tube are computed from the formulas:

$$\left.\begin{array}{l} \sigma_r = \dfrac{pr_1^2}{r_0^2 - r_1^2}\left(\dfrac{r_0^2}{r^2} - 1\right) \\[2mm] \sigma_\theta = -\dfrac{pr_1^2}{r_0^2 - r_1^2}\left(\dfrac{r_0^2}{r^2} + 1\right) \end{array}\right\} \tag{3.11.7}$$

Suppose now that the compound tube is subjected to an internal pressure p_1. The stresses are the same as in a thick tube of internal radius r_i and external radius r_0. To these stresses the shrink fit stresses have to be added.

For the case in which the internal member is a solid shaft instead of a tube, one obtains the following expression by setting $r_i = 0$:

$$p = E\frac{\delta}{r_1}\frac{r_0^2 - r_1^2}{2r_0^2}$$

Example 3.11.1. For a composite steel tube formed from two cylinders having radii $r_1 = 12$ in., $r_i = 8$ in., $r_0 = 16$ in., and $\delta = 0.006$ in. determine the stresses, if the internal pressure is 15,000 psi.

From Eq. (3.11.5) one obtains

$$p = 30(10^6)\frac{0.006}{12}\frac{[(12)^2 - (8)^2][(16)^2 - (12)^2]}{2(12)^2[(16)^2 - (8)^2]} = 2430 \text{ psi}$$

The negative prestresses (due to compression) in the inside tube are

$$(\sigma_\theta)_i = 2(2430)\frac{12^2}{(8)^2 - (12)^2} = -8748 \text{ psi}$$

$$(\sigma_\theta)_1 = 2430 \frac{(12)^2 + (8)^2}{(8)^2 - (12)^2} = -6318 \text{ psi}$$

The prestresses in the outside tube are

$$(\sigma_\theta)_1 = 2430 \frac{(16)^2 + (12)^2}{(16)^2 - (12)^2} = +8675$$

$$(\sigma_\theta)_0 = 2430 \frac{(12)^2}{(16)^2 - (12)^2} = 6221$$

The inside presure of 15,000 psi produces inside the composite tube the same stresses as those computed in the Ex. 3.10.1. The total stresses are

$$\begin{rcases} (\sigma_\theta)_i = 25{,}000 - 8{,}748 = 16{,}252 \text{ psi} \\ (\sigma_\theta)_1 = 13{,}894 - 6{,}318 = 7{,}576 \text{ psi} \end{rcases} \text{(internal tube)}$$

$$\begin{rcases} (\sigma_\theta)_1 = 13{,}894 + 8{,}675 = 22{,}569 \text{ psi} \\ (\sigma_\theta)_0 = 10{,}000 + 6{,}221 = 16{,}221 \text{ psi} \end{rcases} \text{(external tube)}$$

In Fig. 3.11.2, the diagram of σ_θ is represented by a continuous line for the case of the compound tube and by a dotted line for the case of the thick tube.

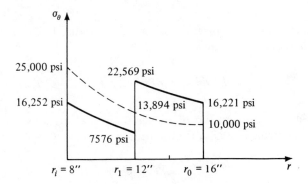

Figure 3.11.2

3.12 Rotating Disks and Cylinders

The problem of determining stresses produced in disks and cylinders rotating at high speed has many important applications, such as in the design of steam and gas turbines. In this section stresses produced in disks and cylinders of constant thickness will be discussed.[15]

[15] Cases of disks of variable thickness and disks of uniform stress are discussed in the books and papers listed in the Bibliography, Sec. 3.12 at the end of the chapter.

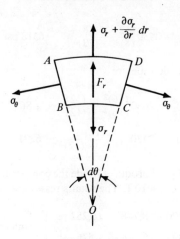

Figure 3.12.1

Consider an element $dr\, r\, d\theta$ cut out from a disk of unit thickness which is rotating with constant angular velocity ω around its axis (see Fig. 3.12.1). Calling ρ the mass density of the disk material, the centrifugal force per unit mass is

$$F_r = \rho \omega^2 r$$

Expressing the condition of equilibrium of the radial forces acting on the element $ABCD$, one finds that

$$\left(\sigma_r + \frac{\partial \sigma_r}{\partial r} dr\right)(r + dr)\, d\theta - \sigma_r r\, d\theta - 2\sigma_\theta dr \sin \frac{d\theta}{2} + F_r r\, dr\, d\theta = 0$$

By neglecting small quantities of higher order and dividing each term by $r\, dr\, d\theta$ the following equation is obtained:

$$\frac{d\sigma_r}{dr} + \frac{\sigma_r - \sigma_\theta}{r} + \rho\omega^2 r = 0 \qquad (3.12.1)$$

This equation can be written in the form

$$\frac{d}{dr}(r\sigma_r) - \sigma_\theta + \rho\omega^2 r^2 = 0 \qquad (3.12.1')$$

Equation (3.12.1') is satisfied by a stress function $\Phi(r)$ such that

$$\left.\begin{aligned}\sigma_r &= \frac{\Phi}{r} \\ \sigma_\theta &= \frac{d\Phi}{dr} + \rho\omega^2 r^2\end{aligned}\right\} \qquad (3.12.2)$$

Because of symmetry, the expressions for strain [Eqs. (3.8.10)] reduce to the form given by Eqs. (3.10.2), i.e.,

$$\left.\begin{aligned} \epsilon_r &= \frac{du}{dr} \\ \epsilon_\theta &= \frac{u}{r} \\ \gamma_{r\theta} &= 0 \end{aligned}\right\} \quad (3.10.2)$$

By eliminating u from Eqs. (3.10.2), the equation

$$\frac{d}{dr}(r\epsilon_\theta) - \epsilon_r = 0$$

or

$$r\frac{d\epsilon_\theta}{dr} + \epsilon_\theta - \epsilon_r = 0 \quad (3.12.3)$$

is obtained. Hooke's law is given by the following equations:

$$\left.\begin{aligned} \epsilon_r &= \frac{1}{E}(\sigma_r - \nu\sigma_\theta) \\ \epsilon_\theta &= \frac{1}{E}(\sigma_\theta - \nu\sigma_r) \end{aligned}\right\} \quad (3.12.4)$$

Combining Eqs. (3.12.2), (3.12.3), and (3.12.4), one obtains the equation

$$\frac{d^2\Phi}{dr^2} + \frac{1}{r}\frac{d\Phi}{dr} - \frac{\Phi}{r^2} + (3+\nu)\rho\omega^2 r = 0$$

which can be written in the form

$$\frac{d}{dr}\left[\frac{1}{r}\frac{d}{dr}(r\Phi)\right] = -(3+\nu)\rho\omega^2 r \quad (3.12.5)$$

By integrating Eq. (3.12.5) directly, one obtains

$$\Phi(r) = -\frac{3+\nu}{8}\rho\omega^2 r^3 + C_1\frac{r}{2} + C_2\frac{1}{r} \quad (3.12.6)$$

where C_1 and C_2 are constants. In view of Eqs. (3.12.2), one obtains the following expressions for the stresses:

$$\left.\begin{aligned} \sigma_r &= \frac{\Phi}{r} = -\frac{3+\nu}{8}\rho\omega^2 r^2 + \frac{C_1}{2} + \frac{C_2}{r^2} \\ \sigma_\theta &= \frac{d\Phi}{dr} + \rho\omega^2 r^2 = -\frac{1+3\nu}{8}\rho\omega^2 r^2 + \frac{C_1}{2} - \frac{C_2}{r^2} \end{aligned}\right\} \quad (3.12.7)$$

(a) Case of the thin solid disk of radius R. For this case, constant C_2 must be zero since the stresses must be of finite magnitude. In the absence of external radial stresses on the free surface of the disk $(\sigma_r)_{r=R} = 0$, and from the first of Eqs. (3.12.7) one obtains for C_1 the expression

$$C_1 = \frac{3+v}{4}\rho\omega^2 R^2$$

where R is the radius of the disk. Substituting the above values of C_1 and C_2 into Eqs. (3.12.7), one obtains the stress components as follows:

$$\left.\begin{array}{l} \sigma_r = \dfrac{3+v}{8}\rho\omega^2(R^2 - r^2) \\[2mm] \sigma_\theta = \dfrac{\rho\omega^2}{8}[(3+v)R^2 - (1+3v)r^2] \end{array}\right\} \quad (3.12.8)$$

The maximum stresses occur at the center of the disk where

$$(\sigma_r)_{r=0} = (\sigma_\theta)_{r=0} = \frac{3+v}{8}\rho\omega^2 R^2 \quad (3.12.9)$$

(b) Case of a thin hollow disk (of external radius R_2 and internal radius R_1). The two constants C_1 and C_2 of Eqs. (3.12.7) are determined in this case by expressing the absence of external radial stresses on the free surfaces of the hollow disk. Thus

$$(\sigma_r)_{r=R_1} = -\frac{3+v}{8}\rho\omega^2 R_1^2 + \frac{C_1}{2} + \frac{C_2}{R_1^2} = 0$$

$$(\sigma_r)_{r=R_2} = -\frac{3+v}{8}\rho\omega^2 R_2^2 + \frac{C_1}{2} + \frac{C_2}{R_2^2} = 0$$

from which it follows that

$$\frac{C_1}{2} = \frac{3+v}{8}\rho\omega^2(R_1^2 + R_2^2)$$

$$C_2 = -\frac{3+v}{8}\rho\omega^2 R_1^2 R_2^2$$

The stresses are

$$\left.\begin{array}{l} \sigma_r = \dfrac{3+v}{8}\rho\omega^2\left(R_1^2 + R_2^2 - \dfrac{R_1^2 R_2^2}{r^2} - r^2\right) \\[2mm] \sigma_\theta = \dfrac{3+v}{8}\rho\omega^2\left(R_1^2 + R_2^2 + \dfrac{R_1^2 R_2^2}{r^2} - \dfrac{1+3v}{3+v}r^2\right) \end{array}\right\} \quad (3.12.10)$$

The maximum value for σ_θ occurs at $r = R_1$ for which

$$(\sigma_\theta)_{r=R_1} = (\sigma_\theta)_{\max} = \frac{3+\nu}{4}\rho\omega^2 R_2^2\left(1 + \frac{1-\nu}{3+\nu}\frac{R_1^2}{R_2^2}\right) \quad (3.12.11)$$

If the quantity R_1/R_2 is negligible as compared with unity, Eq. (3.12.10) becomes

$$(\sigma_\theta)_{\max} = \frac{3+\nu}{4}\rho\omega^2 R_2^2 \quad (3.12.12)$$

Comparison of Eqs. (3.12.9) and (3.12.12) shows that the maximum value of stress σ_θ for this case is twice that at the center of a solid disk. It follows that a small circular hole at the center of a rotating disk doubles the stress at the center of the disk.

(c) **Case of the solid cylinder.** The problem is now one of plane strain, Hooke's law being expressed by

$$\left.\begin{aligned}\epsilon_r &= \frac{1+\nu}{E}[(1-\nu)\sigma_r - \nu\sigma_\theta] \\ \epsilon_\theta &= \frac{1+\nu}{E}[(1-\nu)\sigma_\theta - \nu\sigma_r] \\ \gamma_{r\theta} &= \frac{2(1+\nu)}{E}\tau_{r\theta}\end{aligned}\right\} \quad (3.12.13)$$

By combining Eqs. (3.10.2), (3.12.2), and Eqs. (3.12.13) one obtains the following differential equation for the stress function Φ:

$$\frac{d^2\Phi}{dr^2} + \frac{1}{r}\frac{d\Phi}{dr} - \frac{\Phi}{r^2} + \frac{3-2\nu}{1-\nu}\rho\omega^2 r = 0$$

or

$$\frac{d}{dr}\left[\frac{1}{r}\frac{d}{dr}(r\Phi)\right] = -\frac{3-2\nu}{1-\nu}\rho\omega^2 r \quad (3.12.14)$$

By integrating Eq. (3.12.14) one obtains

$$\Phi = -\frac{1}{8}\frac{3-2\nu}{1-\nu}\rho\omega^2 r^3 + \frac{C_1}{2} + \frac{C_2}{r}$$

The stresses are, in this case

$$\left.\begin{aligned}\sigma_r &= \frac{\Phi}{r} = -\frac{1}{8}\frac{3-2\nu}{1-\nu}\rho\omega^2 r^2 + \frac{C_1}{2} + \frac{C_2}{r^2} \\ \sigma_\theta &= \frac{d\Phi}{dr} + \omega^2\rho r^2 = -\frac{1}{8}\frac{1+2\nu}{1-\nu}\rho\omega^2 r^2 + \frac{C_1}{2} - \frac{C_2}{r^2}\end{aligned}\right\} \quad (3.12.15)$$

Constant C_2 must be zero since the stresses remain finite at $r = 0$. For a solid cylinder with a free surface, the following condition must be satisfied:

$$(\sigma_r)_{r=R} = -\frac{1}{8}\frac{3-2\nu}{1-\nu}\rho\omega^2 R^2 + \frac{C_1}{2} = 0$$

or

$$C_1 = \frac{1}{4}\frac{3-\nu}{1-\nu}\rho\omega^2 R^2$$

Thus, stresses in the case of a solid cylinder are

$$\left.\begin{aligned}\sigma_r &= \frac{1}{8}\frac{3-2\nu}{1-\nu}\rho\omega^2(R^2 - r^2) \\ \sigma_\theta &= \frac{1}{8}\frac{\rho\omega^2}{1-\nu}[(3-2\nu)R^2 - (1+2\nu)r^2]\end{aligned}\right\} \quad (3.12.16)$$

The maximum stresses occur at the center of the cylinder, where

$$(\sigma_r)_{r=0} = (\sigma_\theta)_{r=0} = \frac{3-2\nu}{8(1-\nu)}\rho\omega^2 R^2 \quad (3.12.17)$$

(d) Case of a hollow cylinder (of inner radius R_1 and outer radius R_2). The two constants C_1 and C_2 of Eqs. (3.12.15) are determined in this case by expressing the absence of external radial stresses on the free surfaces of the hollow cylinder. Thus, from Eqs. (3.12.15) the following expressions for the stresses in the case of a hollow cylinder are obtained:

$$\left.\begin{aligned}\sigma_r &= \frac{1}{8}\frac{3-2\nu}{1-\nu}\rho\omega^2\left(R_2^2 + R_1^2 - \frac{R_1^2 R_2^2}{r^2} - r^2\right) \\ \sigma_\theta &= \frac{1}{8}\frac{3-2\nu}{1-\nu}\rho\omega^2\left(R_2^2 + R_1^2 + \frac{R_2^2 R_1^2}{r^2} - \frac{1+2\nu}{3-2\nu}r^2\right)\end{aligned}\right\} \quad (3.12.18)$$

The maximum stress σ_θ occurs at $r = R_1$ for which

$$(\sigma_\theta)_{r=R_1} = (\sigma_\theta)_{\max} = \frac{1}{4}\frac{3-2\nu}{1-\nu}\rho\omega^2 R_2^2\left(1 + \frac{1-2\nu}{3-2\nu}\frac{R_1^2}{R_2^2}\right) \quad (3.12.19)$$

If the quantity R_1/R_2 is negligible as compared with unity, Eq. (3.12.19) becomes

$$(\sigma_\theta)_{\max} = \frac{3-2\nu}{4(1-\nu)}\rho\omega^2 R_2^2 \quad (3.12.20)$$

Comparison of Eqs. (3.12.17) and (3.12.20) shows that the maximum value of stress σ_θ for a hollow cylinder with a small hole is twice that at the center of a solid cylinder.

3.13 Stress Concentration due to a Circular Hole in a Stressed Plate (Kirsch's Problem)

The problem of the distribution of stresses around a hole in a plate is of great practical importance in ship and airplane construction. The problem of *stress concentration* around an elliptic cavity has been discussed by Sir Charles Inglis.[16] In this section, the simpler problem of stress concentration around a circular hole will be discussed. This problem was solved by G. Kirsch in 1898.[17] Consider first an infinite plate referred to a system of orthogonal coordinates Oxy and assume that the plate is subjected to a uniform tensile stress of intensity S in the x-direction (Fig. 3.13.1). If there is no hole in the plate, the state of stress will be given by:

$$\left. \begin{array}{l} (\sigma_x)_1 = S \\ (\sigma_y)_1 = 0 \\ (\tau_{xy})_1 = 0 \end{array} \right\} \quad (3.13.1)$$

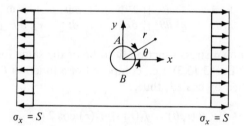

Figure 3.13.1

This state of stress can be derived from the stress function

$$\Phi_1 = \tfrac{1}{2} S y^2 \qquad (3.13.2)$$

Introducing polar coordinates r and θ, then

$$y = r \sin \theta$$

and Eq. (3.13.2) becomes

$$\Phi_1 = \frac{S}{2} r^2 \sin^2 \theta = \frac{S}{4} r^2 (1 - \cos 2\theta) \qquad (3.13.3)$$

and the corresponding stress components are

---

[16] See Bibliography Sec. 3.13, No. 12 at the end of this chapter.
[17] See Bibliography Sec. 3.13, No. 13 at the end of this chapter.

$$(\sigma_r)_1 = \frac{1}{r^2}\frac{\partial^2 \Phi_1}{\partial \theta^2} + \frac{1}{r}\frac{\partial \Phi_1}{\partial r} = \frac{1}{2}S(1 + \cos 2\theta)$$

$$(\sigma_\theta)_1 = \frac{\partial^2 \Phi_1}{\partial r^2} = \frac{1}{2}S(1 - \cos 2\theta) \qquad (3.13.4)$$

$$(\tau_{r\theta})_1 = -\frac{\partial}{\partial r}\left(\frac{1}{r}\frac{\partial \Phi_1}{\partial \theta}\right) = -\frac{1}{2}S\sin 2\theta$$

Now suppose that a hole of radius a with the center at the origin O is drilled through the plate. A new system of stresses σ_r, σ_θ, and $\tau_{r\theta}$ will be produced inside the plate. These stresses must satisfy the following boundary conditions

$$\begin{array}{ll} \text{For } r = a & \sigma_r = 0 \quad \tau_{r\theta} = 0 \\ \text{For } r = \infty & \sigma_r = (\sigma_r)_1 \quad \tau_{r\theta} = (\tau_{r\theta})_1 \quad \sigma_\theta = (\sigma_\theta)_1 \end{array} \qquad (3.13.5)$$

and will be derived from a stress function Φ which must satisfy the compatibility equation

$$\left[\frac{\partial^2}{\partial r^2} + \frac{1}{r}\frac{\partial}{\partial r} + \frac{1}{r^2}\frac{\partial^2}{\partial \theta^2}\right]\left[\frac{\partial^2 \Phi}{\partial r^2} + \frac{1}{r}\frac{\partial \Phi}{\partial r} + \frac{1}{r^2}\frac{\partial^2 \Phi}{\partial \theta^2}\right] = 0 \qquad (3.13.6)$$

Assume this unknown stress function Φ to be of the same form as the stress function given by Eq. (3.13.3), i.e., composed of a function $f_1(r)$ plus a function $f_2(r)$ multiplied by $\cos 2\theta$; thus,

$$\Phi(r, \theta) = f_1(r) + f_2(r)\cos 2\theta \qquad (3.13.7)$$

By substituting Eq. (3.13.7) into Eq. (3.13.6) one obtains

$$\left[\frac{d^2}{dr^2} + \frac{1}{r}\frac{d}{dr}\right]\left[\frac{d^2 f_1}{dr^2} + \frac{1}{r}\frac{df_1}{dr}\right]$$
$$+ \left[\frac{d^2}{dr^2} + \frac{1}{r}\frac{d}{dr} - \frac{4}{r^2}\right]\left[\frac{d^2 f_2}{dr^2} + \frac{1}{r}\frac{df_2}{dr} - \frac{4f_2}{r^2}\right]\cos 2\theta = 0$$

or

$$\left.\begin{array}{l} \left[\dfrac{d^2}{dr^2} + \dfrac{1}{r}\dfrac{d}{dr}\right]\left[\dfrac{d^2 f_1}{dr^2} + \dfrac{1}{r}\dfrac{df_1}{dr}\right] = 0 \\ \left[\dfrac{d^2}{dr^2} + \dfrac{1}{r}\dfrac{d}{dr} - \dfrac{4}{r^2}\right]\left[\dfrac{d^2 f_2}{dr^2} + \dfrac{1}{r}\dfrac{df_2}{dr} - \dfrac{4f_2}{r^2}\right] = 0 \end{array}\right\} \qquad (3.13.8)$$

The general solutions of Eqs. (3.13.8) are[18]

---

[18] The first of Eqs. (3.13.8) is the same as Eq. (3.8.4). It can be shown by substitution that the expression for $f_2(r)$ satisfies the second of Eqs. (3.13.8).

$$f_1(r) = C_1 r^2 \log r + C_2 r^2 + C_3 \log r + C_4$$
$$f_2(r) = C_5 r^2 + C_6 r^4 + \frac{C_7}{r^2} + C_8$$

and the stress function $\Phi(r, \theta)$ is

$$\Phi = [C_1 r^2 \log r + C_2 r^2 + C_3 \log r + C_4] + \left[C_5 r^2 + C_6 r^4 + \frac{C_7}{r^2} + C_8\right] \cos 2\theta$$

which corresponds to the following system of stresses:

$$\sigma_r = C_1(1 + 2\log r) + 2C_2 + \frac{C_3}{r^2} - \left(2C_5 + \frac{6C_7}{r^4} + \frac{4C_8}{r^2}\right) \cos 2\theta$$

$$\sigma_\theta = C_1(3 + 2\log r) + 2C_2 - \frac{C_3}{r^2} + \left(2C_5 + 12C_6 r^2 + \frac{6C_7}{r^4}\right) \cos 2\theta$$

$$\tau_{r\theta} = \left(2C_5 + 6C_6 r^2 - \frac{6C_7}{r^4} - \frac{2C_8}{r^2}\right) \sin 2\theta$$

where $C_1, C_2, C_3, C_5, C_6, C_7$, and C_8 are seven constants to be determined. Since, for $r \to \infty$, σ_r, σ_θ, and $\tau_{r\theta}$ must be finite, it follows from the above equations that $C_1 = C_6 = 0$. The remaining five constants are determined from the five boundary conditions given by Eqs. (3.13.5); thus,

$$2C_2 + \frac{C_3}{a^2} = 0$$

$$2C_5 + \frac{6C_7}{a^4} + \frac{4C_8}{a^2} = 0$$

$$2C_5 - \frac{6C_7}{a^4} - \frac{2C_8}{a^2} = 0$$

$$2C_5 = -\frac{S}{2} \quad 2C_2 = \frac{S}{2}$$

By solving the above equations, one obtains

$$C_2 = \frac{S}{4}, \quad C_3 = -\frac{a^2 S}{2}, \quad C_5 = -\frac{S}{4}, \quad C_7 = -\frac{a^4 S}{4}, \quad C_8 = \frac{a^2 S}{2}$$

and the components of stress are given by the following expressions:

$$\left.\begin{aligned}\sigma_r &= \frac{S}{2}\left[1 - \frac{a^2}{r^2}\right] + \frac{S}{2}\left[1 + \frac{3a^4}{r^4} - \frac{4a^2}{r^2}\right] \cos 2\theta \\ \sigma_\theta &= \frac{S}{2}\left[1 + \frac{a^2}{r^2}\right] - \frac{S}{2}\left[1 + \frac{3a^4}{r^4}\right] \cos 2\theta \\ \tau_{r\theta} &= -\frac{S}{2}\left[1 - \frac{3a^4}{r^4} + \frac{2a^2}{r^2}\right] \sin 2\theta\end{aligned}\right\} \quad (3.13.9)$$

From the second of Eqs. (3.13.9), one sees that

$$(\sigma_\theta)_{\max} = 3S \quad \text{for } \theta = \begin{cases} \dfrac{\pi}{2} \\ \dfrac{3\pi}{2} \end{cases}$$

Thus, due to the presence of the hole one has at the two points A and B (see Fig. 3.13.1), a stress concentration, i.e., the stress is increased three times its average value. This phenomenon is localized near the hole. In fact, from the second of Eqs. (3.13.9), one has

$$(\sigma_\theta)_{\theta=\pi/2, 3\pi/2} = \frac{S}{2}\left[2 + \frac{a^2}{r^2} + \frac{3a^4}{r^4}\right] \simeq S \quad \text{for } r > 10a$$

For $r = 10a$, terms containing r in the above expression, which represent the effect of the hole, have only $\frac{1}{200}$ of the value they had at the edge of the hole. One is justified in considering a distance of five diameters from the center of the hole as practically an infinite distance away.

In order to consider the problem of a small circular hole in a plate subjected to a uniform shearing stress of intensity S, one has to superimpose a uniform tensile stress S in the x-direction and a uniform compression stress S in the y-direction. The resultant stress is a shear S in the diagonal plane (see Fig. 3.13.2). The state of stress due to tension in the x-direction is given by Eq. (3.13.9). The state of stress due to compression in the y-direction is obtained from Eqs. (3.13.9) by introducing $-S$ instead of S and $\theta - \pi/2$ instead of θ; hence, one obtains the following components of stress:

$$\left.\begin{aligned}\sigma_r &= -\frac{S}{2}\left[1 - \frac{a^2}{r^2}\right] - \frac{S}{2}\left[1 + \frac{3a^4}{r^4} - \frac{4a^2}{r^2}\right]\cos 2\left(\theta - \frac{\pi}{2}\right) \\ \sigma_\theta &= -\frac{S}{2}\left[1 + \frac{a^2}{r^2}\right] + \frac{S}{2}\left[1 + \frac{3a^4}{r^4}\right]\cos 2\left(\theta - \frac{\pi}{2}\right) \\ \tau_{r\theta} &= \frac{S}{2}\left[1 - \frac{3a^4}{r^4} + \frac{2a^2}{r^2}\right]\sin 2\left(\theta - \frac{\pi}{2}\right)\end{aligned}\right\} \quad (3.13.10)$$

By superimposing Eqs. (3.13.9) and (3.13.10) one obtains

$$\left.\begin{aligned}\sigma_r &= S\left[1 + \frac{3a^4}{r^4} - \frac{4a^2}{r^2}\right]\cos 2\theta \\ \sigma_\theta &= -S\left[1 + \frac{3a^4}{r^4}\right]\cos 2\theta \\ \tau_{r\theta} &= -S\left[1 - \frac{3a^4}{r^4} + \frac{2a^2}{r^2}\right]\sin 2\theta\end{aligned}\right\} \quad (3.13.11)$$

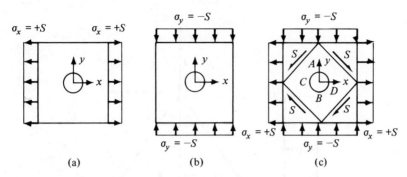

Figure 3.13.2

From the second of Eqs. (3.13.11) one sees that

$$(\sigma_\theta) = (\sigma_\theta)_{\min} = -4S \quad \text{for } \theta = 0, \pi$$

$$(\sigma_\theta) = (\sigma_\theta)_{\max} = +4S \quad \text{for } \theta = \frac{\pi}{2}, \frac{3\pi}{2}$$

or that at points A, B, C, and D (see Fig. 3.13.2), the concentration of stresses is four times the average stress in the plate.

3.14 Concentrated Load Acting on the Vertex of a Wedge (Michell's Problem)[19]

Consider the vertical concentrated load of intensity P acting on the vertex of the wedge of unit thickness shown in Fig. 3.14.1. For this case the stress function is assumed to be

$$\Phi(r, \theta) = CPr\theta \sin \theta \qquad (3.14.1)$$

where C is a constant to be determined from the boundary conditions. Equation (3.14.1) satisfies the compatibility equation:

$$\left[\frac{\partial^2}{\partial r^2} + \frac{1}{r}\frac{\partial}{\partial r} + \frac{1}{r^2}\frac{\partial^2}{\partial \theta^2}\right]\left[\frac{\partial^2 \Phi}{\partial r^2} + \frac{1}{r}\frac{\partial \Phi}{\partial r} + \frac{1}{r^2}\frac{\partial^2 \Phi}{\partial \theta^2}\right] = 0$$

The components of stress in this case are given by:

$$\left.\begin{aligned}\sigma_r &= \frac{1}{r}\frac{\partial \Phi}{\partial r} + \frac{1}{r^2}\frac{\partial^2 \Phi}{\partial \theta^2} = 2CP\frac{\cos \theta}{r} \\ \sigma_\theta &= \frac{\partial^2 \Phi}{\partial r^2} = 0 \\ \tau_{r\theta} &= \frac{1}{r^2}\frac{\partial \Phi}{\partial \theta} - \frac{1}{r}\frac{\partial^2 \Phi}{\partial \theta \partial r} = 0\end{aligned}\right\} \qquad (3.14.2)$$

[19] See Bibliography Sec. 3.14, No. 1 at the end of this chapter.

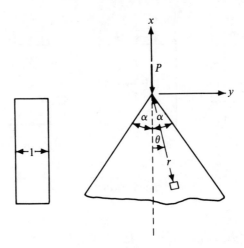

Figure 3.14.1

The boundary conditions which must be satisfied are as follows:

$$\left.\begin{array}{c} \text{for } \theta = \pm\alpha \quad \sigma_\theta = \tau_{r\theta} = 0 \\ 2\int_0^\alpha \sigma_r r \cos\theta\, d\theta = -P \end{array}\right\} \quad (3.14.3)$$

The first two boundary conditions are identically satisfied by the last two of Eqs. (3.14.2). The third of Eqs. (3.14.3) determines the constant C. In fact, by substituting into this equation the expression for σ_r, one obtains

$$-P = 2\int_0^\alpha \sigma_r r \cos\theta\, d\theta = 4CP\int_0^\alpha \cos^2\theta\, d\theta = CP(2\alpha + \sin 2\alpha)$$

from which

$$C = -\frac{1}{2\alpha \sin 2\alpha}$$

By substituting this value of C into the first of Eqs. (3.14.2), one gets for the components of stress, the values given by Michell; thus,

$$\left.\begin{array}{c} \sigma_r = -\frac{2P}{2\alpha + \sin 2\alpha}\frac{\cos\theta}{r} \\ \sigma_\theta = 0 \\ \tau_{r\theta} = 0 \end{array}\right\} \quad (3.14.4)$$

By letting $\alpha = \pi/2$ in Eqs. (3.13.4), the Flamant solution for the case of a vertical, concentrated force of intensity P acting on the free surface of a

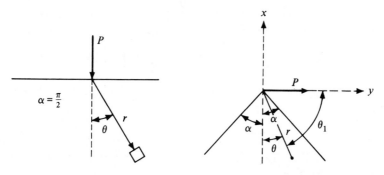

Figure 3.14.2 **Figure 3.14.3**

semi-infinite plate (see Fig. 3.14.2) is obtained. The components of stress are in this case

$$\left.\begin{array}{l}\sigma_r = -\dfrac{2P}{\pi}\dfrac{\cos\theta}{r} \\ \sigma_\theta = 0 \\ \tau_{r\theta} = 0\end{array}\right\} \quad (3.14.5)$$

To study the case of a horizontal, concentrated force of intensity P acting on the vertex of the wedge of unit thickness, shown in Fig. 3.14.3, one assumes the stress function to be

$$\Phi = CPr\theta_1 \sin\theta_1$$

which corresponds to the following stresses:

$$\left.\begin{array}{l}\sigma_r = \dfrac{2CP\cos\theta_1}{r} \\ \sigma_\theta = \tau_{r\theta} = 0\end{array}\right\} \quad (3.14.6)$$

Constant C is determined from the condition that

$$P = \int_{90-\alpha}^{90+\alpha} \sigma_r r \cos\theta_1 d\theta_1 = 2CP\int_{90-\alpha}^{90+\alpha} \cos^2\theta_1\, d\theta_1 = CP\left[\theta_1 + \frac{\sin 2\theta_1}{2}\right]_{90-\alpha}^{90+\alpha}$$
$$= [2\alpha - \sin 2\alpha]CP$$

Thus,

$$C = \frac{1}{2\alpha - \sin 2\alpha}$$

and

$$\sigma_r = \frac{2P}{2\alpha - \sin 2\alpha}\frac{\cos\theta_1}{r}$$

By replacing θ_1 with $90 - \theta$ in Eqs. (3.14.6) one obtains the stress components for the case of a horizontal force; thus,

$$\left. \begin{array}{l} \sigma_r = \dfrac{2P}{2\alpha - \sin 2\alpha} \dfrac{\sin \theta}{r} \\ \sigma_\theta = 0 \\ \tau_{r\theta} = 0 \end{array} \right\} \quad (3.14.7)$$

By letting $\alpha = \pi/2$, the Flamant solution for the case of horizontal, concentrated force of intensity P acting on the free surface of a semi-infinite plate is obtained in the form

$$\left. \begin{array}{l} \sigma_r = \dfrac{2P}{\pi} \dfrac{\sin \theta}{r} \\ \sigma_\theta = 0 \\ \tau_{r\theta} = 0 \end{array} \right\} \quad (3.14.8)$$

Now consider the force shown in Fig. 3.14.4 inclined at the angle β to the vertical. Its horizontal component will be $P \sin \beta$, while its vertical component will be $-P \cos \beta$. By superimposing the effects of the two forces one obtains for the components of stresses

$$\left. \begin{array}{l} \sigma_r = \dfrac{2P \cos \beta}{2\alpha + \sin 2\alpha} \dfrac{\cos \theta}{r} + \dfrac{2P \sin \beta}{2\alpha - \sin 2\alpha} \dfrac{\sin \theta}{r} \\ \sigma_\theta = 0 \\ \tau_{r\theta} = 0 \end{array} \right\} \quad (3.14.9)$$

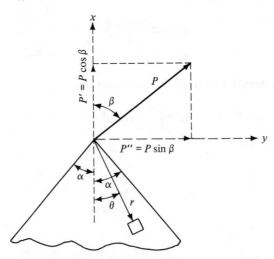

Volterra & Gaines 3.14.4

Sec. 3.14   TWO-DIMENSIONAL ELASTICITY

Example 3.14.1. By the use of Eqs. (3.14.7) compare the results given by the mathematical theory of elasticity with the results given by the strength of materials in the case of the triangular wedge subjected to the concentrated load shown in Fig. 3.14.5.

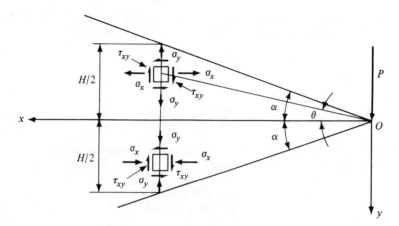

Figure 3.14.5

The rectangular stress components acting on the wedge are

$$\left.\begin{array}{l}\sigma_x = -\dfrac{2P}{(2\alpha - \sin 2\alpha)} \dfrac{x^2 y}{(x^2 + y^2)^2} \\[4pt] \sigma_y = -\dfrac{2P}{(2\alpha - \sin 2\alpha)} \dfrac{y^3}{(x^2 + y^2)^2} \\[4pt] \tau_{xy} = -\dfrac{2P}{(2\alpha - \sin 2\alpha)} \dfrac{xy^2}{(x^2 + y^2)^2}\end{array}\right\} \qquad (3.14.10)[20]$$

From Eqs. (3.14.10) it follows that across a transverse section of the wedge (for which $x =$ constant)[21]

---

[20] Equations (3.14.10) are obtained directly from Eqs. (3.14.7) through the formulas

$$\sigma_x = \sigma_r \cos^2 \theta, \quad \sigma_y = \sigma_r \sin^2 \theta, \quad \tau_{xy} = \sigma_r \sin \theta \cos \theta$$

$$\sin \theta = \frac{y}{r}, \quad \cos \theta = \frac{x}{r}, \quad r = \sqrt{x^2 + y^2}$$

[21] In fact, from Eqs. (3.14.10) it follows that

$$\frac{\partial \sigma_x}{\partial y} = -\frac{2P}{(2\alpha - \sin 2\alpha)} \frac{x^2}{(x^2 + y^2)^3} [(x^2 + y^2) - 4y^2] = 0$$

when $(x^2 + y^2) = r^2 = 4y^2$ or when $y/r = \sin \theta = \pm\frac{1}{2}$, i.e., when $\theta = \pm 30°$. Also

$$\frac{\partial \sigma_y}{\partial y} = -\frac{2P}{(2\alpha - \sin 2\alpha)} \frac{y^2[3(x^2 + y^2) - 4y^2]}{(x^2 + y^2)^3} = 0$$

σ_x is a maximum for $\theta = \pm 30°$

σ_y is a maximum for $\theta = \pm 60°$

τ_{xy} is a maximum for $\theta = \pm 45°$

By introducing the moment of inertia $I = H^3/12$ and the bending moment $M(x) = Px$, for a section located at distance x from vertex O, one obtains for σ_x

$$\sigma_x = -K\frac{My}{I}$$

where

$$K = \frac{4}{3}\frac{\tan^3 \alpha \cos^4 \theta}{(2\alpha - \sin 2\alpha)} \qquad (3.14.11)$$

is the correction factor to the well-known flexure formula $\sigma_x = My/I$ given by the strength of materials.[22] On the extreme fibers of the wedge where $\theta = \pm \alpha$

$$K = \frac{4}{3}\frac{\sin^3 \alpha \cos \alpha}{[2\alpha - \sin 2\alpha]} \qquad (3.14.12)$$

and for small values of α, K is approximately equal to unity.[23]

The lateral stress σ_y given by the second of Eqs. (3.14.10) is ignored in the elementary theory. The last of Eqs. (3.14.10) can be written

$$\tau_{xy} = -K\frac{Py^2}{I} \qquad (3.14.13)$$

where the correction factor K is given by Eq. (3.14.11). On the extreme fibers of the wedge where $\theta = \pm \alpha$ and $y = \pm H/2$ Eq. (3.14.13) becomes

when $3(x^2 + y^2) = 3r^2 = 4y^2$ or when $y/r = \sin \theta = \pm\frac{\sqrt{3}}{2}$, i.e., when $\theta = \pm 60°$. Finally

$$\frac{\partial \tau_{xy}}{\partial y} = -\frac{4P}{(2\alpha - \sin 2\alpha)}\frac{xy[(x^2 + y^2) - 2y^2]}{(x^2 + y^2)^3} = 0$$

when $(x^2 + y^2) = r^2 = 2y^2$ or when $y/r = \sin \theta = \pm\frac{1}{\sqrt{2}}$, i.e., when $\theta = 45°$.

[22] See Sec. 5.1.
[23] In fact, since

$$\sin 2\alpha = 2\alpha - \frac{(2\alpha)^3}{3!} + \frac{((2\alpha)^5}{5!} - \cdots$$

it follows that, by considering only the first two terms of the series, one obtains

$$3(2\alpha - \sin 2\alpha) = 4\alpha^3$$

For small values of α, $\sin^3 \alpha \doteq \alpha^3$ and $\cos \alpha \doteq 1$ and Eq. (3.14.11) becomes

$$K = \frac{4}{3}\frac{\sin^3 \alpha \cos \alpha}{[2\alpha - \sin 2\alpha]} \doteq \frac{4\alpha^3}{4\alpha^3} \doteq 1$$

$$\tau_{xy} = -K\frac{PH^2}{4I} \qquad (3.14.14)$$

where K in this case is given by Eq. (3.14.12). According to the strength of materials the shear stress is given by the formula[24]

$$\tau_{xy} = \frac{VQ}{Ib}$$

which in our notation reaches its maximum value

$$\tau_{xy} = \frac{PH^2}{8I} \qquad (3.14.15)$$

at the neutral axis. Comparison of Eqs. (3.14.14) and (3.14.15) shows that in the case of small angles α, when the correction factor K is close to unity, the numerical value for the maximum shear stress given by the Mathematical Theory of Elasticity is twice as great as that given by the strength of materials and occurs at the extreme fibers instead of at the neutral axis.

3.15 Concentrated Load Acting on the Free Surface of a Plate (Flamant's Problem)

In the case of a concentrated, vertical force acting on the free surface of a plate (see Fig. 3.15.1), the following expressions were found for stresses [see Eqs. (3.14.5)]:

$$\left. \begin{array}{l} \sigma_r = -\dfrac{2P}{\pi r}\cos\theta \\ \sigma_\theta = 0 \\ \tau_{r\theta} = 0 \end{array} \right\} \qquad (3.15.1)$$

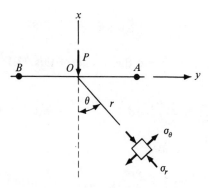

Figure 3.15.1

---

[24] See Sec. 5.1.

Components of the elastic displacement u and v in the r-and θ-directions will now be found starting from the Flamant solution, i.e., Eqs. (3.15.1).

From the equations

$$\left.\begin{aligned}\epsilon_r &= \frac{1}{E}(\sigma_r - v\sigma_\theta) \\ \epsilon_\theta &= \frac{1}{E}(\sigma_\theta - v\sigma_r) \\ \gamma_{r\theta} &= \frac{1}{G}\tau_{r\theta}\end{aligned}\right\} \qquad (3.15.2)$$

and

$$\left.\begin{aligned}\epsilon_r &= \frac{\partial u}{\partial r} \\ \epsilon_\theta &= \frac{u}{r} + \frac{1}{r}\frac{\partial v}{\partial \theta} \\ \gamma_{r\theta} &= \frac{1}{r}\frac{\partial u}{\partial \theta} + \frac{\partial v}{\partial r} - \frac{v}{r}\end{aligned}\right\} \qquad (3.15.3)$$

and in view of Eqs. (3.15.1) one obtains

$$\left.\begin{aligned}\frac{\partial u}{\partial r} &= -\frac{2P\cos\theta}{\pi E\, r} \\ \frac{u}{r} + \frac{1}{r}\frac{\partial v}{\partial \theta} &= \frac{2vP\cos\theta}{\pi E\, r} \\ \frac{1}{r}\frac{\partial u}{\partial \theta} + \frac{\partial v}{\partial r} - \frac{v}{r} &= 0\end{aligned}\right\} \qquad (3.15.4)$$

By integrating the first of Eqs. (3.15.4) one finds that

$$u = -\frac{2P}{\pi E}\cos\theta \log r + f(\theta) \qquad (3.15.5)$$

By substituting Eq. (3.15.5) into the second of Eqs. (3.15.4) and multiplying by r one obtains

$$\frac{\partial v}{\partial \theta} = \frac{2vP}{\pi E}\cos\theta + \frac{2P}{\pi E}\log r \cos\theta - f(\theta)$$

By integrating it is found that

$$v = \frac{2vP}{\pi E}\sin\theta + \frac{2P}{\pi E}\log r \sin\theta - \int f(\theta)d\theta + F(r) \qquad (3.15.6)$$

From Eqs. (3.15.5) and (3.15.6) one obtains

$$\left.\begin{array}{l}\dfrac{\partial u}{\partial \theta}=\dfrac{2P}{\pi E}\log r \sin \theta + \dfrac{d}{d\theta}f(\theta)\\[2mm] \dfrac{\partial v}{\partial r}=\dfrac{2P}{\pi E}\dfrac{\sin \theta}{r}+\dfrac{d}{dr}F(r)\end{array}\right\} \quad (3.15.7)$$

By substituting Eqs. (3.15.6) and (3.15.7) into the last of Eqs. (3.15.4), one obtains the equation

$$\dfrac{2P}{\pi E}\sin\theta(1-v)+\dfrac{d}{d\theta}f(\theta)+\int f(\theta)d\theta+r\dfrac{d}{dr}F(r)-F(r)=0$$

The above equation yields two equations, one a function of variable θ

$$\dfrac{2P}{\pi E}(1-v)\sin\theta+\dfrac{d}{d\theta}f(\theta)+\int f(\theta)d\theta=0 \quad (3.15.8)$$

and one a function of variable r

$$r\dfrac{d}{dr}F(r)-F(r)=0 \quad (3.15.9)$$

By integrating Eq. (3.15.8) one obtains

$$f(\theta)=-\dfrac{(1-v)P}{\pi E}\theta\sin\theta+A\sin\theta+B\cos\theta \quad (3.15.10)$$

By integrating Eq. (3.1.5.9) one obtains

$$F(r)=Cr \quad (3.15.11)$$

A, B, and C are constants of integration which will be shown to be zero. Equations (3.15.5) and (3.15.6) can now be written as

$$\left.\begin{array}{l}u=-\dfrac{2P}{\pi E}\cos\theta\log r-\dfrac{(1-v)P}{\pi E}\theta\sin\theta+A\sin\theta+B\cos\theta\\[2mm] v=-\dfrac{(1-v)}{\pi E}P\theta\cos\theta+\dfrac{(1+v)}{\pi E}P\sin\theta+\dfrac{2P}{\pi E}\log r\sin\theta\\[2mm] \quad+A\cos\theta-B\sin\theta+Cr\end{array}\right\} \quad (3.15.12)$$

Since $v=0$ for $\theta=0$ and for any value of r, it follows from the second of Eqs. (3.15.12) that

$$A+Cr=0 \quad \text{or} \quad A=C=0$$

It can be shown that constant B represents a rigid displacement of every point of the plate in the x-direction and can therefore be assumed to be equal to zero. In fact, if one lets

$$u' = B \cos \theta$$
$$v' = -B \sin \theta$$

then the resultant is

$$\sqrt{(u')^2 + (v')^2} = \sqrt{(B \cos \theta)^2 + (-B \sin \theta)^2} = B$$

and its direction is parallel to the x-direction. From Eqs. (3.15.12), for the components of the elastic displacement u and v, it follows that

$$\left. \begin{aligned} u &= \frac{P}{\pi E}[-(1 - v)\theta \sin \theta - 2 \cos \theta \log r] \\ v &= \frac{P}{\pi E}[-(1 - v)\theta \cos \theta + (1 + v) \sin \theta + 2 \sin \theta \log r] \end{aligned} \right\} \quad (3.15.13)$$

Displacements on the free straight boundary (i.e., on the y-axis) can be derived from Eqs. (3.15.13). The radial displacement u is obtained by letting $\theta = \pm \pi/2$ in the first of Eqs. (3.15.13); thus,

$$(u)_{\theta = \pm \pi/2} = -\frac{(1 - v)P}{2E} \quad (3.15.14)$$

Since $v < 1$, Eq. (3.15.14) shows that the displacement u is always negative, i.e., every point A or B on the free boundary (see Fig. 3.15.1) moves with a translatory motion toward the origin O. The vertical displacement v is obtained by letting $\theta = \pi/2$ (or $\theta = -\pi/2$) in the second of Eqs. (3.15.13); thus,

$$\left. \begin{aligned} (v)_{\theta = \pi/2} &= \frac{P}{\pi E}[(1 + v) + 2 \log r] \quad \text{on } OA \\ (v)_{\theta = -\pi/2} &= -\frac{P}{\pi E}[(1 + v) + 2 \log r] \quad \text{on } OB \end{aligned} \right\} \quad (3.15.15)$$

(see Fig. 3.15.2). At the origin ($r = 0$) Eqs. (3.15.15) give infinitely large displacements v, while for other points of the free boundary the displacements are finite. This corresponds to physical reality since plastic deformation occurs under the concentrated load.

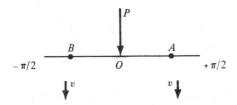

Figure 3.15.2

Sec. 3.15 TWO-DIMENSIONAL ELASTICITY

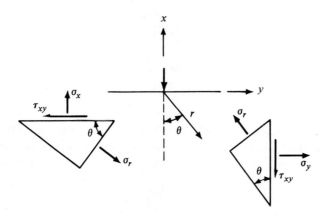

Figure 3.15.3

Equations (3.15.1) can be transformed from polar (r, θ) to cartesian (x, y) coordinates (see Fig. 3.15.3); thus, since

$$\left. \begin{aligned} \sigma_x &= \sigma_r \cos^2 \theta = -\frac{2P \cos^3 \theta}{\pi r} \\ \sigma_y &= \sigma_r \sin^2 \theta = -\frac{2P \sin^2 \theta \cos \theta}{\pi r} \\ \tau_{xy} &= \sigma_r \sin \theta \cos \theta = -\frac{2P \sin \theta \cos^2 \theta}{\pi r} \\ \cos \theta &= \frac{x}{r} \quad \sin \theta = \frac{y}{r} \quad r = \sqrt{x^2 + y^2} \end{aligned} \right\} \quad (3.15.16)$$

one obtains the following expressions for the components of stress σ_x, σ_y, and τ_{xy} in cartesian coordinates (see Fig. 3.15.4):

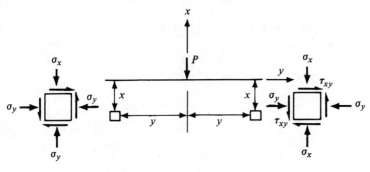

Figure 3.15.4

$$\left.\begin{array}{l}\sigma_x = -\dfrac{2Px^3}{\pi(x^2+y^2)^2} \\[6pt] \sigma_y = -\dfrac{2Pxy^2}{\pi(x^2+y^2)^2} \\[6pt] \tau_{xy} = -\dfrac{2Px^2y}{\pi(x^2+y^2)^2}\end{array}\right\} \quad (3.15.17)$$

Suppose now that a uniformly distributed load of intensity p_0 is applied on the free surface (see Fig. 3.15.5). Let $dP = p_0 dy$ denote the load acting on an infinitesimal length dy of the free surface. From Fig. 3.15.6 one has

$$dy = \frac{r d\theta}{\cos \theta}$$

so that

$$p_0 dy = dP = \frac{p_0 r d\theta}{\cos \theta}$$

and Eqs. (3.15.16) become

$$\left.\begin{array}{l}d\sigma_x = -\dfrac{2p_0}{\pi} \cos^2 \theta \, d\theta \\[6pt] d\sigma_y = -\dfrac{2p_0}{\pi} \sin^2 \theta \, d\theta \\[6pt] d\tau_{xy} = -\dfrac{p_0}{\pi} \sin 2\theta \, d\theta\end{array}\right\} \quad (3.15.18)$$

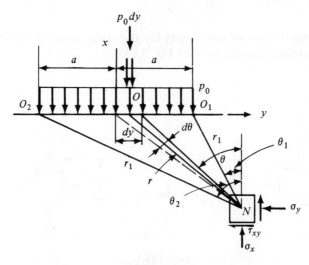

Figure 3.15.5

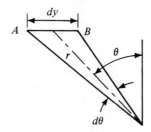

Figure 3.15.6

and the components of stress at a generic point are obtained by integration of Eqs. (3.15.18); hence,

$$\sigma_x = -\frac{2p_0}{\pi}\int_{\theta_1}^{\theta_2}\cos^2\theta\,d\theta = -\frac{p_0}{2\pi}[2(\theta_2-\theta_1)+(\sin 2\theta_2-\sin 2\theta_1)]$$

$$\sigma_y = -\frac{2p_0}{\pi}\int_{\theta_1}^{\theta_2}\sin^2\theta\,d\theta = -\frac{p_0}{2\pi}[2(\theta_2-\theta_1)-(\sin 2\theta_2-\sin 2\theta_1)]$$

$$\tau_{xy} = -\frac{p_0}{\pi}\int_{\theta_1}^{\theta_2}\sin 2\theta\,d\theta = -\frac{p_0}{2\pi}[\cos 2\theta_1-\cos 2\theta_2]$$

3.16 Moment Acting on the Vertex of a Wedge (Inglis' Problem)

Assume that the moment M is acting on the vertex of the wedge shown in Fig. 3.16.1. This problem was solved by Sir Charles Inglis by assuming the following stress function:

$$\Phi = -\frac{M(\sin 2\theta - 2\theta\cos 2\alpha)}{2(\sin 2\alpha - 2\alpha\cos 2\alpha)} \tag{3.16.1}$$

This is the unique solution of the problem since Eq. (3.16.1) satisfies the compatibility equation [Eq. (3.8.1)] and the components of stress

$$\left.\begin{array}{l}\sigma_r = \dfrac{1}{r}\dfrac{\partial\Phi}{\partial r}+\dfrac{1}{r^2}\dfrac{\partial^2\Phi}{\partial\theta^2}=\dfrac{M}{2(\sin 2\alpha-2\alpha\cos 2\alpha)}\dfrac{4}{r^2}\sin 2\theta \\[6pt] \sigma_\theta = \dfrac{\partial^2\Phi}{\partial r^2}=0 \\[6pt] \tau_{r\theta} = \dfrac{1}{r^2}\dfrac{\partial\Phi}{\partial\theta}-\dfrac{1}{r}\dfrac{\partial^2\Phi}{\partial r\,\partial\theta}=\dfrac{M}{2(\sin 2\alpha-2\alpha\cos 2\alpha)}\dfrac{2}{r^2}(\cos 2\theta-\cos 2\alpha)\end{array}\right\} \tag{3.16.2}$$

satisfy the boundary conditions that

$$\sigma_\theta = \tau_{r\theta} = 0 \quad \text{for} \quad \theta = \pm\alpha$$

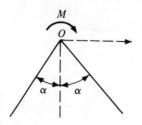

Figure 3.16.1

For $\alpha = \pi/2$ Eqs. (3.16.2) become

$$\left. \begin{array}{l} \sigma_r = \dfrac{2M \sin 2\theta}{\pi r^2} \\[2mm] \sigma_\theta = 0 \\[2mm] \tau_{r\theta} = \dfrac{2M \cos^2 2\theta}{\pi r^2} \end{array} \right\} \quad (3.16.3)$$

Example 3.16.1. Obtain Eqs. (3.16.3) directly from Flamant's solution discussed in Sec. 3.15.

From Flamant's formulas [Eq. (3.15.1)], we obtain by superposition the results due to the two forces P shown in Fig. 3.16.2; thus

$$\left. \begin{array}{l} \sigma_r = \dfrac{2P}{\pi}\left[-\dfrac{\cos\theta}{r} + \dfrac{\cos\theta_1}{r_1}\cos^2\alpha\right] \\[2mm] \sigma_\theta = \dfrac{2P\cos\theta_1}{\pi r_1}\sin^2\alpha \\[2mm] \tau_{r\theta} = \dfrac{P}{\pi}\dfrac{\cos\theta_1}{r_1}\sin 2\alpha \end{array} \right\} \quad (3.16.4)$$

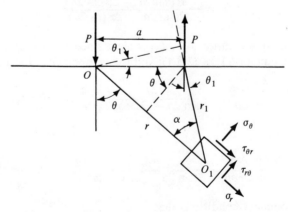

Figure 3.16.2

Assume now that in the expression for the moment $M = Pa$ produced by the couple, a is getting smaller and smaller while P is getting proportionally greater so that their product remains constant. This will in the limit correspond to a couple acting on an infinitesimal area around point O. From Fig. 3.16.2, the following relationships are derived:

$$r_1 \cos \theta_1 = r \cos \theta$$
$$r \cos \alpha = r_1 + a \sin \theta_1$$
$$r_1 \sin \alpha = a \cos \theta$$

Equations (3.16.4) may be rewritten as

$$\left. \begin{array}{l} \sigma_r = \dfrac{2P \cos \theta}{\pi r}\left[-1 + \dfrac{r^2}{r_1^2}\cos^2 \alpha\right] = \dfrac{2Pa}{\pi r}\cos\theta\left[\dfrac{2\sin\theta_1}{r_1} + \dfrac{a\sin^2\theta_1}{r_1^2}\right] \\[2mm] \sigma_\theta = \dfrac{2Pa^2 \cos^3 \theta_1}{\pi r_1 r^2} \\[2mm] \tau_{r\theta} = \dfrac{2P \cos \theta_1 \cos \alpha \sin \alpha}{\pi r_1} \end{array} \right\} \quad (3.16.5)$$

When a approaches the limit zero then $\alpha \to 0$, $Pa \to M$, $r_1 \to r$, $\theta \to \theta_1$, and $r_1 \sin \alpha = a \cos \theta$ and Eqs. (3.16.5) reduce to Eqs. (3.16.3) of the Inglis' solution.

3.17 Disk Subjected to Two Opposite Concentrated Forces (Hertz's Problem)

In order to study the problem of the circular disk of diameter D subjected to two diametral loads P (see Fig. 3.17.1), consider first the disk of diameter D subjected to the uniform radial tension of intensity σ_n as shown in Fig. 3.17.2. The stress function for this case is assumed to be

$$\Phi = Cr^2 \quad (3.17.1)$$

C being an arbitrary constant to be determined. Equation (3.17.1) satisfies the compatibility equation [Eq. (3.8.1)]. In view of Eqs. (3.8.3), the stress components are

$$\sigma_r = 2C$$
$$\sigma_\theta = 2C$$
$$\tau_{r\theta} = 0$$

From the boundary condition that $\sigma_r = \sigma_n$ at $r = D/2$, one obtains $C = \sigma_n/2$; thus, the stress components can be written as

178 TWO-DIMENSIONAL ELASTICITY Sec. 3.17

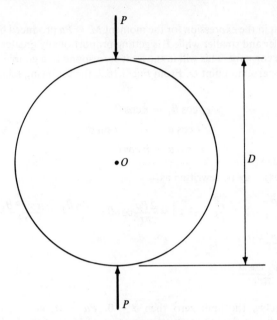

Figure 3.17.1

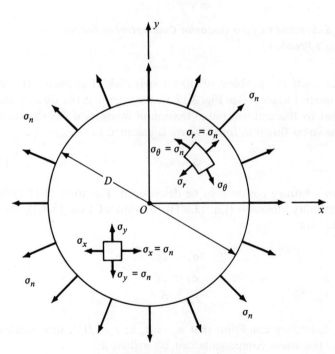

Figure 3.17.2

$$\left.\begin{array}{l}\sigma_r = \sigma_n \\ \sigma_\theta = \sigma_n \\ \tau_{r\theta} = 0\end{array}\right\} \quad (3.17.2)$$

From Eqs. (3.2.2) and (3.2.4) it follows that $\sigma_x = \sigma_y = \sigma_n$ and $\tau_{xy} = 0$ (see Fig. 3.17.2).

Consider now the Flamant solution for the concentrated load of intensity P acting on the surface of a plate. The stress distribution is given by Eqs. (3.15.1), i.e.,

$$\left.\begin{array}{l}\sigma_r = -\dfrac{2P}{\pi r}\cos\theta \\ \sigma_\theta = 0 \\ \tau_{r\theta} = 0\end{array}\right\} \quad (3.15.1)$$

On the circle of diameter D, shown in Fig. 3.17.3, $r/\cos\theta = D$; thus, it follows from the first of Eqs. (3.15.1) that

$$\sigma_r = -\frac{2P}{\pi D}$$

which is constant at each point of the circle. To isolate the disk of diameter D inside the plate, a stress q must be applied at each point of the surface of the disk to equilibrate the stress σ_r (see Fig. 3.17.4). It follows that $q\,\overline{AB}$

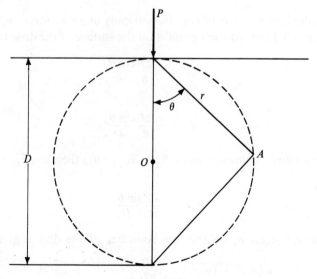

Figure 3.17.3

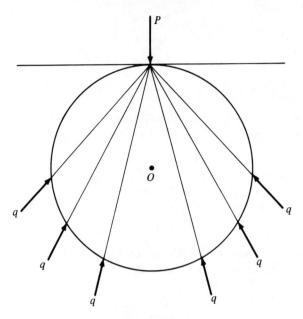

Figure 3.17.4

$= \sigma_r \overline{AC}$ (see Fig. 3.17.5). Since OA is perpendicular to AB and O_1A is perpendicular to AC, angle $CAB = \theta$ and

$$q = \sigma_r \frac{AC}{AB} = \sigma_r \cos\theta = -\frac{2P}{\pi D}\cos\theta$$

Consider now the case of two diametrically opposite forces acting on the disk (see Fig. 3.17.6). At each point A on the surface of the disk, two stresses

$$q_1 = -\frac{2P}{\pi}\frac{\cos\theta_1}{D}$$

and

$$q_2 = -\frac{2P}{\pi}\frac{\cos\theta_2}{D}$$

must be applied. However, since $\theta_2 = 90° - \theta_1$, then $\cos\theta_2 = \sin\theta_1$ and thus,

$$q_2 = -\frac{2P}{\pi}\frac{\sin\theta_1}{D}$$

The resultant stress σ_n on the free boundary of the disk is given by

$$\sigma_n = -\sqrt{(q_1)^2 + (q_2)^2} = -\frac{2P}{\pi D}\sqrt{\cos^2\theta_1 + \sin^2\theta_1} = -\frac{2P}{\pi D}$$

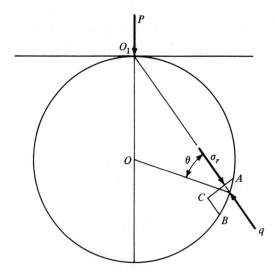

Figure 3.17.5

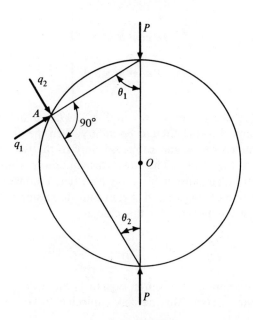

Figure 3.17.6

which is a constant radial compression of magnitude $\sigma_n = -2P/\pi D$. Thus, the disk is subjected to two concentrated forces P and the distributed force σ_n as shown in Fig. 3.17.7. The state of stress in the disk is the resultant of the following states of stress [see Fig. 3.17.8(a)]:

$$\sigma'_r = -\frac{2P}{\pi}\frac{\cos\theta_1}{r_1}, \qquad \sigma'_\theta = \tau'_{r\theta} = 0 \tag{3.17.3}$$

and

$$\sigma''_r = -\frac{2P}{\pi}\frac{\cos\theta_2}{r_2}, \qquad \sigma''_\theta = \tau''_{r\theta} = 0 \tag{3.17.4}$$

The rectangular components for the above states of stress are, respectively [see Fig. 3.17.8(b)],

$$\left.\begin{aligned}\sigma'_x &= -\frac{2P}{\pi}\frac{\cos\theta_1 \sin^2\theta_1}{r_1} \\ \sigma'_y &= -\frac{2P}{\pi}\frac{\cos^3\theta_1}{r_1} \\ \tau'_{xy} &= \frac{2P}{\pi}\frac{\cos^2\theta_1 \sin\theta_1}{r_1}\end{aligned}\right\} \tag{3.17.5}$$

and

$$\left.\begin{aligned}\sigma''_x &= -\frac{2P}{\pi}\frac{\cos\theta_2 \sin^2\theta_2}{r_2} \\ \sigma''_y &= -\frac{2P}{\pi}\frac{\cos^3\theta_2}{r_2} \\ \tau''_{xy} &= -\frac{2P}{\pi}\frac{\cos^2\theta_2 \sin\theta_2}{r_2}\end{aligned}\right\} \tag{3.17.6}$$

If, in addition to the forces shown in Fig. 3.17.7, a uniform normal tension $2P/\pi D$ is applied to the circular boundary, then the disk will be free of externally distributed forces and subjected only to the two concentrated forces P as shown in Fig. 3.17.1. The stress distribution in a disk subjected to a uniformly distributed normal force on its circular boundary was discussed at the beginning of this section. From that discussion it follows that the rectangular stress components for this case are

$$\sigma'''_x = \frac{2P}{\pi D} \qquad \sigma'''_y = \frac{2P}{\pi D} \qquad \tau'''_{xy} = 0 \tag{3.17.7}$$

By adding the stress components given in Eqs. (3.17.5), (3.17.6), and (3.17.7), one obtains the state of stress for the disk subjected to two opposite concentrated forces; thus,

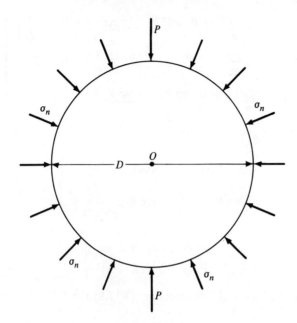

Figure 3.17.7

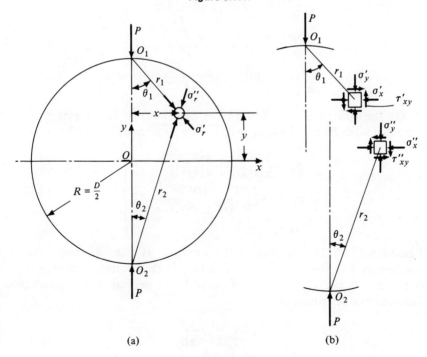

Figure 3.17.8

$$\left.\begin{array}{l}\sigma_x = -\dfrac{2P}{\pi}\left[\dfrac{\cos\theta_1 \sin^2\theta_1}{r_1} + \dfrac{\cos\theta_2 \sin^2\theta_2}{r_2} - \dfrac{1}{D}\right] \\[6pt] \sigma_y = -\dfrac{2P}{\pi}\left[\dfrac{\cos^3\theta_1}{r_1} + \dfrac{\cos^3\theta_2}{r_2} - \dfrac{1}{D}\right] \\[6pt] \tau_{xy} = \dfrac{2P}{\pi}\left[\dfrac{\cos^2\theta_1 \sin\theta_1}{r_1} - \dfrac{\cos^2\theta_2 \sin\theta_2}{r_2}\right]\end{array}\right\} \quad (3.17.8)$$

From Fig. 3.17.8(a) one finds that

$$\left.\begin{array}{l}\sin\theta_1 = \dfrac{x}{r_1}, \quad \cos\theta_1 = \dfrac{R-y}{r_1} \\[6pt] \sin\theta_2 = \dfrac{x}{r_2}, \quad \cos\theta_2 = \dfrac{R+y}{r_2}\end{array}\right\} \quad (3.17.9)$$

where

$$\left.\begin{array}{l}r_1^2 = x^2 + (R-y)^2 \\ r_2^2 = x^2 + (R+y)^2\end{array}\right\} \quad (3.17.10)$$

By introducing Eqs. (3.17.9) into Eqs. (3.17.8) the following equations are derived:

$$\left.\begin{array}{l}\sigma_x = -\dfrac{2P}{\pi}\left[\dfrac{(R-y)x^2}{r_1^4} + \dfrac{(R+y)x^2}{r_2^4} - \dfrac{1}{D}\right] \\[6pt] \sigma_y = -\dfrac{2P}{\pi}\left[\dfrac{(R-y)^3}{r_1^4} + \dfrac{(R+y)^3}{r_2^4} - \dfrac{1}{D}\right] \\[6pt] \tau_{xy} = \dfrac{2P}{\pi}\left[\dfrac{(R-y)^2 x}{r_1^4} - \dfrac{(R+y)^2 x}{r_2^4}\right]\end{array}\right\} \quad (3.17.11)$$

On the x-axis (i.e., on the diameter perpendicular to the external, concentrated loads P) $y = 0$ and $r_1 = r_2 = \sqrt{x^2 + R^2}$ and Eqs. (3.17.11) become

$$\left.\begin{array}{l}\sigma_x = \dfrac{2P}{\pi D}\left[\dfrac{D^2 - 4x^2}{D^2 + 4x^2}\right]^2 \\[6pt] \sigma_y = -\dfrac{2P}{\pi D}\left[\dfrac{4D^4}{(D^2 + 4x^2)^2} - 1\right] \\[6pt] \tau_{xy} = 0\end{array}\right\} \quad (3.17.12)$$

Equations (3.17.12) show that on the x-axis, σ_x is a tensile stress while σ_y is a compressive stress. For $x = \pm D/2$, i.e., on the circumference, $\sigma_x = \sigma_y = 0$. For $x = 0$, i.e., at the center of the disk, the normal stresses reach their maximum values which are

$$(\sigma_x)_{\substack{x=0 \\ y=0}} = \dfrac{2P}{\pi D}$$

$$(\sigma_y)_{\substack{x=0 \\ y=0}} = -\dfrac{6P}{\pi D}$$

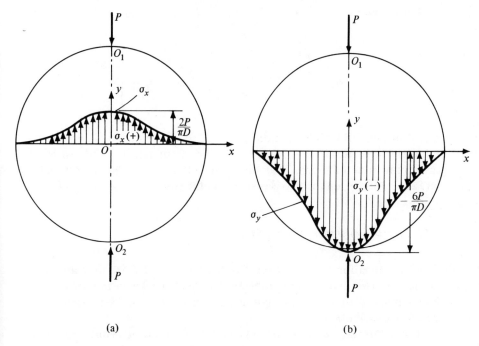

(a)   (b)

Figure 3.17.9

The maximum value of σ_y is approximately twice the average compression value $-P/D$. The distribution of stresses σ_x and σ_y on the x-axis is shown in the two graphs of Fig. 3.17.9.

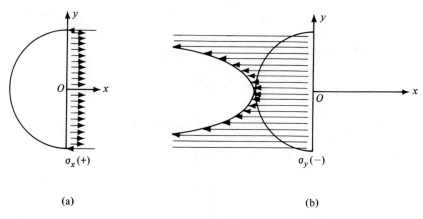

(a)   (b)

Figure 3.17.10

On the y-axis, $x = 0$, $r_1 = R - y$, $r_2 = R + y$ and Eqs. (3.17.11) become

$$\left. \begin{array}{l} \sigma_x = \dfrac{2P}{\pi D} \\[6pt] \sigma_y = -\dfrac{2P}{\pi}\left[\dfrac{2}{D - 2y} + \dfrac{2}{D + 2y} - \dfrac{1}{D}\right] \\[6pt] \tau_{xy} = 0 \end{array} \right\} \quad (3.17.13)$$

Equations (3.17.13) show that on the y-axis, σ_x is constant while σ_y becomes infinitely large for $y = \pm D/2$, i.e., on the circumference, and reaches its minimum value of $-6P/\pi D$ at the center of the disk. The distribution of stresses σ_x and σ_y on the y-axis is shown in the two graphs of Fig. 3.17.10.

3.18 Two-Dimensional Thermal Stresses

The equations for thermal stresses derived in Sec. 1.7 are now applied to elementary two-dimensional problems. Thermal stresses in thin disks and in long circular cylinders will be considered for the case of a symmetrical temperature distribution.

For a thin disk $\sigma_z = 0$ and Eqs. (1.7.2), in terms of cylindrical coordinates, become

$$\left. \begin{array}{l} \epsilon_r = \dfrac{1}{E}(\sigma_r - \nu\sigma_\theta) + \alpha T \\[6pt] \epsilon_\theta = \dfrac{1}{E}(\sigma_\theta - \nu\sigma_r) + \alpha T \end{array} \right\} \quad (3.18.1)$$

The equilibrium equation is

$$\frac{d\sigma_r}{dr} + \frac{\sigma_r - \sigma_\theta}{r} = 0 \quad (3.12.1')$$

and the compatibility equation is

$$\frac{d\epsilon_\theta}{dr} + \frac{\epsilon_\theta - \epsilon_r}{r} = 0 \quad (3.12.3)$$

Equation (3.12.1') is satisfied by the stress function Φ if

$$\left. \begin{array}{l} \sigma_r = \dfrac{\Phi}{r} \\[6pt] \sigma_\theta = \dfrac{d\Phi}{dr} \end{array} \right\} \quad (3.18.2)$$

By substituting Eqs. (3.18.1) and (3.18.2) into Eq. (3.12.3) one obtains

$$\frac{d^2\Phi}{dr^2} + \frac{1}{r}\frac{d\Phi}{dr} - \frac{\Phi}{r^2} = -\alpha E \frac{dT}{dr}$$

or

$$\frac{d}{dr}\left[\frac{1}{r}\frac{d}{dr}(r\Phi)\right] = -\alpha E \frac{dT}{dr} \qquad (3.18.3)$$

from which it follows that

$$\Phi(r) = -\frac{\alpha E}{r}\int_{R_1}^{R_2} Tr\,dr + \frac{C_1 r}{2} + \frac{C_2}{r} \qquad (3.18.4)$$

and in view of the first of Eqs. (3.18.2)

$$\left.\begin{aligned}\sigma_r &= -\frac{\alpha E}{r^2}\int_{R_1}^{R_2} Tr\,dr + \frac{C_1}{2} + \frac{C_2}{r^2} \\ \sigma_\theta &= \frac{\alpha E}{r^2}\int_{R_1}^{R_2} Tr\,dr - \alpha ET + \frac{C_1}{2} - \frac{C_2}{r^2}\end{aligned}\right\} \qquad (3.18.5)$$

C_1 and C_2 are constants to be determined by the boundary conditions on the free borders of the disk. In the case of a solid disk of radius R, $C_2 = 0$ since $(\sigma_r)_{r=0}$ must remain a finite quantity. If there are no external forces applied at the free boundary then from the condition that $(\sigma_r)_{r=R} = 0$ it follows that

$$C_1 = \frac{2\alpha E}{R^2}\int_0^R Tr\,dr$$

and Eqs. (3.18.2) become

$$\left.\begin{aligned}\sigma_r &= \alpha E\left[\frac{1}{R^2}\int_0^R T(r)r\,dr - \frac{1}{r^2}\int_0^r T(r)r\,dr\right] \\ \sigma_\theta &= \alpha E\left[-T + \frac{1}{R^2}\int_0^R T(r)r\,dr + \frac{1}{r^2}\int_0^r T(r)r\,dr\right]\end{aligned}\right\} \qquad (3.18.6)$$

In the case of a hollow disk (the radius of the inner hole being R_1 and the outside radius of the disk being R_2), if the edges are free of external forces, then from the conditions that

$$(\sigma_r)_{r=R_1} = (\sigma_r)_{r=R_2} = 0$$

it follows that

$$C_1 = \frac{2\alpha E}{(R_2^2 - R_1^2)}\int_{R_1}^{R_2} T(r)r\,dr$$

$$C_2 = -\frac{R_1^2 E}{(R_2^2 - R_1^2)}\int_{R_1}^{R_2} T(r)r\,dr$$

and Eqs. (3.18.2) become

$$\begin{aligned}
\sigma_r &= \alpha E\left[-\frac{1}{r^2}\int_{R_1}^{R_2} T(r)r\,dr + \frac{d}{R_2^2 - R_1^2}\int_{R_1}^{R_2} Tr(r)r\,dr \right. \\
&\qquad \left. - \frac{R_1^2}{r^2(R_2^2 - R_1^2)}\int_{R_1}^{R_2} T(r)r\,dr\right] \\
\sigma_\theta &= \alpha E\left[-T + \frac{1}{r^2}\int_{R_1}^{R_2} T(r)r\,dr + \frac{1}{R_2^2 - R_1^2}\int_{R_1}^{R_2} T(r)r\,dr \right. \\
&\qquad \left. + \frac{R_1^2}{r^2(R_2^2 - R_1^2)}\int_{R_1}^{R_2} T(r)r\,dr\right]
\end{aligned} \qquad (3.18.7)$$

Example 3.18.1. Determine the thermal stresses in a solid disk for the case in which the temperature varies according to the law

$$T(r) = (T_0 - T_1) - (T_0 - T_1)\frac{r^2}{R^2} \qquad (3.18.8)$$

where T_1 is the temperature at the free edge and T_0, the temperature at the center of the disk.

By substituting Eq. (3.18.8) into Eqs. (3.18.7) one obtains

$$\begin{aligned}
\sigma_r &= \frac{\alpha E}{4}(T_1 - T_0)\left(1 - \frac{r^2}{R^2}\right) \\
\sigma_\theta &= \frac{\alpha E}{4}(T_1 - T_0)\left(1 - \frac{3r^2}{R^2}\right)
\end{aligned} \qquad (3.18.9)$$

For a long, circular cylinder, Eqs. (1.7.2), in terms of cylindrical coordinates, become

$$\begin{aligned}
\epsilon_r &= \frac{1}{E}[\sigma_r - \nu(\sigma_\theta + \sigma_z)] + \alpha T \\
\epsilon_\theta &= \frac{1}{E}[\sigma_\theta - \nu(\sigma_z + \sigma_r)] + \alpha T \\
\epsilon_z &= \frac{1}{E}[\sigma_z - \nu(\sigma_r + \sigma_\theta)] + \alpha T
\end{aligned} \qquad (3.18.10)$$

In the case of plane strain, $\epsilon_z = 0$ and

$$\sigma_z = \nu(\sigma_r + \sigma_\theta) - \alpha T \qquad (3.18.11)$$

By substituting Eq. (3.18.11) in the first two of Eqs. (3.18.10), one obtains

$$\begin{aligned}
\epsilon_r &= \frac{1+\nu}{E}[(1-\nu)\sigma_r - \nu\sigma_\theta + \alpha ET] \\
\epsilon_\theta &= \frac{1+\nu}{E}[(1-\nu)\sigma_\theta - \nu\sigma_r + \alpha ET]
\end{aligned} \qquad (3.18.12)$$

By substituting Eqs. (3.18.12) and (3.18.2) into Eq. (3.12.3), one obtains

$$\frac{d}{dr}\left[\frac{1}{r}\frac{d}{dr}(r\Phi)\right] = -\frac{\alpha E}{(1-v)}\frac{dT}{dr} \tag{3.18.13}$$

Equation (3.18.13) is identical to Eq. (3.18.3) except for the coefficient which multiplies dT/dr ($\alpha E/(1-v)$ instead of αE). The solution of Eq. (3.18.13) is

$$\Phi = -\frac{\alpha E}{(1-v)}\frac{1}{r}\int_R^r T(r) r\, dr + \frac{C_1}{2}r + \frac{C_2}{r}$$

In the case of a solid cylinder of radius R, $C_2 = 0$ since $(\sigma_r)_{r=0}$ must remain a finite quantity. If there are no external forces applied at the free boundary then from the condition that $(\sigma_r)_{r=R} = 0$ it follows that

$$C_1 = \frac{2\alpha E}{(1-v)}\frac{1}{R^2}\int_0^R T(r) r\, dr$$

The stress components are

$$\left.\begin{aligned}\sigma_r &= \frac{\alpha E}{(1-v)}\left[\frac{1}{R^2}\int_0^R T(r) r\, dr - \frac{1}{r^2}\int_0^r T(r) r\, dr\right] \\ \sigma_\theta &= \frac{\alpha E}{(1-v)}\left[-T + \frac{1}{R^2}\int_0^R T(r) r\, dr + \frac{1}{r^2}\int_0^r T(r) r\, dr\right] \\ \sigma_z &= \frac{\alpha E}{(1-v)}\left[\frac{2v}{R^2}\int_0^R T(r) r\, dr - T\right]\end{aligned}\right\} \tag{3.18.14}$$

The last of Eqs. (3.18.14) is obtained from Eq. (3.18.11) and corresponds to the condition that

$$\epsilon_z = 0$$

For a cylinder with free ends, a uniform constant axial stress $\sigma'_z = C$ has to be superimposed in order that the resultant force P at the end be zero. The condition

$$P = \int_0^R \sigma_z 2\pi r\, dr = 0$$

gives

$$C = \frac{2\alpha E}{R^2}\int_0^R T(r) r\, dr$$

and

$$\sigma_z = \frac{\alpha E}{(1-v)}\left[\frac{2}{R^2}\int_0^R T(r) r\, dr - T\right] \tag{3.18.15}$$

In the case of a hollow cylinder (the radius of the inner hole being R_1 and the outside radius of the cylinder being R_2), the conditions that

$$(\sigma_r)_{r=R_1} = (\sigma_r)_{r=R_2} = 0$$

give for constants C_1 and C_2 the values

$$C_1 = \frac{2\alpha E}{(1-\nu)} \frac{1}{R_2^2 - R_1^2} \int_{R_1}^{R_2} T(r)\, r\, dr$$

$$C_2 = -\frac{\alpha E}{(1-\nu)} \frac{R_1^2}{R_2^2 - R_1^2} \int_{R_1}^{R_2} T(r)\, r\, dr$$

The stress components are

$$\sigma_r = \frac{\alpha E}{(1-\nu)} \frac{1}{r^2} \left[\frac{r^2 - R_1^2}{R_2^2 - R_1^2} \int_{R_1}^{R_2} T(r)\, r\, dr - \int_{R_1}^{r} T(r)\, r\, dr \right] \quad (3.18.16)$$

$$\sigma_\theta = \frac{\alpha E}{(1-\nu)} \frac{1}{r^2} \left[\frac{r^2 + R_1^2}{R_2^2 - R_1^2} \int_{R_1}^{R_2} T(r)\, r\, dr + \int_{R_1}^{r} T(r)\, r\, dr - T r^2 \right]$$

Example 3.18.2. Determine the thermal stresses in a solid, circular cylinder of radius R, if the temperature varies according to the law (steady heat flow)

$$T(r) = \frac{T_0 - T_1}{\log \frac{R_2}{R_1}} \log \frac{R_2}{r} \quad (3.18.17)$$

where T_0 is the temperature on the inner surface and T_1 the temperature on the outer surface of the cylinder.

By substituting Eq. (3.18.17) into Eqs. (3.18.14), one finds for the stress components

$$\sigma_r = \frac{\alpha E(T_0 - T_1)}{2(1-\nu) \log \frac{R_2}{R_1}} \left[-\log \frac{R_2}{r} - \frac{R_1^2(r^2 - R_2^2)}{r^2(R_2^2 - R_1^2)} \log \frac{R_2}{R_1} \right]$$

$$\sigma_\theta = \frac{\alpha E(T_0 - T_1)}{2(1-\nu) \log \frac{R_2}{R_1}} \left[1 - \log \frac{R_2}{r} - \frac{R_1^2(r^2 + R_2^2)}{r^2(R_2^2 - R_1^2)} \log \frac{R_2}{R_1} \right]$$

$$\sigma_z = \frac{\alpha E(T_0 - T_1)}{2(1-\nu) \log \frac{R_2}{R_1}} \left[1 - \log \frac{R_2}{r} - \frac{2R_1^2}{R_2^2 - R_1^2} \log \frac{R_2}{R_1} \right]$$

If $T_0 > T_1$ then $\sigma_r < 0$, i.e., there is compression. Stresses σ_θ and σ_z attain their maximum values on the outside and inside surfaces of the cylinder.

APPENDIX 3.A

Expression of the Laplacian Operator in Polar Coordinates[25]

To express the Laplacian operator

$$\Delta = \frac{\partial^2}{\partial x^2} + \frac{\partial^2}{\partial y^2}$$

in polar coordinates r and θ, consider the function $f(x, y)$ expressed in terms of variables r and θ i.e.,

$$f(x, y) = f[r(x, y); \theta(x, y)] \quad (3.A.1)$$

From the relationships

$$r^2 = x^2 + y^2, \quad \theta = \tan^{-1}\frac{y}{x}$$

one obtains

$$\frac{\partial r}{\partial x} = \frac{x}{r} = \cos\theta, \qquad \frac{\partial r}{\partial y} = \frac{y}{r} = \sin\theta$$

$$\frac{\partial \theta}{\partial x} = -\frac{y}{r^2} = -\frac{\sin\theta}{r}, \qquad \frac{\partial \theta}{\partial y} = \frac{x}{r^2} = \frac{\cos\theta}{r}$$

By differentiating Eq. (3.A.1), one has

$$\frac{\partial f}{\partial x} = \frac{\partial f}{\partial r}\frac{\partial r}{\partial x} + \frac{\partial f}{\partial \theta}\frac{\partial \theta}{\partial x} = \frac{\partial f}{\partial r}\cos\theta - \frac{1}{r}\frac{\partial f}{\partial \theta}\sin\theta \quad (3.A.2)$$

The expression for the second derivative $\partial^2 f/\partial x^2$ can be obtained by replacing f in Eq. (3.A.2) with

$$\frac{\partial f}{\partial x} = \frac{\partial f}{\partial r}\cos\theta - \frac{1}{r}\frac{\partial f}{\partial \theta}\sin\theta$$

[25] Reprinted from *Dynamics of Vibrations* by Enrico Volterra and E. C. Zachmanoglou, 1965, with the permission of Charles E. Merrill Books, Inc., Columbus, Ohio.

which gives

$$\frac{\partial^2 f}{\partial x^2} = \frac{\partial}{\partial r}\left[\frac{\partial f}{\partial r}\cos\theta - \frac{1}{r}\frac{\partial f}{\partial \theta}\sin\theta\right]\cos\theta - \frac{1}{r}\frac{\partial}{\partial \theta}\left[\frac{\partial f}{\partial r}\cos\theta - \frac{1}{r}\frac{\partial f}{\partial \theta}\sin\theta\right]\sin\theta$$

$$= \frac{\partial^2 f}{\partial r^2}\cos^2\theta - \frac{\partial^2 f}{\partial r\partial\theta}\frac{\sin 2\theta}{r} + \frac{\partial^2 f}{\partial \theta^2}\frac{\sin^2\theta}{r^2} + \frac{\partial f}{\partial r}\frac{\sin^2\theta}{r} + \frac{\partial f}{\partial \theta}\frac{\sin 2\theta}{r^2} \quad (3.A.3)$$

Similarly, one obtains

$$\frac{\partial^2 f}{\partial y^2} = \frac{\partial^2 f}{\partial r^2}\sin^2\theta + \frac{\partial^2 f}{\partial r\partial\theta}\frac{\sin 2\theta}{r} + \frac{\partial^2 f}{\partial \theta^2}\frac{\cos^2\theta}{r^2} + \frac{\partial f}{\partial r}\frac{\cos^2\theta}{r} - \frac{\partial f}{\partial \theta}\frac{\sin 2\theta}{r^2} \quad (3.A.4)$$

and finally

$$\Delta f = \frac{\partial^2 f}{\partial x^2} + \frac{\partial^2 f}{\partial y^2} = \frac{\partial^2 f}{\partial r^2} + \frac{1}{r}\frac{\partial f}{\partial r} + \frac{1}{r^2}\frac{\partial^2 f}{\partial \theta^2}$$

Therefore, the Laplacian operator Δ expressed in terms of polar coordinates r and θ is

$$\Delta = \frac{\partial^2}{\partial r^2} + \frac{1}{r}\frac{\partial}{\partial r} + \frac{1}{r^2}\frac{\partial^2}{\partial \theta^2} \quad (3.A.5)$$

APPENDIX 3.B

Integration of Euler's Equation

Euler's equation may be written as

$$\left(\frac{d^2}{dr^2} + \frac{1}{r}\frac{d}{dr}\right)\left(\frac{d^2 f(r)}{dr^2} + \frac{1}{r}\frac{df(r)}{dr}\right) = 0 \quad (3.B.1)$$

To integrate Eq. (3.B.1), which may also be written in the form

$$\frac{d^4 f(r)}{dr^4} + \frac{2}{r}\frac{d^3 f(r)}{dr^3} - \frac{1}{r^2}\frac{d^2 f(r)}{dr^2} + \frac{1}{r^3}\frac{df(r)}{dr} = 0 \quad (3.B.2)$$

let
$$r = e^x \tag{3.B.3}$$

where x represents a new independent variable. Since

$$\frac{dr}{dx} = e^x \quad \text{and} \quad \frac{dx}{dr} = e^{-x}$$

it follows that

$$\frac{df(r)}{dr} = \frac{df(r)}{dx}\frac{dx}{dr} = \frac{df(x)}{dx}e^{-x}$$

$$\frac{d^2f(r)}{dr^2} = \frac{d}{dx}\left[\frac{df(r)}{dr}\right]\frac{dx}{dr} = e^{-x}\frac{d}{dx}\left[\frac{df(x)}{dx}e^{-x}\right] = \left[\frac{d^2f(x)}{dx^2} - \frac{df(x)}{dx}\right]e^{-2x}$$

$$\frac{d^3f(r)}{dr^3} = \frac{d}{dx}\left[\frac{d^2f(r)}{dr^2}\right]\frac{dx}{dr} = e^{-x}\frac{d}{dx}\left\{\left[\frac{d^2f(x)}{dx^2} - \frac{df(x)}{dx}\right]e^{-2x}\right\}$$

$$= \left[\frac{d^3f(x)}{dx^3} - 3\frac{d^2f(x)}{dx^2} + 2\frac{df(x)}{dx}\right]e^{-3x}$$

$$\frac{d^4f(r)}{dr^4} = \frac{d}{dx}\left[\frac{d^3f(r)}{dr^3}\right]\frac{dx}{dr} = e^{-x}\frac{d}{dx}\left\{\left[\frac{d^3f(x)}{dx^3} - 3\frac{d^2f(x)}{dx^2} + 2\frac{df(x)}{dx}\right]e^{-3x}\right\}$$

$$= \left[\frac{d^4f(x)}{dx^4} - 6\frac{d^3f(x)}{dx^3} + 11\frac{d^2f(x)}{dx^2} - 6\frac{df(x)}{dx}\right]e^{-4x}$$

By substituting the above values into Eq. (3.B.2), the following linear differential equation with constant coefficients is obtained:

$$\frac{d^4f(x)}{dx^4} - 4\frac{d^3f(x)}{dx^3} + 4\frac{d^2f(x)}{dx^2} = 0 \tag{3.B.4}$$

To integrate Eq. (3.B.4) let

$$f(x) = e^{\lambda x} \tag{3.B.5}$$

Euler, Leonhard. Swiss mathematician (b. Basel 1707; d. St. Petersburg 1783) who was a pupil of Johann Bernoulli. He was invited to Russia by Empress Catherine where he became professor of physics and later succeeded Daniel Bernoulli in the chair of mathematics at the Russian Academy. In 1741, at the invitation of King Frederick the Great, he went to Berlin where he became director of the Academy. Later, from 1766 until his death he lived again in St. Petersburg. He left outstanding evidence of his genius in every branch of pure and applied mathematics and physics. [Reproduction from the Vito Volterra collection, Villa Volterra, Ariccia (Rome).]

By substituting Eq. (3.B.5) into Eq. (3.B.4), the characteristic equation

$$\lambda^4 - 4\lambda^3 + 4\lambda^2 = 0 \quad \text{or} \quad \lambda^2(\lambda - 2)^2 = 0$$

having the two double roots

$$\lambda_{1,2} = 0 \quad \lambda_{3,4} = +2$$

is obtained. It follows that the general solution of Eq. (3.B.4) is

$$f(x) = a + bx + ce^{2x} + dx\, e^{2x}$$

where a, b, c, and d are constants. In view of Eq. (3.B.3), the general solution of Eq. (3.B.1) is

$$f(r) = A + B \log r + Cr^2 + Dr^2 \log r$$

where A, B, C, and D are constants.

BIBLIOGRAPHY

Sections 3.1, 3.2, 3.3, 3.4, 3.5. PLANE STRESS; MOHR'S CIRCLE FOR STRESS; PLANE STRAIN; MOHR'S CIRCLE FOR STRAIN AND STRAIN ROSETTES; AIRY'S STRESS FUNCTION

1. Airy, G. B., "On the Strains in the Interior of Beams," *Rept. Brit. Assoc. Advan. Sci.*, 1862, p. 82.
2. ———, "On the Strains in the Interior of Beams," *Phil. Trans. Roy. Soc., London*, Vol. **153**, 1863, p. 49.
3. Almansi, E., "Sull'integrazione dell'equazione $\Delta^2\Delta^2 u = 0$," *Atti Accad. Torino*, Vol. **31**, 1895–1896.
4. ———, "Sull'integrazione dell'equazione differenziale $\Delta^{2n} = 0$," *Ann. Mat. Pura Appl.* Vol. **2**, 1898, p. 1.
5. ———, "Sulla integrazione dell'equazione $\Delta^2\Delta^2 u = 0$," *Atti Accad. Torino*, Vol. **34**, 1898–1899.
6. ———, "Sull'integrazione dell'equazione differenziale $\Delta^2\Delta^2 u = 0$," *Atti Accad. Nazl. Lincei, Rend., Serie 5*, Vol. **8**, 1899, p. 1.
7. ———, "Sulle ricerche delle funzioni poliarmoniche in un'area piana semplicemente connessa per date condizioni al contorno," *Rend. Circolo Mat. Palermo*, Vol. **13**, 1899, p. 225.
8. ———, "Integrazione della doppia equazione di Laplace," *Atti Accad. Lincei, Rend., Serie 5*, Vol. **9**, 1900, p. 1.
9. Biezeno, C. B., and R. Grammel, "Engineering Dynamics," Vol. **1**, Blackie & Son, Ltd., Glasgow, 1955.

10. Boggio, T., "Integrazione dell'Equazione $\Delta^2\Delta^2 u = 0$ in un'Area Ellittica," *Atti Ist. Veneto*, Vol. **62**, 1900–1901, pp. 591–609.

11. Bricas, M., "La Théorie de l'élasticité bidimensionelle," Athens, Greece, 1937.

12. Butty E., "Tratado de elasticidad teorico-tecnico," Vol. **1**, Centro Estud. Ing., Buenos Aires, Argentina, 1957.

13. Coker, E. G., and L. N. G. Filon, "A Treatise on Photoelasticity," Cambridge University Press, London, England, 1957.

14. Durelli, A. J., E. A. Phillips, and C. H. Tsao, "Introduction to the Theoretical and Experimental Analysis of Stress and Strain," McGraw-Hill Book Company, New York, N.Y., 1958.

15. Filon, L. N. G., "On the Relation between Corresponding Problems in Plane Stress and in Generalized Plane Stress," *Quart. J.*, Vol. **1**, 1930, pp. 289–299.

16. Frocht, M. M., "Photoelasticity," Vol. **2**, John Wiley & Sons, Inc., New York, N.Y., 1948.

17. Goursat, E., "Sur l'Equation $\Delta_2\Delta_2 u = 0$," *Bull. Soc. Math. France*, Vol. **26**, 1898, p. 236.

18. Grioli, G., "Structura della funzione di Airy nei sistemi molteplicemente connessi," *Giorn. Mat. Battaglini*, 1947, p. 119.

19. ———, "Mathematical Theory of Elastic Equilibrium," Academic Press, Inc., New York, N.Y., 1962.

20. Hetenyi, M., "Handbook on Experimental Stress Analysis," John Wiley & Sons, Inc., New York, N.Y., 1950.

21. Inglis, Sir Charles, "Stresses in a Plate Due to the Presence of Cracks and Sharp Corners," *Trans. Inst. Naval Arch., London*, Vol. **55**, 1913, pp. 219–230.

22. Jeffery, G. B., "Plane Stress and Plane Strain in Bipolar Coordinates," *Phil. Trans. Roy. Soc., London*, Vol. **221**, 1921, pp. 265–293.

23. Klein, F., and K. Wieghardt, "Über Spannungsflächen und reziproke Diagramme, mit besonderer Berücksichtigung der MaxwellschenArbeiten," *Arch. Math. Phys.* **3, Reihe VIII**, 1904.

24. Lauricella, G., "Integrazione dell'Equazione $\Delta_2\Delta_2 u = 0$ in un Campo di Forma Circolare," *Atti Accad. Torino*, Vol. **31**, 1896.

25. ———, "Sull'integrazione delle equazioni dell'equilibrio dei corpi elastici istropi," *Atti Accad. Nazl. Lincei, Rend., Serie 5*, Vol. **15**, 1906.

26. Le Boiteux, H., and R. Bourssart, "Elasticité et photoélasticimétrie," Hermann, Paris, France, 1940.

27. L'Hermite, R., "Résistance des matériaux théorique et expérimentale," Vol. **1**, Dunod, Paris, France, 1958.

28. Levi-Civita, T., "Sulla Integrazione dell'Equazione $\Delta_2\Delta_2 u = 0$" *Atti Accad. Torino*, Vol. **33**, 1897–1898, pp. 932–956. See also Vol. **1** of Levi-Civita's "Opere Matematiche," N. Zanichelli, Bologna, Italy, 1954.

29. ———, "Sopra una trasformazione in sé stessa della equazione," *Atti Ist. Veneto, Serie 7*, T. **9**, 1897–1898. See also Vol. **1** of Levi-Civita's" Opere Matematiche," N. Zanichelli, Bologna, Italy, 1954.

30. Love, A. E. H., "A Treatise on the Mathematical Theory of Elasticity," Cambridge University Press, London, England, 1934.

31. Mathieu, E., "Mémoire sur l'Equation aux Différences Partielles...," *J. Math.*, Serie 2, **T. 14**, 1869.

32. Maxwell, J. C., "On the Equilibrium of Elastic Solids," *Trans. Roy. Soc., Edinburgh*, **Vol. 20**, 1853, p. 87.

33. ———, "On Reciprocal Figures, Diagrams in Space and Their Relation to Airy's Function of Stress," *Proc. Math. Soc., London*, 1868, p. 58.

34. Michell, J. H., "Elementary Distributions of Plane Stress," *Proc. Math. Soc., London*, **Vol. 31**, 1901, pp. 35–61.

35. ———, "The Inversion of Plane Stress," *Proc. Math. Soc., London*, **Vol. 34**, 1902, pp. 134–142.

36. Mikhlin, S. G., "Le Problème fondamental biharmonique à deux dimensions," *Compt. Rend.*, **Vol. 197**, 1933, p. 608.

37. ———, "Reduction of the Fundamental Problems on the Plane Theory of Elasticity to Fredholm Integral Equations," *Dokl. Akad. Nauk, SSSR, New Series*, **Vol. 1**, 1934, p. 295.

38. ———, "La Solution du problème plan biharmonique et des problèmes de la théorie statique d'élasticité à deux dimensions," *Tr. Seismologiczesk Inst., A.H., SSSR*, 1934.

39. ———, "The Method of Successive Approximation in the Biharmonic Problem," *Tr. Seismologiczesk Inst., A.H., SSSR*, **No. 39**, 1934.

40. ———, "The Plane Problem of the Theory of Elasticity," *Tr. Seismologiczesk Inst. A.H., SSSR*, **No. 65**, 1935.

41. Milne-Thomson, L. M., "Plane Elastic Systems," Julius Springer, Berlin, Germany, 1960.

42. Mohr, O., "Abhandlungen aus dem Gebiete der Technischen Mechanik," 2nd ed., W. Ernst und Sohn, Berlin, Germany, 1914.

43. Muskhelishvili, N. I., "Some Basic Problems of the Mathematical Theory of Elasticity," P. Nooordhoff, Ltd., Groningen, Holland, 1953.

44. ———, "Singular Integral Equations," P. Noordhoff, Ltd., Groningen, Holland, 1953

45. Poritsky, H., "Application of Analytic Functions to Two-Dimensional Biharmonic Analysis," *Trans. Am. Math. Soc.*, **Vol. 24**, No. 2, 1946, pp. 248–279.

46. Reissner, E., "On the Calculation of Three-Dimensional Corrections for the Two-Dimensional Theory of Plane Stress," *Proc. Fifteenth Eastern Photoelastic Conf.*, 1942.

47. Sadovski, M. A., "Zwei dimensionale Probleme der Elastizitätstheorie," *Z. Angew. Math. Mech.*, **Vol. 8**, 1928, pp. 107–121.

48. Sen, B., "Two-Dimensional Boundary Value Problems of Elasticity," *Proc. Roy. Soc., London*, 1946, p. 87.

49. Sobrero, L., "Delle funzioni analoghe al potenziale intervenienti nella fisica-matematica," *Atti Accad. Nazl. Lincei, Rend.*, **Vol. 21**, 1935, p. 448.

50. Stevenson, A. C., "Some Boundary-Value Problems of the Two-Dimensional Elasticity," *Phil. Mag.*, **Vol. 34**, 7, 1943, p. 766.

51. Tedone, O., "Sui Problemi di Equilibrio Elastico a Due Dimensioni," *Atti Accad. Torino*, Vol. **41**, 1905–1906, pp. 86–101.

52. Theodorescu, P. P., "One Hundred Years of Investigation in the Plane Problem of the Theory of Elasticity," *Appl. Mech. Rev.*, Vol. **17**, No. **3**, pp. 175–186. 1964

53. Timoshenko, S. P., and J. N. Goodier, "Theory of Elasticity," McGraw-Hill Book Company, New York, 1951.

54. Timpe, A., "Probleme der Spannungsverteilung in ebenen Systemen, einfach gelost mit Hilfe der Airyschen Funktion," *Z. Math. Phys.*, **Bd. 52**, 1905, pp. 348–383.

55. Venske, O., "Zur Integration der Gleichung $\Delta_2\Delta_2 u = 0$ fur Ebene Bereiche," Gottingen Nachrichten, 1891.

56. Wang, C. T., "Applied Elasticity," McGraw-Hill Book Company, New York, N.Y., 1953.

57. Westergard, H. M., "Theory of Elasticity and Plasticity," Harvard University Press, Cambridge, Mass., 1952.

Section 3.6. SOLUTION OF TWO-DIMENSIONAL PROBLEMS BY THE USE OF POLYNOMIALS

1. Mesnager, A.,[26] "Sur l'Application de la théorie de l'élasticité au calcul des pieces rectangulaires fléchies," *Compt. Rend.* Vol. **132**, 1901, p. 1475.

Section 3.7. TRIANGULAR AND RECTANGULAR WALLS SUBJECTED TO HYDROSTATIC PRESSURES (MAURICE LEVY'S PROBLEMS)

1. Levy, Maurice, "Sur la Légitimité de la Règle Dite du Trapèze dans l'Étude de la Résistance des Barrages en Maconnerie," *Compt. Rend,* Vol. **126, No. 18**, Première Semestre, 1878, pp. 1235–1240.

2. Muller, J., "Etude de Trois Profiles de Murs Encastrés sollicités a la compression et a la flexion," Publications du Laboratoire de Photo-élasticité de l'Ecole Polytechnique Fédérale de Zurich, Switzerland, 1930.

3. Osgood, W. R., "Rectangular Plate Loaded Along Two Adjacent Edges by Couples in Its Own Plane," *Natl. Bur. Std. (U.S.), Res. Paper*, **1450**; *J. Res. Natl. Bur. Std.*, Vol. **28**, 1942, pp. 159–163.

Sections 3.8, 3.9, 3.10, 3.11, 3.12, 3.13, 3.15, 3.17. USE OF POLAR COORDINATES; BENDING OF A CIRCULAR BAR (GOLOVIN-RIBIÈRE PROBLEM); THICK TUBE SUBJECTED TO EXTERNAL AND INTERNAL UNIFORMLY DISTRIBUTED PRESSURES (LAMÉ'S PROBLEM); SHRINK FITS; ROTATING DISKS AND CYLINDERS; STRESS CONCENTRATION DUE TO A CIRCULAR HOLE IN A STRESSED PLATE (KIRSCH'S PROBLEM); CONCENTRATED LOAD ACTING ON THE FREE SURFACE OF A PLATE (FLAMANT'S PROBLEM); DISK SUBJECTED TO TWO OPPOSITE CONCENTRATED FORCES (HERTZ'S PROBLEM)

1. Biezeno, C. B., and R. Grammel, "Engineering Dynamics," Vol. **1**, Blackie & Son, Ltd., Glasgow, 1955.

2. Billevicz, V., Doctorate Thesis, University of Michigan, Ann Arbor, Mich., 1931.

---

[26] **Mesnager, Augustin.** French engineer and mechanicist (b. Paris 1862; d. Paris 1933). He was successively professor at the Ecole des Ponts et Chaussées and at the Conservatoire National des Arts et Métiers.

3. Coker, E. G., and L. N. G. Filon, "A Treatise on Photoelasticity," Cambridge University Press, London, England, 1957.
4. Den Hartog, J. P., "Advanced Strength of Materials," McGraw-Hill Book Company, New York, N.Y., 1952.
5. Durelli, A. J., and W. M. Murray, "Stress distribution around an elliptical discontinuity in any two-dimensional, uniform and axial system of combined stresses," *Exptl. Stress Anal.*, **Vol. 1, No. 1,** 1943.
6. Durelli, A. J., E. A. Phillips, and C. H. Tsao, "Introduction to the Theoretical and Experimental Analysis of Stress and Strain," McGraw-Hill Book Company, New York, N.Y., 1958.
7. Foppl, A., "Vorlesungen uber Technische Mechanik," Vol. 5, 1907, p. 72.
8. Frocht, M. M., "Photoelasticity," **Vol. II,** John Wiley & Sons, Inc., New York, N.Y., 1948.
9. Golovin, H., *Trans. Inst. Tech., St. Petersburg,* 1881.
10. Greenspan, M., "Effect of a Small Hole on the Stresses in an Uniformly Loaded Plate," *Quart. Appl. Math.,* 1944.
11. Hertz, H., "Ueber die Vertheilung der Druckkräfte in einen Elastischen Dreiscylinder," *Z. Math. Phys.,* **XXVIII Jahrgang,** 1883, pp. 125–128; see also Hertz's "Gesammelte Werke," J. A. Barth, Leipzig, Germany, 1894.
12. Inglis, Sir Charles, "Stresses in a Plate Due to the Presence of Cracks and Sharp Corners," *Trans. Inst. Naval Arch.,* London, **Vol. LV,** 1913, pp. 219–230.
13. Kirsch, G., "Die Theorie der Elastizität und die Bedürfnisse der Festigkeitslehre," *Ver. Deut. Ing.,* **Vol. 42,** 1898, p. 797.
14. Lamé, G., "Leçons sur la théorie mathématique de l'élasticité des corps solides," Paris, France, Bachelier, 1852.
15. L'Hermite, R., "Résistance des matériauz théorique et experimentale," **Vol. 1,** Dunod, Paris, France, 1958.
16. Morkovin, D., "Effect of a Small Hole on the Stresses in an Uniformly Loaded Plate," *Quart. Appl. Math,* 1945, p. 350.
17. Pearson, K., "History of the Theory of Elasticity," **Vol. 2, Pt. 1,** Cambridge University Press, 1893, London, England, p. 422.
18. Ribiére, C., "Sur l'Equilibre d'élasticité des voutes en arc de cercle," *Compt. Rend.* **Vol. 108,** 1889, pp. 561–563.
19. ———, "Sur les voutes en arc de cercle encastrées aux naissances," *Compt. Rend.,* **Vol. 132,** 1901, pp. 315–317.
20. Southwell, Sir Richard, "An Introduction to the Theory of Elasticity for Engineers and Physicists," Oxford University Press, London, England, 1941.
21. Sternberg, E., and M. A. Sadowsky, "Three-Dimensional Solution for the Stress Concentration around a Circular Hole in a Plate of Arbitrary Thickness," *J. Appl. Mech.,* **Vol. 16,** 1949, p. 27.
22. Timoshenko, S. P., "On Stresses in a Plate with a Circular Hole," *J. Franklin Inst.* **197, No. 4,** 1924, pp. 505–516.
23. Timoshenko, S. P., and J. N. Goodier, "Theory of Elasticity," McGraw-Hill Book Company, New York, N.Y., 1951.

24. Timpe, A., "Probleme der Spannungsverteilung in Ebenen Systemen, Einfach Gelost mit Hilfe der Airyschen Funktion," *Z. Math. Phys.*, **Bd. 52**, 1905, pp. 348–383.

25. Winkler, E., "Formänderung und Festigkeit Gekrümmter Körper, Insbesondere der Ringe," *Civiling.*, **Bd. IV**, 1858, pp. 232–246.

26. ———, "Die Lehre von der Elastizität und Festigkeit," Chap. 15, Prague, 1867.

Section 3.14. CONCENTRATED LOAD ACTING ON THE VERTEX OF A WEDGE (MICHELL'S PROBLEM)

1. Michell, J. H., "The Collected Mathematical Works of J. H. and A. G. M. Michell," P. Nordhoff, Ltd., Groningen, Holland, 1964.

Section 3.16. MOMENT ACTING ON THE VERTEX OF A WEDGE (INGLIS' PROBLEM)

1. Carothers, S. D., "Plane Strain in a Wedge," *Proc. Roy. Soc., Edinburgh*, **Vol. 13**, 1912, p. 292.

2. Inglis, Sir Charles, "Some Special Cases of Two-Dimensional Stress or Strain," *Trans. Inst. Naval Arch., London*, **Vol. LXIV**, 1922.

3. Sternberg, E., and W. Koiter, "The Wedge under a Concentrated Couple. A Paradox in the Two-Dimensional Theory of Elasticity," *J. Appl. Mech.*, **Vol. 525**, 1958, p. 575.

Section 3.18. TWO-DIMENSIONAL THERMAL STRESSES

1. Boley, B., and J. Weiner, "Theory of Thermal Stresses," John Wiley & Sons, Inc., New York, N.Y., 1960.

2. Nowacki, W., "Thermoelasticity," Pergamon Press, Inc., New York, 1962.

3. Strutt, J. W., Baron Rayleigh, "On the Stresses in Solid Bodies Due to Unequal Heating, and on the Double Refraction Resulting Therefrom," *Phil. Mag.*, Series 6, **Vol. 1**, 1901, p. 169.

4. Timoshenko, S. P., and J. N. Goodier, "Theory of Elasticity," McGraw-Hill Book Company, New York, N.Y., 1951.

Love, August Edward Hough. English mathematician (b. Weston super Mare 1863; d. Oxford 1940). He was Sedleian Professor of Natural Philosophy at Oxford University and the author of many important papers in elasticity and geophysics. His classical book, *A Treatise on the Mathematical Theory of Elasticity*, was first published in 1892. His works on geophysics were collected in the volume entitled *Some Problems of Geodynamics*, published in 1911 in Cambridge. (Reproduction by permission of the Royal Society of London.)

Strutt, John William, 3rd Baron Rayleigh. English physicist and chemist (b. Langford Grove, Essex 1842; d. Witham, Essex 1919). From 1879 to 1884 he was Cavendish Professor of Experimental Physics at Cambridge University, and from 1887 to 1905 he was Professor of Natural Philosophy at the Royal Institution in London. Lord Rayleigh also served as Secretary and, later, as President of the Royal Society, and in late years was Chancellor of Cambridge University. He left outstanding contributions in hydrodynamics, optics, and acoustics (his book, *Theory of Sound*, was first published in 1877). Together with Sir William Ramsay he was Nobel Laureate in chemistry in 1904 for their discovery of argon. His complete works were published in seven volumes. [Reproduction from the Vito Volterra collection, Villa Volterra, Ariccia (Rome).]

PROBLEMS

3-1 to 3-6. By use of Mohr's circle and by use of formulas find graphically and analytically the magnitudes of the normal, shearing, and resultant stresses acting on:

(a) The principal planes.
(b) The planes of maximum shearing stress.
(c) The plane whose normal makes an angle of $+30°$ with the x-axis in the case where the stresses at a point in a body are

1. $\sigma_x = 800$ psi, $\sigma_y = 200$ psi, $\tau_{xy} = 300$ psi
2. $\sigma_x = -400$ psi, $\sigma_y = -100$ psi, $\tau_{xy} = 0$ psi
3. $\sigma_x = 300$ psi, $\sigma_y = -200$ psi, $\tau_{xy} = -200$ psi
4. $\sigma_x = 0$ psi, $\sigma_y = 300$ psi, $\tau_{xy} = 0$ psi
5. $\sigma_x = -200$ psi, $\sigma_y = 200$ psi, $\tau_{xy} = 300$ psi
6. $\sigma_x = 100$ psi, $\sigma_y = 800$ psi, $\tau_{xy} = -500$ psi

3-7. The equilateral triangle wedge of Fig. P.3.1. is loaded on two adjacent faces by a uniform normal stress of 800 psi and by uniform shear stress of 400 psi. Determine σ_y and τ_{xy}.

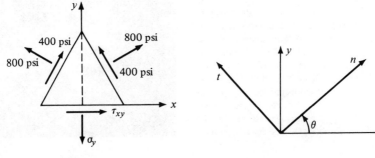

Figure P.3.1   Figure P.3.2

3-8. If n and t are two perpendicular axes in the xy plane (see Fig. P.3.2) prove the following relationships:

$$\sigma_n + \sigma_t = \sigma_x + \sigma_y$$
$$\sigma_n \sigma_t - \tau_{nt}^2 = \sigma_x \sigma_y - \tau_{xy}^2$$

3-9. In a state of plane stress in the xy-plane, the maximum principal stress $\sigma_{max} = 8000$ psi, $\sigma_y = 3500$ psi, and $\tau_{xy} = 6000$ psi. Determine the value of σ_x and show the principal directions in the xy-plane.

3-10. to 3-13. Sketch Mohr's circle for each of the states of plane stress shown in Figs. P.3.3 through P.3.6.

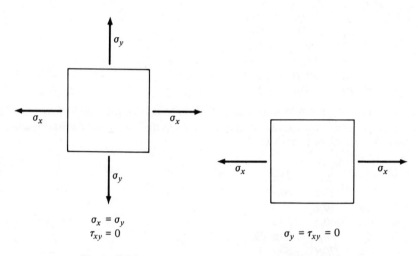

Figure P.3.3

Figure P.3.4

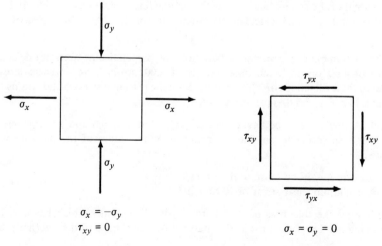

Figure P.3.5

Figure P.3.6

3-14. A small cube of an elastic material rests at the bottom of the sea where the pressure of water is 50,000 psi. What are the normal and shear stresses on any plane of the cube?

3-15. The propeller shaft of a ship is subjected to a longitudinal thrust of 6 tons/in.^{2} and (due to torsion) there is also a shear stress of 3 ton/in.^{2} Find the principal stresses and the maximum shear stress.

3-16. to 3-21. The following states of strain have been measured by means of a 45° strain rosette. Find the magnitudes and directions of the principal strains analytically and by use of Mohr's circle.

	$\epsilon_{0°}$	$\epsilon_{45°}$	$\epsilon_{90°}$
1.	80μ	0μ	30μ
2.	-60μ	40μ	20μ
3.	100μ	30μ	40μ
4.	60μ	-100μ	20μ
5.	-200μ	300μ	-200μ
6.	150μ	50μ	-20μ

3-22. to 3-27. The following states of strain have been measured by means of a 60° strain rosette. Find the magnitudes and directions of the principal strains analytically and by use of Mohr's circle.

	$\epsilon_{0°}$	$\epsilon_{60°}$	$\epsilon_{120°}$
1.	30μ	-20μ	40μ
2.	20μ	40μ	60μ
3.	30μ	10μ	25μ
4.	70μ	70μ	50μ
5.	40μ	60μ	80μ
6.	30μ	20μ	-50μ

3-28. A flat bar 4 × 0.5 in. is subjected to an axial pull of 16 tons. One side of the bar is polished and fine lines are ruled on it to form a square with 2 in. sides, one diagonal of the square being along the middle line of the polished side. If $E = 30(10^6)$ psi and $\nu = 0.28$, calculate the alterations of the sides and angles of the square.

3-29. A thin plate is subjected to two mutually perpendicular stresses, one compressive of 4.00 tons/in.^{2}, the other tensile of 5.00 tons/in.^{2}, and a shear stress, parallel to these directions of 3.00 tons/in.^{2} Find the principal stresses and strains by assuming that $E = 30(10^6)$ psi and $\nu = 0.3$.

3-30. A thick steel tube having, respectively, 3 in. internal and 12 in. external diameters is subjected to an internal pressure. What values of the pressure will produce:

1. A maximum hoop tension of 15 tons/in.^{2}
2. A maximum shear-stress of 10 tons/in.^{2}?

3-31. Find the thickness of a cast iron hydraulic cylinder which has a 12 in. inside diameter. The safe stress in the material is 4 tons/in.^{2} and the internal pressure is 1000 psi. Also sketch the curves of circumferential and radial stresses existing under these conditions.

3-32. A steel tube 10 ft long with 3 in. internal and 9 in. external diameter respectively, is to be subjected to an internal hydraulic pressure. Assuming that the tube remains elastic, estimate the amount of water that must be forced into it to raise the internal pressure to 15 tons/in.^{2} Assume that for water $K = 3(10^5)$ psi.

3-33. The cylinder of an engine which has to withstand an internal pressure of 2.5 tons/in.^{2} has an internal diameter of 8 in. The safety factor is to be 2 and the material being steel has an elastic limit of 20 tons/in.^{2} in simple tension. Calculate the outside diameter based on

1. the maximum shear-stress theory.
2. the strain-energy theory.

3-34. Calculate the initial difference between the internal diameter of a steel tube of 12 in. external diameter and a solid shaft of 8 in. diameter if the shrink fit pressure between the tube and the shaft is 2 tons/in.^{2}

3-35. Calculate the maximum safe speed of a flat circular disk 12 in. in diameter with a hole 2 in. in diameter at the center, the thickness of the disk being 1.5 in. The elastic limit of the material in simple tension is 15 tons/in.^{2}

3-36. Calculate the safe speed of the disk of the above problem, if there is no hole at the center.

3-37. Compare the components of stress given by the Michell formula for the case of a triangular wedge subjected to a concentrated load at the apex with the components of stress given by the strength of materials for the case in which the angle $\alpha = 10°$.

3-38. Solve problem 3-37 for the case of $\alpha = 20°$.

3-39. Solve problem 3-37 for the case of $\alpha = 30°$.

3-40. Solve problem 3-37 for the case of $\alpha = 45°$.

CHAPTER

4

ELEMENTARY PROBLEMS IN THREE-DIMENSIONAL ELASTICITY

4.0 Introduction

In this chapter, elementary problems in three-dimensional elasticity will be discussed starting with Lamé's solution of the problem of the thick spherical shell under uniform internal and external pressures. It is one of the rare cases in which an elasticity problem can be solved by elementary means. This problem is the equivalent in three-dimensions of the two-dimensional problem of the thick cylinder under uniform internal and external pressures whose solution was discussed in the previous chapter.

Next, the problem of pure bending of a prismatic bar will be considered. To solve this problem the semi-inverse method will be used with the stress components assumed to be the same as those obtained from the elementary beam theory. It is then shown that this assumed solution satisfies all of the equations of the theory of elasticity (stress-strain equations, strain-displacement equations, equilibrium equations, and boundary conditions). By the uniqueness theorem it follows that this is the only correct solution of the problem.

The problem of torsion of a bar is then discussed by starting with the theory of torsion of circular shafts derived by Coulomb in 1784. Navier's theory is then presented and it is shown that this solution satisfies the boundary conditions only in the case in which the cross-section of the twisted shaft is circular.

Then Saint-Venant's theory of torsion is presented. To solve the torsion problem, Saint-Venant used the semi-inverse method, i.e., he assumed that some features of the displacement are known and he determined the remaining features so as to satisfy all the equations of elasticity. Saint-Venant's theory of torsion is then applied to bars of circular, elliptic, and rectangular cross sections.

Prandtl's theory of torsion is then discussed and applied to the study of the torsion of shafts having elliptical and equilateral triangular cross sections. Prandtl's membrane analogy is then presented and the experiments by Taylor and Griffith[1] using this analogy for solving torsion problems is briefly described. Prandtl's analogy is then applied to the study of torsion of thin-walled sections. The chapter ends with the application of Ritz's method to torsion problems.

4.1 Lamé's Solution of the Problem of the Thick Spherical Shell under Uniform Internal and External Pressures

Lamé's equations of equilibrium

$$\left. \begin{array}{l} G\Delta u + (\lambda + G)\dfrac{\partial e}{\partial x} + X = 0 \\[4pt] G\Delta v + (\lambda + G)\dfrac{\partial e}{\partial y} + Y = 0 \\[4pt] G\Delta w + (\lambda + G)\dfrac{\partial e}{\partial z} + Z = 0 \end{array} \right\} \qquad (1.4.3)$$

[1] **Griffith, Alan Arnold.** English physicist and applied mechanicist (b. London 1893; d. 1963). His papers with G. I. (Sir Geoffrey) Taylor, read before the Institution of Mechanical Engineers, on the estimation of torsional stresses in sections of complicated shapes by the use of soap films gained for the authors the Thomas Hawksley Gold Medal. In 1920, he published in the Philosophical Transactions of the London Royal Society his classical paper, "The Phenomena of Rupture and Flow in Solids," in which he pointed out that cracks could cause fracture. This theory was the forerunner of all later theories based on the presence of flaws accounting for loss of strength.

become, in the absence of body forces,

$$\left.\begin{aligned}\Delta u + \frac{1}{1-2\nu}\frac{\partial e}{\partial x} &= 0 \\ \Delta v + \frac{1}{1-2\nu}\frac{\partial e}{\partial y} &= 0 \\ \Delta w + \frac{1}{1-2\nu}\frac{\partial e}{\partial z} &= 0\end{aligned}\right\} \quad (4.1.1)$$

Now, with the origin O of the system of cartesian coordinates x, y, and z at the center of the shell (see Fig. 4.1.1), the components of the elastic displacement are assumed to be

$$\left.\begin{aligned}u &= \frac{x}{r}\phi(r) \\ v &= \frac{y}{r}\phi(r) \\ w &= \frac{z}{r}\phi(r)\end{aligned}\right\} \quad (4.1.2)$$

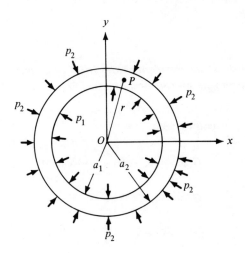

Figure 4.1.1

where $\phi(r)$ is the total (unknown) radial displacement, and

$$r = \sqrt{x^2 + y^2 + z^2} \quad (4.1.3)$$

is the distance of generic point P from the center of the sphere. Since the deformation of the shell is assumed to be symmetrical with respect to the origin, it seems logical to assume the components of the elastic displacement in the form expressed by Eqs. (4.1.2). From Eqs (4.1.2) and (4.1.3), the following relationships are derived

$$e = \frac{\partial u}{\partial x} + \frac{\partial v}{\partial y} + \frac{\partial w}{\partial z} = \frac{2}{r}\phi(r) + \frac{d\phi(r)}{dr}$$

$$\Delta u = \frac{\partial e}{\partial x}$$

$$\Delta v = \frac{\partial e}{\partial y}$$

$$\Delta w = \frac{\partial e}{\partial z}$$

and Eqs. (4.1.1) reduce to the single equation

$$\frac{d}{dr}\left[\frac{2}{r}\phi(r) + \frac{d\phi(r)}{dr}\right] = 0 \qquad (4.1.4)$$

whose general integral, as can be shown by substitution, is given by

$$\phi(r) = C_1 r + C_2 r^{-2} \qquad (4.1.5)$$

where C_1 and C_2 are two constants.

The radial stress σ_r, which is normal to the free concentric surfaces of the shell, is given by

$$\sigma_r = \sigma_x l^2 + \sigma_y m^2 + \sigma_z n^2 + 2\tau_{xy} lm + 2\tau_{yz} mn + 2\tau_{zx} nl \qquad (4.1.6)$$

where

$$\left.\begin{aligned} l &= \frac{x}{r} \\ m &= \frac{y}{r} \\ n &= \frac{z}{r} \end{aligned}\right\} \qquad (4.1.7)$$

and

$$\left.\begin{aligned} \sigma_x &= \lambda e + 2G\epsilon_x \\ \sigma_y &= \lambda e + 2G\epsilon_y \\ \sigma_z &= \lambda e + 2G\epsilon_z \end{aligned}\right\} \qquad (1.3.4)$$

and

$$\left.\begin{array}{l}\tau_{xy} = G\gamma_{xy}\\ \tau_{yz} = G\gamma_{yz}\\ \tau_{zx} = G\gamma_{zx}\end{array}\right\} \quad (1.3.5)$$

Let $\bar{X}$, $\bar{Y}$, and $\bar{Z}$ denote the components of the surface forces in the x-, y-, z-directions and l, m, and n the direction cosines of the normal stress σ_r on the element of surface (see Fig. 4.1.2). To derive Eq. (4.1.6) one then notes that

$$\sigma_r = \bar{X}l + \bar{Y}m + \bar{Z}n \quad (4.1.8)$$

But

$$\left.\begin{array}{l}\bar{X} = \sigma_x l + \tau_{xy} m + \tau_{xz} n\\ \bar{Y} = \tau_{yx} l + \sigma_y m + \tau_{yz} n\\ \bar{Z} = \tau_{zx} l + \tau_{zy} m + \sigma_z n\end{array}\right\} \quad (1.6.2)$$

By substituting Eqs. (1.6.2) into Eq. (4.1.8), Eq. (4.1.6) follows immediately. By combining Eqs. (4.1.2), (4.1.5), (4.1.7), (1.3.4) and (1.3.5) with Eq. (4.1.6) one gets

$$\sigma_r = (3\lambda + 2G)C_1 - \frac{4GC_2}{r^3} \quad (4.1.9)$$

Considering the case of a thick shell of inside radius a_1 and outside radius a_2 (see Fig. 4.1.3), the following boundary conditions, which will determine constants C_1 and C_2, must be satisfied:

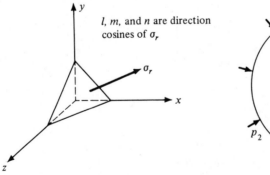

Figure 4.1.2

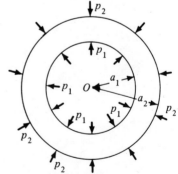

Figure 4.1.3

Sec. 4.1 ELEMENTARY PROBLEMS IN THREE-DIMENSIONAL ELASTICITY

$$\left.\begin{array}{l}\sigma_r = -p_1 \quad \text{for } r = a_1 \\ \sigma_r = -p_2 \quad \text{for } r = a_2\end{array}\right\} \quad (4.1.10)$$

By substituting Eq. (4.1.9) into Eqs. (4.1.10) one obtains

$$\left.\begin{array}{l}C_1 = \dfrac{p_1 a_1^3 - p_2 a_2^3}{(3\lambda + 2G)(a_2^3 - a_1^3)} \\ C_2 = \dfrac{a_1^3 a_2^3 (p_1 - p_2)}{4G(a_2^3 - a_1^3)}\end{array}\right\} \quad (4.1.11)$$

By substituting Eqs. (4.1.11) into Eq. (4.1.9) one gets

$$\sigma_r = \frac{p_1 a_1^3 - p_2 a_2^3}{a_2^3 - a_1^3} - \frac{a_1^3 a_2^3}{r^3} \frac{p_1 - p_2}{a_2^3 - a_1^3} \quad (4.1.12)$$

To determine the stress σ_θ, an element of volume is isolated by the cone having a small angle $d\theta$ at the vertex (which is at the origin O) and the two spherical surfaces of radii r and $r + dr$ (see Fig. 4.1.4). The internal face of radius r has an area

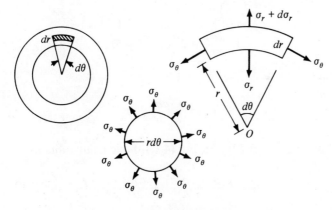

Figure 4.1.4

$$\frac{\pi (r d\theta)^2}{4}$$

and is subjected to the force

$$\frac{\pi}{4} (r^2 \sigma_r)(d\theta)^2$$

The force which is acting on the external face has the same expression with the quantity

$$(r^2\sigma_r) + d(r^2\sigma_r)$$

in place of $(r^2\sigma_r)$. The difference between the two forces is therefore

$$\frac{\pi}{4} d(r^2\sigma_r)(d\theta)^2 = \frac{\pi}{4}\left[2r\sigma_r + r^2\frac{d\sigma_r}{dr}\right]dr\,(d\theta)^2$$

The contour of the element subjected to the stress σ_θ has area $\pi r\,d\theta\,dr$. The radial component of the resultant force on this contour is

$$(\sigma_\theta \pi r\,d\theta\,dr)\sin\frac{d\theta}{2} = \sigma_\theta \pi r\,d\theta\,dr\,\frac{d\theta}{2}$$

For equilibrium in the radial direction, the following equation must be satisfied

$$\frac{\pi}{4}\left(2r\sigma_r + r^2\frac{d\sigma_r}{dr}\right)dr\,(d\theta)^2 - \sigma_\theta \pi r\,d\theta\,dr\,\frac{d\theta}{2} = 0$$

or

$$\frac{d\sigma_r}{dr} + 2\frac{\sigma_r - \sigma_\theta}{r} = 0$$

from which

$$\sigma_\theta = \frac{r}{2}\frac{d\sigma_r}{dr} + \sigma_r$$

By substituting Eq. (4.1.12) into Eq. (4.1.13) one finally obtains

$$\sigma_\theta = \frac{p_1 a_1^3 - p_2 a_2^3}{a_2^3 - a_1^3} + \frac{a_1^3 a_2^3}{2r^3}\frac{p_1 - p_2}{a_2^3 - a_1^3} \tag{4.1.14}$$

Equations (4.1.12) and (4.1.14) are Lamé's formulas for the spherical shell. Some special cases of these formulas are now considered.

(a) *The external pressure $p_2 = 0$.* For this case

$$\sigma_r = \frac{p_1 a_1^3}{a_2^3 - a_1^3}\left(1 - \frac{a_2^3}{r^3}\right) \leq 0 \quad \text{since} \quad r \leq a_2$$

$$\sigma_\theta = \frac{p_1 a_1^3}{a_2^3 - a_1^3}\left(1 + \frac{a_2^3}{2r^3}\right) > 0$$

It follows that the maximum value of σ_θ, which occurs at $r = a_1$, i.e., at the inner surface of the shell, is

$$(\sigma_\theta)_{\max} = \frac{p_1}{2} \frac{2a_1^3 + a_2^3}{a_2^3 - a_1^3} \qquad (4.1.15)$$

If the shell is very thin ($t = a_2 - a_1$ is a very small quantity), then from Eq. (4.1.15), the approximate formula

$$(\sigma_\theta)_{\max} = \frac{p_1 a}{2t} \qquad (4.1.16)$$

is derived, where a and t are, respectively, the mean radius and thickness of the shell.[2]

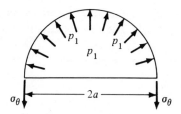

Figure 4.1.5

[2] To derive Eq. (4.1.16) from Eq. (4.1.15) let

$$a_1 = \left(a - \frac{t}{2}\right)$$

$$a_2 = \left(a + \frac{t}{2}\right)$$

and Eq. (4.1.15) becomes

$$(\sigma_\theta)_{\max} = \frac{p_1}{2} \frac{2\left[a - \frac{t}{2}\right]^3 + \left[a + \frac{t}{2}\right]^3}{\left[a + \frac{t}{2}\right]^3 - \left[a - \frac{t}{2}\right]^3}$$

$$= \frac{p_1}{2} \frac{2\left[a^3 + 3\frac{at^2}{4} - 3a^2\frac{t}{2} - \frac{t^3}{8}\right] + \left[a^3 + 3a\frac{t^2}{4} + 3a^2\frac{t}{2} + \frac{t^3}{8}\right]}{\left[a^3 + 3\frac{at^2}{4} + 3\frac{a^2 t}{2} + \frac{t^3}{8}\right] - \left[a^3 + 3a\frac{t^2}{4} - 3a^2\frac{t}{2} - \frac{t^3}{8}\right]}$$

$$= \frac{p_1}{2} \frac{3a^3}{6a^2 \frac{t}{2}} = \frac{p_1 a}{2t}$$

Equation (4.1.16) can be obtained directly by assuming that the wall of the shell is very thin and that the stress σ_θ is uniformly distributed (see Fig. 4.1.5). Then $2\pi a \sigma_\theta t = \pi p_1 a^2$ from which Eq. (4.1.16) follows.

(b) Spherical cavity inside an elastic solid. This situation is obtained, if $a_2 = \infty$. Then Eqs. (4.1.12) and (4.1.14) become[3]

$$\left. \begin{array}{l} \sigma_r = -p_1 \dfrac{a_1^3}{r^3} \\ \sigma_\theta = +p_1 \dfrac{a_1^3}{2r^3} \end{array} \right\} \qquad (4.1.17)$$

(c) Case in which the internal pressure $p_1 = 0$ and the external pressure $p_2 \neq 0$. Equation (4.1.14) gives, in this case,

$$(\sigma_\theta)_{\max} = (\sigma_\theta)_{r=r_1} = \frac{3p_2 a_2^3}{2(a_2^3 - a_1^3)}$$

(d) Case of a sphere subjected to an external pressure p. This situation is obtained, if $a_1 = 0$ and $p_1 = 0$. Then Eqs. (4.1.12) and (4.1.14) become

$$\sigma_r = \sigma_\theta = -p$$

4.2 Pure Bending of a Prismatic Bar

Consider a prismatic beam which is bent in one of its principal planes by two equal and opposite couples M as shown in Fig. 4.2.1. The origin of the coordinates is taken at the centroid of a cross section, and the xz plane is the principal plane of bending. Assume that the stress components are the same values as those given by the elementary theory of bending. Then

---

[3] To derive Eqs. (4.1.17) divide numerator and denominator of the equations for case (a), i.e.,

$$\sigma_r = \frac{p_1 a_1^3}{a_2^3 - a_1^3} \left(1 - \frac{a_2^3}{r^3} \right)$$

$$\sigma_\theta = \frac{p_1 a_1^3}{a_2^3 - a_1^3} \left(1 + \frac{a_2^3}{2r^3} \right)$$

by a_2^3. One then obtains

$$\sigma_r = \frac{p_1 a_1^3}{1 - \left(\frac{a_1}{a_2}\right)^3} \left(\frac{1}{a_2^3} - \frac{1}{r^3} \right) \quad \text{and for} \quad a_2 = \infty \quad \sigma_r = -\frac{p_1 a_1^3}{r^3}$$

$$\sigma_\theta = \frac{p_1 a_1^3}{1 - \left(\frac{a_1}{a_2}\right)^3} \left(\frac{1}{a_2^3} + \frac{1}{2r^3} \right) \quad \text{and for} \quad a_2 = \infty \quad \sigma_\theta = +\frac{p_1 a_1^3}{2r^3}$$

ELEMENTARY PROBLEMS IN THREE-DIMENSIONAL ELASTICITY

$$\left.\begin{array}{c} \sigma_x = \sigma_y = \tau_{xy} = \tau_{yz} = \tau_{zx} = 0 \\ \sigma_z = \dfrac{Ex}{R} \end{array}\right\} \quad (4.2.1)$$

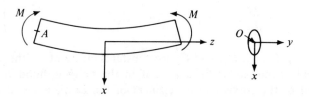

Figure 4.2.1

where R represents the radius of curvature of the bent beam. The bending moment is

$$M = \iint_\Omega \sigma_z x \, dx \, dy = \frac{E}{R} \iint_\Omega x^2 \, dx \, dy = \frac{EI}{R} \quad (4.2.2)$$

where Ω is the area of the cross section of the beam and I is the moment of inertia of the cross section of the beam with respect to the y-axis. In view of Eq. (4.2.2), Eqs. (4.2.1) can be written in the following form:

$$\left.\begin{array}{c} \sigma_x = \sigma_y = \tau_{xy} = \tau_{yz} = \tau_{zx} = 0 \\ \sigma_z = \dfrac{Mx}{I} \end{array}\right\} \quad (4.2.3)$$

By substituting Eqs. (4.2.3) into the equations of equilibrium, the compatibility equations, and the boundary conditions on the lateral surface of the beam it can be shown that these equations are satisfied. From Eqs. (4.2.3) the components of strain are derived; thus,

$$\epsilon_x = \epsilon_y = -\nu\epsilon_z = -\frac{\nu Mx}{EI}, \quad \epsilon_z = \frac{Mx}{EI}, \quad \gamma_{xy} = \gamma_{yz} = \gamma_{zx} = 0 \quad (4.2.4)$$

It follows from Eqs. (1.2.1) and (4.2.4) that

$$\left.\begin{array}{c} \dfrac{\partial u}{\partial x} = \dfrac{\partial v}{\partial y} = -\nu\dfrac{Mx}{EI}, \quad \dfrac{\partial w}{\partial z} = \dfrac{Mx}{EI}, \quad \dfrac{\partial u}{\partial y} + \dfrac{\partial v}{\partial x} = 0 \\ \dfrac{\partial v}{\partial z} + \dfrac{\partial w}{\partial y} = 0, \quad \dfrac{\partial w}{\partial x} + \dfrac{\partial u}{\partial z} = 0 \end{array}\right\} \quad (4.2.5)$$

After integration of Eqs. (4.2.5), one obtains

$$\left.\begin{aligned} u &= -\frac{M}{2EI}[z^2 + v(x^2 - y^2)] - C_1 z + C_3 y + C_5 \\ v &= -\frac{vM}{EI}xy - C_2 z - C_3 x + C_6 \\ w &= \frac{M}{EI}zx + C_1 x + C_2 y + C_4 \end{aligned}\right\} \quad (4.2.6)$$

where $C_1, C_2, C_3, C_4, C_5,$ and C_6 are constants. If point A, the centroid of the left end of the bar, is fixed and, if in the neighborhood of this point an element of the z-axis and an element of the zx-plane are also fixed, the following conditions must be satisfied:

$$\text{for } x = y = z = 0 \quad \left\{ \begin{aligned} u &= v = w = 0 \\ \frac{\partial u}{\partial z} &= \frac{\partial v}{\partial z} = \frac{\partial v}{\partial x} = 0 \end{aligned} \right\} \quad (4.2.7)$$

Conditions (4.2.7) imply that

$$C_1 = C_2 = C_3 = C_4 = C_5 = C_6 = 0$$

and Eqs. (4.2.6) become

$$\left.\begin{aligned} u &= -\frac{M}{2EI}[z^2 + v(x^2 - y^2)] \\ v &= -\frac{vM}{EI}xy \\ w &= \frac{M}{EI}zx \end{aligned}\right\} \quad (4.2.8)$$

In order to discuss the deformation of the bar, consider a general point P which prior to the deformation has coordinates x, y, and z. After the bar is bent, point P will move to point P' of coordinates x', y', and z', such that

$$\left.\begin{aligned} \text{(a)} \quad & x' = x + u = x - \frac{M}{2EI}[z^2 + v(x^2 - y^2)] \\ \text{(b)} \quad & y' = y + v = y - \frac{vM}{EI}xy \\ \text{(c)} \quad & z' = z + w = z + \frac{M}{EI}zx \end{aligned}\right\} \quad (4.2.9)$$

Consider now the plane which in the undeformed state has equation $z = a$. Substituting a for z into Eq. (c) one obtains

$$z' = a + \frac{M}{EI}ax \qquad (4.2.10)$$

But from Eq. (a)

$$x = x' + \frac{M}{2EI}[a^2 + v(x^2 - y^2)] \qquad (4.2.11)$$

By substituting Eq. (4.2.11) into Eq. (4.2.10), one finds that

$$z' = a + \frac{Ma}{EI}x' + \left(\frac{M}{EI}\right)^2\left(\frac{a}{2}\right)[a^2 + v(x^2 - y^2)] \qquad (4.2.12)$$

The quantity $M/EI = 1/R$, being the inverse of the radius of curvature R, is a very small quantity and its square can be neglected in Eq. 4.2.12. Thus, Eq. (4.2.12) reduces to

$$z' = a + \frac{Ma}{EI}x' \qquad (4.2.13)$$

which is the equation of a plane. This agrees with the elementary Bernoulli-Navier hypothesis that transverse cross sections of a bent beam will remain plane.

Consider now the plane which in the undeformed state has the equation $x = b$. By substituting b for x in Eqs. (4.2.9) and then eliminating y and z from Eq. (a) by use of Eqs. (b) and (c), one obtains

$$x' = b - \frac{M}{2EI}\left\{\left(z' - \frac{Mzb}{EI}\right)^2 + v\left[b^2 - \left(y' + \frac{Mv}{EI}by\right)^2\right]\right\} \qquad (4.2.14)$$

By neglecting terms containing higher powers of M/EI, Eq. (4.2.14) reduces to the equation

$$x' = b - \frac{M}{2EI}[z'^2 + v(b^2 - y'^2)] \qquad (4.2.15)$$

Equation (4.2.15) shows that points at the same level ($x = b$) and on the same hyperbola ($z^2 - vy^2 = $ constant) will displace to new points at the same distance from the original level and constitute a contour line. It therefore follows that the contour lines of the deformed surface form a family of hyperbolas of equation $z^2 - \nu y^2 =$ constant as shown in Fig. 4.2.2. The slope of the asymptotes is given by the equation

$$\tan \alpha = \frac{1}{\sqrt{\nu}}$$

or

$$\nu = \text{cotangent}^2 \alpha$$

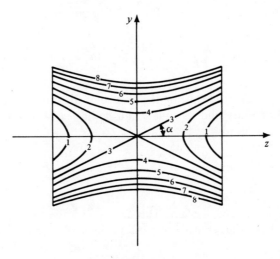

Figure 4.2.2

A. Cornu determined experimentally Poisson's ratio of a given material from the above theory. By polishing the upper or lower surface of a beam subjected to pure bending and by putting a glass plate over it, the contour lines of the deformed surface can be obtained optically. The technique used takes advantage of the existence of a variable thickness of air gap between the glass and the polished surface of the bent beam. This variable thickness can be measured by means of a beam of monochromatic light directed perpendicular to the glass plate. The light will be reflected partially by the plate and partially by the surface of the beam. At points where the thickness of the air gap is such that the difference in the paths of the reflected rays is equal to an uneven number of half wavelengths of light, the light rays interfere with one another. These points appear as dark points and the locus of these points indicate contour lines of the deformed beam surface. The photograph shown in Fig. 4.2.3 was obtained in this manner.

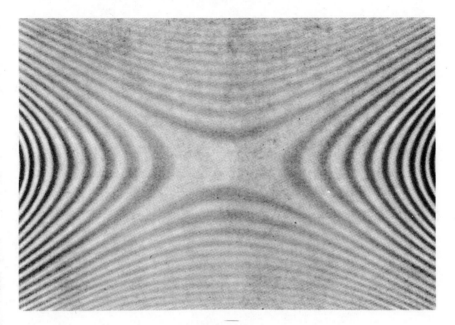

Figure 4.2.3

4.3 Coulomb's Theory of Torsion of a Circular Shaft

Consider the shaft of length L having circular cross section of radius R and polar moment of inertia J and subjected to a torque M_T (see Fig. 4.3.1). According to Coulomb's theory,[4] the shear stress at a distance r from the z-axis of the shaft is

$$\tau = \frac{M_T r}{J} \qquad (4.3.1)$$

while the angle of twist of the free end with respect to the fixed end is

$$\phi = \frac{M_T L}{GJ} \qquad (4.3.2)$$

where G is the shear modulus of the material of the shaft.

[4] **Coulomb, Charles Augustin.** French natural philosopher (b. Angoûleme 1736; d. Paris 1806). He was the author of important works in mechanics as well as in electricity and magnetism. His study of the laws of friction was published in 1779, and his work on fluid resistance in 1799. A military engineer by profession, in the last years of his life he was inspector of public instruction at the University of Paris. The practical unit of quantity of electricity, the *coulomb*, is named for him.

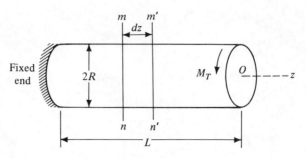

Figure 4.3.1

In order to derive Eqs. (4.3.1) and (4.3.2), consider the element of shaft of length dz shown in Fig. 4.3.2. Call A a generic point on the free surface of the shaft before torque M_T is applied, A' the position of the same point after deformation, γ the shearing strain, and ds the arc AA'. The following relationship is obtained from Fig. 4.3.2:

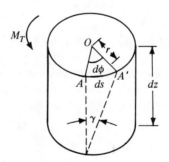

Figure 4.3.2

$$\tan \gamma \simeq \gamma = \frac{ds}{dz} = \frac{R\,d\phi}{dz}$$

During the torsional deformation of the circular shaft, a rigid rotation occurs such that the cross sections remain plane and the points of the shaft, which before deformation were on a diameter, remain on the same diameter after deformation. Thus, if τ_r and τ_R represent shear stresses at distances r and R from the shaft axis then the following relationship will be obtained:

$$\frac{\tau_r}{\tau_R} = \frac{r}{R}$$

or

$$\tau_r = \tau_R \frac{r}{R} \tag{4.3.3}$$

Also,

$$\tau_R = G\gamma = G\frac{R\,d\phi}{dz} \tag{4.3.4}$$

The condition of equilibrium between torque M_T and the internal moment is expressed by

$$M_T = \int_0^R \tau_r r\,dA = \int_0^R \tau_r r(2\pi r\,dr) = \frac{2\pi}{R}\tau_R \int_0^R r^3\,dr = \frac{\tau_R}{R}J$$

where

$$J = 2\pi \int_0^R r^3\,dr = \frac{\pi R^4}{2}$$

Finally

$$\tau_R = \frac{M_T R}{J} \tag{4.3.5}$$

By substituting Eq. (4.3.5) into Eq. (4.3.3), Eq. (4.3.1) is derived. By substituting Eq. (4.3.5) into Eq. (4.3.4) the equation

$$d\phi = \frac{M_T\,dz}{GJ}$$

is derived; then by substituting the total angle of twist in place of $d\phi$ and the total length L in place of dz, Eq. 4.3.2 is derived.

Equations (4.3.1) and (4.3.2) apply also in the case of a hollow circular shaft. In the case of a solid circular shaft of radius R and diameter D

$$J = \frac{\pi R^4}{2} = \frac{\pi D^4}{32}$$

In the case of a hollow shaft of inside radius R_1 and diameter D_1 and outside radius R_2 and diameter D_2

$$J = \frac{\pi}{2}[R_2^4 - R_1^4] = \frac{\pi}{32}[D_2^4 - D_1^4]$$

4.4 Navier's Theory of Torsion

According to Navier's theory, a generic section S of a prismatic solid (see Fig. 4.4.1) does not change its form due to torsion but only rotates around its centroidal axis by an angle θ proportional to the distance z of section S from the initial section S_0. The components of the elastic displacement will, in accordance with this assumption, be

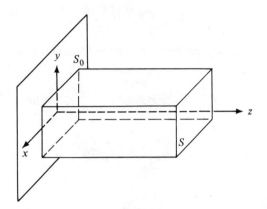

Figure 4.4.1

$$\left. \begin{array}{l} u = -\theta z y \\ v = \theta z x \\ w = 0 \end{array} \right\} \quad (4.4.1)$$

It follows that

$$\left. \begin{array}{l} \epsilon_x = \dfrac{\partial u}{\partial x} = 0, \quad \epsilon_y = \dfrac{\partial v}{\partial y} = 0, \quad \epsilon_z = \dfrac{\partial w}{\partial z} = 0 \\[4pt] \gamma_{xy} = \dfrac{\partial u}{\partial y} + \dfrac{\partial v}{\partial x} = 0 \\[4pt] \gamma_{yz} = \dfrac{\partial v}{\partial z} + \dfrac{\partial w}{\partial y} = x\theta \\[4pt] \gamma_{zx} = \dfrac{\partial w}{\partial x} + \dfrac{\partial u}{\partial z} = -y\theta \end{array} \right\} \quad (4.4.2)$$

which correspond to the following components of stress:

$$\left. \begin{array}{l} \sigma_x = \sigma_y = \sigma_z = \tau_{xy} = 0 \\ \tau_{yz} = Gx\theta \\ \tau_{zx} = -Gy\theta \end{array} \right\} \quad (4.4.3)$$

Equations (4.4.3) satisfy the equilibrium equations

$$\frac{\partial \sigma_x}{\partial x} + \frac{\partial \tau_{xy}}{\partial y} + \frac{\partial \tau_{xz}}{\partial z} = 0$$

$$\frac{\partial \tau_{yx}}{\partial x} + \frac{\partial \sigma_y}{\partial y} + \frac{\partial \tau_{yz}}{\partial z} = 0$$

$$\frac{\partial \tau_{zx}}{\partial x} + \frac{\partial \tau_{zy}}{\partial y} + \frac{\partial \sigma_z}{\partial z} = 0$$

By computing the resultant forces and moments on a cross-sectional area of the solid (parallel to the xy-plane), and using the equilibrium conditions, one finds

(a) *for the resultant forces.*

$$\iint_\Omega \tau_{xz} dx dy = -G\theta \iint_\Omega y \, dx dy = 0 \text{ (component in the } x\text{-direction)}$$

$$\iint_\Omega \tau_{yz} dx dy = G\theta \iint_\Omega x \, dx dy = 0 \text{ (component in the } y\text{-direction)}$$

$$\iint_\Omega \sigma_z dx dy = 0 \text{ (component in the } z\text{-direction)}$$

(b) *for the resultant moments.*

$$\iint_\Omega y\sigma_z dx dy = 0 \text{ (component about the } x \text{ axis)}$$

$$\iint_\Omega x\sigma_z dx dy = 0 \text{ (component about the } y \text{ axis)}$$

$$\iint_\Omega (x\tau_{zy} - y\tau_{zx}) dx dy = G\theta \iint_\Omega (x^2 + y^2) dx dy = M_T \text{ (applied torque)}$$

By calling J the polar moment of inertia of the cross section where

$$J = \iint_\Omega (x^2 + y^2) dx dy$$

the above equation can be written in the form

$$\theta = \frac{M_T}{GJ} \tag{4.4.4}$$

The weak point in Navier's theory of torsion lies in the fact that in general, the boundary conditions, i.e., the equations

$$\begin{aligned}
\bar{X} &= \sigma_x \cos{(\widehat{nx})} + \tau_{xy} \cos{(\widehat{ny})} + \tau_{xz} \cos{(\widehat{nz})} \\
\bar{Y} &= \tau_{yx} \cos{(\widehat{nx})} + \sigma_y \cos{(\widehat{ny})} + \tau_{yz} \cos{(\widehat{nz})} \\
\bar{Z} &= \tau_{zx} \cos{(\widehat{nx})} + \tau_{zy} \cos{(\widehat{ny})} + \sigma_z \cos{(\widehat{nz})}
\end{aligned} \qquad (4.4.5)$$

are not satisfied. In fact, for the present case, the following conditions exist

$$\bar{X} = \bar{Y} = \bar{Z} = \sigma_x = \sigma_y = \sigma_z = \tau_{xy} = \cos{(\widehat{nz})} = 0$$

and the first two of Eqs. (4.4.5) are identically satisfied, while the third becomes

$$\tau_{zx} \cos{(\widehat{nx})} + \tau_{zy} \cos{(\widehat{ny})} = 0$$

or, in view of Eqs. (4.4.3),

$$-G\theta y \cos{(\widehat{nx})} + G\theta x \cos{(\widehat{ny})} = 0 \qquad (4.4.6)$$

But, from Fig. 4.4.2

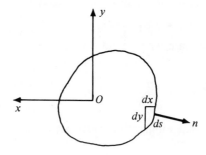

Figure 4.4.2

$$\left.\begin{aligned}
\cos{(\widehat{nx})} &= -\frac{dy}{ds} \\
\cos{(\widehat{ny})} &= \frac{dx}{ds}
\end{aligned}\right\} \qquad (4.4.7)$$

and Eq. (4.4.6) becomes

$$ydy + xdx = 0$$

or

$$x^2 + y^2 = \text{constant} \tag{4.4.8}$$

Equation (4.4.8) shows that with the exception of the cases in which the cross section of the bar is a circle or two concentric circles (hollow cylinder), the boundary conditions are not satisfied by Navier's assumptions. In the cases in which the cross section of the bar is a solid circular cylinder or a hollow circular cylinder, Navier's theory of torsion coincides with Coulomb's theory.

4.5 Saint-Venant's Semi-Inverse Method for Solving Torsion Problems

From Timoshenko's *History of Strength of Materials*[5] we learn that: "In 1853, Saint-Venant presented his epoch making memoir on torsion to the French Academy. The committee composed of Cauchy, Poncelet, Piobert, and Lamé were very impressed by this work and recommended its publication. The memoir contains not only the author's theory of torsion but also gives an account of all that was known at that time in the theory of elasticity together with many important additions developed by its author..."

"Saint-Venant proposes the semi-inverse method by which he assumes only some features of the displacement and of the forces and determines the remaining features of those quantities so as to satisfy all the equations of elasticity. He remarks that an engineer, guided by the approximate solutions of the elementary strength of materials, can obtain rigorous solutions of practical importance in this way."

In solving the torsion problem, Saint-Venant assumed that the component w of the displacement is different from zero, depending on the variables x and y. Saint-Venant assumed for the components of the elastic displacement, the following expressions:

$$\left.\begin{aligned} u &= -\theta zy \\ v &= \theta zx \\ w &= \theta \phi(x, y) \end{aligned}\right\} \tag{4.5.1}$$

---

[5] See Bibliography, Sec. 4.5, No. 39 at the end of this chapter.

where ϕ is a function, a priori unknown, of the variables x and y. It is called the warping function. From Eqs. (4.5.1) it follows that

$$\epsilon_x = \epsilon_y = \epsilon_z = \gamma_{xy} = 0$$
$$\gamma_{zy} = \theta\left(x + \frac{\partial \phi}{\partial y}\right)$$
$$\gamma_{zx} = \theta\left(-y + \frac{\partial \phi}{\partial x}\right)$$

and for the components of stress that

$$\sigma_x = \sigma_y = \sigma_z = \tau_{xy} = 0$$
$$\tau_{zy} = G\theta\left(x + \frac{\partial \phi}{\partial y}\right) \qquad (4.5.2)$$
$$\tau_{zx} = G\theta\left(-y + \frac{\partial \phi}{\partial x}\right)$$

By substituting Eqs. (4.5.2) into the equations of equilibrium, i.e.,

$$\frac{\partial \sigma_x}{\partial x} + \frac{\partial \tau_{xy}}{\partial y} + \frac{\partial \tau_{xz}}{\partial z} = 0$$
$$\frac{\partial \tau_{yx}}{\partial x} + \frac{\partial \sigma_y}{\partial y} + \frac{\partial \tau_{yz}}{\partial z} = 0$$
$$\frac{\partial \tau_{zx}}{\partial x} + \frac{\partial \tau_{zy}}{\partial y} + \frac{\partial \sigma_z}{\partial z} = 0$$

we find that the first two equations are identically satisfied and that the third equation becomes

$$\frac{\partial^2 \phi}{\partial x^2} + \frac{\partial^2 \phi}{\partial y^2} = 0 \qquad (4.5.3)$$

By substituting Eqs. (4.5.2) into the boundary equations

$$\bar{X} = \sigma_x \cos(\widehat{nx}) + \tau_{xy} \cos(\widehat{ny}) + \tau_{xz} \cos(\widehat{nz})$$
$$\bar{Y} = \tau_{yx} \cos(\widehat{nx}) + \sigma_y \cos(\widehat{ny}) + \tau_{yz} \cos(\widehat{nz})$$
$$\bar{Z} = \tau_{zx} \cos(\widehat{nx}) + \tau_{zy} \cos(\widehat{ny}) + \sigma_z \cos(\widehat{nz})$$

where

$$\bar{X} = \bar{Y} = \bar{Z} = \cos(\widehat{nz}) = 0$$

we find that the first two equations are identically satisfied, while the third equation becomes

$$\tau_{zx} \cos(\widehat{nx}) + \tau_{zy} \cos(\widehat{ny}) = 0 \qquad (4.5.4)$$

Equation (4.5.4) expresses the fact that on the boundary of the cross section of the shaft, the resultant shearing stress has zero component in the normal direction or that the resultant stress is directed along the tangent to the boundary.

By use of Eqs. (4.4.7) and (4.5.2), Eq. (4.5.4) can be written in the form

$$\left[\frac{\partial \phi}{\partial x} - y\right]\frac{dy}{ds} - \left[\frac{\partial \phi}{\partial y} + x\right]\frac{dx}{ds} = 0 \qquad (4.5.5)$$

According to Saint-Venant's theory of torsion of prismatic shafts, the torsion problem is reduced to the problem of finding the function $\phi(x, y)$ satisfying Eq. (4.5.3) and the boundary equation [Eq. (4.5.5)]. Mathematicians call this problem "Neumann's problem." It can be shown that a function $\phi(x, y)$ satisfying Eqs. (4.5.3) and (4.5.5) exists and is uniquely defined in a given region.[6]

The resultant twisting moment is obtained from the equation

$$M_T = \iint_\Omega (x\tau_{yz} - y\tau_{xz})dxdy = G\theta \iint_\Omega \left[x^2 + y^2 + x\frac{\partial \phi}{\partial y} - y\frac{\partial \phi}{\partial x}\right]dxdy$$

$$(4.5.6)$$

The integral

$$J = \iint_\Omega \left[x^2 + y^2 + x\frac{\partial \phi}{\partial y} - y\frac{\partial \phi}{\partial x}\right]dxdy \qquad (4.5.7)$$

of Eq. (4.5.6) depends on the warping function $\phi(x, y)$ and therefore on the cross section of the bar. It is called the torsional constant of the bar. Equation (4.5.6) can be written

$$M_T = G\theta J \qquad (4.5.8)$$

The product GJ is called the torsional rigidity of the bar.

Saint-Venant's theory of torsion will now be applied to the study of the torsion of bars having

---

[6] Here, as in every section of this book, unless differently specified, we refer only to simply-connected regions.

1. Circular cross sections
2. Elliptical cross sections
3. Rectangular cross sections

(a) Case of a circular cross section. Assume

$$\phi = C = \text{constant} \tag{4.5.9}$$

Equation (4.5.3) is satisfied while Eq. (4.5.5) becomes in this case

$$-y\frac{dy}{ds} - x\frac{dx}{ds} = 0$$

or

$$x^2 + y^2 = \text{constant} \tag{4.5.10}$$

In Eq. (4.5.10) x and y represent the coordinates of a generic point of the boundary. Since this equation represents the equation of a circle, it follows that Eq. (4.5.9) represents the solution of the torsion problem for a bar of circular cross section. If $w = 0$ for $z = 0$ then $C = 0$ and the solution coincides with the solution of Sec. 4.3 in accordance with Coulomb's theory for shafts of circular cross sections.

(b) Case of an elliptic cross section. Assume the ellipse has semi-axes a and b and that the warping function is of the form

$$\phi(x, y) = \frac{b^2 - a^2}{a^2 + b^2} xy \tag{4.5.11}$$

Equation (4.5.11) satisfies Eq. (4.5.3). By substituting Eq. (4.5.11) into Eq. (4.5.5) one obtains

$$\left[\frac{b^2 - a^2}{a^2 + b^2} - 1\right] y\frac{dy}{ds} - \left[\frac{b^2 - a^2}{a^2 + b^2} + 1\right] x\frac{dx}{ds} = \frac{-2a^2}{a^2 + b^2} y\frac{dy}{ds} - \frac{2b^2}{a^2 + b^2} x\frac{dy}{ds} = 0$$

or

$$a^2 \frac{d(2y)}{ds} + b^2 \frac{d(2x)}{ds} = 0$$

Thus,

$$a^2 y^2 + b^2 x^2 = C = \text{constant}$$

and finally, by dividing by a^2b^2 and letting $C/a^2b^2 = 1$, one obtains

$$\frac{x^2}{a^2} + \frac{y^2}{b^2} = 1$$

which is the equation of an ellipse with its center at the origin and semi-axes a and b. The torsional constant of the bar [see Eq. (4.5.7)] is

$$J = \iint_\Omega \left[x^2 + y^2 + x\frac{\partial \phi}{\partial y} - y\frac{\partial \phi}{\partial x} \right] dxdy$$

$$= \iint_\Omega \left[x^2 + y^2 + \frac{b^2 - a^2}{a^2 + b^2} x^2 - \frac{b^2 - a^2}{a^2 + b^2} y^2 \right] dxdy$$

$$= \frac{2b^2}{a^2 + b^2} \iint_\Omega x^2 dxdy + \frac{2a^2}{a^2 + b^2} \iint_\Omega y^2 dxdy$$

$$= \frac{2b^2}{a^2 + b^2} I_y + \frac{2a^2}{a^2 + b^2} I_x = \frac{\pi a^3 b^3}{a^2 + b^2} \qquad (4.5.12)$$

since the moments of inertia of the cross-sectional area with respect to the x- and y-axes are, respectively,

$$I_x = \frac{\pi ab^3}{4}, \quad I_y = \frac{\pi a^3 b}{4}$$

The angle of twist in accordance with Eq. (4.5.8) is given by

$$\theta = \frac{M_T}{GJ} = \frac{M_T(a^2 + b^2)}{G\pi a^3 b^3} \qquad (4.5.13)$$

The components of the elastic displacement are

$$\left. \begin{array}{l} u = -\theta zy = -\dfrac{M_T(a^2 + b^2)}{G\pi a^3 b^3} zy \\[2mm] v = \theta zx = \dfrac{M_T(a^2 + b^2)}{G\pi a^3 b^3} zx \\[2mm] w = \theta \phi = \dfrac{M_T(b^2 - a^2)}{G\pi a^3 b^3} xy \end{array} \right\} \qquad (4.5.14)$$

The last of Eqs. (4.5.14) shows that the contour lines defined by the equation

$$w = \text{constant}$$

are hyperbolas having the x- and y-axes (principal axes of the ellipse) as asymptotes (see Fig. 4.5.1). The stresses are given by the formulas

$$\left.\begin{aligned} \tau_{yz} &= G\theta\left[x + \frac{\partial \phi}{\partial y}\right] = \frac{2M_T x}{\pi a^3 b} \\ \tau_{xz} &= G\theta\left[-y + \frac{\partial \phi}{\partial x}\right] = -\frac{2M_T y}{\pi a b^3} \end{aligned}\right\} \quad (4.5.15)$$

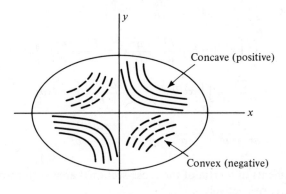

Figure 4.5.1

The resultant shear stress is

$$\tau_{\text{resultant}} = \sqrt{(\tau_{xz})^2 + (\tau_{yz})^2} = \frac{2M_T}{\pi a^2 b^2}\sqrt{\frac{b^2 x^2}{a^2} + \frac{a^2 y^2}{b^2}} \quad (4.5.16)$$

It can be shown that the maximum value of the resultant shear stress is

$$\tau_{\max} = \frac{2M_T}{\pi a b^2} \quad (4.5.17)$$

and occurs at the extremities of the minor axis, i.e., at points A and B of Fig. 4.5.2. To prove this, consider an ellipse of semi-axes a_1 and b_1 such that

$$\frac{a_1}{a} = \frac{b_1}{b} < 1$$

The equation of this ellipse, in parametric form will be

$$\left.\begin{aligned} x &= a_1 \cos \beta \\ y &= b_1 \sin \beta \end{aligned}\right\} \quad (4.5.18)$$

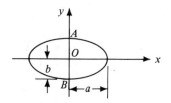

Figure 4.5.2

By substituting Eqs. (4.5.18) into Eq. (4.5.16) one obtains

$$\tau_{\text{resultant}} = \frac{2M_T}{\pi a^2 b^2} \sqrt{b^2 \left(\frac{a_1}{a}\right)^2 \cos^2 \beta + a^2 \left(\frac{b_1}{b}\right)^2 \sin^2 \beta}$$

$$= \frac{2M_T}{\pi a^2 b^2} \frac{a_1}{a} \sqrt{b^2 \cos^2 \beta + a^2 \sin^2 \beta}$$

$$= \frac{2M_T}{\pi a^2 b^2} \frac{a_1}{a} \sqrt{b^2 + (a^2 - b^2) \sin^2 \beta}$$

Since $a > b$, $\tau_{\text{resultant}}$ is a maximum when $a_1 = a$ and $\beta = \pm\pi/2$, i.e., at the extremities of the minor axis. The value of $\tau_{\max}$ is then given by Eq. (4.5.17).

In the case $a = b$, the cross section reduces to a circle and Eq. (4.5.17) becomes Coulomb's formula, i.e.,

$$\tau_{\max} = \frac{2M_T}{\pi a^3} = \frac{M_T a}{J}$$

The direction of $\tau_{\text{resultant}}$ is given by

$$\frac{\tau_{xz}}{\tau_{yz}} \propto \frac{y}{x}$$

and therefore is constant along a line going through the center O as, for instance, along the straight line OP of Fig. 4.5.3. The direction of the resultant coincides with the direction of the tangent PP' to the border of the cross section at the point of intersection of line OP with the border.

(c) **Case of a rectangular cross section.** Consider the rectangular cross section of sides $2a$ and $2b$ as shown in Fig. 4.5.4. The coordinate axes x and y are parallel to the sides, and the origin of coordinates O is at the centroid of the cross section. In order to solve the torsion problem for this rectangular cross section in accordance with the Saint-Venant theory of torsion, a warping function $\phi(x, y)$ must be found which satisfies the equation

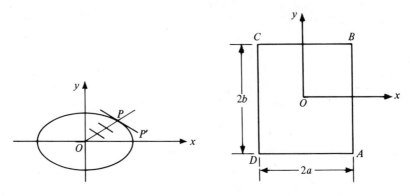

Figure 4.5.3   Figure 4.5.4

$$\frac{\partial^2 \phi}{\partial x^2} + \frac{\partial^2 \phi}{\partial y^2} = 0 \qquad (a)$$

and the boundary conditions

$$\left(\frac{\partial \phi}{\partial x} - y\right)\frac{dy}{ds} - \left(\frac{\partial \phi}{\partial y} + x\right)\frac{dx}{ds} = 0 \qquad (b)$$

On lines AB and CD ($x = \pm a$) of Fig. 4.5.4, $dy/ds = 1$ while $dx/ds = 0$; on lines BC and AD ($y = \pm b$) $dy/ds = 0$ while $dx/ds = 1$. Equation (b) can therefore be written in the following form:

$$\left.\begin{aligned}\frac{\partial \phi}{\partial x} &= y \quad \text{for } x = \pm a \\ \frac{\partial \phi}{\partial y} &= -x \quad \text{for } y = \pm b\end{aligned}\right\} \qquad (4.5.19)$$

Now assume a new warping function

$$\phi_1(x, y) = xy - \phi(x, y) \qquad (4.5.20)$$

which satisfies Eq. (a), i.e.,

$$\frac{\partial^2 \phi_1}{\partial x^2} + \frac{\partial^2 \phi_1}{\partial y^2} = 0 \qquad (4.5.21)$$

Equations (4.5.19) become

$$\frac{\partial \phi_1}{\partial x} = 0 \quad \text{for } x = \pm a$$
$$\frac{\partial \phi_1}{\partial y} = 2x \quad \text{for } y = \pm b \qquad (4.5.22)$$

Let

$$\phi_1(x,y) = \sum_{n=0}^{\infty} X_n(x) Y_n(y) \qquad (4.5.23)$$

By substituting Eq. (4.5.23) into Eq. (4.5.21) one obtains

$$\frac{d^2 X_n(x)}{dx^2} Y_n(y) + X_n(x) \frac{d^2 Y(y)}{dy^2} = 0$$

or

$$\frac{1}{X_n(x)} \frac{d^2 X(x)}{dx^2} = -\frac{1}{Y_n(y)} \frac{d^2 Y_n(y)}{dy^2} \qquad (4.5.24)$$

The left-hand side of Eq. (4.5.24) is a function of x only while the right-hand side is a function of y only. Equation (4.5.24) can be satisfied only if

$$\frac{1}{X_n(x)} \frac{d^2 X(x)}{dx^2} = -\frac{1}{Y_n(y)} \frac{d^2 Y_n(y)}{dy^2} = -K_n^2 \qquad (4.5.25)$$

where K_n is a real quantity; the minus sign must be taken before K_n^2 in order to satisfy the boundary conditions.

From Eq. (4.5.25) the two differential equations

$$\frac{d^2 X_n(x)}{dx^2} + K_n^2 X_n(x) = 0$$
$$\frac{d^2 Y_n(y)}{dy^2} - K_n^2 Y_n(y) = 0 \qquad (4.5.26)$$

are derived. The solutions of these equations are

$$X_n(x) = C_1 \sin K_n x + C_3 \cos K_n x$$
$$Y_n(y) = C_2 \sinh K_n y + C_4 \cosh K_n y \qquad (4.5.27)$$

where C_1, C_2, C_3, and C_4, are constants. To satisfy the second boundary condition [see Eq. (4.5.22)], i.e.,

$$\frac{\partial \phi_1(x,y)}{\partial y} = \sum_{n=0}^{\infty} X_n(x) Y_n'(y) = 2x \quad \text{for } y = \pm b$$

$Y'(y)$ must be a symmetric function of y, while $X(x)$ must be an antisymmetric function of x; thus, it follows that $C_3 = C_4 = 0$ and Eqs. (4.5.27) reduce to

$$\left.\begin{array}{l} X_n(x) = C_1 \sin K_n x \\ Y_n(y) = C_2 \sinh K_n y \end{array}\right\} \quad (4.5.28)$$

The first boundary condition, i.e.,

$$\frac{\partial \phi_1(x, y)}{\partial x} = 0 \quad \text{for } x = \pm a$$

is satisfied if

$$K_n \cos K_n a = 0$$

from which it follows that

$$K_n = \frac{(2n+1)\pi}{2a} \quad (4.5.29)$$

Since C_1 and C_2 are arbitrary constants, the function $\phi_1(x, y)$ can be written in the form

$$\phi_1(x, y) = \sum_{n=0}^{\infty} A_n \sin K_n x \sinh K_n y \quad (4.5.30)$$

where K_n is given by Eq. (4.5.29) and the constants A_n have to be determined so as to satisfy the second of the boundary conditions (4.5.22), i.e.,

$$\frac{\partial \phi_1(x, y)}{\partial y} = 2x \quad \text{for } y = \pm b$$

or

$$\sum_{n=0}^{\infty} A_n K_n \cosh K_n b \sin K_n x = \sum_{n=0}^{\infty} B_n \sin K_n x = 2x \quad (4.5.31)$$

where the constant

$$B_n = A_n K_n \cosh K_n b$$

To compute the coefficients B_n, both terms of Eq. (4.5.31) are multiplied by

$$\sin \frac{(2m+1)\pi x}{2a} dx$$

and integrated from $-a$ to $+a$. Since the orthogonality property of the trigometric functions is

$$\int_{-a}^{a} \sin K_n x \sin K_m x \, dx = \begin{cases} = 0 & \text{for } m \neq n \\ = a & \text{for } m = n \end{cases}$$

one obtains

$$B_n = \frac{16(-1)^n a}{\pi^2 (2n+1)^2}$$

or

$$A_n = \frac{32(-1)^n a^2}{\pi^3 (2n+1)^3 \cosh K_n b}$$

and Eq. (4.5.20) gives for the function $\phi(x, y)$, the value

$$\phi(x, y) = xy - \phi_1(x, y)$$
$$= xy - \frac{32 a^2}{\pi^3} \sum_{n=0}^{\infty} \frac{(-1)^n}{(2n+1)^3} \frac{1}{\cosh K_n b} \sin K_n x \sinh K_n y \quad (4.5.32)$$

The torsional constant J is, in this case,

$$J = \int_{-b}^{b} dy \int_{-a}^{a} \left[x^2 + y^2 + x \frac{\partial \phi}{\partial y} - y \frac{\partial \phi}{\partial x} \right] dx$$
$$= \frac{8 a^3 b}{3} \left[1 + \frac{96}{\pi^4} \sum_{n=0}^{\infty} \frac{1}{(2n+1)^4} - \frac{384 a}{\pi^5 b} \sum_{n=0}^{\infty} \frac{\tanh K_n b}{(2n+1)^5} \right]$$

But

$$\sum_{n=0}^{\infty} \frac{1}{(2n+1)^4} = \frac{\pi^4}{96}$$

and thus

$$J = 16 a^3 b \left[\frac{1}{3} - \frac{64 a}{\pi^5 b} \sum_{n=0}^{\infty} \frac{\tanh K_n b}{(2n+1)^5} \right] = k_1 a^3 b \quad (4.5.33)$$

In Table 4.5.1, values of constant k_1 are given for various values of the ratio b/a. It can, however, be shown,[7] that for practical purposes, within an approximation of 0.5 percent, the formula

$$J = 16 a^3 b \left[\frac{1}{3} - \frac{64}{\pi^5} \frac{a}{b} \tanh \frac{\pi b}{2a} \right] \quad (4.5.34)$$

can be used instead of Eq. (4.5.33). The shearing stresses are given by

[7] See Bibliography Sec. 4.5, No. 42 at the end of this chapter.

TABLE 4.5.1

b/a	k_1	k_2	k_3
1.0	2.250	1.350	0.600
1.2	2.656	1.518	0.571
1.5	3.136	1.696	0.541
2.0	3.664	1.860	0.508
2.5	3.984	1.936	0.484
3.0	4.208	1.970	0.468
4.0	4.496	1.994	0.443
5.0	4.656	1.998	0.430
10.0	4.992	2.000	0.401
∞	5.328	2.000	0.375

$$\left.\begin{aligned}\tau_{xz} &= \frac{M_T}{J}\left(\frac{\partial \phi}{\partial x} - y\right) = -\frac{16 M_T a}{\pi^2 J}\sum_{n=0}^{\infty}\frac{(-1)^n}{(2n+1)^2}\frac{\sinh K_n y}{\cosh K_n b}\cos K_n x \\ \tau_{yz} &= \frac{M_T}{J}\left(\frac{\partial \phi}{\partial y} + x\right) = \frac{M_T}{J}\left[2x - \frac{16a}{\pi^2}\sum_{n=0}^{\infty}\frac{(-1)^n \cosh K_n y}{(2n+1)^2 \cosh K_n b}\sin K_n x\right]\end{aligned}\right\}$$
(4.5.35)

In the case in which $b > a$, the maximum shear stress occurs at the midpoints of the long sides $x = \pm a$ of the rectangle. Letting $x = a$ and $y = 0$ in Eqs. (4.5.35) gives

$$\left.\begin{aligned}\tau_{xz} &= 0 \\ \tau_{yz} &= \tau_{max} = \frac{2 M_T a}{J}\left[1 - \frac{8}{\pi^2}\sum_{n=0}^{\infty}\frac{1}{(2n+1)^2 \cosh K_n b}\right] = k_2 \frac{M_T a}{J}\end{aligned}\right\}$$
(4.5.36)

In Table 4.5.1, values of constant k_2 are given for various values of the ratio b/a. By substituting the expression for J given by Eq. (4.5.33) into Eq. (4.5.36) one obtains

$$\tau_{max} = k_3 \frac{M_T}{a^2 b} \tag{4.5.37}$$

Values of constant k_3 are given in Table 4.5.1 for various values of the ratio b/a.

In the case of a narrow, rectangular cross section (see Fig. 4.5.5) Eqs. (4.5.33) and (4.5.36) can be simplified. Writing t instead of $2a$ and H instead of 2b we have that H/t is a large number. In this case, $\tanh H/t$ is near unity and the second term inside the bracket of Eq. (4.5.33) becomes negligible. It follows that

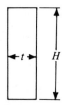

Figure 4.5.5

$$J = \frac{16a^3b}{3} = \frac{t^3 H}{3} \qquad (4.5.38)$$

It also follows that

$$\cosh K_n H = \cosh \frac{(2n+1)\pi}{2} \frac{H}{t}$$

is a large number and the second term inside the bracket of Eq. (4.5.36) becomes negligible; therefore, Eq. (4.5.36) reduces to

$$\tau_{yz} = \tau_{max} = \frac{M_T t}{J} \qquad (4.5.39)$$

Once the torsional constant J is determined, the angle of twist is calculated from the formula

$$\theta = \frac{M_T}{GJ} \qquad (4.5.40)$$

4.6 Prandtl's Theory of Torsion

An alternative to the Saint-Venant method for solving torsion problems was introduced by Ludwig Prandtl. Consider the components of the elastic displacements in the form

$$\left. \begin{array}{l} u = -\theta z y \\ v = \theta z x \\ w = \theta \phi(x, y) \end{array} \right\} \qquad (4.6.1)$$

Prandtl, Ludwig. German physicist and applied mechanicist (b. Freising, Bavaria, 1875; d. Gottingen 1953). He was professor of applied mechanics and director of the Kaiser Wilhelm Institute for Fluid Mechanics at Gottingen University. He did work in the mathematical theory of elasticity and in plasticity. His most important contributions were in aerodynamics with his discovery in 1904 of the "boundary layer" in the theories of supersonic flow and of turbulence. He also contributed to the development of wind tunnels and other aerodynamic techniques. Theodore von Kármán was one of his students. [Reproduction from the Vito Volterra collection, Villa Volterra, Ariccia (Rome).]

From Eqs. (4.6.1) the components of stress are derived; thus,

$$\left.\begin{array}{l} \sigma_x = \sigma_y = \sigma_z = \tau_{xy} = 0 \\ \tau_{xz} = G\theta\left(-y + \dfrac{\partial \phi}{\partial x}\right) \\ \tau_{yz} = G\theta\left(x + \dfrac{\partial \phi}{\partial y}\right) \end{array}\right\} \qquad (4.6.2)$$

By substituting Eqs. (4.6.2) into the equations of equilibrium [Eqs. (1.4.1)] which in absence of body forces are

$$\left.\begin{array}{l} \dfrac{\partial \sigma_x}{\partial x} + \dfrac{\partial \tau_{xy}}{\partial y} + \dfrac{\partial \tau_{xz}}{\partial z} = 0 \\ \dfrac{\partial \tau_{yx}}{\partial x} + \dfrac{\partial \sigma_y}{\partial y} + \dfrac{\partial \tau_{yz}}{\partial z} = 0 \\ \dfrac{\partial \tau_{zx}}{\partial x} + \dfrac{\partial \tau_{zy}}{\partial y} + \dfrac{\partial \sigma_z}{\partial z} = 0 \end{array}\right\} \qquad (1.4.1)$$

one finds that

$$\left.\begin{array}{l} \dfrac{\partial \tau_{xz}}{\partial z} = 0 \\ \dfrac{\partial \tau_{yz}}{\partial z} = 0 \\ \dfrac{\partial \tau_{zx}}{\partial x} + \dfrac{\partial \tau_{zy}}{\partial y} = 0 \end{array}\right\} \qquad (4.6.3)$$

Since the components of stress given by Eqs. (4.6.2) are independent of variable z, the first two of Eqs. (4.6.3) are satisfied. In order to satisfy the third of Eqs. (4.6.3), Prandtl introduces a stress function $\psi(x, y)$ such that

$$\left.\begin{array}{l}\tau_{xz} = \dfrac{\partial \psi}{\partial y} \\[1em] \tau_{yz} = -\dfrac{\partial \psi}{\partial x}\end{array}\right\} \quad (4.6.4)$$

In view of Eqs. (4.6.2), Eqs. (4.6.4) can be written

$$\left.\begin{array}{l}\tau_{xz} = \dfrac{\partial \psi}{\partial y} = G\theta\left(-y + \dfrac{\partial \phi}{\partial x}\right) \\[1em] \tau_{yz} = -\dfrac{\partial \psi}{\partial x} = G\theta\left(x + \dfrac{\partial \phi}{\partial y}\right)\end{array}\right\} \quad (4.6.5)$$

By differentiating the first of Eqs. (4.6.5) with respect to y, the second with respect to x, and subtracting the first from the second we get

$$\frac{\partial^2 \psi}{\partial x^2} + \frac{\partial^2 \psi}{\partial y^2} = F = -2G\theta \quad (4.6.6)$$

Now consider the boundary conditions:

$$\left.\begin{array}{l}\bar{X} = \sigma_x \cos{(\widehat{nx})} + \tau_{xy} \cos{(\widehat{ny})} + \tau_{xz} \cos{(\widehat{nz})} \\ \bar{Y} = \tau_{yx} \cos{(\widehat{nx})} + \sigma_y \cos{(\widehat{ny})} + \tau_{yz} \cos{(\widehat{nz})} \\ \bar{Z} = \tau_{zx} \cos{(\widehat{nx})} + \tau_{zy} \cos{(\widehat{ny})} + \sigma_z \cos{(\widehat{nz})}\end{array}\right\} \quad (4.6.7)$$

Since in the present case $\bar{X} = \bar{Y} = \bar{Z} = \sigma_x = \sigma_y = \sigma_z = \tau_{xy} = \cos{(\widehat{nz})} = 0$, the first two of Eqs. (4.6.7) are satisfied, while the third becomes

$$\tau_{xz} \cos{(\widehat{nx})} + \tau_{yz} \cos{(\widehat{ny})} = \frac{\partial \psi}{\partial y}\frac{dy}{ds} + \frac{\partial \psi}{\partial x}\frac{dx}{ds} = \frac{d\psi}{ds} = 0 \quad (4.6.8)$$

Equation (4.6.8) shows that the stress function $\psi(x, y)$ must be constant along the boundary of the cross section. This constant will, in the following discussion, be assumed to be zero, i.e.,

$$\psi(x, y) = 0 \quad (4.6.9)$$

on the boundary.

According to Prandtl's theory, determination of the stress distribution over a cross section of a twisted bar is reduced to the problem of finding the function $\psi(x, y)$ which satisfies Eq. (4.6.6) and which is identically zero along the boundary.

The conditions at the free ends of the twisted bar, shown in Fig. 4.6.1, will now be considered, and it will be shown that the action of all the forces acting on an end is a couple. At the free ends, the following conditions exist:

$$\cos(\widehat{nx}) = \cos(\widehat{ny}) = 0, \quad \cos(\widehat{nz}) = \pm 1$$
$$\sigma_x = \sigma_y = \sigma_z = \tau_{xy} = 0$$

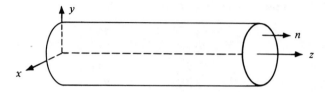

Figure 4.6.1

The boundary equations [Eqs. (1.6.2)], i.e.,

$$\left.\begin{aligned}\bar{X} &= \sigma_x \cos(\widehat{nx}) + \tau_{xy}\cos(\widehat{ny}) + \tau_{xz}\cos(\widehat{nz}) \\ \bar{Y} &= \tau_{yx}\cos(\widehat{nx}) + \sigma_y \cos(\widehat{ny}) + \tau_{yz}\cos(\widehat{nz}) \\ \bar{Z} &= \tau_{zx}\cos(\widehat{nx}) + \tau_{zy}\cos(\widehat{ny}) + \sigma_z \cos(\widehat{nz})\end{aligned}\right\} \quad (1.6.2)$$

become in this case

$$\left.\begin{aligned}\bar{X} &= \pm\tau_{xz} \\ \bar{Y} &= \pm\tau_{yz} \\ \bar{Z} &= 0\end{aligned}\right\} \quad (4.6.10)$$

By substituting Eqs. (4.6.4) into Eqs. (4.6.10) and computing the resultant force, one finds that

$$\iint_\Omega \bar{X}dxdy = \iint_\Omega \tau_{xz}dxdy = \iint_\Omega \frac{\partial \psi}{\partial y}dxdy = \int dx \int \frac{\partial \psi}{\partial y}dy = \int dx[\psi] = 0$$

since ψ is zero at the boundary of the cross section. In the same way one finds that

$$\iint_\Omega \bar{Y} dx dy = \iint_\Omega \tau_{yz} dx dy = -\iint_\Omega \frac{\partial \psi}{\partial x} dx dy = -\int dy \int \frac{\partial \psi}{\partial x} dx$$
$$= \int dy [\psi] = 0$$

Calling M_T the couple acting on a free end, one has that

$$M_T = \iint_\Omega (\bar{Y}x - \bar{X}y) dx dy = -\iint_\Omega \frac{\partial \psi}{\partial x} x dx dy - \iint_\Omega \frac{\partial \psi}{\partial y} y dx dy$$

Since $\psi = 0$ at the boundary, one finds upon integrating this equation by parts that

$$M_T = 2 \iint_\Omega \psi(x, y) \, dx dy \qquad (4.6.11)$$

Prandtl's theory of torsion will now be applied to the study of torsion of bars having

1. an elliptical cross section
2. an equilateral triangular cross section

(a) **Case of an elliptical cross section.** The ellipse of semi-axes a and b shown in Fig. 4.6.2 will be considered. It's equation is

$$\frac{x^2}{a^2} + \frac{y^2}{b^2} = 1$$

For the stress function $\psi(x, y)$, the expression

$$\psi(x, y) = m \left[\frac{x^2}{a^2} + \frac{y^2}{b^2} - 1 \right] \qquad (4.6.12)$$

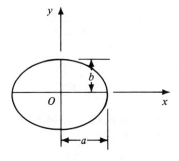

Figure 4.6.2

will be assumed, where m is a constant to be determined. Equation (4.6.12) satisfies the boundary condition $\psi = 0$. By substituting Eq. (4.6.12) into Eq. (4.6.6) one gets

$$\frac{\partial^2 \psi}{\partial x^2} + \frac{\partial^2 \psi}{\partial y^2} = F = -2G\theta = 2m\left(\frac{1}{a^2} + \frac{1}{b^2}\right) = 2\frac{a^2 + b^2}{a^2 b^2} m$$

or

$$m = \frac{a^2 b^2}{2(a^2 + b^2)} F \qquad (4.6.13)$$

By substituting Eq. (4.6.13) into Eq. (4.6.12) one obtains

$$\psi = \frac{a^2 b^2 F}{2(a^2 + b^2)}\left[\frac{x^2}{a^2} + \frac{y^2}{b^2} - 1\right] \qquad (4.6.14)$$

The magnitude of the constant F is determined by use of Eq. (4.6.11). Upon substituting Eq. (4.6.14) into Eq. (4.6.11) one obtains

$$M_T = \frac{a^2 b^2 F}{a^2 + b^2}\left[\frac{1}{a^2}\iint_\Omega x^2 \, dxdy + \frac{1}{b^2}\iint_\Omega y^2 \, dxdy - \iint_\Omega dxdy\right] \qquad (4.6.15)$$

But

$$\iint_\Omega x^2 \, dxdy = \frac{\pi a^3 b}{4} = I_y$$

$$\iint_\Omega y^2 \, dxdy = \frac{\pi a b^3}{4} = I_x$$

$$\iint_\Omega dxdy = \pi a b$$

and Eq. (4.6.15) becomes

$$M_T = -\frac{\pi a^3 b^3 F}{2(a^2 + b^2)}$$

from which

$$F = -\frac{2M_T(a^2 + b^2)}{\pi a^3 b^3}$$

By substituting the above value of F into Eq. (4.6.14) one finally obtains for the stress function the expression

$$\psi(x, y) = -\frac{M_T}{\pi ab}\left[\frac{x^2}{a^2} + \frac{y^2}{b^2} - 1\right] \qquad (4.6.16)$$

The stress components are

$$\tau_{xz} = \frac{\partial \psi}{\partial y} = -\frac{2M_T y}{\pi a b^3}$$
$$\tau_{yz} = -\frac{\partial \psi}{\partial x} = \frac{2M_T x}{\pi a^3 b}$$
(4.6.17)

and the angle of twist is

$$\theta = -\frac{1}{2G}\left[\frac{\partial^2 \psi}{\partial x^2} + \frac{\partial^2 \psi}{\partial y^2}\right] = M_T \frac{a^2 + b^2}{\pi a^3 b^3 G} \qquad (4.6.18)$$

(b) Case of an equilateral triangular cross section. Consider the equilateral triangle *ABC* referred to the system of coordinates *xy* with the origin *O* at the centroid of the triangle as shown in Fig. 4.6.3. The equations of lines *AB*, *AC*, and *BC* are, respectively,

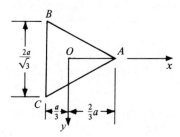

Figure 4.6.3

$$x - \sqrt{3}\, y - \tfrac{2}{3}a = 0$$
$$x + \sqrt{3}\, y - \tfrac{2}{3}a = 0$$
$$x + \tfrac{1}{3}a = 0$$

For the stress function, the expression

$$\psi(x, y) = m[(x - \sqrt{3}\, y - \tfrac{2}{3}a)(x + \sqrt{3}\, y - \tfrac{2}{3}a)(x + \tfrac{1}{3}a)]$$

or

$$\psi(x, y) = m\left[\tfrac{1}{2}(x^2 + y^2) - \frac{1}{2a}(x^3 - 3xy^2) - \tfrac{2}{27}a^2\right] \qquad (4.6.19)$$

is assumed. This function satisfies the boundary condition $\psi(x, y) = 0$. Moreover, by substituting Eq. (4.6.19) into Eq. (4.6.6), one finds that

$$m = -G\theta \qquad (4.6.20)$$

In view of Eq. (4.6.20), Eq. (4.6.19) can be written as

$$\psi(x, y) = -G\theta\left[\tfrac{1}{2}(x^2 + y^2) - \frac{1}{2a}(x^3 - 3xy^2) - \tfrac{2}{27}a^2\right] \quad (4.6.21)$$

It follows that

$$M_T = 2\iint_\Omega \psi(x, y)\,dx\,dy = \frac{G\theta a^4}{15\sqrt{3}}$$

and

$$\left. \begin{aligned} \tau_{xz} &= \frac{\partial \psi}{\partial y} = -G\theta\left[y + \frac{3xy}{a}\right] \\ \tau_{yz} &= -\frac{\partial \psi}{\partial x} = G\theta\left[x - \frac{3x^2}{2a} + \frac{3y^2}{2a}\right] \end{aligned} \right\} \quad (4.6.22)$$

From Eq. (4.6.22) it follows that $\tau_{xz} = \tau_{yz} = 0$ at the three corners A, B, and C and at the centroid O, since the coordinates of these points are, respectively,

$$A(\tfrac{2}{3}a, 0), \quad B\left(-\frac{a}{3}, -\frac{a}{\sqrt{3}}\right), \quad C\left(-\frac{a}{3}, \frac{a}{\sqrt{3}}\right), \quad O(0, 0)$$

Also $\tau_{xz} = 0$ for $y = 0$, i.e., on the x-axis. The maximum stress occurs at $x = -a/3$, $y = 0$ for which

$$\tau_{yz}\left(-\frac{a}{3}, 0\right) = G\theta\left[-\frac{a}{3} - \frac{3}{2a}\left(\frac{a^2}{9}\right)\right] = -\frac{G\theta a}{2} = \tau_{\max}$$

Figure 4.6.4 shows the distribution of the stress τ_{yz} along the x-axis.

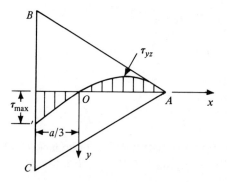

Figure 4.6.4

4.7 Prandtl's Membrane Analogy

The membrane analogy, introduced by Prandtl, has proved of very great value in solving torsion problems. Consider a homogeneous, perfectly flexible membrane, bounded by a curve C in the xy-plane and having the same outline as that of the cross section of a twisted bar. The membrane, which is subjected to a uniform tension S and to a uniform lateral pressure p, is ref-

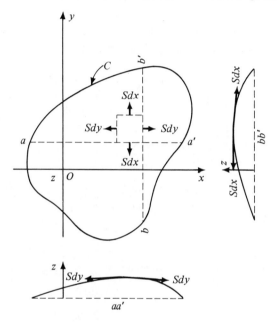

Volterra & Gaines 4.7.1

erred to a system of orthogonal cartesian coordinates xyz (see Fig. 4.7.1). The following assumptions will be made:

1. The undeformed position of the membrane is in the xy-plane. Each point P of the membrane is identified by its coordinates at the undeformed position. Each point $P(x, y)$ can move only in the z-direction and its displacement is denoted by $w = w(x, y)$. It is assumed that the values of w are very small as compared to the dimensions of the membrane and that $w(x, y)$ is zero on the boundary.
2. The tension S is sufficiently great so that the plane, tangent to the membrane at P, makes very small angles with the x- and y-axes.

To obtain the equation of equilibrium of the membrane under the distributed load $p(x, y)$ per unit area, we consider a small element of area $dxdy$.

On its sides are acting forces, $S dy$ and $S dx$. The sum of the projections of the two $S dy$ forces on the z-axis is

$$S dy \left[\frac{\partial w}{\partial x} + \frac{\partial}{\partial x}\left(\frac{\partial w}{\partial x}\right)dx\right] - S dy \frac{\partial w}{\partial x} = S \frac{\partial^2 w}{\partial x^2} dx dy$$

Similarly, the sum of the projections of the two $S dx$ forces on the z-axis is

$$S \frac{\partial^2 w}{\partial y^2} dx dy$$

The external force acting in the z-direction on the area $dxdy$ is $p(x, y)dxdy$. It follows from the equation of equilibrium that

$$-p(x, y) = S \left[\frac{\partial^2 w(x, y)}{\partial x^2} + \frac{\partial^2 w(x, y)}{\partial y^2}\right]$$

or

$$\frac{\partial^2 w(x, y)}{\partial x^2} + \frac{\partial^2 w(x, y)}{\partial y^2} = -\frac{p(x, y)}{S} \qquad (4.7.1)$$

If we compare these results with the results obtained in the previous section from Prandtl's theory of torsion [Eqs. (4.6.6) and (4.6.9)] we see that the following analogy exists between the stretched membrane and the twisted rod:

Membrane Problem		Torsion Problem
The displacement $w(x, y)$	corresponds to	the stress function $\psi(x, y)$
Equation $\Delta w = -\dfrac{p}{S}$		$\Delta \psi = -2G\theta$
Boundary condition $w = 0$		$\psi = 0$
Slope of a contour line $\dfrac{dw}{ds} = 0$		$\dfrac{d\psi}{ds} = 0$

By use of Prandtl's membrane analogy, it can be shown that the direction of the resultant shearing stress at a point in the cross section of a twisted bar has the same direction as the tangent to the contour line at the same point on the membrane. In fact, for a contour line of the membrane

$$\frac{\partial w}{\partial s} = 0 = \frac{\partial w}{\partial x}\frac{dx}{ds} + \frac{\partial w}{\partial y}\frac{dy}{ds}$$

The corresponding expression for the twisted bar is

$$\frac{\partial \psi}{\partial x}\frac{dx}{ds} + \frac{\partial \psi}{\partial y}\frac{dy}{ds} = 0 \qquad (4.7.2)$$

From Fig. 4.7.2 one finds that

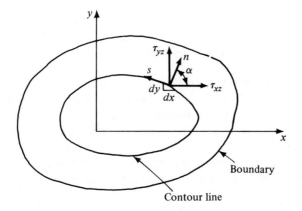

Figure 4.7.2

$$-\frac{dx}{ds} = \sin \alpha, \quad \frac{dy}{ds} = \cos \alpha$$

By use of these relations and Eqs. (4.6.4), it follows that Eq. (4.7.2) may be expressed as

$$\frac{\partial \psi}{\partial x}\frac{dx}{ds} + \frac{\partial \psi}{\partial y}\frac{dy}{ds} = \tau_{yz} \sin \alpha + \tau_{xz} \cos \alpha = 0 \qquad (4.7.3)$$

Equation (4.7.3) shows that the projection of the resultant shearing stress at a point on the normal n to the contour line is zero (see Fig. 4.7.2); therefore, the resultant shearing stress at a point has the same direction as the tangent to the contour line at that point on the deflected membrane. The curves drawn in the cross section of a twisted bar in such a way that the tangent to the curve at a generic point has the direction of the resultant shear stress at the same point are called shearing stress lines.

The magnitude of the resultant shearing stress is given by

$$\tau_{\text{resultant}} = \tau_{yz} \cos \alpha - \tau_{xz} \sin \alpha$$

But since

$$\tau_{yz} = -\frac{\partial \psi}{\partial x} \quad \tau_{xz} = \frac{\partial \psi}{\partial y} \quad \cos \alpha = \frac{dx}{dn} \quad \sin \alpha = \frac{dy}{dn}$$

it follows that

$$\tau_{\text{resultant}} = -\left[\frac{\partial \psi}{\partial x}\frac{dx}{dn} + \frac{\partial \psi}{\partial y}\frac{dy}{dn}\right] = -\frac{\partial \psi}{\partial n}$$

or

$$\tau_{\text{resultant}} \propto \frac{\partial w}{\partial n} \qquad (4.7.4)$$

In Eq. (4.7.4), n is measured positive along the outward normal to the contour line (see Fig. 4.7.2). Equation (4.7.4) shows that $\tau_{\text{resultant}}$ is proportional to the maximum slope of the membrane at each point. The actual value of the resultant stress is obtained by replacing p/S with $2G\theta$. Obviously the greatest values for the shear stress (i.e., the maximum stress concentration) will occur where the contour lines are closest together (i.e., where there is the maximum concentration of contour lines). From Eq. (4.6.11), i.e.,

$$M_T = 2 \iint_\Omega \psi(x, y)\, dx\, dy \qquad (4.6.11)$$

it is concluded that the torque is given by twice the volume bounded by the deflected membrane and the xy-plane with p/S replaced by $2G\theta$.

By applying Saint-Venant's theory of torsion to a narrow, rectangular cross section, the following formulas were derived:

$$J = \frac{t^3 H}{3} \qquad (4.5.38)$$

$$\tau_{\max} = \frac{M_T t}{J} \qquad (4.5.39)$$

$$\theta = \frac{M_T}{GJ} \qquad (4.5.40)$$

Equations (4.5.38), (4.5.39), and (4.5.40) can be derived directly by the use of Prandtl's membrane analogy [see Figs. 4.7.3 (a) and (b)]. When H/t is very large the effects of the short sides of the rectangle are negligible; therefore, the surface of the deflected membrane can be assumed to be cylindrical and it follows that Eq. (4.7.1) reduces to

$$\frac{\partial^2 w}{\partial x^2} = -\frac{p}{S}$$

and upon integration

$$w = -\frac{px^2}{2S} + C_1 x + C_2$$

The two constants C_1 and C_2 are determined by the two boundary conditions, $w = 0$ for $x = \pm t/2$, and have values $C_1 = 0$ and $C_2 = pt^2/8S$. It follows that

ELEMENTARY PROBLEMS IN THREE-DIMENSIONAL ELASTICITY

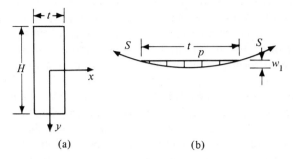

Figure 4.7.3

$$w(x) = -\frac{px^2}{2S} + \frac{pt^2}{8S}$$

$$w(0) = w_1 = \frac{pt^2}{8S}$$

The volume of the membrane is

$$V = \tfrac{2}{3}Htw_1 = \frac{pt^3 H}{12S} \tag{4.7.5}$$

while the maximum slope of the parabola is

$$\left(\frac{\partial w}{\partial x}\right)_{\max} = \frac{4w_1}{t} = \frac{pt}{2S} \tag{4.7.6}$$

Now using the membrane analogy, $\tau_{\max}$ is substituted for $(\partial w/\partial x)_{\max}$, $2G\theta$ is substituted for p/S and M_T is substituted for $2V$. It follows that

$$\tau_{\max} = tG\theta \tag{4.7.7}$$

$$M_T = \frac{Ht^3 G\theta}{3} \tag{4.7.8}$$

From Eqs. (4.7.7) and (4.7.8) one obtains

$$\tau_{\max} = \frac{3M_T}{Ht^2} = \frac{M_T t}{J} \tag{4.5.39}$$

$$\theta = \frac{3M_T}{Ht^3 G} = \frac{M_T}{GJ} \tag{4.5.40}$$

$$J = \frac{M_T}{G\theta} = \frac{Ht^3}{3} \tag{4.5.38}$$

The practical importance of Prandtl's analogy was recognized by A. A. Griffith and G. I. Taylor, who used the soap film method for studying twisting of bars of various complicated cross sections. If a thin membrane, such as one made of a thin rubber sheet or a soap film, is stretched over a hole having the same form as that of the cross section of a twisted bar and is subjected to a pressure p, it will deflect according to Eq. (4.7.1). It is then possible with a probe to determine the deflections of the membrane and to plot contour lines (i.e., lines joining points having the same deflection). Once the shape of the deflected membrane is determined, the volume and the corresponding torque can be obtained by numerical integration. Similarly, approximate values of the slopes and the corresponding stresses can be obtained from numerical expressions for the derivatives of functions.

4.8 Ritz's Method Applied to Torsion Problems

The internal strain energy per unit length of the twisted bar shown in Fig. 4.8.1 is

$$U' = \frac{1}{2} \iint_\Omega [\tau_{xz}\gamma_{xz} + \tau_{yz}\gamma_{yz}]\,dxdy = \frac{1}{2G} \iint_\Omega [\tau_{xz}^2 + \tau_{yz}^2]\,dxdy \qquad (4.8.1)$$
$$= \frac{1}{2G} \iint_\Omega \left[\left(\frac{\partial \psi}{\partial x}\right)^2 + \left(\frac{\partial \psi}{\partial y}\right)^2\right] dxdy$$

since, in accordance with Prandtl's theory,

$$\left. \begin{array}{l} \tau_{xz} = \dfrac{\partial \psi}{\partial y} \\[6pt] \tau_{yz} = -\dfrac{\partial \psi}{\partial x} \end{array} \right\} \qquad (4.6.4)$$

The stress function $\psi(x, y)$ must satisfy the condition

$$\psi(x, y) = 0 \qquad (4.6.9)$$

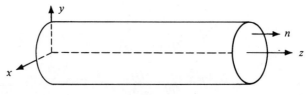

Figure 4.8.1

on the boundary and the torque is given by

$$M_T = 2 \iint_\Omega \psi(x, y)\, dx\, dy \qquad (4.6.11)$$

It follows from the minimum energy theorem (Menabrea's theorem), that of all the stress functions $\psi(x, y)$ which satisfy Eq. (4.6.9), the one which corresponds to the true stress components is the one which makes the variation of the strain energy zero. Equation (4.8.1) gives the strain energy per unit length of the shaft, since the function ψ is independent of variable z. Calling l the length of the shaft, the total strain energy will be expressed by

$$U = \frac{l}{2G} \iint_\Omega \left[\left(\frac{\partial \psi}{\partial x}\right)^2 + \left(\frac{\partial \psi}{\partial y}\right)^2 \right] dx\, dy \qquad (4.8.2)$$

In accordance with Menabrea's theorem, the torsion problem is therefore reduced to the following problem: to determine among all the functions $\psi = \psi(x, y)$, which become zero on the border of the cross sectional area of the twisted bar, the one which makes Eq. (4.8.2) a minimum. In order to satisfy this condition, Ritz's method will be followed. Suppose that n functions $\psi_i(x, y)$ are known which satisfy the condition $\psi_i(x, y) = 0$ on the border of the cross section of the twisted shaft. Any linear combination of these n functions, i.e.,

$$p_1 \psi_1(x, y) + p_2 \psi_2(x, y) + \ldots + p_n \psi_n(x, y) \qquad (4.8.3)$$

will be zero on the boundary. By substituting Eq. (4.8.3) into Eq. (4.6.11) one has

$$M_T = 2 \left\{ p_1 \iint_\Omega \psi_1(x, y)\, dx\, dy + p_2 \iint_\Omega \psi_2(x, y)\, dx\, dy + \ldots \right. $$
$$\left. + p_n \iint_\Omega \psi_n(x, y)\, dx\, dy \right\} \qquad (4.8.4)$$

while Eq. (4.8.2) becomes

$$U = \frac{l}{2G} \left\{ p_1^2 \iint_\Omega \left[\left(\frac{\partial \psi_1}{\partial x}\right)^2 + \left(\frac{\partial \psi_1}{\partial y}\right)^2 \right] dx\, dy \right.$$
$$+ p_2^2 \iint_\Omega \left[\left(\frac{\partial \psi_2}{\partial x}\right)^2 + \left(\frac{\partial \psi_2}{\partial y}\right)^2 \right] dx\, dy + \ldots$$
$$\left. + 2 p_1 p_2 \iint_\Omega \left[\frac{\partial \psi_1}{\partial x} \frac{\partial \psi_2}{\partial x} + \frac{\partial \psi_1}{\partial y} \frac{\partial \psi_2}{\partial y} \right] dx\, dy + \ldots \right\} \qquad (4.8.5)$$

The torsion problem is therefore reduced to the problem of determining constants p_i ($i = 1, 2, \ldots, n$) so as to render the function U of Eq. (4.8.5) a minimum. Once the parameters $p_i(i = 1, 2, \ldots, n)$ are determined, the function

$$\psi(x, y) = \sum_{i=1}^{n} p_i \psi_i(x, y)$$

is also determined, and from this function the stresses τ_{xz} and τ_{yz} are derived. The angle of twist θ is determined by the use of Clapeyron's theorem, which states that the strain energy is equal to half the work produced by the external forces if they remained constant during the elastic deformation of the shaft. Since, in the present case, the external forces reduce to the twisting couple M_T applied at the extreme section of the shaft ($z = l$) and this section rotates through the angle $l\theta$, one half of the work L of the external forces is

$$\frac{L}{2} = \frac{M_T l \theta}{2} \tag{4.8.6}$$

Example 4.8.1. Ritz's[8] method will now be applied to a shaft of rectangular cross section[9] of sides $2a$ and $2b$ (see Fig. 4.8.2). Let

$$\psi(x, y) = \sum_{i=1}^{3} p_i \psi_i(x, y)$$

with

$$\left.\begin{aligned}
\psi_1(x, y) &= (a^2 - x^2)(b^2 - y^2) \\
\psi_2(x, y) &= x^2(a^2 - x^2)(b^2 - y^2) \\
\psi_3(x, y) &= y^2(a^2 - x^2)(b^2 - y^2)
\end{aligned}\right\} \tag{4.8.7}$$

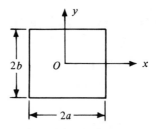

Figure 4.8.2

Equations (4.8.7) and therefore the function ψ satisfy the zero condition on the border of the rectangle ($x = \pm a, y = \pm b$). Equation (4.8.4) becomes in this case

---

[8] **Ritz, Walter.** Swiss applied mathematician and physicist (b. Sion in Valais 1878; d. Gottingen 1909). His complete works were published by the Société Suisse de Physique in 1911.

[9] This example is taken from August Föppl's "Drang und Zwang," Vol. 2, 1920, R. Oldenbourg, p. 77.

$$M_T = \frac{32a^3b^3}{45}[5p_1 + a^2p_2 + b^2p_3] \qquad (4.8.8)$$

while Eq. (4.8.5) becomes

$$U = \frac{64a^3b^3}{45G}l\left\{p_1^2(a^2 + b^2) + p_2^2 a^4\left(\frac{a^2}{21} + \frac{11b^2}{35}\right)\right.$$
$$+ p_3^2 b^4\left(\frac{b^2}{21} + \frac{11a^2}{35}\right) + 2p_1 p_2 a^2\left(\frac{a^2}{7} + \frac{b^2}{5}\right)$$
$$\left. + 2p_1 p_3 b^3\left(\frac{b^2}{7} + \frac{a^2}{5}\right) + 2p_2 p_3 a^2 b^2 \frac{a^2 + b^2}{35}\right\}$$

By use of the equation

$$\frac{\partial U}{\partial p_i} = 0 \quad (i = 1, 2, 3)$$

and Eq. (4.8.8), one obtains

$$p_1 = \frac{45M_T(9a^4 + 130a^2b^2 + 9b^4)}{256a^3b^3(a^2 + 9b^2)(b^2 + 9a^2)}$$

$$p_2 = \frac{135M_T}{256a^3b^3(a^2 + 9b^2)}$$

$$p_3 = \frac{135M_T}{256a^3b^3(9a^2 + b^2)}$$

In the case of a square cross section ($a = b$), the above formulas simplify. One finds, for instance, that the maximum shear stress occurs at the midpoints of the sides and has the value

$$\tau_{max} = 0.618 \frac{M_T}{a^3}$$

while the angle of twist is given by the formula

$$\theta = 0.445 \frac{M_T}{a^4 G}$$

The exact values found with more accurate methods (by development in series) are given by

$$\tau_{max} = 0.591 \frac{M_T}{a^3}$$

$$\theta = 0.438 \frac{M_T}{a^4 G}$$

BIBLIOGRAPHY

Section 4.1. LAMÉ'S SOLUTION OF THE PROBLEM OF THE THICK SPHERICAL SHELL UNDER UNIFORM INTERNAL AND EXTERNAL PRESSURES

1. Almansi, E., "Sulla Deformazione della Sfera Elastica," *Atti Accad. Nazl. Lincei, Rend.*, Serie 5, Vol. 6, 1st Sem., 1897, pp. 61–64.

2. ——, "Sulla Deformazione della Sfera Elastica," *Mem. Accad. Torino, Serie 2*, T. 47, 1897, pp. 103–125.

3. Arzelà, C., "Deformazione di un Ellissoide Omogeneo Elastico Isotropo...," *Giorn. Mat.*, Vol. 12, 1874, pp. 339–347.

4. Boggio, T., "Sulla Deformazione di una Sfera Elastica Isotropa," *Atti Accad. Torino*, Vol. 41, 1905–1906, pp. 579–587.

5. ——, "Sulla Deformazione di un Ellissoide Elastico," *Atti Accad. Nazl. Lincei, Rend.*, Serie 5, Vol. 15, 1st Semestre, 1906, pp. 104–111.

6. Cerruti, V., "Sur la Déformation d'une sphère homogène isotrope," *Assoc. Franc. Avan. Sci. Compt. Rend.*, *14 Session*, Grenoble, France, 1885, pp. 68–79.

7. ——, "Sulla Deformazione d'una Sfera Omogenea Isotropa," *Atti Accad. Nazl. Lincei, Rend.*, Serie 4, Vol. 2, 1st Semestre, 1885–1886, pp. 461–469, 586–593.

8. ——, "Sulla Deformazione di una Sfera Omogenea Isotropa per dati Spostamenti dei Punti della Superficie," *Nuovo Cimento, Serie 3*, T. 32, 1892, pp. 121–133.

9. ——, "Sulla Deformazione di un Involucro Sferico Isotropo per dati Spostamenti dei Punti delle Due Superfici limiti," *Atti Accad. Lincei, Rend.*, Serie 4, Vol. 5, 2nd Semestre, 1889, pp. 189–201; *Nuovo Cimento, Serie 3*, T. 33, 1893, pp. 5–14, 49–56.

10. ——, "Sulla Deformazione di un Involucro Sferico Isotropo per date Forze agenti sulle Due Superfici Limiti," *Mem. Accad. Nazl. Lincei, Serie 4*, Vol. 7, 1890, pp. 25–44; *Nuovo Cimento, Serie 3*, T. 33, 1893, pp. 97–115, 145–151, 202–208, 259–268.

11. Cesáro, E., "Introduzione alla Teoria Matematica della Elasticità," Fratelli Bocca Editori, Torino, Italy, 1894.

12. Fung, Y. C., "Foundations of Solid Mechanics," Prentice-Hall, Inc., Englewood Cliffs, N.J., 1965.

13. Lamé, G., "Leçons sur la théorie mathématique de l'élasticité," Bachelier, Paris, France, 1852.

14. Lauricella, G., "Sulla Deformazione di una Sfera Elastica per dati Spostamenti in Superficie," *Ann. Mat., Serie 3*, T. 6, 1900, pp. 289–299.

Cerruti, Valentino. Italian mathematician and engineer (b. Croce Mosso, near Novara 1850; d. Croce Mosso 1909). He was Professor of Rational Mechanics at the University of Rome from 1877 until his death. In 1903, he succeeded Luigi Cremona as director of the Rome School of Engineering. Cerruti left important contributions in mechanics, thermodynamics, analysis, and, above all, in the mathematical theory of elasticity. [Reproduction from the Vito Volterra collection, Villa Volterra, Ariccia (Rome).]

15. ——, "Sulla Deformazione di una Sfera Isotropa per date Tensioni in Superficie," *Nuovo Cimento, Serie 5*, T. **5**, 1903, pp. 5-26.

16. ——, "Sulle Formule che danno la Deformazione di una Sfera Elastica Isotropa," *Atti Accad. Nazl. Lincei, Rend., Serie 5*, Vol. **13**, 2nd Semestre, 1904, pp. 583-590.

17. Love, A. E. H., "A Treatise on the Mathematical Theory of Elasticity," 4th ed., Dover Publications, Inc., New York, N.Y., 1944.

18. Marcolongo, R., "Sulla Deformazione di una Sfera Omogenea Isotropa per speciali Condizioni ai Limiti," *Atti Accad. Nazl. Lincei, Rend., Serie 4*, Vol. **5**, 1889, pp. 349-357.

19. ——, "Risoluzione di Due Problemi relativi alla Deformazione di una Sfera Omogenea Isotropa," *Atti Accad. Nazl. Lincei, Rend., Serie 5*, Vol. **1**, 1st Semestre, 1892, pp. 335-343.

20. ——, "Deformazione di una Sfera Isotropa," *Ann. Mat., Serie 2*, T. **23**, 1895, pp. 111-152.

21. ——, "Teoria Matematica dell' Equilibrio dei Corpi Elastici," Hoepli, Milano, Italy, 1904.

22. Somigliana, C., "Sopra l'Equilibrio di un Corpo Elastico Isotropo limitato da una o due Superfici Sferiche," *Ann. Scuola Normale Superiore Pisa*, Vol. **4**, 1887, pp. 100-172.

23. Southwell, Sir Richard, "An Introduction to the Theory of Elasticity for Engineers and Physicists," Oxford University Press, London, England, 1941.

24. Tedone, O., "Sulle Equazioni dell'Equilibrio Elastico per un Corpo Isotropo con speciale riguardo alle Forze di Massa e su alcuni Problemi relativi alla Sfera Elastica," *Rend. Circolo Mat.*, T. **17**, 1903, pp. 241-274.

25. ——, "Sul Problema dell'Equilibrio Elastico di un Ellissoide di Rotazione," *Atti Accad. Nazl. Lincei, Rend., Serie 5*, Vol. **14**, 1st Semestre, 1905, pp. 76-84.

26. Westergaard, H. M., "Theory of Elasticity and Plasticity," Harvard University Press, Cambridge, Mass., 1952.

Sections 4.2, 4.3, 4.4, 4.5, 4.6, 4.7, 4.8. PURE BENDING OF A PRISMATIC BAR; COULOMB'S THEORY OF TORSION OF A CIRCULAR SHAFT; NAVIER'S THEORY OF TORSION; SAINT-VENANT'S SEMI-INVERSE METHOD FOR SOLVING TORSION PROBLEMS; PRANDTL'S THEORY OF TORSION; PRANDTL'S MEMBRANE ANALOGY; TAYLOR AND GRIFFITH'S APPLICATIONS; RITZ'S METHOD APPLIED TO TORSION PROBLEMS

1. Almansi, E., "Sulla Torsione dei Cilindri Cavi a Spessore piccolissimo," *Atti Accad. Torino*, Vol. **35**, 1899-1900, pp. 39-53.

2. ——, "Sulla Flessione dei Cilindri," *Rend. Circolo Mat. Palermo*, T. **21**, 1906, pp. 36-55.

3. Boresi, A. P., "Elasticity in Engineering Mechanics," Prentice-Hall, Inc., Englewood Cliffs, N.J., 1965.

4. Cesáro, E., "Introduzione alla Teoria Matematica della Elasticità," Fratelli Bocca Editori, Torino, Italy, 1894.

5. Clebsh, A., "Théorie de l'élasticité des corps solides," Leipzig, 1861, French translation by Barré de Saint-Venant and Flamant, Dunod, Paris, France, 1883.

6. Colonnetti, G., "La Statica delle Costruzioni," Vol. **1**, Unione Tipografica Editrice Torinese, Torino, Italy, 1928.

7. Coulomb, C. A., "Recherches théoriques et éxpérimentales sur les forces de torsion et sur l'élasticité des fils de métal," *Mem. Acad. Sci.*, 1784, and also "Collection de Mémoires," Vol. 1, Gauthier-Villars, Paris, France, 1884.

8. Den Hartog, J. P., "Advanced Strength of Materials," McGraw-Hill Book Company, New York, N.Y., 1952.

9. Diaz, J. B., and A. Weinstein, "The Torsional Rigidity and Variational Methods," *Am. J. Math.*, Vol. 70, 1948.

10. Föppl A., and L., "Drang und Zwang," Oldenbourg, Munich, Germany, 1924.

11. Fung, Y. C., "Foundations of Solid Mechanics," Prentice-Hall, Inc., Englewood Cliffs, N.J., 1965.

12. Goodier, J. N., "Torsion," in *Handbook of Engineering Mechanics*, W. Flugge, Ed., McGraw-Hill Book Company, New York, N.Y., 1962.

13. Grossmann, G., "Experimentalle Durchführung einer neuen Hydrodynamischen Analogie fur das Torsion Problem," *Ing. Arch.*, Vol. 25, 1957.

14. Higgins, T. J., "The Approximate Mathematical Methods of Applied Physics as Exemplified by Application to Saint-Venant's Torsion Problem," *J. Appl. Phys.*, Vol. 13, No. 7, 1942, p. 457.

15. Huth, J. H., "Torsional Stress Concentration in Angle and Square Tube Fillets," *J. Appl. Mech.*, 1950, p. 388.

16. L'Hermite, R., "Résistance des matériaux théorique et expérimentale," T. 1, Dunod, Paris, France, 1959,

17. Ling, C. B., "Torsion of a Circular Tube with Longitudinal Circular Holes," *Quart. Appl. Math.*, 1947, p. 168.

18. Marcolongo, R., "Teoria Matematica dello Equilibrio dei Corpi Elastici," U. Hoepli, Milan, Italy, 1904.

19. Mindlin, R. D., and M. G. Salvadori, "Analogies," in *Handbook of Experimental Stress Analysis*, M. Hetenyi, Ed., John Wiley & Sons, Inc., New York, N.Y., 1950.

20. Navier, L. M. H., "Applications de la mécanique," Paris, 1833.

21. Padova, E., "Il Problema di De St. Venant per un Prisma Rettangolare," *Nuovo Cimento*, Serie 3, T. 10, 1881, pp. 102–113.

22. ——, "Estensione del Problema di De St. Venant," *Atti Accad. Nazl. Lincei, Rend.*, Serie 4, Vol. 6, 2nd Semestre, 1890, pp. 95–102.

23. Pestel, E., "Ein neues Strömungsgleichnis der Torsion," *Z. Angew. Met. Mech.*, Vol. 34, 1954.

24. ——, "Eine Neue Hydrodynamische Analogie zur Torsion Prismatischer Stabe," *Ing. Arch.*, Vol. 23, 1955.

25. Picone, M., "Sulla Torsione di un Prisma Elastico Cavo secondo la Teoria di Saint-Venant," *Ist. Applicazioni Calcolo C.N.R.*, Rome, No. 222, 1948.

26. Polya, G., "Sur la Fréquence Fondamentale des Membranes Vibrantes et la Résistance Elastique des Tiges à la Torsion," *Compt. Rend.*, 1947, p. 346.

27. ——, "Torsional Rigidity, Principal Frequency Electrostatic Capacity and Symmetrisation," *Quart. Appl. Math.*, 1948, p. 267.

28. Prandtl, L., "Zur Torsion von Prismatischen Staben," *Physik. Z.*, **Vol. 4,** 1903, pp. 758–759.

29. ——, "Eine neue Darstellung von Torsionsspannungen bei prismatischen Staben von beliebigem Querschnitt," *Jahresber. Math. Ver.*, **Bd. 13,** 1904, pp. 31–36.

30. Saint-Venant, Barré de A. J. C., "Mémoire sur la torsion des prismes, avec des considérations sur leur flexion, ainsi que sur l'équilibre intérieur des solides élastiques en général et des formules pratiques pour le calcul de leur résistance à divers efforts s'exerçant simultanément," *Mem. Savants Etrangers*, **T. 14,** 1855, pp. 233–560.

31. ——, "Mémoire sur la flexion des prismes, sur le glissements transversaux et longitudinaux qui l'accompagnent lorsqu'elle ne s'opère pas uniformement ou en arc de cercle, et sur la forme courbe affectée alors par leurs sections transversales primitivement planes," *J. Math. Liouville, Serie 2*, **T. 1,** 1856, pp. 89–189.

32. Sokolnikoff, I. S., "Mathematical Theory of Elasticity," McGraw-Hill Book Company, New York, N.Y., 1956.

33. Sommerfeld, A., "Mechanics of Deformable Bodies," *Lectures on Theoretical Physics*, **Vol. 2,** Academic Press, Inc., New York, N.Y., 1950.

34. Southwell, Sir Richard, "An Introduction to the Theory of Elasticity for Engineers and Physicists," Oxford University Press, London, England, 1941.

35. Stevenson, A. C., "The Centre of Flexure of a Hollow Shaft," *Proc. London Math. Soc.*, **Vol. 50,** 1949, p. 536.

36. Taylor, Sir Geoffrey, "The Use of Soap Films in Solving Torsion Problems" (with A. A. Griffith), from the Scientific Papers of Sir Geoffrey Taylor, Cambridge, University Press, London, England, 1958, **Vol. 1,** pp. 1–23.

37. ——, "The Problem of Flexure and Its Solution by the Soap-Film Method" (with A. A. Griffith), from the Scientific Papers of Sir Geoffrey Taylor, Cambridge, University Press, London, England, 1958, **Vol. 1,** pp. 24–45.

38. ——, "The Application of Soap Films to the Determination of the Torsion and Flexure of Hollow Shafts" (with A. A. Griffith), from the Scientific Papers of Sir Geoffrey Taylor, Cambridge University Press, London, England, 1958, **Vol. 1,** pp. 46–60.

39. Timoshenko, S. P., "History of Strength of Materials," McGraw-Hill Book Company, New York, N.Y., 1953.

40. Timoshenko, S. P., and J. N. Goodier, "Theory of Elasticity," McGraw-Hill Book Company, New York, N.Y., 1951.

41. Todhunter, I., and K. Pearson, "A History of the Theory of Elasticity and of the Strength of Materials," Cambridge University Press, **Vol. 1,** 1886; **Vol. 2,** 1893. Reprinted by Dover Publications, Inc., New York, N.Y., 1960.

42. Wang, C. T., "Applied Elasticity," McGraw-Hill Book Company, New York, N.Y., 1953.

43. Weber, C., and W. Gunther, "Torsiontheorie," F. Vieweg und Sohn, Berlin, Germany, 1958.

44. Weinstein, A., "New Methods for the Estimation of Torsional Rigidity," *Third Symposium on Applied Mathematics*, McGraw-Hill Book Company, New York, N.Y., 1950.

PROBLEMS

4-1. Calculate the thickness of a spherical shell having an inside diameter of 4 in., if the allowable working stress is 8000 psi and the internal pressure is 2 ton/sq. in.

4-2. Two bars made of the same material and having the same length are subjected to the same torque. One bar has a square cross section of side a and the other bar has a circular cross section of diameter a. Which bar has the greater angle of twist?

4-3. Apply Ritz's method to the problem of torsion of a bar of rectangular cross section of sides $2a$ and $2b$ by assuming the stress function to be

$$\psi(x, y) = p_1 \cos \frac{\pi x}{2a} \cos \frac{\pi y}{2b} + p_2 \cos \frac{3\pi x}{2a} \cos \frac{3\pi y}{2b}$$

Compare the values of $\tau_{\max}$ and of θ in the case of the square with those obtained by using other methods of solution described in the text.

CHAPTER

5

BENDING OF STRAIGHT BEAMS

5.0 Introduction

In this chapter bending of straight beams is discussed, with both linearly elastic and inelastic behaviors being considered. Fundamental formulas of the strength of materials are first derived; these include the differential equations of equilibrium, Navier's flexure formula, Jourawski's shearing stress formula, and the Bernoulli-Euler differential equation for the bending deflections of elastic beams. Then various methods for studying deflections of elastic beams are presented, among which are direct integration methods, use of singularity functions, and Taylor's and Fourier's series representations. Mohr's conjugate beam method is discussed and applied to derive Clapeyron's three-moment equation for solving continuous elastic beam problems. The problem of elastic-plastic bending of beams is then considered in the cases of perfectly plastic materials and linearly hardening materials. The chapter ends with a brief discussion of limit or plastic design of beams.

5.1 Differential Equations of Equilibrium; Navier's Flexure Formula; Jourawski's Shearing Stress Formula

Consider the horizontal beam shown in Fig. 5.1.1. It is referred to a system of cartesian coordinates $Oxyz$, in which the x-axis coincides with the undeformed axis of the beam and the y- and z-axes coincide with the principal axes of inertia of the cross-sectional area. The xy-plane is assumed to be a plane of

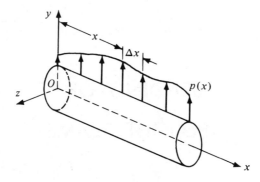

Figure 5.1.1

symmetry of the beam and the distributed force $p(x)$ and any concentrated force or moment acting on the beam are assumed to be applied in this plane of symmetry. Forces like distributed forces $p(x)$ or concentrated forces $P_i (i = 1, 2, 3, \ldots, n)$ will be assumed to be positive when directed in the positive y-direction.

Assume the beam of Fig. 5.1.1 to be cut at a distance x from the origin O by a plane parallel to the yz-plane. Each portion of the beam, shown in Fig. 5.1.2, remains in equilibrium under the action of the applied load, the reactions, and the shear force and bending moment acting at the cut. The shear force V is assumed to be positive on a section if it tends to rotate the free body diagram of a portion of the beam in a clockwise sense with respect to a point located within the free body diagram of the portion. For example, in Fig. 5.1.2 the shear force is positive since it tends to rotate the left-hand

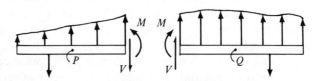

Figure 5.1.2

portion of the beam in a clockwise sense about point P; also, it tends to rotate the right-hand portion in a clockwise sense about point Q.

The bending moment M is assumed to be positive if it tends to produce compression at the top side of the beam. Thus, the bending moment as it is shown in Fig. 5.1.2 is assumed positive. The segment of the beam of length Δx of Fig. 5.1.1 is shown isolated in Fig. 5.1.3 with all forces and moments

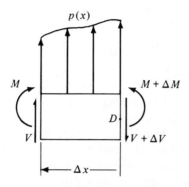

Figure 5.1.3

acting in the positive directions. In general, the moment and shear force will vary with x, the distance measured along the axis of the beam. These changes will be indicated by ΔV and ΔM. Let $\bar{p}$ represent the average value of the distributed load acting on the portion of beam of length Δx. For equilibrium, the following equations must be satisfied:

$$\sum F_y = 0 = V + \bar{p}\,\Delta x - (V + \Delta V)$$
$$\sum M_D = 0 = M + \bar{p}\,\Delta x\,(k\,\Delta x) + V\,\Delta x - (M + \Delta M)$$

where $0 \leq k \leq 1$. The above relations may be written as

$$\frac{\Delta V}{\Delta x} = \bar{p}$$
$$\frac{\Delta M}{\Delta x} = V + \bar{p}k(\Delta x)$$

which in the limit, as Δx approaches zero, become

$$\frac{dV}{dx} = p \qquad (5.1.1)$$

$$\frac{dM}{dx} = V \qquad (5.1.2)$$

These equations relate the loading, shear force, and bending moment and are valid for continuous loading ranges, i.e., portions of the beam where no concentrated forces or moments are acting. Equation (5.1.1) states that the rate of change of shear force at a point is equal to the intensity of the loading at that point. Equation (5.1.2) states that the rate of change of bending moment at a point is equal to the shear force at that point. If Eqs. (5.1.1) and (5.1.2) are integrated, the following relations are obtained:

$$V(b) - V(a) = \int_a^b p(x)\, dx \tag{5.1.3}$$

$$M(b) - M(a) = \int_a^b V(x)\, dx \tag{5.1.4}$$

where $x = a$ and $x = b$ are two points in the same continuous loading range. Equation (5.1.3) states that the change in shear between points a and b is equal to the area under the p-x curve between a and b. Equation (5.1.4) states that the change in bending moment between points a and b is equal to the area under the V-x curve between points a and b. Equations (5.1.1), (5.1.2), (5.1.3), and (5.1.4) are useful in obtaining shear and moment distributions for a beam. Such applications are thoroughly discussed in elementary books on the strength of materials.

Example 5.1.1. Determine the shear force and bending moment equations for the simply supported beam of length l subjected to a uniformly distributed load of intensity p_0 as shown in Fig. 5.1.4.

In view of Eq. (5.1.1)

$$V = -\int p_0\, dx + C_1 = -p_0 x + C_1$$

where $C_1 = V(0) = p_0 l/2$ is the reaction at the left support of the beam. In view of Eq. (5.1.2)

$$M = \int \left(-p_0 x + \frac{p_0 l}{2}\right) dx + C_2 = -p_0 \frac{x^2}{2} + \frac{p_0 l x}{2} + C_2$$

where $C_2 = M(0) = 0$ is the moment at the left support of the beam. It follows that

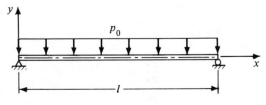

Figure 5.1.4

the shear force equation is

$$V(x) = -\frac{p_0 x}{2} + \frac{p_0 l}{2}$$

and the bending moment equation is

$$M(x) = -\frac{p_0 x^2}{2} + \frac{p_0 l x}{2}$$

At the center of the beam i.e., at $x = l/2$

$$V = 0$$

and

$$M\left(\frac{l}{2}\right) = M_{\max} = \frac{p_0 l^2}{8}$$

To derive Navier's flexure formula for a beam subjected to pure bending, consider the initially straight beam of uniform cross section shown in Fig. 5.1.5 (a) whose axis $PBFQ$ coincides with the x-axis. Suppose that two external bending couples M are acting in the longitudinal plane of symmetry of the beam as shown in Fig. 5.1.5 (b). Bending will occur in the xy-plane and the axis of the beam will be deformed into the axis $P'B'F'Q'$.

Navier's theory of bending is based on the following assumptions:

1. There is no lateral shearing stress and no lateral pressure between the longitudinal fibers; thus every longitudinal fiber acts as if separate from any other fiber.
2. Each transverse section of the bar, initially plane, remains plane and normal to the longitudinal fibers of the bar after bending.

Consider in Fig. 5.1.5 (a) lines AD and EH which represent two adjacent cross sections separated by distance Δx. Under the action of the applied couples M, these lines will rotate through an angle $\Delta\theta$ relative to each other [see Fig. 5.1.5 (b)]. Let R be the radius of curvature of the axis of the beam at point B. As the beam bends, the fibers above the $PBFQ$ axis, such as DH and CG, will decrease in length, whereas the fibers below this axis, such as AE, will increase in length. Fiber $PBFQ$, however, will not change in length. This fiber which lies in the plane of symmetry and remains unstrained is called the neutral axis of the beam.

Let us determine the strain after bending of the fiber CG located at a distance y above the axis. This strained element is shown as line $C'G'$ in Fig. 5.1.5 (c). In this figure, line $F'G''$ is drawn parallel to $B'C'$. The change in length of the fiber is $G'G'' = y\,\Delta\theta$ while its original length was

$$CG = BF = B'F' = R\,\Delta\theta$$

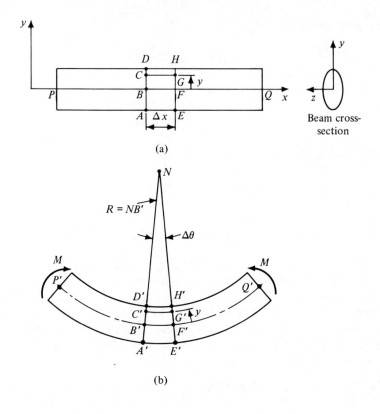

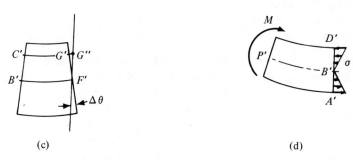

Figure 5.1.5

The strain of CG will, therefore, be

$$\epsilon = -\frac{G'G''}{CG} = -\frac{y\,\Delta\theta}{R\,\Delta\theta}$$

or

$$\epsilon = -\frac{y}{R} \tag{5.1.5}$$

The minus sign in Eq. (5.1.5) indicates that the fibers above the neutral axis decrease in length while those below the neutral axis (where y is negative) increase in length.

Assuming Hooke's law to be valid, the stress in fiber CG will be

$$\sigma = E\epsilon = -\frac{Ey}{R} \qquad (5.1.6)$$

By expressing the conditions of equilibrium at a generic section $A'D'$ [shown in Fig. 5.1.5 (d)], the two equations

$$\left. \begin{array}{l} F_x = 0 = \displaystyle\int_A \sigma \, dA \\[2mm] M_{B'} = 0 = M + \displaystyle\int_A y\sigma \, dA \end{array} \right\} \qquad (5.1.7)$$

are obtained, where A represents the area of the cross section of the beam. By substituting Eq. (5.1.6) into Eqs. (5.1.7) one finds that

$$\left. \begin{array}{l} 0 = \displaystyle\int_A \left(-\frac{Ey}{R}\right) dA = -\frac{E}{R}\displaystyle\int_A y \, dA \\[2mm] M = -\displaystyle\int_A y\left(-\frac{Ey}{R}\right) dA = \frac{E}{R}\displaystyle\int_A y^2 \, dA \end{array} \right\} \qquad (5.1.8)$$

Since the constant term E/R is different from zero the first of Eqs. (5.1.8) can be written in the form

$$\int_A y \, dA = \bar{y}A = 0 \qquad (5.1.9)$$

where $\bar{y}$ is the distance from the neutral axis to the centroid of the area A. Since A is not zero, it follows from Eq. (5.1.9) that $\bar{y} = 0$; therefore, the neutral axis passes through the centroid of the section. The integral $\int_A y^2 \, dA$ is the moment of inertia of the cross section of the beam with respect to the centroidal z-axis. By letting

$$I = \int_A y^2 \, dA$$

the second of Eqs. (5.1.8) becomes

$$M = \frac{EI}{R} \qquad (5.1.10)$$

By combining Eqs. (5.1.6) and (5.1.10) one finally obtains

$$\sigma = -\frac{My}{I} \qquad (5.1.11)$$

which is known as Navier's flexure formula.

264   BENDING OF STRAIGHT BEAMS   Sec. 5.1

To derive Jourawski's shearing stress formula, let us consider the equilibrium conditions for the element of beam of length Δx and cross-sectional area A^*. The area A^* is the shaded area shown in Fig. 5.1.6 (a). The stresses acting in the x-direction are shown in Fig. 5.1.6 (b). There are, of course, also shear stresses acting in the y-direction on sections S and S' and the shear stress on the vertical plane at point P is equal to the shear stress on the horizontal plane at that point.

The resultant force in the x-direction due to the flexural stress acting on area A^* at section S' is

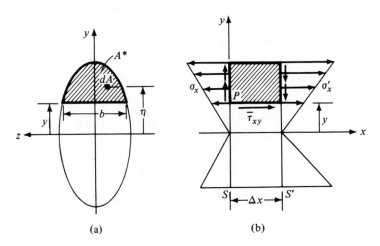

Figure 5.1.6

$$\int_{A^*} \sigma'_x \, dA = \int_{A^*} \left[-\frac{(M + \Delta M)\eta}{I} \right] dA = -\frac{(M + \Delta M)}{I} \int_{A^*} \eta \, dA$$

and the corresponding force acting on area A^* at section S is

$$\int_{A^*} \sigma_x \, dA = \int_{A^*} \left[-\frac{M\eta}{I} \right] dA = -\frac{M}{I} \int_{A^*} \eta \, dA$$

The resultant force in the x-direction due to the shear stress on the lower side of the element is

$$\bar{\tau}_{xy} b \, \Delta x$$

where b is the width of the beam at distance y above the neutral axis and $\bar{\tau}_{xy}$ is the average value of the shear stress on the area $b\Delta x$. For equilibrium

$$\Sigma F_x = 0 = \bar{\tau}_{xy} b \, \Delta x - \left(-\frac{M}{I} \int_{A^*} \eta \, dA \right) + \left(-\frac{M + \Delta M}{I} \int_{A^*} \eta \, dA \right)$$

from which

$$\bar{\tau}_{xy} = \frac{\Delta M}{\Delta X} \frac{1}{Ib} \int_{A^*} \eta \, dA$$

In the limit as Δx approaches zero, the above equation becomes

$$\tau_{xy} = \frac{dM}{dx} \frac{1}{Ib} \int_{A^*} \eta \, dA \tag{5.1.12}$$

or, in view of Eq. (5.1.2),

$$\tau_{xy} = \frac{VQ}{Ib} \tag{5.1.13}$$

where τ_{xy} is the shear stress on a horizontal or vertical plane at point P [see Fig. 5.1.6 (b)] and $Q = \int_{A^*} \eta \, dA$ is the static moment of the area A^* taken with respect to the z-axis which passes through the centroid of the cross sectional area.[1]

Example 5.1.2. Determine the distribution of shearing stresses on the rectangular cross section of breadth B and height H shown in Fig. 5.1.7 (a).

For this section

$$b = B, \quad Q = \frac{B}{2}\left[\left(\frac{H}{2}\right)^2 - y^2\right], \quad \frac{V}{BH} = \tau_m, \quad I = \frac{BH^3}{12}$$

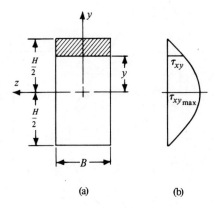

Figure 5.1.7

---

[1] According to Eq. (5.1.13) the shear stress τ_{xy} is positive, if directed down (i.e., in the negative-y-direction) on the positive x face. This is opposite to the shear stress sign convention adopted in Chapter 1. If one desires to use the sign convention of Chapter 1, then a minus sign should be used on the right-hand side of Eq. (5.1.13).

and it follows from Eq. (5.1.13) that

$$\tau_{xy} = \frac{VQ}{Ib} = \frac{3V}{2BH}\left[1 - \frac{4y^2}{H^2}\right] = \frac{3\tau_m}{2}\left[1 - \frac{4y^2}{H^2}\right] \quad (5.1.14)$$

where τ_m is the average uniformly distributed shear stress. Equation (5.1.14) shows that in a rectangular cross section, the shearing stress varies according to a parabolic law [see Fig. 5.1.7 (b)].

Example 5.1.3 Determine the distribution of shearing stresses on the circular cross section of radius r shown in Fig. 5.1.8 (a).

For this section

$$b = 2\sqrt{r^2 - y^2}, \quad Q = \int_y^r \eta(2\sqrt{r^2 - \eta^2}\,d\eta) = \frac{2}{3}(r^2 - y^2)^{3/2}, \quad I = \frac{\pi r^4}{4}$$

and it follows that

$$\tau_{xy} = \frac{VQ}{Ib} = \frac{4V}{3\pi r^4}(r^2 - y^2) = \frac{4V}{3A}\left(1 - \frac{y^2}{r^2}\right) \quad (5.1.15)$$

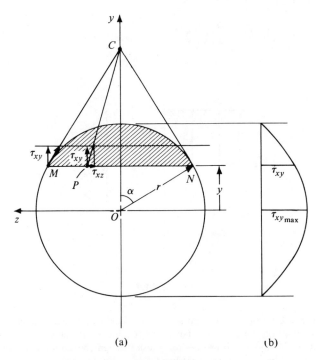

Figure 5.1.8

where A represents the area of the circular cross section. Equation (5.1.15) shows that in a circular cross section, the shearing stress varies according to a parabolic law [see Fig. 5.1.8(b)] and the maximum value

$$\tau_{max} = \frac{4V}{3A}$$

occurs at the neutral axis. It must be pointed out, however, that the results obtained from Eq. (5.1.13) are open to question in this case. For instance, the shear stress is not uniform along a horizontal line MN [Fig. 5.1.8 (a)] as indicated by the shear stress equation. Instead, the shear stress at points M and N must be directed tangent to the circle at these points as indicated in Fig. 5.1.8 (a). In fact, the shear stress equation only provides an approximation to the vertical component of shear stress at points along line MN. If one assumes that the resultant shear stress, at any point P on line MN, passes through point C then the component τ_{xz} and the resultant shear stress at P can be computed, but the results are only approximations.

The results for this example indicate that the shear stress has the value $1.33 V/A$ along the horizontal diameter. However, a more exact analysis based on the theory of elasticity[2] shows that the shear stress on this line varies from $1.23 V/A$ at the ends of the diameter to $1.38 V/A$ at the center of the cross section.[3] Along the horizontal diameter then, the error of the approximate solution is only about 4 percent.

Example 5.1.4. Determine the distribution of shearing stresses for the I beam section [see Fig. 5.1.9 (a)] of height H and breadth B whose flanges have thickness h while the web has breadth b.

The diagram of shearing stress τ_{xy} [see Fig. 5.1.9 (b)] will still be parabolic but it will be discontinuous at the intersection of the web and flange, i.e. at $y = \pm y_2$. In fact, in the flanges, the breadth of the beam is B, while in the web the breadth is reduced to b. In the flanges at a distance y from the neutral axis of the beam, the shearing stress will be:

$$\tau_{xy} = \frac{V}{IB}\frac{B}{2}(y_1^2 - y^2) = \frac{V}{2I}(y_1^2 - y^2) \tag{5.1.16}$$

At the lower or upper border of the flange where $y = \pm y_2^+$ (the symbol y_2^+ represents a value of slightly larger than y_2) the shearing stress will be

$$\tau_f = \frac{V}{2I}(y_1^2 - y_2^2) \tag{5.1.17}$$

while at the upper or lower border of the web, the shearing stress will jump to the value

$$\tau_w = \frac{V}{2I}\frac{B}{b}(y_1^2 - y_2^2)$$

---

[2] See Bibliography, Sec. 5.1, No. 15 at the end of this chapter.
[3] These results are based on a value of 0.3 for Poisson's ratio.

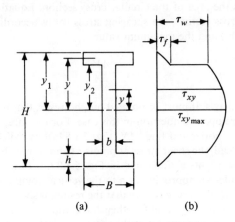

Figure 5.1.9

In the web, the shearing stress varies according to the parabolic law

$$\tau_{xy} = \frac{V}{Ib}\left[\frac{B}{2}(y_1^2 - y_2^2) + \frac{b}{2}(y_2^2 - y^2)\right] = \tau_w + \frac{V}{2I}(y_2^2 - y^2) \quad (5.1.18)$$

The maximum value of the shearing stress occurs at the neutral axis and is obtained by letting $y = 0$ in Eq. (5.1.18).

Example 5.1.5. Consider the I section, shown in Fig. 5.1.9, and determine the shear stress distribution on a cross section, assuming that the total height H of the beam is very large as compared with the thickness h of the flanges and the breadth b of the web.

If Ω represents the area of one of the flanges it follows with a good approximation that

$$Q = \Omega\frac{H}{2}, \quad I = 2\Omega\left(\frac{H}{2}\right)^2 = \frac{\Omega H^2}{2}, \quad \text{and } \frac{I}{y_1} = \Omega H$$

It follows from the formulas derived in this section that the normal stress

$$\sigma_{max} = \frac{M}{\Omega H}$$

is uniformly distributed in the flanges of the I beam, while the shear stress

$$\tau_{xy} = \frac{V}{bH}$$

is uniformly distributed in the web of the beam.

5.2 Differential Equations for Deflections of Elastic Beams According to the Bernoulli-Euler Theory

Let the curve of Fig. 5.2.1 represent the neutral axis of a beam deformed under the action of externally applied loads. Calling $v(x)$ the vertical or transverse displacement of the deformed axis of the beam from the x-axis, the radius of curvature R of the deformed axis will be expressed by the equation

$$\frac{1}{R} = \frac{\frac{d^2v}{dx^2}}{\left[1 + \left(\frac{dv}{dx}\right)^2\right]^{3/2}} \tag{5.2.1}$$

Figure 5.2.1

Equation (5.1.10) can be expressed as

$$\frac{1}{R} = \frac{M}{EI} \tag{5.2.2}$$

By combining Eqs. (5.2.1) and (5.2.2), the equation

$$M = \frac{EI\frac{d^2v}{dx^2}}{\left[1 + \left(\frac{dv}{dx}\right)^2\right]^{3/2}} \tag{5.2.3}$$

is obtained. If the elastic deflections of a beam are small, then the quantity dv/dx is small and the quantity $(dv/dx)^2$ can be neglected in comparison with unity. Equation (5.2.3) then becomes

$$EI\frac{d^2v(x)}{dx^2} = M \tag{5.2.4}$$

Equation (5.2.4) is the differential equation of the elastic curve of the beam.

Equations (5.1.11) and (5.2.4) have been derived for the case of a uniform beam subjected to a state of pure bending. However, these equations are often applied to cases where the bending moment varies and also to beams of variable cross sections. For long, slender beams the results, obtained by use of Eqs. (5.1.11) and (5.2.4), are found to be in good agreement with experimental results.

Direct relations between deflection and shear force and between deflection and loading may be obtained by differentiating Eq. (5.2.4) and using Eqs. (5.1.1) and (5.1.2). Thus, one finds that

$$\frac{d}{dx}\left(EI\frac{d^2v}{dx^2}\right) = \frac{dM}{dx} = V \tag{5.2.5}$$

$$\frac{d^2}{dx^2}\left(EI\frac{d^2v}{dx^2}\right) = \frac{d^2M}{dx^2} = p \tag{5.2.6}$$

If the quantity EI is constant, the above equations may be written as follows:

$$EI\frac{d^3v}{dx^3} = V \tag{5.2.7}$$

$$EI\frac{d^4v}{dx^4} = p \tag{5.2.8}$$

Equation (5.2.6) or (5.2.8) is often used as the differential equation of the elastic curve instead of Eq. (5.2.4).

5.3 Solution of Beam Deflection Problems by Direct Integration

The application of Eq. (5.2.4) or Eq. (5.2.8) is quite simple. For instance, in using Eq. (5.2.4), the bending moment is expressed in terms of the applied loads and the distance x along the beam. The bending moment is then substituted into Eq. (5.2.4) which is integrated twice to obtain the equation for the elastic curve. This equation contains two constants of integration which are evaluated by using the boundary conditions on deflection and slope. If the beam has concentrated forces or moments then a bending moment expression is usually written for each range between concentrated loads. Thus a differential equation is obtained for each range. Each equation is integrated twice and the constants of integration are evaluated by use of boundary conditions and continuity conditions. For example, consider the beam shown in Fig. 5.3.1. The bending moment equations for the ranges $0 \leq x \leq a$ and $a \leq x \leq l$ are substituted into Eq. (5.2.4) The two differential equations are integrated twice to obtain the equations of the elastic curve for each range.

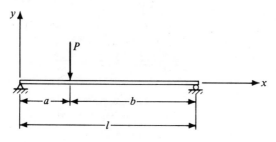

Figure 5.3.1

These equations contain four constants of integration which are evaluated by using the boundary conditions

$$v_1(0) = v_2(l) = 0$$

and the continuity conditions

$$\frac{dv_1(a)}{dx} = \frac{dv_2(a)}{dx}, \quad v_1(a) = v_2(a)$$

where v_1 is the deflection for $0 \le x \le a$ and v_2 is the deflection for $a \le x \le l$. The evaluation of the constants, therefore, requires that a set of four algebraic equations be solved. The solution becomes even more cumbersome, if more concentrated forces act along the span of the beam. Similar difficulties are encountered when Eq. (5.2.6) or (5.2.8) is used as the differential equation of the elastic curve. Such difficulties can be circumvented, however, and a single equation can be derived which is valid at all points along the beam. This is accomplished by the use of singularity functions which are described in the next sections.

5.4 Macaulay's Use of Singularity Functions for Studying Deflections of Beams

As a means of introducing singularity functions consider the beam shown in Fig. 5.4.1 (a). To determine the equation of the elastic curve from the second order differential equation [Eq. (5.2.4)], we must first write the bending moment expressions for the ranges between concentrated loads. This is accomplished by isolating segments to the left of a section at distance x from the left end of the beam [see Figs. 5.4.1 (b), (c), (d), (e)] and writing the moment equilibrium equations. Then, by substituting the bending moment expressions into Eq. (5.2.4), the following differential equations are obtained:

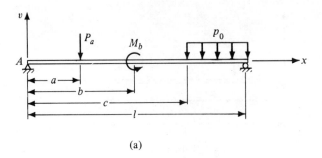

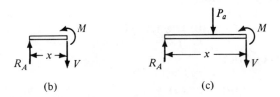

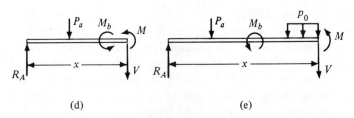

Figure 5.4.1

$$\left.\begin{aligned}
EI\frac{d^2v}{dx^2} &= R_A x, \quad 0 \leq x \leq a \\
EI\frac{d^2v}{dx^2} &= R_A x - P_a(x-a), \quad a \leq x < b \\
EI\frac{d^2v}{dx^2} &= R_A x - P_a(x-a) - M_b, \quad b < x \leq c \\
EI\frac{d^2v}{dx^2} &= R_A x - P_a(x-a) - M_b - \frac{p_0}{2}(x-b)^2, \quad c \leq x \leq l
\end{aligned}\right\} \quad (5.4.1)$$

In the above equations, R_A is the support reaction at the left end of the beam and can be determined from the equilibrium conditions. To complete the solution of the above equations one must integrate each of them twice and evaluate the constants of integration by using the boundary conditions and the continuity conditions as mentioned in Sec. 5.3. However, the solution can be determined much more efficiently by observing that the last of Eqs. (5.4.1) contains all the terms common to the other three equations and introducing a function $\langle x - a \rangle^n$ with the following characteristics:

$$\left.\begin{array}{l} \langle x - a \rangle^n = \begin{cases} 0 \text{ if } x < a \\ (x - a)^n \text{ if } x > a \end{cases} \quad n \geq 0 \\ \int_{-\infty}^{x} \langle x - a \rangle^n \, dx = \frac{\langle x - a \rangle^{n+1}}{n + 1} \quad n \geq 0 \\ \frac{d}{dx} \langle x - a \rangle^n = n \langle x - a \rangle^{n-1} \quad n \geq 1 \end{array}\right\} \quad (5.4.2)$$

The function $\langle x - a \rangle^n$ defined above behaves just as the function $(x - a)^n$, except that it is zero for $x < a$. This peculiarity is emphasized by the pointed brackets. Two examples of singularity functions are given in Fig. 5.4.2. The function shown in Fig. 5.4.2 (a) is called the "unit step" and the function shown in Fig. 5.4.2 (b) is called the "unit ramp." These particular functions are frequently used in mathematical physics textbooks.

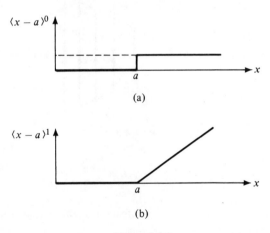

Figure 5.4.2

By using the singularity functions, a single equation can now be written which encompasses all four of Eqs. (5.4.1); thus, one may easily verify that the equation

$$EI \frac{d^2v}{dx^2} = R_A \langle x - 0 \rangle - P_a \langle x - a \rangle - M_b \langle x - b \rangle^0 - \frac{p_0}{2} \langle x - c \rangle^2 \quad (5.4.3)$$

contains all four of Eqs. (5.4.1). It is now apparent that we can simply write the bending moment equation for the last range (i.e., for $c \leq x \leq l$ in this particular case) and express it in terms of singularity functions. When Eq. (5.4.3) is integrated there will be only two constants of integration. These are evaluated by use of the boundary conditions $v(0) = 0$ and $v(l) = 0$.

In order to use singularity functions to express the bending moment for a beam loaded as shown in Fig. 5.4.3 (a), we let the loading continue past $x = b$ and introduce the opposite loading as indicated in Fig. 5.4.3 (b). Thus, due to the loading shown, the moment equation would be

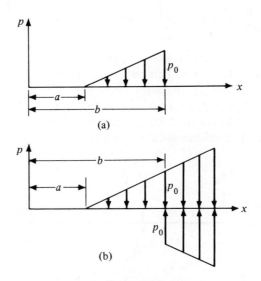

Figure 5.4.3

$$M = -\frac{p_0}{6(b-a)}\langle x - a\rangle^3 + \frac{p_0}{2}\langle x - b\rangle^2 + \frac{p_0}{6(b-a)}\langle x - b\rangle^3 \quad (5.4.4)$$

Other load distributions, which terminate before the end of the beam, can be handled in a similar manner.

Example 5.4.1. Determine the equation of the elastic curve for the beam shown in Fig. 5.4.4 (a). Use the second order differential equation to obtain the solution.

From the equilibrium conditions for the beam, the support reactions are found to be $R_A = 200$ lb $\uparrow$ and $R_B = 2320$ lb $\uparrow$. The bending moment is determined from the free body diagram shown in Fig. 5.4.4 (b). The uniformly distributed load introduced at $x = 10$ ft cancels the effect of the other distributed load which continues past $x = 10$ ft. The bending moment is

$$M = 200\langle x - 0\rangle - \tfrac{200}{2}\langle x - 4\rangle^2 + \tfrac{200}{2}\langle x - 10\rangle^2 + 2320\langle x - 15\rangle \quad (5.4.5)$$

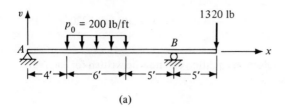

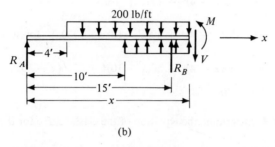

Figure 5.4.4

By substituting the above relation into Eq. (5.2.4) and noting that $\langle x - 0 \rangle = x$, we obtain

$$EI\frac{d^2v}{dx^2} = 200x - \tfrac{200}{2}\langle x - 4 \rangle^2 + \tfrac{200}{2}\langle x - 10 \rangle^2 + 2320\langle x - 15 \rangle \quad (5.4.6)$$

By integrating Eq. (5.4.6) it follows that

$$EI\frac{dv}{dx} = \tfrac{200}{2}x^2 - \tfrac{200}{6}\langle x - 4 \rangle^3 + \tfrac{200}{6}\langle x - 10 \rangle^3 + \tfrac{2320}{2}\langle x - 15 \rangle + C_1 \quad (5.4.7)$$

and

$$EIv = \tfrac{200}{6}x^3 - \tfrac{200}{24}\langle x - 4 \rangle^4 + \tfrac{200}{24}\langle x - 10 \rangle^4 + \tfrac{2320}{6}\langle x - 15 \rangle^3 + C_1 x + C_2 \quad (5.4.8)$$

Constants C_1 and C_2 are determined from the boundary conditions $v(0) = 0$ and $v(15) = 0$; hence

$$EIv(0) = 0 = C_2$$
$$EIv(15) = 0 = \tfrac{200}{6}(15)^3 - \tfrac{200}{24}(11)^4 + \tfrac{200}{24}(5)^4 + 15C_1 + C_2$$

from which

$$C_1 = \tfrac{1720}{6} \text{ and } C_2 = 0$$

The equation of the deflection curve is now obtained by substituting the above values into Eq. (5.4.8); thus,

$$v = \frac{1}{6EI}[200x^3 - 50\langle x - 4\rangle^4 + 50\langle x - 10\rangle^4 + 2320\langle x - 15\rangle^3 + 1720x] \tag{5.4.9}$$

If desired, the deflection equation can now be written for each range; thus, from Eq. (5.4.9) it follows that

$$v = \frac{1}{6EI}[200x^3 + 1720x] \quad \text{for } 0 \leq x \leq 4$$

$$v = \frac{1}{6EI}[200x^3 - 50(x-4)^4 + 1720x] \quad \text{for } 4 \leq x \leq 10$$

$$v = \frac{1}{6EI}[200x^3 - 50(x-4)^4 + 50(x-10)^4 + 1720x] \quad \text{for } 10 \leq x \leq 15$$

$$v = \frac{1}{6EI}[200x^3 - 50(x-4)^4 + 50(x-10)^4 + 2320(x-15)^3 + 1720x]$$
$$\text{for } 15 \leq x \leq 20$$

Example 5.4.2. Determine the equation of the elastic curve for the beam shown in Fig. 5.4.5 (a).

The beam in this case is statically indeterminate with one redundant. The redundant reaction is assumed to be R_A, the reaction at the roller support. The bending moment is expressed in terms of the applied couple and the redundant reaction by applying the equilibrium conditions to the free body diagram in Fig. 5.4.5 (b); thus,

$$M = R_A x - M_0 \langle x - l \rangle^0 \tag{5.4.10}$$

By substituting the bending moment into Eq. (5.2.4) one obtains

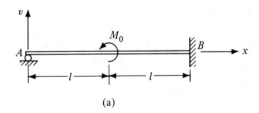

(a)

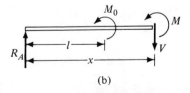

(b)

Figure 5.4.5

$$EI\frac{d^2v}{dx^2} = R_A x - M_0 \langle x - l \rangle^0 \tag{5.4.11}$$

The slope and deflection equations are

$$EI\frac{dv}{dx} = R_A \frac{x^2}{2} - M_0 \langle x - l \rangle + C_1 \tag{5.4.12}$$

and

$$EIv = R_A \frac{x^3}{6} - M_0 \frac{\langle x - l \rangle^2}{2} + C_1 x + C_2 \tag{5.4.13}$$

The boundary conditions in this case are

$$v(0) = 0$$
$$v(2l) = 0$$
$$\frac{dv(2l)}{dx} = 0$$

By using these conditions and Eqs. (5.4.12) and (5.4.13) we find that

$$EIv(0) = 0 = C_2$$
$$EIv(2l) = 0 = R_A \frac{(2l)^3}{6} - M_0 \frac{l^2}{2} + C_1(2l) + C_2$$
$$EI\frac{dv(2l)}{dx} = 0 = R_A \frac{(2l)^2}{2} + M_0 l + C_1$$

from which

$$C_1 = -\frac{M_0 l}{8}$$
$$C_2 = 0$$
$$R_A = \frac{9M_0}{16l}$$

By substituting the above values into Eq. (5.4.13), the equation of the deflection curve takes the following form:

$$v = \frac{M_0}{32EI}[3x^3 - 16l\langle x - l \rangle^2 - 4l^2 x] \tag{5.4.14}$$

We have now determined the redundant force R_A and the equation of the elastic curve. The reactions at B can be determined by use of the equilibrium equations for the beam.

In the preceding discussion, we employed singularity functions and the second order differential equation [Eq. (5.2.4)] to obtain elastic curve equations. The fourth order differential equation can also be used to obtain deflection curves. The use of Eq. (5.2.6) or (5.2.8) requires that the load be expressed as a function of the distance along the beam. The singularity functions introduced earlier are very convenient for expressing distributed loads in terms of x; however, those functions are not sufficient for treating concentrated forces or couples. For this reason, we shall now consider a method for representing concentrated forces and couples as distributed forces. This will lead to the introduction of two additional singularity functions.

A unit couple is shown acting on a beam in Fig. 5.4.6 (a) and an equivalent distributed load $p_2(x)$ is shown in Fig. 5.4.6 (d). The resultant R of the uniformly distributed force in the interval ϵ to the left of $x = a$ acts vertically up. The resultant for the interval ϵ to the right of $x = a$ acts vertically down and has the same magnitude. These two forces form a clockwise couple $R\epsilon = 1$. We now define the distributed force as

$$p_2(x) = \begin{cases} = 0 & \text{for } x < (a - \epsilon) \\ = \lim_{\epsilon \to 0} \frac{1}{\epsilon^2} & \text{for } (a - \epsilon) < x < a \\ = -\lim_{\epsilon \to 0} \frac{1}{\epsilon^2} & \text{for } a < x < (a + \epsilon) \\ = 0 & \text{for } x > (a + \epsilon) \end{cases} \qquad (5.4.15a)$$

with the condition that

$$\lim_{\epsilon \to 0} R\epsilon = 1 \qquad (5.4.15b)$$

Thus, the distributed force becomes very large in the vicinity of $x = a$ as ϵ approaches zero; however, the product $R\epsilon$ is always a unit couple. The function defined above is called the "unit doublet." Because of the unusual behavior of this function we shall use the bracket notation hereafter; thus,

$$\langle x - a \rangle_{-2} = p_2(x) \qquad (5.4.16)$$

represents a clockwise unit couple at $x = a$.

A unit concentrated force is shown acting on a beam in Fig. 5.4.6 (b) and an equivalent distributed force $p_1(x)$ is shown in Fig. 5.4.6 (e). The distributed force is defined as

$$p_1(x) = \begin{cases} 0 & \text{for } x < a - \epsilon \\ \lim_{\epsilon \to 0} \frac{\epsilon + (x - a)}{\epsilon^2} & \text{for } a - \epsilon < x < a \\ \lim_{\epsilon \to 0} \frac{\epsilon - (x - a)}{\epsilon^2} & \text{for } a < x < a + \epsilon \\ 0 & \text{for } x > a + \epsilon \end{cases} \qquad (5.4.17a)$$

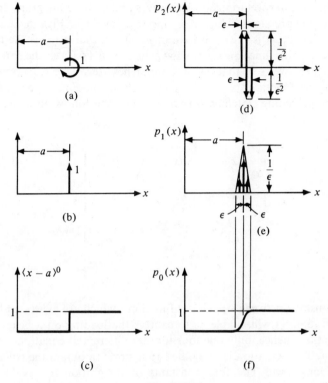

Figure 5.4.6

with the condition that

$$\lim_{\epsilon \to 0} \int_{-\infty}^{\infty} p_1(x)\,dx = 1 \qquad (5.4.17b)$$

Thus, the distributed force becomes very large as ϵ approaches zero, but the resultant of the distribution is always a unit force. The bracket notation will be used for the function $p_1(x)$ also. In this case then

$$\langle x - a \rangle_{-1} = p_1(x) \qquad (5.4.18)$$

represents a unit force acting up at $x = a$. This function is often called the "unit impulse."

We have now defined functions which enable us to represent concentrated forces and couples as distributed forces. These functions will be integrated when used in the fourth order differential equation of the elastic curve; therefore, we must determine the integrals of the new functions. Because of the way in which these functions have been defined we can, on the basis of the geometrical interpretations of derivatives and definite integrals, arrive at the necessary expressions. Thus, referring once again to Figs. 5.4.6 (d) and (e), we see that $p_2(x)$ is the derivative of $p_1(x)$, or, that $p_1(x)$ is the integral of $p_2(x)$. Also, the function $p_0(x)$ shown in Fig. 5.4.6 (f) is the integral of $p_1(x)$. Further, we see that as ϵ approaches zero the function $p_0(x)$ approaches the function $\langle x - a \rangle^0$ which is shown in Fig. 5.4.6 (c). On the basis of the preceding statements we now define the following integrals and derivatives:

$$\int_{-\infty}^{x} \langle x - a \rangle_{-2} dx = \langle x - a \rangle_{-1} \quad (5.4.19a)$$

$$\int_{-\infty}^{x} \langle x - a \rangle_{-1} dx = \langle x - a \rangle^0 \quad (5.4.19b)$$

$$\frac{d}{dx}\langle x - a \rangle^0 = \langle x - a \rangle_{-1} \quad (5.4.19c)$$

$$\frac{d}{dx}\langle x - a \rangle_{-1} = \langle x - a \rangle_{-2} \quad (5.4.19d)$$

The relationships between singularity functions, which are expressed by Eqs. (5.4.2) and (5.4.19) will enable us to express the loading on a beam with a single equation; hence, only one fourth order differential equation is written for a beam. This equation is integrated four times to obtain the equation of the deflection curve. The four constants of integration are evaluated by using the boundary conditions on shear, bending moment, slope, and deflection.

Example 5.4.3. Determine the equation of the elastic curve for the beam shown in Fig. 5.4.7. Use the fourth order differential equation [Eq. (5.2.8)] to obtain the solution.

As usual, when using singularity functions, the origin $x = 0$ is taken at the left end of the beam. The beam loading for $0 < x < 20$ ft can be expressed as follows:

$$p(x) = -100\langle x - 5 \rangle_{-1} - 10\langle x - 10 \rangle^0 \quad (5.4.20)$$

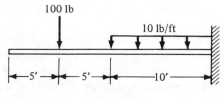

Figure 5.4.7

Substituting the above relation into Eq. (5.2.8) gives

$$EI\frac{d^4v}{dx^4} = -100\langle x-5\rangle_{-1} - 10\langle x-10\rangle^0 \quad (5.4.21)$$

By integration the following expressions are obtained:

$$EI\frac{d^3v}{dx^3} = V = -100\langle x-5\rangle^0 - 10\langle x-10\rangle^1 + C_1 \quad (5.4.22)$$

$$EI\frac{d^2v}{dx^2} = M = -100\langle x-5\rangle^1 - \tfrac{10}{2}\langle x-10\rangle^2 + C_1 x + C_2 \quad (5.4.23)$$

$$EI\frac{dv}{dx} = -\tfrac{100}{2}\langle x-5\rangle^2 - \tfrac{10}{6}\langle x-10\rangle^3 + \frac{C_1 x^2}{2} + C_2 x + C_3 \quad (5.4.24)$$

$$EIv = -\tfrac{100}{6}\langle x-5\rangle^3 - \tfrac{10}{24}\langle x-10\rangle^4 + \frac{C_1 x^3}{6} + \frac{C_2 x^2}{2} + C_3 x + C_4 \quad (5.4.25)$$

The boundary conditions in this case are

$$M = EI\frac{d^2v}{dx^2} = 0 \quad \text{and} \quad V = EI\frac{d^3v}{dx^3} = 0 \quad \text{at} \quad x = 0$$

$$v = 0 \quad \text{and} \quad \frac{dv}{dx} = 0 \quad \text{at} \quad x = 20 \text{ ft}$$

Thus, by use of Eqs. (5.4.22) through (5.4.25) and the boundary conditions we find that

$$V(0) = 0 = C_1$$
$$M(0) = 0 = C_2$$
$$EI\frac{dv(20)}{dx} = 0 = -\tfrac{100}{2}(15)^2 - \tfrac{10}{6}(10)^3 + C_3$$
$$EIv(20) = 0 = -\tfrac{100}{6}(15)^3 - \tfrac{10}{24}(10)^4 + C_3(20) + C_4$$

which yield the following values for the constants of integration:

$$C_1 = C_2 = 0, \quad C_3 = \tfrac{155000}{12}, \quad C_4 = -\tfrac{2375000}{12}$$

By substituting these values into Eqs. (5.4.22), (5.4.23), (5.4.24), and (5.4.25), respectively, the equations for shear force, bending moment, slope, and deflection can be obtained. The deflection equation is

$$v = \frac{1}{12EI}[-200\langle x-5\rangle^3 - 5\langle x-10\rangle^4 + 155000x - 2375000] \quad (5.4.26)$$

An alternate approach to solving beam problems with the aid of singularity functions utilizes the concept of an infinitely long beam. For instance, beam AB shown in Fig. 5.4.8 (a) can be replaced by the beam shown in Fig.

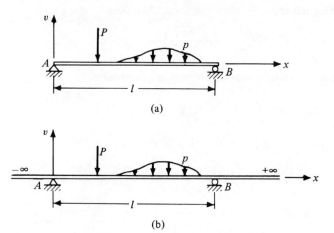

Figure 5.4.8

5.4.8 (b). The infinitely long beam is loaded in the range $0 \leq x \leq l$ and the following conditions are valid:

$$V(0^-) = 0 \quad V(l^+) = 0$$
$$M(0^-) = 0 \quad M(l^+) = 0$$
$$v(0) = 0 \quad v(l) = 0$$

where 0^- indicates a point just to the left of $x = 0$ and l^+ indicates a point just to the right of $x = l$. Conditions $V(0^-) = 0$ and $M(0^-) = 0$ indicate that the shear and bending moment are zero in the unloaded portion to the left of A. Similarly, conditions $V(l^+) = 0$ and $M(l^+) = 0$ indicate that the shear and bending moment are zero in the unloaded portion to the right of B. The remaining conditions, $v(0) = v(l) = 0$, are the deflection boundary conditions.

Example 5.4.4. The method of solution discussed in the preceding paragraph will now be used to determine the deflection curve for the beam shown in Fig. 5.4.9 (a).

This is the same configuration which was studied in Example 5.4.1. As before an upward uniform load is introduced at $x = 10$ ft to counteract the downward uniform load which, when expressed in terms of singularity functions, continues past its actual termination point [see Fig. 5.4.9 (b)]. For $0 \leq x \leq 20$ ft the beam loading is

$$p(x) = R_A \langle x - 0 \rangle_{-1} - 200 \langle x - 4 \rangle^0 + 200 \langle x - 10 \rangle^0 + R_B \langle x - 15 \rangle_{-1}$$
$$- 1320 \langle x - 20 \rangle_{-1} \tag{5.4.27}$$

Substituting the above relation into Eq. (5.2.8) gives

$$EI \frac{d^4 v}{dx^4} = R_A \langle x \rangle_{-1} - 200 \langle x - 4 \rangle^0 + 200 \langle x - 10 \rangle^0 + R_B \langle x - 15 \rangle_{-1}$$
$$- 1320 \langle x - 20 \rangle_{-1} \tag{5.4.28}$$

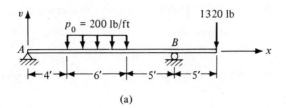

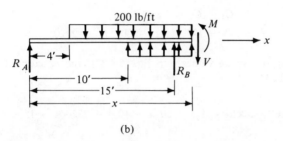

Figure 5.4.9

By integration, the following relations are obtained:

$$V = EI\frac{d^3v}{dx^3} = R_A\langle x\rangle^0 - 200\langle x-4\rangle^1 + 200\langle x-10\rangle^1 + R_B\langle x-15\rangle^0$$
$$- 1320\langle x-20\rangle^0 + C_1 \tag{5.4.29}$$

$$M = EI\frac{d^2v}{dx^2} = R_A\langle x\rangle^1 - \tfrac{200}{2}\langle x-4\rangle^2 + \tfrac{200}{2}\langle x-10\rangle^2 + R_B\langle x-15\rangle^1$$
$$- 1320\langle x-20\rangle^1 + C_1 x + C_2 \tag{5.4.30}$$

$$EI\frac{dv}{dx} = \frac{R_A\langle x\rangle^2}{2} - \tfrac{200}{6}\langle x-4\rangle^3 + \tfrac{200}{6}\langle x-10\rangle^3 + \frac{R_B}{2}\langle x-15\rangle^2$$
$$- \tfrac{1320}{2}\langle x-20\rangle^2 + \frac{C_1 x^2}{2} + C_2 x + C_3 \tag{5.4.31}$$

$$EIv = \frac{R_A\langle x\rangle^3}{6} - \tfrac{200}{24}\langle x-4\rangle^4 + \tfrac{200}{24}\langle x-10\rangle^4 + \frac{R_B}{6}\langle x-15\rangle^3$$
$$- \tfrac{1320}{6}\langle x-20\rangle^3 + \frac{C_1 x^3}{6} + \frac{C_2 x^2}{2} + C_3 + C_4 \tag{5.4.32}$$

Assuming that the reactions at A and B are unknown, then there are six constants ($R_A, R_B, C_1, C_2, C_3,$ and C_4) to be evaluated in the above equations. The six boundary conditions are $V(0^-) = M(0^-) = 0$, $V(20^+) = M(20^+) = 0$, $v(0) = 0$, and $v(15) = 0$; therefore, the following relations are obtained:

$$V(0^-) = 0 = C_1 \tag{5.4.33a}$$
$$M(0^-) = 0 = C_2 \tag{5.4.33b}$$
$$V(20^+) = 0 = R_A - 200(16) + 200(10) + R_B - 1320 + C_1 \tag{5.4.33c}$$

$$M(20^+) = 0 = R_A(20) - \tfrac{200}{2}(16)^2 + \tfrac{200}{2}(10)^2 + R_B(5) + 20C_1 + C_2 \tag{5.4.33d}$$

$$EIv(0) = 0 = C_4 \tag{5.4.33e}$$

$$EIv(15) = 0 = \tfrac{R_A}{6}(15)^3 - \tfrac{200}{24}(11)^4 + \tfrac{200}{24}(5)^4 + \tfrac{C_1}{6}(15)^3 + \tfrac{C_2}{6}(15)^2$$
$$+ 15C_3 + C_4 \tag{5.4.33f}$$

From the above relations, the constants are found to be

$$C_1 = C_2 = C_4 = 0$$
$$C_3 = \tfrac{1720}{6}, \quad R_A = 200, \quad R_B = 2320$$

By substituting the above values into Eq. (5.4.32), the equation for the elastic curve is found to be

$$v = \frac{1}{6EI}[200x^3 - 50\langle x-4\rangle^4 + 50\langle x-10\rangle^4 + 2320\langle x-15\rangle^3 + 1720x] \tag{5.4.34}$$

Since the term involving $\langle x - 20 \rangle^3$ is zero over the entire length of the beam, it has been dropped from the deflection equation.

If the origin of coordinates is at the left end of the beam then the boundary conditions $V(0^-) = 0$ and $M(0^-) = 0$ will be valid regardless of the type of surport; therefore, these conditions will always require that the integration constants C_1 and C_2 be equal to zero. Conditions $V(L^+) = 0$ and $M(L^+) = 0$, where L is the length of the beam, will also be valid regardless of the type of support. These two conditions will yield the equations of equilibrium for the beam. This last statement is true providing, of course, that all forces and moments are included in the loading function.

Example 5.4.5. Determine the equation of the elastic curve for the beam shown in Fig. 5.4.10 (a). This is the same configuration which was considered in Example 5.4.2. The corresponding infinite beam is shown in Fig. 5.4.10 (b). The loading for $0 \leq x \leq 2l$ is expressed as

$$p(x) = R_A\langle x - 0\rangle_{-1} - M_0\langle x - l\rangle_{-2} + R_B\langle x - 2l\rangle_{-1} + M_B\langle x - 2l\rangle_{-2}$$

where R_A, R_B, and M_B are the support reactions. Substituting the above expression into Eq. (5.2.8) gives

$$EI\frac{d^4v}{dx^4} = R_A\langle x\rangle_{-1} - M_0\langle x - l\rangle_{-2} + R_B\langle x - 2l\rangle_{-1} + M_B\langle x - 2l\rangle_{-2} \tag{5.4.35}$$

By integrating the above equation and remembering that, in view of the boundary conditions $V(0^-) = M(0^-) = 0$, the first two constants of integration are zero, we obtain

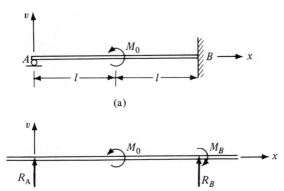

Figure 5.4.10

$$V = EI\frac{d^3v}{dx^3} = R_A\langle x\rangle^0 - M_0\langle x - l\rangle_{-1} + R_B\langle x - 2l\rangle^0 + M_B\langle x - 2l\rangle_{-1}$$
(5.4.36)

$$M = EI\frac{d^2v}{dx^2} = R_A\langle x\rangle^1 - M_0\langle x - l\rangle^0 + R_B\langle x - 2l\rangle^1 + M_B\langle x - 2l\rangle^0$$
(5.4.37)

$$EI\frac{dv}{dx} = \frac{R_A}{2}\langle x\rangle^2 - M_0\langle x - l\rangle^1 + \frac{R_B}{2}\langle x - 2l\rangle^2 + M_B\langle x - 2l\rangle^1 + C_3$$
(5.4.38)

$$EIv = \frac{R_A}{6}\langle x\rangle^3 - \frac{M_0}{2}\langle x - l\rangle^2 + \frac{R_B}{6}\langle x - 2l\rangle^3 + \frac{M_B}{2}\langle x - 2l\rangle^2 + C_3 x + C_4$$
(5.4.39)

The boundary conditions require that

$$EIv(0) = 0 = C_4 \tag{5.4.40a}$$
$$V(2l^+) = 0 = R_A + R_B \tag{5.4.40b}$$
$$M(2l^+) = 0 = R_A(2l) - M_0 + M_B \tag{5.4.40c}$$
$$EI\frac{dv(2l)}{dx} = 0 = \frac{R_A}{2}(2l)^2 - M_0 l + C_3 \tag{5.4.40d}$$
$$EIv(2l) = 0 = \frac{R_A}{6}(2l)^3 - \frac{M_0}{2}l^2 + C_3(2l) \tag{5.4.40e}$$

which yield the following values

$$\left.\begin{array}{l} C_3 = -\dfrac{M_0 l}{8}, \quad C_4 = 0 \\[1ex] R_A = \dfrac{9M_0}{16l}, \quad R_B = -\dfrac{9M_0}{16l}, \quad M_B = -\dfrac{M_0}{8} \end{array}\right\} \tag{5.4.41}$$

From the analysis of this indeterminate beam we have determined the constants of integration and the support reactions from the boundary conditions. An inspection of Eqs. (5.4.40b) and (5.4.40c) shows that they are simply the equations of equilibrium for the beam. By substituting values from Eqs. (5.4.41) into Eqs. (5.4.36) through (5.4.39), expressions for shear force, bending moment, slope, and deflection are determined. The deflection equation, for instance, is

$$v = \frac{M_0}{32EI}[3x^3 - 16\langle x - l\rangle^3 - 4l^2 x] \tag{5.4.42}$$

Since the terms involving $\langle x - 2l\rangle^2$ and $\langle x - 2l\rangle^3$ are zero over the entire length of the beam they have been deleted from the deflection equation.

The shear equation can be written as follows:

$$V = \frac{9M_0}{16l} - M_0\langle x - l\rangle_{-1} - \frac{M_0}{8}\langle x - 2l\rangle_{-1} \tag{5.4.43}$$

for $0 < x < 2l$. Terms $\langle x - l\rangle_{-1}$ and $\langle x - 2l\rangle_{-1}$ are zero at all points, except $x = l$ and $x = 2l$, respectively, where they become very large. Thus, the last two terms in Eq. (5.4.43) indicate the presence of large indeterminate values of shear force at $x = l$ and $x = 2l$. When the shear is written for the separate ranges we have

$$V = \frac{9M_0}{16l} = R_A \text{ for } 0 < x < l$$

$$V = \frac{9M_0}{16l} = R_A \text{ for } l < x < 2l$$

5.5 Use of Singularity Functions in the Case of Beams of Variable Cross Sections

The analysis of beams of variable cross sections can be facilitated by the use of singularity functions. Such applications result in products of singularity functions as well as products of continuous functions and singularity functions; thus, a few comments concerning the behavior of such products are necessary.

First, consider the integral of the product of a step function and a continuous function $f(x)$, i.e.,

$$g(x) = \int_{-\infty}^{\xi} \langle x - a\rangle^0 f(x)\, dx \tag{5.5.1}$$

If $\xi \leq a$ then the integral $g(x)$ is zero since the integrand is zero. For $\xi > a$, the only contribution to $g(x)$ is that due to integration from $x = a$ to $x = \xi$; therefore, Eq. (5.5.1) may be written in the form

$$g(x) = \int_a^{\xi} \langle x - a\rangle^0 f(x)\, dx$$

Sec. 5.5   BENDING OF STRAIGHT BEAMS   287

or, since $\langle x - a \rangle^0$ is constant in the above equation,

$$g(x) = \int_{-\infty}^{\xi} \langle x - a \rangle^0 f(x)\, dx = \langle \xi - a \rangle^0 \int_{a}^{\xi} f(x)\, dx \qquad (5.5.2)$$

In view of the above discussion Eq. (5.5.2) is obviously valid for all values of ξ.

Another useful relation is the product of two step functions, i.e.,

$$\langle x - a \rangle^0 \langle x - b \rangle^0 \qquad (5.5.3)$$

where $b \geq a$. The above product is zero for $x < b$; therefore, the function $\langle x - b \rangle^0$ is the controlling function and the product may be written as follows:

$$\langle x - a \rangle^0 \langle x - b \rangle^0 = \langle x - b \rangle^0 \qquad (5.5.4)$$

Equations (5.5.2) and (5.5.4) are quite useful as will be demonstrated in the examples that follow.

Example 5.5.1. The beam shown in Fig. 5.5.1 has constant thickness t and linearly varying depth. Determine the equation of the elastic curve for the beam. The loading function is

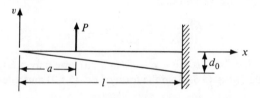

Figure 5.5.1

$$p(x) = P\langle x - a \rangle_{-1}$$

for $0 \leq x < l$. Substituting the above function into Eq. (5.2.6) gives the differential equation of the elastic curve; thus,

$$\frac{d^2}{dx^2}\left(EI\frac{d^2v}{dx^2}\right) = p(x) = P\langle x - a \rangle_{-1} \qquad (5.5.5)$$

By sucessive integrations of the above equation we find

$$\frac{d}{dx}\left(EI\frac{d^2v}{dx^2}\right) = V(x) = P\langle x - a \rangle^0 \qquad (5.5.6)$$

$$EI\frac{d^2v}{dx^2} = M(x) = P\langle x - a \rangle^1 \qquad (5.5.7)$$

For this case, depth d varies with x and the moment of inertia of any cross section is

$$I(x) = \frac{td^3}{12} = \frac{td_0^3 \, x^3}{12 \, l^3} = \frac{I_0}{l^3} x^3$$

where I_0 is the moment of inertia of the cross section at $x = l$. Substituting the above relation into Eq. (5.5.7) gives

$$\frac{EI_0}{l^3} x^3 \frac{d^2v}{dx^2} = P\langle x - a \rangle^1$$

or

$$\frac{EI_0}{l^3} \frac{d^2v}{dx^2} = P\frac{\langle x - a \rangle^1}{x^3} \tag{5.5.8}$$

If we note that

$$\langle x - a \rangle^1 = \langle x - a \rangle^0 (x - a)$$

then Eq. (5.5.8) may be written as

$$\frac{EI_0}{l^3} \frac{d^2v}{dx^2} = P\langle x - a \rangle^0 \frac{(x - a)}{x^3} \tag{5.5.9}$$

The right side of the above equation is the type of product represented in Eq. (5.5.2); therefore, by integrating we find that

$$\frac{EI_0}{l^3} \frac{dv}{dx} = \frac{P\langle x - a \rangle^2}{2ax^2} + C_1 \tag{5.5.10}$$

Again, by noting that $\langle x - a \rangle^2 = \langle x - a \rangle^0 (x - a)^2$, we may put the first term on the right-hand side of Eq. (5.5.10) in the form of Eq. (5.5.2) and integrate to find

$$\frac{EI_0}{l^3} v = P\langle x - a \rangle^0 \left(\frac{x^2 - a^2}{2ax} - \log \frac{x}{a} \right) + C_1 x + C_2 \tag{5.5.11}$$

The boundary conditions are $dv/dx = v = 0$ at $x = l$. From Eqs. (5.5.10) and (5.5.11) and the boundary conditions we find

$$\frac{EI_0}{l^3} \frac{dv(l)}{dx} = 0 = \frac{P(l - a)^2}{2al^2} + C_1$$

$$\frac{EI_0}{l^3} v(l) = 0 = P\left(\frac{l^2 - a^2}{2al} - \log \frac{l}{a} \right) + C_1 l + C_2$$

which give

$$C_1 = -\frac{P(l - a)^2}{2al^2}$$

$$C_2 = -\frac{P(l - a)}{2l} + P \log \frac{l}{a}$$

The equation of the elastic curve is obtained by substituting the above values of C_1 and C_2 into Eq. (5.5.11); hence

$$v(x) = \frac{Pl^3}{EI_0}\left[\langle x - a\rangle^0\left(\frac{x^2 - a^2}{2ax} - \log\frac{x}{a}\right) - \frac{x(l - a)^2}{2al^2} - \frac{(l - 2)}{l} + \log\frac{l}{a}\right]$$
(5.5.12)

Example 5.5.2. The beam in Fig. 5.5.2 has an abrupt change in cross section at $x = b$ and is loaded by concentrated forces as shown. Determine the equation of the elastic curve.

The loading function for this case is

$$p(x) = P_1\langle x - a_1\rangle_{-1} + P_2\langle x - a_2\rangle_{-1}$$

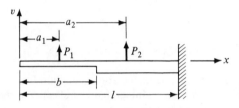

Figure 5.5.2

for $0 \leq x < l$. By substituting the above expression into Eq. (5.2.6), the differential equation of the elastic curve is obtained; thus,

$$\frac{d^2}{dx^2}\left(EI\frac{d^2v}{dx^2}\right) = p(x) = P_1\langle x - a_1\rangle_{-1} + P_2\langle x - a_2\rangle_{-1} \quad (5.5.13)$$

By integration the shear and moment relations are found to be

$$\frac{d}{dx}\left(EI\frac{d^2v}{dx_2}\right) = V(x) = P_1\langle x - a_1\rangle^0 + P_2\langle x - a_2\rangle^0 \quad (5.5.14)$$

$$EI\frac{d^2v}{dx^2} = M(x) = P_1\langle x - a_1\rangle^1 + P_2\langle x - a_2\rangle^1 \quad (5.5.15)$$

where the constants of integration are zero since $V(0^-) = M(0^-) = 0$, as explained previously. Equation (5.5.15) may be rewritten as

$$\frac{d^2v}{dx^2} = \frac{1}{EI}(P_1\langle x - a_1\rangle^1 + P_2\langle x - a_2\rangle^1) \quad (5.5.16)$$

By the use of singularity functions, the quantity $1/EI$ may be written as

$$\frac{1}{EI} = \frac{1}{EI_1} - \frac{\langle x - b\rangle^0}{EI_1} + \frac{\langle x - b\rangle^0}{EI_2} = \frac{1}{EI_1} + \langle x - b\rangle^0\left(-\frac{1}{EI_1} + \frac{1}{EI_2}\right) \quad (5.5.17)$$

By combining Eqs. (5.5.16) and (5.5.17), we find that

$$\frac{d^2v}{dx^2} = \frac{P_1}{EI_1}\langle x - a_1\rangle^1 + \frac{P_2}{EI_1}\langle x - a_2\rangle^1 + \left(-\frac{P_1}{EI_1} + \frac{P_1}{EI_2}\right)\langle x - b\rangle^0\langle x - a_1\rangle^1$$
$$+ \left(-\frac{P_2}{EI_1} + \frac{P_2}{EI_2}\right)\langle x - b\rangle^0 \langle x - a_2\rangle^1 \qquad (5.5.18)$$

The product $\langle x - b\rangle^0\langle x - a_2\rangle^1$ in the last term of the above equation may be written more concisely as

$$\langle x - b\rangle^0\langle x - a_2\rangle^1 = \langle x - a_2\rangle^1 \qquad (5.5.19)$$

since $a_2 > b$. The product $\langle x - b\rangle^0\langle x - a_1\rangle^1$ may be rewritten as

$$\langle x - b\rangle^0\langle x - a_1\rangle^1 = \langle x - b\rangle^0\langle x - a_1\rangle^0(x - a_1)$$

or

$$\langle x - b\rangle^0\langle x - a_1\rangle^1 = \langle x - b\rangle^0(x - a_1) \qquad (5.5.20)$$

since $b > a_1$. The right-hand side of the above equation has the same form as the integrand in Eq. (5.5.2) and can be integrated as indicated in that equation. Alternately, the last equation may be written as

$$\langle x - b\rangle^0\langle x - a_1\rangle^1 = \langle x - b\rangle^0(x - b + b - a_1)$$
$$= \langle x - b\rangle^0(x - b)^1 + \langle x - b\rangle^0(b - a_1)$$

or finally

$$\langle x - b\rangle^0\langle x - a_1\rangle^1 = \langle x - b\rangle^1 + \langle x - b\rangle^0(b - a_1) \qquad (5.5.21)$$

The right-hand side of the last equation is easily integrated. By substituting Eqs. (5.5.19) and (5.5.21) into Eq. (5.5.18) one finds that

$$\frac{d^2v}{dx^2} = \frac{P_1}{EI_1}\langle x - a_1\rangle^1 + \frac{P_2}{EI_1}\langle x - a_1\rangle^1 + \left(-\frac{P_1}{EI_1} + \frac{P_1}{EI_2}\right)(\langle x - b\rangle^1$$
$$+ (b - a_1)\langle x - b\rangle^0) + \left(-\frac{P_2}{EI_1} + \frac{P_2}{EI_2}\right)\langle x - a_2\rangle^1 \qquad (5.5.22)$$

The equations for slope and deflection are determined by integrating the last equation; thus,

$$\frac{dv}{dx} = \frac{P_2}{2EI_1}\langle x - a_1\rangle^2 + \frac{P_2}{2EI_2}\langle x - a_2\rangle^2 + \left[-\frac{P_1}{EI_1} + \frac{P_1}{EI_2}\right]\left[\frac{\langle x - b\rangle^2}{2}\right.$$
$$\left. + (b - a_1)\langle x - b\rangle^1\right] + \left(-\frac{P_2}{EI_1} + \frac{P_2}{EI_2}\right)\frac{\langle x - a_2\rangle^2}{2} + C_1 \qquad (5.5.23)$$

$$v = \frac{P_1}{6EI_1}\langle x - a_1\rangle^3 + \frac{P_2}{6EI_2}\langle x - a_2\rangle^3 + \left(-\frac{P_1}{EI_1} + \frac{P_1}{EI_2}\right)\left(\frac{\langle x-b\rangle^3}{6}\right.$$

$$\left. + (b-a_1)\frac{\langle x-b\rangle^2}{2}\right) + \left(-\frac{P_2}{EI_1} + \frac{P_2}{EI_2}\right)\frac{\langle x-a_2\rangle^3}{6} + C_1 x + C_2 \quad (5.5.24)$$

The constants of integration may be determined by use of the boundary conditions $dv/dx = v = 0$ at $x = l$. These conditions give

$$\frac{dv(l)}{dx} = 0 = \frac{P_1}{2EI_1}(l-a_1)^2 + \frac{P_2}{2EI_2}(l-a_2)^2 + \left[-\frac{P_1}{EI_1} + \frac{P_1}{EI_2}\right]\left[\frac{(l-b)^2}{2}\right.$$

$$\left. + (b-a_1)(l-b)\right] + \left(-\frac{P_2}{EI_1} + \frac{P_2}{EI_2}\right)\frac{(l-a_2)^2}{2} + C_1$$

and

$$v(l) = 0 = \frac{P_1}{6EI_1}(l-a_1)^3 + \frac{P_2}{6EI_2}(l-a_1)^3 + \left[-\frac{P_1}{EI_1} + \frac{P_1}{EI_2}\right]\left[\frac{(l-b)^3}{6}\right.$$

$$\left. + (b-a_1)\frac{(l-b)^2}{2}\right] + \left(-\frac{P_2}{EI_1} + \frac{P_2}{EI_2}\right)\frac{(l-a_2)^3}{6} + C_1 l + C_2$$

which can be solved to determine C_1 and C_2.

5.6 Use of Taylor's and Maclaurin's Series for Studying Deflections of Beams

The use of Taylor's[4] and Maclaurin's series often provides a simple method of directly obtaining the equations of elastic curves of bent beams.

If a function $v(x)$ and all its derivatives are continuous for $c < x < d$, then the function can be expanded into a Taylor's series, i.e.,

$$v(x) = v_a + v_a^{(1)}(x-a) + v_a^{(2)}\frac{(x-a)^2}{2!} + v_a^{(3)}\frac{(x-a)^3}{3!} + \cdots \quad (5.6.1)$$

where $c < a < d$ and $v_a^{(n)}$ represents the nth derivative of $v(x)$ evaluated at $x = a$. If $v(x)$ represents the deflection of a uniform beam then in accordance with Eqs. (5.2.4), (5.2.7), and (5.2.8) it follows that

$$v_a^{(2)} = \frac{M_a}{EI} \quad (5.6.2)$$

$$v_a^{(3)} = \frac{V_a}{EI} \quad (5.6.3)$$

$$v_a^{(4)} = \frac{p_a}{EI} \quad (5.6.4)$$

---

[4] Taylor, Brook. English mathematician (b. Edmonton, Middlesex 1685; d. London 1731). He was a Fellow and Secretary of the Royal Society. His "Methodus Incrementorum Directa et Inversa," which contained the celebrated Taylor's series, was published in London in 1715.

Maclaurin, Colin. Scottish mathematician (b. Kilmodan, Argyllshire 1698; d. Edinburgh 1746). When only nineteen years old, Maclaurin was appointed Professor of Mathematics at Marichal College, Aberdeen. At twenty-one, he was elected a Fellow of the Royal Society and there made the acquaintance of Sir Isaac Newton. In 1725, upon Newton's recommendation, he was elected Professor of Mathematics at Edinburgh University. His scientific activity continued many aspects of Newton's work. Especially well-known among his works are the *Geometria Organica, Sive Descriptio Linearum Curvarum Universalis* (1720), and his *Treatise on Fluxions* (1742). (Text from *Dynamics of Vibrations* by Enrico Volterra and E. C. Zachmanoglou. Reprinted with permission of Charles E. Merrill Books, Inc., Columbus, Ohio.) [Reproduction from the Vito Volterra collection, Villa Volterra, Ariccia (Rome).]

By substituting the above relations into Eq. (5.6.1) and letting $\theta_a = v_n^{(1)}$ one obtains:

$$v(x) = v_a + \theta_a(x-a) + \frac{M_a(x-a)^2}{2!\,EI} + \frac{V_a(x-a)^3}{3!\,EI} + \frac{p_a(x-a)^4}{4!\,EI}$$
$$+ \frac{p_a^{(1)}(x-a)^5}{5!\,EI} + \cdots \tag{5.6.5}$$

where $p_a^{(n)}$ represents the nth derivative of the load $p(x)$ evaluated at $x = a$.

Example 5.6 1. Determine the equation of the elastic curve for the beam shown in Fig. 5.6.1.

Taking $a = 0$, it follows from the boundary conditions at the fixed end that $v_0 = \theta_0 = 0$. Also, the loading and its derivatives are

$$p(x) = -\frac{p_0 x}{l}$$

$$p^{(1)}(x) = -\frac{p_0}{l}$$

$$p^{(n)}(x) = 0 \quad n = 2, 3, 4, \ldots$$

In view of the above relations, Eq. (5.6.5) may be written as

$$v = \frac{M_0 x^2}{2!\,EI} + \frac{V_0 x^3}{3!\,EI} - \frac{p_0 x^5}{5!\,EI} \tag{5.6.6}$$

From the equilibrium conditions for the beam we find that

$$V_0 = \frac{p_0 l}{2} \quad \text{and} \quad M_0 = -\frac{p_0 l^2}{3} \tag{5.6.7}$$

By substituting Eqs. (5.6.7) into Eq. (5.6.6), the equation of the elastic curve is found to be

$$v = \frac{p_0 x^2}{120 EIl}(-20l^3 + 10l^2 x - x^3) \tag{5.6.8}$$

Example 5.6.2. Determine the equation of the elastic curve for the cantilever beam shown in Fig. 5.6.2.

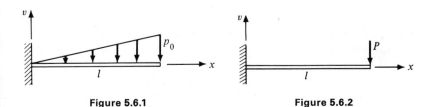

Figure 5.6.1 Figure 5.6.2

By taking the coordinates as indicated in Fig. 5.6.2 and letting $a = 0$ Eq. (5.6.5) becomes

$$v(x) = \frac{M_0 x^2}{2EI} + \frac{V_0 x^3}{6EI} \tag{5.6.9}$$

since $v_0 = \theta_0 = 0$ at the fixed end. By use of the equilibrium conditions for the beam, one finds that

$$M_0 = -Pl \quad \text{and} \quad V_0 = P$$

By substituting the above values into Eq. (5.6.9) it follows that

$$v(x) = -\frac{Px^2}{6EI}(3l - x) \tag{5.9.10}$$

Example 5.6.3. Determine the equation of the elastic curve and the end reactions for the statically indeterminate beam shown in Fig. 5.6.3 (a).

Taking $a = 0$, the boundary conditions at the end $x = 0$ require that $v_0 = \theta_0 = 0$. Since the loading is constant, the derivatives $p^{(n)} = 0$ for $n = 1, 2, 3, \ldots$. By assuming that M_0 and V_0 are positive then Eq. (5.6.5) can be written as follows:

$$v = \frac{M_0 x^2}{2EI} + \frac{V_0 x^3}{6EI} - \frac{p_0 x^4}{24EI} \tag{5.6.11}$$

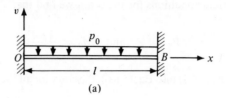

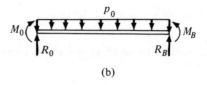

Figure 5.6.3

The quantities M_0 and V_0 are determined from the boundary conditions $v = v^{(1)} = 0$ at $x = l$. Thus,

$$v(l) = 0 = \frac{M_0 l^2}{2EI} + \frac{V_0 l^3}{6EI} - \frac{p_0 l^4}{24EI}$$

$$v^{(1)}(l) = 0 = \frac{2M_0 l}{2EI} + \frac{3V_0 l^2}{6EI} - \frac{4p_0 l^3}{24EI}$$

By solving the above equations for V_0 and M_0 the values

$$M_0 = -\frac{p_0 l^2}{12} \quad \text{and} \quad V_0 = \frac{p_0 l}{2} \tag{5.6.12}$$

are obtained. By use of the free body diagram, shown in Fig. 5.6.3 (b), and the equations of equilibrium we find that the reactions at B are

$$M_B = -\frac{p_0 l^2}{12} \quad \text{and} \quad V_B = -\frac{p_0 l}{6} \tag{5.6.13}$$

The reactive forces R_0 and R_B are shown in the proper direction in Fig. 5.6.3 (b); the reactive moments M_0 and M_b, however, act in directions opposite to those indicated in the figure. By substituting the values of M_0 and V_0 from Eq. (5.6.12) into Eq. (5.6.11) we find that

$$v = \frac{p_0 x^2}{24EI}(-l^2 + 2lx - x^2) \tag{5.6.14}$$

The series method is modified and applied to more complicated beam loadings in the next section.

5.7 General Deflection Equation for Beams of Uniform Cross Section

The method presented in Sec. 5.6 can be generalized so as to include concentrated forces and moments applied to the beam. Consider a function $f(x)$ such that

$$f(x) = \begin{cases} f_1(x) & \text{for } x < a \\ f_2(x) & \text{for } x > a \end{cases} \tag{5.7.1}$$

where $f_1(x), f_2(x), f_1^{(n)}(x)$, and $f_2^{(n)}(x)$ are continuous for all n and

$$\begin{aligned} f_1(a) &= f_2(a) \\ f_1^{(n)}(a) &= f_2^{(n)}(a) \end{aligned} \tag{5.7.2}$$

for all n except $n = j$ where

$$f_2^{(j)}(a) - f_1^{(j)}(a) = \delta \tag{5.7.3}$$

Then it can be shown that

$$f_2(x) = f_1(x) + \frac{\delta(x-a)^j}{j!} \tag{5.7.4}$$

By expanding $f_1(x)$ in a Taylor's series at $a = 0$ (i.e., in a Maclaurin's series), it follows that

$$f_1(x) = f_1(0) + x f_1^{(1)}(0) + \frac{x^2}{2!} f_1^{(2)}(0) + \cdots \tag{5.7.5}$$

Then according to Eq. (5.7.4), one finds that

$$f_2(x) = f_1(0) + x f_1^{(1)}(0) + \frac{x^2}{2!} f_1^{(2)}(0) + \cdots + \frac{\delta(x-a)^j}{j!} \tag{5.7.6}$$

By use of the singularity function, the last two equations may be written more concisely as

$$f(x) = f(0) + x f^{(1)}(0) + \frac{x^2}{2!} f^{(2)}(0) + \cdots + \frac{\delta \langle x-a \rangle^j}{j!} \tag{5.7.7}$$

In the case of the deflection curve for a uniform beam, the conditions specified by Eqs. (5.7.1), (5.7.2), and (5.7.3) are satisfied and the above relation becomes

$$v(x) = v_0 + x v_0^{(1)} + \frac{x^2}{2!} v_0^{(2)} + \cdots + \frac{\delta \langle x-a \rangle^j}{j} \tag{5.7.8}$$

If the relations (see Sec. 5.6)

$$v^{(2)} = \frac{M}{EI} \tag{5.6.2}$$

$$v^{(3)} = \frac{V}{EI} \tag{5.6.3}$$

$$v^{(4)} = \frac{p}{EI} \tag{5.6.4}$$

are used then Eq. (5.7.8) may be written in the form

$$v(x) = v_0 + \theta_0 x + \frac{M_0 x^2}{2!EI} + \frac{V_0 x^3}{3!EI} + \frac{p_0 x^4}{4!EI} + \frac{p_0^{(1)} x^5}{5!EI} + \cdots + \sum_{i=1}^{k} \frac{\Delta_i \langle x - a_i \rangle^j}{j!EI} \tag{5.7.9}$$

where $\Delta = EI\delta$. The last term has been written as a summation, thus allowing for the possibility of k discontinuities Δ_i at points $x = a_i$ along the beam.

For the case where a clockwise concentrated moment M_a is applied at $x = a$, the bending moment $M = EIv^{(2)}$ is discontinuous and the last term in Eq. (5.7.9) becomes

$$\frac{M_a \langle x - a \rangle^2}{2!EI} \tag{5.7.10}$$

If instead, a force P_a acting upward is applied at $x = a$, the discontinuity is in the shear (i.e., in $EIv^{(3)}$) and the last term Eq. (5.7.9) becomes

$$\frac{P_a \langle x - a \rangle^3}{3!EI} \tag{5.7.11}$$

For a uniformly distributed load p_a acting upward and starting at $x = a$, the last term becomes

$$\frac{p_a \langle x - a \rangle^4}{4!EI} \tag{5.7.12}$$

It follows that the term arising from a discontinuity in the loading (or derivative of the loading) will be of the same form as the corresponding term in the initial part of the deflection equation [Eq. (5.7.9)].

Example 5.7.1. Determine the deflection equation for the beam of constant cross section shown in Fig. 5.7.1.

For this case

$$v_0 = M_0 = p_0 = p_0^{(n)} = 0$$

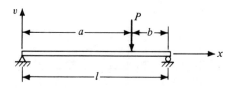

Figure 5.7.1

for all n and from the equilibrium condition one finds that $V_0 = Pb/l$; therefore, Eq. (5.7.9) reduces to

$$v(x) = \theta_0 x + \frac{Pbx^3}{3!EIl} + \frac{\Delta \langle x - a \rangle^j}{j!EI} \qquad (5.7.13)$$

The applied force at $x = a$ causes a shear discontinuity and the last term in the above equation is of the form indicated in Eq. (5.7.11). Consequently, Eq. (5.7.13) becomes

$$v(x) = \theta_0 x + \frac{Pbx^3}{3!EIl} - \frac{P\langle x - a \rangle^3}{3!EI} \qquad (5.7.14)$$

where the minus sign indicates that the applied force acts downward. The unknown value θ_0 can be determined from the condition that $v(l) = 0$; hence,

$$v(l) = 0 = \theta_0 l + \frac{Pbl^3}{3!EIl} - \frac{P(l-a)^3}{3!EI}$$

from which

$$\theta_0 = -\frac{Pb(l^2 - b^2)}{6EIl}$$

Substituting this expression into Eq. (5.7.14) gives

$$v(x) = \frac{P}{6EIl}[-b(l^2 - b^2)x + bx^3 - l\langle x - a \rangle^3] \qquad (5.7.15)$$

Example 5.7.2. Determine the equation of the elastic curve for the beam shown in Fig. 5.7.2 (a). The maximum value of the uniformly varying load is w lb/ft.
 For this beam $v_0 = M_0 = 0$ and, from the equilibrium conditions, $V_0 = wl/4$. For $0 \leq x \leq l/2$

$$p = \frac{2wx}{l}, \quad p^{(1)} = -\frac{2w}{l}, \quad p^{(n)} = 0 \quad \text{for} \quad n > 1$$

and consequently

$$p_0 = 0, \quad p_0^{(1)} = -\frac{2w}{l}$$

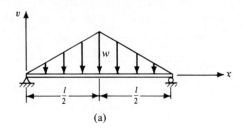

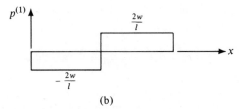

Figure 5.7.2

Substituting the above values into Eq. (5.6.9) gives

$$v(x) = \theta_0 x + \frac{wl}{4}\frac{x^3}{3!EI} - \frac{2w}{l}\frac{x^5}{5!EI} + \frac{\Delta\langle x - a\rangle^j}{j!EI} \tag{5.7.16}$$

For the given loading, the only discontinuity for $0 < x < l$ occurs in $p^{(1)}$ at $x = l/2$ [see the curve $p^{(1)}$-x shown in Fig. 5.7.2(b)]. The discontinuity is $\Delta = 4w/l$ and, therefore, the last term in the above equation has the form

$$\frac{4w}{l}\frac{\langle x - \frac{l}{2}\rangle^5}{5!EI}$$

and Eq. (5.7.16) becomes

$$v(x) = \theta_0 x + \frac{wlx^3}{24EI} - \frac{wx^5}{60lEI} + \frac{w\langle x - \frac{l}{2}\rangle^5}{30lEI} \tag{5.7.17}$$

The unknown θ_0 is determined from the condition $v(l) = 0$; thus,

$$v(l) = 0 = \theta_0 l + \frac{wl^4}{24EI} - \frac{wl^4}{60EI} + \frac{w\left(\frac{l}{2}\right)^5}{30lEI}$$

from which it follows that

$$\theta_0 = -\frac{5wl^3}{192EI}$$

Equation (5.7.17) can now be written as follows:

$$v(x) = \frac{w}{960lEI}\left(25l^4 x - 40l^2 x^3 + 16x^5 + 32\left\langle x - \frac{l}{2}\right\rangle^5\right) \tag{5.7.18}$$

Example 5.7.3. Determine the equation of the elastic curve and the support reactions for the statically indeterminate beam shown in Fig. 5.7.3 (a).

For this configuration $M_0 = V_0 = 0$, $p_0 = -60$ lb/ft, and $p_0^{(n)} = 0$ for $n > 0$. There are discontinuities in V at $x = 4$ ft and in p at $x = 10$ ft which lead, respectively, to the terms

$$\frac{R_A\langle x - 4\rangle^3}{3!EI} \quad \text{and} \quad \frac{60\langle x - 10\rangle^4}{4!EI}$$

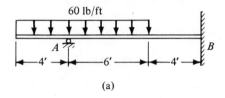

(a)

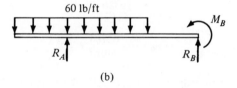

(b)

Figure 5.7.3

By substituting the above values into Eq. (5.7.9), one obtains

$$v(x) = v_0 + \theta_0 x - \frac{60x^4}{24EI} + \frac{R_A\langle x - 4\rangle^3}{6EI} + \frac{60\langle x - 10\rangle^4}{24EI} \tag{5.7.19}$$

The equation for the slope of the elastic curve is needed in order to evaluate the unknown quantities. By differentiating Eq. (5.7.19), the slope is found to be

$$v^{(1)}(x) = \theta_0 - \frac{60x^3}{6EI} + \frac{R_A\langle x - 4\rangle^2}{2EI} + \frac{60\langle x - 10\rangle^3}{6EI} \tag{5.7.20}$$

There are five unknown quantities which must be evaluated in order to complete the solution; these are v_0, θ_0, and the support reactions [see Fig. 5.7.3 (b)] R_A, R_B, and M_B. The boundary conditions

$$v(4) = v(14) = v^{(1)}(14) = 0$$

and the conditions of equilibrium

$$\Sigma F = 0 \quad \text{and} \quad \Sigma M = 0$$

provide a sufficient number of equations to determine the unknowns. The use of Eqs. (5.7.19) and (5.7.20) and the above conditions gives

$$v(4) = 0 = v_0 + 4\theta_0 - \frac{(4)^4(60)}{24EI}$$

$$v(14) = 0 = v_0 + 14\theta_0 - \frac{(14)^4(60)}{24EI} + \frac{R_A(10)^3}{6EI} + \frac{(4)^4(60)}{24EI}$$

$$v'(14) = 0 = \theta_0 - \frac{(14)^3(60)}{6EI} + \frac{R_A(10)^2}{2EI} + \frac{(4)^3(60)}{6EI}$$

$$\sum F = 0 = R_A + R_B - 600$$

$$\sum M_B = 0 = -R_A(10) + (9)(500) + M_B$$

By solving the above equations one finds that

$$R_A = 519.72 \text{ lb} \uparrow$$
$$R_B = 80.28 \text{ lb} \uparrow$$
$$M_B \doteq 202.8 \text{ lb-ft} \circlearrowright$$
$$\theta_0 = \frac{19536}{24EI}$$
$$v_0 = -\frac{62784}{240EI}$$

By substituting the above values into Eq. (5.7.19) the equation of the elastic curve is found to be

$$v = \frac{1}{240EI}(-62784 + 195360x - 600x^4 + 20788.8\langle x - 4\rangle^3 + 600\langle x - 10\rangle^4) \quad (5.7.21)$$

The equations for slope, bending moment, shear, and loading can be derived by sucessive differentiations of the deflection equation.

5.8 Mohr's Conjugate Beam Method[5]

In this section, special analogies existing among the functions which appear in the equations derived in Secs. (5.1) and (5.2) will be investigated. A method for determining deflections and slopes of a bent beam, developed by Otto Mohr follows. This method is particularly useful in the analysis of beams with variable cross sections.

In Sec. (5.2), the following relationship was derived for a bent beam:

$$\frac{d^2v}{dx^2} = \frac{d\theta}{dx} = \frac{M}{EI} \quad (5.8.1)$$

---

[5] See Bibliography Sec. 5.8, No. 17 at the end of this chapter.

where $\theta = dv/dx$ is the slope of the elastic curve. Consider now another beam, which is called the conjugate beam, and let the loading, shear force, and bending moment for this beam be denoted by p_c, V_c, and M_c, respectively. Then, from Eqs. (5.1.1) and (5.1.2) it follows that

$$\frac{d^2 M_c}{dx^2} = \frac{dV_c}{dx} = p_c \tag{5.8.2}$$

By comparing Eqs. (5.8.1) and (5.8.2) one notes that

v is analogous to M_c

θ is analogous to V_c

$\dfrac{M}{EI}$ is analogous to p_c

and x has the same meaning in both equations. From these analogies, it may be concluded that if the conjugate beam is subjected to a load which at each point is equal to the M/EI value for the actual beam at the same point, then the shear and bending moment for the conjugate beam are equal, respectively, to the slope and deflection in the actual beam. This is true, of course, provided that the conjugate beam has the proper boundary conditions and the same length as the actual beam. The boundary conditions for the conjugate beam are readily determined. For example, for a beam having a simple support at one end, then $M = 0$ and $v = 0$ which means that $v_c = 0$ and $M_c = 0$, respectively, for the conjugate beam. These are the force and moment conditions corresponding to a simple support; therefore, one can conclude that the support conditions for a simply supported beam and its conjugate beam

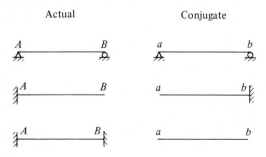

Figure 5.8.1

are identical. Equivalent boundary conditions for various support conditions are given as follows:

Actual Beam	Conjugate Beam
simple support ($M = 0$, $v = 0$)	simple support ($v_c = 0$, $M_c = 0$)
free end ($V = 0$, $M = 0$)	fixed end ($\theta_c = 0$, $v_c = 0$)
fixed end ($\theta = 0$, $v = 0$)	free end ($V_c = 0$, $M_c = 0$)

Some examples of actual beams and their conjugates are shown in Fig. 5.8.1.

Example 5.8.1. For the simply supported beam with an applied end moment [see Fig. 5.8.2 (a)] determine the slope of the elastic curve at points A and B. Determine also the equation of the elastic curve.

The moment diagram is shown in Fig. 5.8.2 (b) and the conjugate beam with its loading ($p_c = M/EI$) is shown in Fig. 5.8.2 (c). The loading p_c is positive and acts

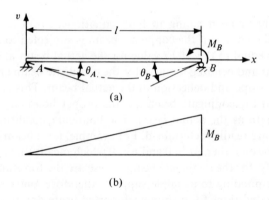

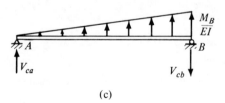

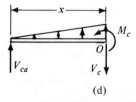

Figure 5.8.2

upward since the bending moment for the actual beam is positive for $0 < x < l$. The slope for the actual beam is equal to the shear force in the conjugate beam; therefore, we shall solve for the shear forces V_{ca} and V_{cb} which are shown in the positive directions according to the shear force sign convention [see Fig. 5.8.2 (c)]. Thus,

$$\sum M_b = 0 = V_{ca}l + \left(\frac{l}{2}\right)\left(\frac{M_B}{EI}\right)\left(\frac{l}{3}\right)$$

$$\sum M_a = 0 = V_{cb}l - \left(\frac{l}{2}\right)\left(\frac{M_B}{EI}\right)\left(\frac{2l}{3}\right)$$

From the above equations one finds that

$$V_{ca} = \theta_A = -\frac{M_B l}{6EI} \tag{5.8.3}$$

$$V_{cb} = \theta_B = \frac{M_B l}{3EI} \tag{5.8.4}$$

To determine the deflection equation we shall refer to Fig. 5.8.2 (d) and solve for the moment M_c at point O located at a distance x from the left end of the conjugate beam. The moment M_c is equal to the actual deflection at point x. Thus

$$\sum M_0 = 0 = V_{ca}x + \left(\frac{x}{2}\right)\left(\frac{M_B x}{EIl}\right)\left(\frac{x}{3}\right) - M_c$$

where V_{ca} is given by Eq. (5.8.3). Solving for M_c we find that

$$M_c = v = \frac{M_B}{6EIl}(x^3 - l^2 x) \tag{5.8.5}$$

Example 5.8.2. Determine the deflection and slope at the free end of the cantilever beam shown in Fig. 5.8.3 (a).

The moment diagram for this case is shown in Fig. 5.8.3 (b) and the M/EI diagram is shown in Fig. 5.8.3 (c) while the conjugate beam appears in Fig. 5.8.3 (d). The loading p_c acts downward since the bending moment is negative for $0 < x < l$. Since the slope and deflection at the free end of the actual beam are equal, respectively, to the shear force and bending moment at the fixed end of the conjugate beam, we shall solve for quantities V_{cb} and M_{cb} shown in Fig. 5.8.3 (d). These quantities are shown in their positive directions.

$$\sum F_v = 0 = V_{cb} + \left(\frac{l}{2}\right)\left(\frac{Pl}{EI}\right) + \left(\frac{l}{2}\right)\left(\frac{Pl}{2EI}\right) + (l)\left(\frac{Pl}{2EI}\right)$$

$$\sum M_b = 0 = M_{cb} + \left(\frac{l}{2}\right)\left(\frac{Pl}{EI}\right)\left(\frac{2l}{3}\right) + \left(\frac{l}{2}\right)\left(\frac{Pl}{2EI}\right)\left(\frac{5l}{3}\right) + (l)\left(\frac{Pl}{2EI}\right)\left(\frac{3l}{3}\right)$$

From the above equations one finds that

$$V_{cb} = \theta_B = -\frac{5Pl^2}{4EI}$$

$$M_{cb} = v_B = -\frac{3Pl^3}{2EI}$$

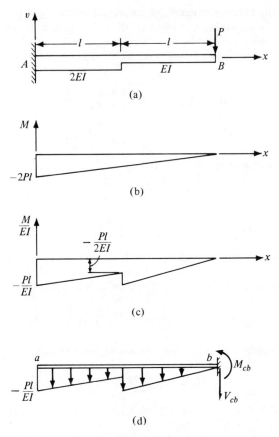

Figure 5.8.3

5.9 Clapeyron's Three-Moment Equation for Solving Continuous Beams

Consider a continuous beam resting on n supports. The supports are rollers except one, which is usually an intermediate fixed pin support so that the horizontal displacements of the beam are prevented. The system is $(n-2)$ times statically indeterminate since at each of the $(n-2)$ intermediate supports there is an unknown vertical reaction. However, if the number of supports n is greater than four, it is more convenient to take the bending moments at the supports as unknowns instead of the vertical reactions. These moments satisfy the so-called three-moment equation, which is an equation involving the bending moments at three successive supports. In the case of a continuous beam having constant cross section, the three-moment equation for the three adjacent supports A, B, and C of Fig. 5.9.1 (a) is

$$M_A l_1 + 2M_B(l_1 + l_2) + M_C l_2 = -\frac{6A_1 a_1}{l_1} - \frac{6A_2 b_2}{l_2} \quad (5.9.1)$$

In Eq. (5.9.1) M_A, M_B, and M_C are the three unknown bending moments at supports A, B, and C [see Fig. 5.9.1(b)]; l_1 and l_2 are the lengths of two successive spans; A_1 is the area of the bending moment diagram for the left-hand span 1 considered as a simply supported beam, and a_1 and b_1 are the horizontal distances of the centroid G_1 of this moment diagram from supports A and B, respectively [see Fig. 5.9.1(c).] In the same way A_2, a_2, b_2, and G_2 apply to similar quantities for the right-hand span 2 [see Fig. 5.9.1 (c)]. Due to the loading on span 1, the slope at B in the left span of the beam is[6]

$$\theta'_1 = \frac{A_1 a_1}{l_1 EI} \quad (5.9.2)$$

The rotation of the cross section at support B in span 1 produced by the couples M_A and M_B [see Figs. 5.9.1 (d) and (e)] is

$$\theta''_1 = \frac{M_B l_1}{3EI} + \frac{M_A l_1}{6EI} \quad (5.9.3)$$

By adding Eqs. (5.9.2) and (5.9.3), one obtains the following expression for the total angle of rotation of the cross section at support B in span 1:

$$\theta_1 = \frac{M_B l_1}{3EI} + \frac{M_A l_1}{6EI} + \frac{A_1 a_1}{l_1 EI} \quad (5.9.4)$$

In a similar manner one obtains for span 2, the following expression for the slope at B:

$$\theta_2 = \frac{M_B l_2}{3EI} + \frac{M_C l_2}{6EI} + \frac{A_2 b_2}{l_2 EI} \quad (5.9.5)$$

To satisfy the condition of continuity of the deflection curve at support B, the angles of rotation of the ends at B must be equal in magnitude and both clockwise or both counterclockwise. It follows that

$$\theta_1 = -\theta_2 \quad (5.9.6)$$

By substituting Eqs. (5.9.4) and (5.9.5) into Eq. (5.9.6), Eq. (5.9.1) is obtained.

---

[6] Equation (5.9.2) is easily derived by referring to Fig. 5.9.1 (c) and using Mohr's conjugate beam method. Similarly, Eqs. (5.9.3), (5.9.4), and (5.9.5) can be derived by referring to Figs. 5.9.1 (c), (d), and (e).

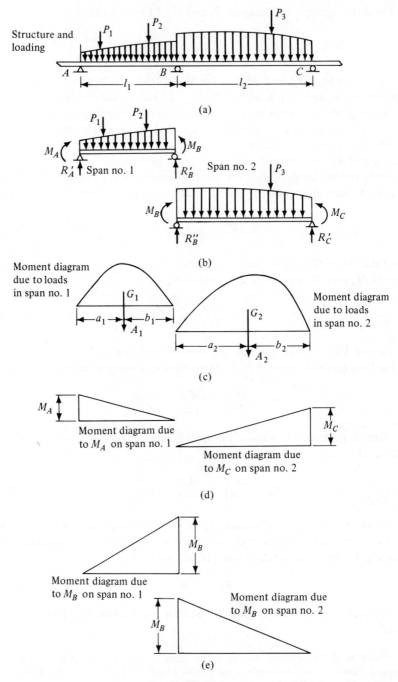

Figure 5.9.1

Example 5.9.1. Construct the bending moment and shear force diagrams for the continous beam resting on four equidistant supports and subjected to the uniformly distributed load of intensity p_0 as shown in Fig. 5.9.2 (a).

The moments at the supports are in this case

$$M_A = M_D = 0$$
$$M_B = M_C$$

Also, one finds that

$$A_1 = A_2 = A_3 = \frac{2}{3}\frac{p_0 l^2}{8}l = \frac{p_0 l^3}{12}$$

$$a_1 = a_2 = a_3 = b_1 = b_2 = b_3 = \frac{l}{2}$$

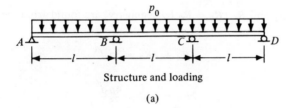

Structure and loading

(a)

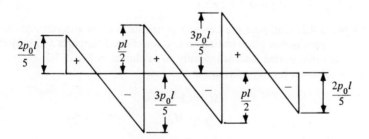

Shearing force diagram

(b)

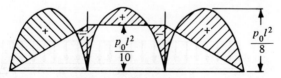

Bending moment diagram

(c)

Figure 5.9.2

By applying Eq. (5.9.1) to the moments at supports A, B, and C, the equation

$$5M_B l = -\frac{p_0 l^3}{2}$$

is obtained and it follows that $M_B = -p_0 l^2/10$. To calculate the reactions R_A and R_D ($R_A = R_D$ by symmetry) one writes the expression for the bending moment M_B at support B based on forces acting to the left of support B, i.e.,

$$M_B = \frac{p_0 l^2}{10} = R_A l - \frac{p_0 l^2}{2}$$

from which one finds that

$$R_A = \frac{2p_0 l}{5}$$

The reactions R_B and R_C ($R_B = R_C$ by symmetry) are found from the equilibrium equation

$$2R_A + 2R_B = 3p_0 l$$

or

$$2R_B = 3p_0 l - \frac{4p_0 l}{5}$$

which gives $R_B = 11p_0 l/10$. The shearing force and bending moment diagrams are shown in Figs. 5.9.2 (b) and 5.9.2 (c).

Example 5.9.2. Compute the bending moments and the reactions at the supports for the continuous beam resting on five equidistant supports and subjected to the uniformly distributed load of intensity p_0 as shown in Fig. 5.9.3.

The moments at the supports are in this case:

$$M_A = M_E = 0$$
$$M_B = M_D$$

Also, one has that

$$A_1 = A_2 = A_3 = A_4 = \frac{p_0 l^3}{12}$$

$$a_1 = a_2 = a_3 = a_4 = b_1 = b_2 = b_3 = b_4 = \frac{l}{2}$$

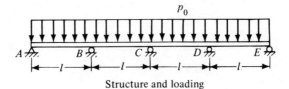

Structure and loading

Figure 5.9.3

By applying Eq. (5.9.1) successively to the moments at supports A, B, C, and B, C, D one finds

$$4M_Bl + M_Cl = -\frac{p_0l^3}{2}$$

$$2M_Bl + 4M_Cl = -\frac{p_0l^3}{2}$$

from which it follows that

$$M_B = M_D = -\tfrac{3}{28}p_0l^2$$

$$M_C = -\frac{p_0l^2}{14}$$

To calculate the reactions $R_A = R_E$, $R_B = R_D$, and R_C the following equilibrium equations are used:

$$M_B = -\tfrac{3}{28}p_0l^2 = R_Al - \frac{p_0l^2}{2}$$

$$M_C = -\frac{p_0l^2}{14} = R_A(2l) - 2p_0l^2 + R_Bl$$

$$2R_A + 2R_B + R_C = 4p_0l$$

From the above equations it follows that

$$R_A = \tfrac{11}{28}p_0l, \quad R_B = \tfrac{8}{7}p_0l, \quad R_C = \tfrac{13}{14}p_0l$$

Example 5.9.3. Construct the shear force and bending moment diagrams for the continuous beam resting on four equidistant supports and subjected to the concentrated loads of intensity P acting at the midspans as shown in Fig. 5.9.4 (a).

The moments at the supports in this case are

$$M_A = M_D = 0$$

$$M_B = M_C$$

Also, one has that

$$A_1 = A_2 = A_3 = \frac{Pl^2}{8}$$

$$a_1 = a_2 = a_3 = b_1 = b_2 = b_3 = \frac{l}{2}$$

By applying Eq. (5.9.1) to the moments at supports A, B, and C one obtains the relation

$$5M_Bl = -\tfrac{3}{4}Pl^2$$

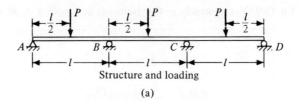
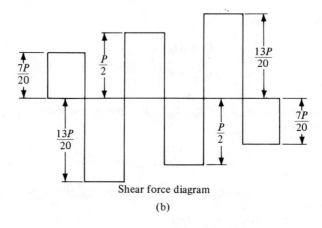
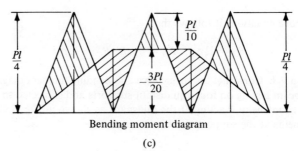

Figure 5.9.4

from which it follows that

$$M_B = M_C = -\tfrac{3}{20}Pl$$

The reaction R_A ($R_D = R_A$) is found from the equation

$$M_B = -\frac{3}{20}Pl = R_A l - \frac{Pl}{2}$$

which gives

$$R_A = R_D = \tfrac{7}{20}P$$

The reaction R_B ($R_C = R_B$) is found from the equation

$$2R_A + 2R_B = 3P$$

which gives

$$R_B = R_C = \tfrac{3}{2}P - R_A = \tfrac{3}{2}P - \tfrac{7}{20}P = \tfrac{23}{20}P$$

The shear force and bending moment diagrams are shown in Figs. 5.9.4 (b) and 5.9.4 (c).

5.10 Application of Trigonometric Series to the Study of Deflections of Beams[7]

Consider a beam of constant cross section and length l which is simply supported at its ends and subjected to the distributed load

$$p(x) = \sum_{n=1}^{\infty} p_n \sin \frac{n\pi x}{l} \qquad (5.10.1)$$

From Eq. (5.2.8), i.e.,

$$\frac{d^4 v(x)}{dx^4} = \frac{p(x)}{EI} \qquad (5.8.2)$$

it follows that

$$\frac{d^4 v(x)}{dx^4} = \frac{1}{EI} \sum_{n=1}^{\infty} p_n \sin \frac{n\pi x}{l} \qquad (5.10.2)$$

The solution of Eq. (5.10.2) which satisfies the conditions that the beam is simply supported at its ends, i.e.,

$$v(0) = v(l) = \frac{d^2 v(0)}{dx^2} = \frac{d^2 v(l)}{dx^2} = 0$$

is

$$v(x) = \frac{l^4}{\pi^4 EI} \sum_{n=1}^{\infty} \frac{1}{n^4} p_n \sin \frac{n\pi x}{l} \qquad (5.10.3)$$

Equation (5.10.3) is a very rapidly convergent series and the first few terms

[7] For a discussion of the representation of periodic functions by means of trigonometric series, see Appendix 5A at the end of this chapter.

of it are in general sufficient to represent, with a good approximation, the deflection curve of the beam.

Example 5.10.1. Derive the sine series representation of the deflection curve of a simply supported beam of length l which is subjected to a uniformly distributed load p_0 acting over its entire length.

Assume the load to act vertically downward, i.e.,

$$p(x) = -p_0 \quad \text{for} \quad 0 < x < l$$

Since the sine series expansion of the function[8]

$$f(x) = 1 \quad \text{for} \quad 0 < x < l$$

is

$$f(x) = \sum_{n=1}^{\infty} b_n \sin \frac{n\pi x}{l}$$

where

$$b_n = \frac{4}{n\pi}, \quad n = 1, 3, 5, \ldots$$

the coefficients p_n of the present example are

$$p_n = -p_0 b_n = -\frac{4p_0}{n\pi}, \quad n = 1, 3, 5, \ldots \tag{5.10.4}$$

By substituting Eqs. (5.10.4) into Eq. (5.10.3) it follows that

$$v(x) = -\frac{4p_0 l^4}{\pi^5 EI} \sum_{n=1,3,5,\ldots}^{\infty} \frac{1}{n^5} \sin \frac{n\pi x}{l} \tag{5.10.5}$$

Equation (5.10.5) is a very rapidly convergent series. By taking only the first term of the series, one finds that

$$v\left(\frac{l}{2}\right) = -\frac{4p_0 l^4}{\pi^5 EI} = -\frac{p_0 l^4}{76.505 EI}$$

whereas the exact value derived from the strength of materials is

$$v\left(\frac{l}{2}\right) = -\frac{5p_0 l^4}{384 EI} = -\frac{p_0 l^4}{76.8 EI}$$

Thus, by neglecting all but the first term of the series the error is only 0.38 percent.

[8] See Example 5.A.3 in the Appendix at the end of this chapter.

Example 5.10.2. Determine the sine series representation of the deflection curve of a simply supported beam subjected to a concentrated load applied at the section $x = a$ (see Fig. 5.10.1).

In this case, the loading is given by the expression

$$p(x) = -P\langle x - a\rangle_{-1}$$

and the coefficients p_n of Eq. (5.10.1) are given by

$$p_n = -\frac{2}{l}\int_0^l p(x)\sin\frac{n\pi x}{l}\,dx = -\frac{2P}{l}\int_0^l \langle x-a\rangle_{-1}\sin\frac{n\pi x}{l}\,dx$$

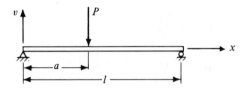

Figure 5.10.1

The integrand in the above equation is zero everywhere except in the small interval $a - \epsilon < x < a + \epsilon$ [see the definition of the function $\langle x - a \rangle_{-1}$ given by Eq. (5.4.19)]. In this interval, the sine function approaches the value $\sin n\pi a/l$ as ϵ approaches zero; therefore, the last equation may be written as

$$p_n = -\frac{2P}{l}\sin\frac{n\pi a}{l}\int_0^l \langle x-a\rangle_{-1}\,dx$$

$$= -\frac{2P}{l}\sin\frac{n\pi a}{l}\langle l - a\rangle^0$$

Since $l > a$, the function $\langle l - a \rangle^0$ is unity and thus

$$p_n = -\frac{2P}{l}\sin\frac{n\pi a}{l}$$

By substituting the above relation into Eq. (5.10.3), the sine series representing the deflection curve for the uniform, simply supported beam subjected to a concentrated force at $x = a$ is obtained; thus,

$$v = -\frac{2Pl^3}{\pi^4 EI}\sum_{n=1}^\infty \frac{1}{n^4}\sin\frac{n\pi a}{l}\sin\frac{n\pi x}{l} \qquad (5.10.6)$$

which is the same result that was obtained in Example 2.4.3 [see Eq. (2.4.16)].

In Chapter VIII, in discussing bending of rectangular, simply supported plates of sides a and b, the deflection $w(x, y)$ will be represented by a double trigonometric series

$$w(x, y) = \sum_{m=1}^\infty \sum_{n=1}^\infty A_{mn}\sin\left(\frac{m\pi x}{a}\right)\sin\left(\frac{n\pi y}{b}\right) \qquad (5.10.7)$$

To calculate coefficients A_{mn} of the double series, both sides of Eq. (5.10.7) are multiplied by

$$\sin\left(\frac{m'\pi n}{a}\right)\sin\left(\frac{n'\pi y}{b}\right)dx\,dy$$

and integrated from 0 to a and from 0 to b. Since

$$\int_0^a \sin\left(\frac{m\pi x}{a}\right)\sin\left(\frac{m'\pi x}{a}\right)dx = \begin{cases} =0 & \text{when } m \neq m' \\ =\dfrac{a}{2} & \text{when } m = m' \end{cases}$$

and

$$\int_0^b \sin\left(\frac{n\pi y}{b}\right)\sin\left(\frac{n'\pi y}{b}\right)dy = \begin{cases} =0 & \text{when } n \neq n' \\ =\dfrac{b}{2} & \text{when } n = n' \end{cases}$$

the coefficients A_{mn} are given by the following formula:

$$A_{mn} = \frac{4}{ab}\int_0^a \int_0^b w(x,y)\sin\left(\frac{m\pi x}{a}\right)\sin\left(\frac{n\pi y}{b}\right)dx\,dy$$

5.11 Elastic-Plastic Bending of Beams[9]

If a metallic beam of cross section similar to that shown in Fig. 5.11.1 (a) is subjected to a bending moment M which increases from zero to a final value under which the beam yields, the extreme fibers of the beam quickly become highly strained. The cross section of the beam is referred to its principal axes of inertia (y, z) and is symmetrical with respect to the y-axis, which lies in the plane of the bending moment, and in the more general case, is not symmetrical with respect to the z-axis.

The stress-strain diagram obtained from a specimen cut from the beam may be represented as in Fig. 5.11.1 (b). The proportional limit of the material is represented by σ_p, the upper yield-stress by σ_y', the lower yield-stress by σ_y^*, and the ultimate strength by σ_m. In order to simplify the stress-strain diagram for the purpose of calculation, it may be assumed that beyond the yield point, the material is increasingly deformed under a constant stress [see Fig. 5.11.1 (c)]. However, the phenomenon is more accurately described for some special materials by taking into account the effect of hardening.

---

[9] Some of Section 5.11 is from the paper entitled "Results of Experiments on Metallic Beams Bent Beyond the Elastic Limit" by E. Volterra (see Bibliography Sec. 5.11, No. 3, at the end of this chapter) and is reproduced with the permission of the London Institution of Civil Engineers.

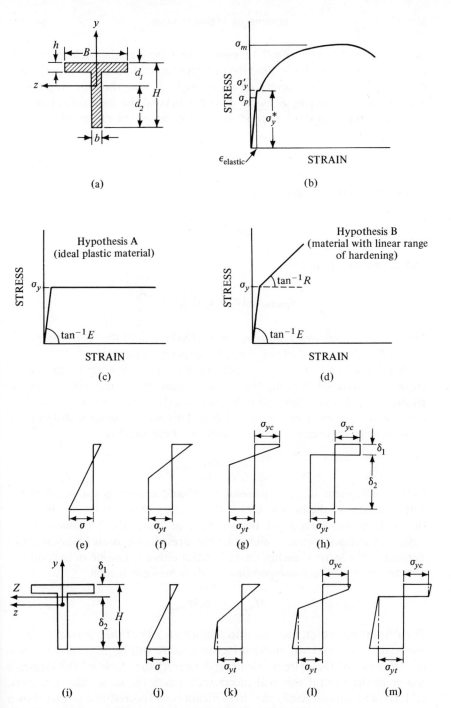

Reproduced with permission of the London Institution of Civil Engineers.

Figure 5.11.1

In this case, it is convenient to assume that beyond the yield point, the increase of strain is a linear function of stress and that the material maintains its characteristic of not returning to its original state when the stress is removed. Thus, by taking R as the modulus beyond the yield point [see Fig. 5.11.1 (d)], the total strain beyond the yield point will be expressed by:

$$\epsilon_{\text{total}} = \epsilon_{\text{elastic}} + \epsilon_{\text{plastic}} = \frac{\sigma}{R} - \sigma_y\left(\frac{1}{R} - \frac{1}{E}\right)$$

The elastic strain is expressed by

$$\epsilon_{\text{elastic}} = \frac{\sigma}{E}$$

and the plastic strain by

$$\epsilon_{\text{plastic}} = (\sigma - \sigma_y)\left(\frac{1}{R} - \frac{1}{E}\right)$$

Melan[10] calls these two hypotheses, respectively, that of the "perfectly plastic material" (*A*), and that of the "linearly hardening material" (*B*).

If the bending moment on the section of the metallic beam under consideration is increased, keeping the greatest stress on the material within the elastic limit, that is, below the yield point σ_y, then the stress distribution on the section is linear [see Fig. 5.11.1 (e)]. This normal stress distribution is equivalent to an internal bending moment whose value is:

$$M = \sigma W_{\text{elastic}}$$

In this expression, W_{elastic} represents the elastic section modulus calculated with reference to the neutral axis z, the position of which is determined by the well-known methods of the strength of materials. The linear distribution of stresses remains valid until the stress in the most highly strained element of the section reaches the yield point either in tension or in compression (σ_{yt} or σ_{yc}). The corresponding bending moment is then

$$M_{\text{elastic}} = \sigma_y W_{\text{elastic}}$$

Then from that point, according to hypothesis (*A*) of the "perfectly plastic material," if the external moment increases there will be no further increase in the stress of the extreme element; however, the stress of the elements nearer to the neutral axis will successively reach the same value [see Figs. 5.11.1 (f) and (g)] and finally the stress diagram will correspond to that shown

---

[10] See Bibliography Sec. 5.11, No. 1 at the end of this chapter.

in Fig. 5.11.1 (h), all of the section being in a plastic condition. On the assumption that sections remain plane throughout deformation, the external bending moment to which the section is subjected at the final stage is

$$M_{\text{plastic}} = \sigma_y W_{\text{plastic}}$$

It must be calculated with reference to a new neutral axis Z [Fig. 5.11.1 (i)] which, in the limiting case, divides the section into two areas A_1 and A_2, such that

$$\sigma_{yc} A_1 = \sigma_{yt} A_2$$

The distances δ_1 and δ_2 in Figs. 5.11.1(h) and (i) can be determined from the above equation and the relation $\delta_1 + \delta_2 = H$ where H is the beam depth.

If instead of hypothesis (A), hypothesis (B) of the "linearly hardening material" were used, the stress diagrams of the section would be as shown in Figs. 5.11.1 (j), (k), (l), and (m).

In the case where $\sigma_{yc} = \sigma_{yt}$, it is possible to obtain simple expressions for the values of W_{elastic} and W_{plastic} for the most common cross sections. The ratio $W_{\text{plastic}}/W_{\text{elastic}}$, which is denoted by w and called the "coefficient of plasticity of the section," is a measure of the capacity of the section to resist plastic deformation. The values of W_{elastic}, W_{plastic}, and w are given for eight different sections in Fig. 5.11.2.

In the elastic-plastic stage of strain, it is possible to distinguish in the total curvature (c_{total}) of a section of a bent beam an elastic fraction (c_{elastic}), which is the curvature that the section would assume if it could resist elastically the bending moment M, and also a fraction which will be designated plastic curvature (c_{plastic}) representing the difference between the two. Thus

$$c_{\text{total}} = c_{\text{elastic}} + c_{\text{plastic}}$$

where

$$c_{\text{elastic}} = \frac{M}{EI}$$

For a rectangular section [see Fig. 5.11.3 (a)], assuming the material to have the same resistance to compression as to tension ($\sigma_{yc} = \sigma_{yt}$), it is very simple to determine the relation between total curvature and bending moment. For this purpose, it is necessary to express the condition of equilibrium of the section subjected to the bending moment M. Thus, considering the stress diagram [Fig. 5.11.3 (b)] of the bent section, it is possible to deduce the following relation between the two bending moments, M_{plastic} and M:

$$M_{\text{plastic}} - M = \tfrac{1}{3} B d^2 \sigma_y \tag{5.11.1}$$

Type of section	Moment of inertia, I	Elastic section modulus, W elastic	Plastic section modulus, W plastic	$w = \dfrac{W \text{ plastic}}{W \text{ elastic}}$
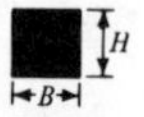 I. Square $H = B$	$\dfrac{H^4}{12} =$ $= 0.0833 H^4$	$\dfrac{H^3}{6} =$ $= 0.1666 H^3$	$\dfrac{H^3}{4} =$ $= 0.25 H^3$	1.50
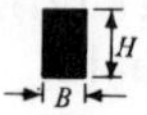 II. Rectangle $H \neq B$	$\dfrac{BH^3}{12} =$ $= 0.0833 BH^3$	$\dfrac{BH^2}{6} =$ $= 0.1666 BH^2$	$\dfrac{BH^2}{4} =$ $= 0.25 BH^2$	1.50
 III. Diamond $H = B$	$\dfrac{H^4}{48} =$ $= 0.0208 H^4$	$\dfrac{H^3}{24} =$ $= 0.0416 H^3$	$\dfrac{H^3}{12} =$ $= 0.0833 H^3$	2.00
 IV. Circle $H = B$	$\dfrac{\pi H^4}{64} =$ $= 0.0491 H^4$	$\dfrac{\pi H^3}{32} =$ $= 0.0982 H^3$	$\dfrac{0.2122 \pi H^3}{4} =$ $0.0167 H^3$	1.70
V. Triangle	 $\dfrac{BH^3}{36} =$ $= 0.0278 BH^3$	$\dfrac{BH^2}{24} =$ $= 0.0416 BH^2$	$\dfrac{BH^2}{10.126}$ $= 0.0987 BH^2$	2.37
VI.	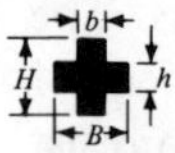 $\dfrac{BH^3}{12}[\beta + (1-\beta)\alpha^3]$	$\dfrac{BH^2}{6}[\beta + (1-\beta)\alpha^3]$	$\dfrac{BH^2}{4}[\beta + (1-\beta)\alpha^2]$	$1.5 \dfrac{\beta + (1-\beta)\alpha^2}{\beta + (1-\beta)\alpha^3}$
VII.	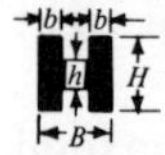 $\dfrac{BH^3}{12}[2\beta + (1-2\beta)\alpha^3]$	$\dfrac{BH^2}{6}[2\beta + (1-2\beta)\alpha^3]$	$\dfrac{BH^2}{4}[2\beta + (1-2\beta)\alpha^2]$	$1.5 \dfrac{2\beta + (1-2\beta)\alpha^2}{2\beta + (1-2\beta)\alpha^3}$
VIII.	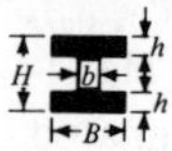 $\dfrac{BH^3}{12}[1 - (1-\beta)(1-2\alpha)^3]$	$\dfrac{BH^2}{6}[1 - (1-\beta)(1-2\alpha)^3]$	$\dfrac{BH^2}{4}[1 - (1-\beta)(1-2\alpha)^2]$	$1.5 \dfrac{1 - (1-\beta)(1-2\alpha)^2}{1 - (1-\beta)(1-2\alpha)^3}$

$\sigma_{yt} = \sigma_{yc}, \alpha = \dfrac{h}{H}, \beta = \dfrac{b}{B}$

For symmetric sections, $d_1 = d_2 = \delta_1 = \delta_2 = \dfrac{H}{2}$

For the triangular sections, $d_1 = \dfrac{2}{3}H; d_2 = \dfrac{1}{3}H; \delta_1 = \dfrac{H}{\sqrt{2}}; \delta_2 = H\left(1 - \dfrac{1}{\sqrt{2}}\right)$

Reproduced with permission of the London Institution of Civil Engineers.

Figure 5.11.2

where d denotes the distance from the neutral axis to the plane whose fibers have just reached the yield point. However, from the strain diagram [Fig. 5.11.3 (c)] of this same section

$$c_{\text{total}} = \frac{\sigma_y}{Ed}$$

Then

$$d = \frac{\sigma_y}{Ec_{\text{total}}}$$

and Eq. (5.11.1) becomes

$$M_{\text{plastic}} - M = \frac{1}{3} \frac{B\sigma_y^3}{E^2 c_{\text{total}}^2}$$

Sec. 5.11 BENDING OF STRAIGHT BEAMS 319

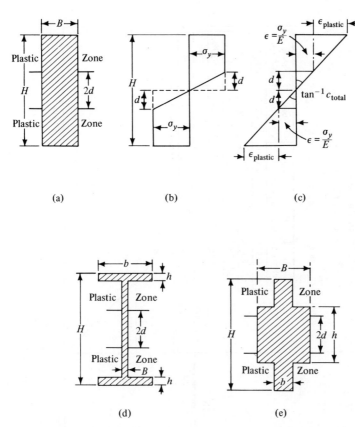

Reproduced with permission of the London Institution of Civil Engineers.

Figure 5.11.3

from which the following expression for c_{total} is derived:

$$c_{\text{total}} = \frac{1}{E}\sqrt{\frac{B\sigma_y^3}{3(M_{\text{plastic}} - M)}} \qquad (5.11.2)$$

Equation (5.11.2) may be applied without alteration to any bent beam having a cross section which is symmetrical with respect to the y- and z-axes and for which the elastic zone of stress lies wholly within a rectangular portion of the section, the breadth of this portion being B [see Figs. 5.11.3 (d) and (e)]. Figure 5.11.4 gives the curves for calculated values of total curvature versus bending moment for two different types of beams; namely, a beam with a cross-shaped section and a beam with a rectangular section. The calculated values are compared with those derived from experiments. As the diagram shows, the experimental curve is always above the calculated one. The difference of the two curves becomes more and more pronounced as the bending

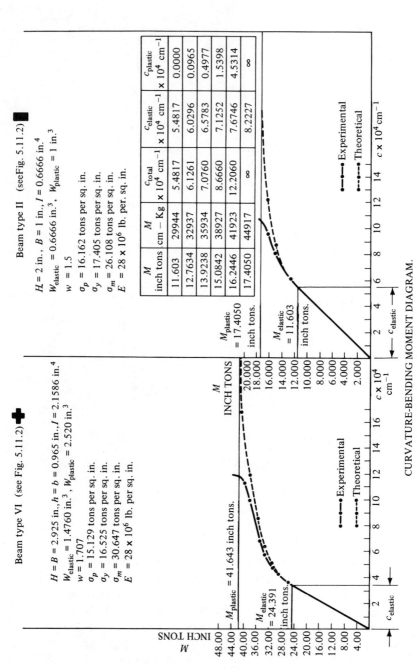

CURVATURE-BENDING MOMENT DIAGRAM.

Figure 5.11.4

Reproduced with permission of the London Institution of Civil Engineers.

moment is increased above the value M_{elastic} and the phenomenon of hardening at the edges of the section is of greater significance.

As is well known in the case of elastic bending of beams, the approximate equation for the deflection curve is obtained by integrating the differential equation

$$\frac{d^2v(x)}{dx^2} = c_{\text{elastic}} = \frac{M}{EI} \qquad (5.2.4)$$

Equation (5.2.4) can be used beyond the yield point for the deflection curve of the beam, provided that the total curvature (c_{total}) is substituted for the elastic curvature (c_{elastic}); thus,

$$\frac{d^2v(x)}{dx^2} = +c_{\text{total}} = +(c_{\text{elastic}} + c_{\text{plastic}}) \qquad (5.11.3)$$

If it is assumed that the bending moment M after having reached a value $M > M_{\text{elastic}}$ returns to zero, the elastic curvature will also become zero; but the plastic curvature will not vanish and Eq. (5.11.3) becomes

$$\frac{d^2v(x)}{dx^2} = c_{\text{plastic}} \qquad (5.11.4)$$

With the aid of Eq. (5.11.4), the permanent deflection of a beam bent beyond the yield point can be calculated.

Mohr's method for calculating elastic deflections of beams can be extended to calculate deflections of beams in plastic conditions by considering the total curvature diagram instead of the elastic curvature diagram. In the same way, the three-moment equation for solving continuous beams can be extended from the elastic to the plastic case. This will now be shown in the case of a continuous beam on four supports under the conditions of loading shown in Figs. 5.11.5, cases A and B.

Suppose that the central span is bent beyond the elastic limit and that plastic deformations have occurred. For case A, the angles of rotation at one of the inner supports will be

$$\theta_1 = \frac{1}{3} \frac{M_A l_1}{EI}$$

$$\theta_2 = \frac{Pl_2^2}{16EI} + \frac{M_A l_2}{2EI} + \frac{1}{2} \int c_{\text{plastic}} \, ds$$

The condition of continuity requires that $\theta_1 = -\theta_2$; thus, the following relation is obtained:[11]

---

[11] G. Colonnetti has obtained the same result as that expressed by Eq. (5.11.5) by the use of his theorem of Least Energy.

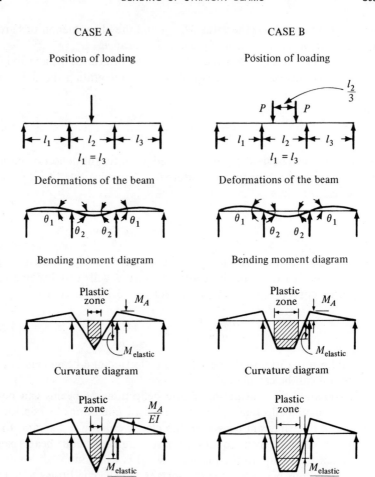

Reproduced with permission of the London Institution of Civil Engineers.

Figure 5.11.5

$$\frac{M_A}{EI}\left(\frac{2}{3}l_1 + l_2\right) + \frac{Pl_2^2}{8EI} + \int c_{\text{plastic}}\, ds = 0 \qquad (5.11.5)$$

The plastic curvature (c_{plastic}) is a function of the bending moment at the center of the beam M_m and also of the hyperstatical unknown M_A, upon which that bending moment depends. From Eq. (5.11.5) it is always possible to deduce the relation between the hyperstatic bending moment M_A and the external force P.

For case B, the angles of rotation at one of the inner supports are

$$\theta_1 = \frac{1}{3}\frac{M_A l_1}{EI}$$

$$\theta_2 = \frac{Pl_2^2}{6EI} + \frac{M_A l_2}{2EI} + \frac{1}{2}\int c_{\text{plastic}}\,ds$$

But here again the condition of continuity, $\theta_1 = -\theta_2$, must be satisfied; thus,

$$\frac{M_A}{EI}\left(\frac{2}{3}l_1 + l_2\right) + \frac{Pl_2^2}{3EI} + \int c_{\text{plastic}}\,ds = 0 \qquad (5.11.6)$$

If the bending moment M_m in the central span is less than the limiting value of the elastic bending moment (that is, $M < M_{\text{elastic}}$) then, for both cases A and B, $c_{\text{plastic}} = 0$ and Clapeyron's equations are obtained; thus,

$$\frac{M_A}{EI}\left(\frac{2}{3}l_1 + l_2\right) + \frac{Pl_2^2}{8EI} = 0 \qquad (5.11.5')$$

$$\frac{M_A}{EI}\left(\frac{2}{3}l_1 + l_2\right) + \frac{Pl_2^2}{3EI} = 0 \qquad (5.11.6')$$

The above discussion has been limited to metallic beams bent beyond the elastic limit. Similar considerations can be employed, at least for a limited period of the deformation, to reinforced concrete beams.[12]

Example 5.11.1. The cantilever beam shown in Figs. 5.11.6 (a) and (b) is made of structural steel and is subjected to a statically increasing concentrated load applied

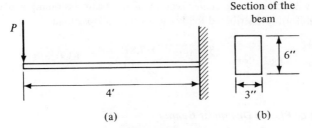

Figure 5.11.6

at its free end. Assuming the material to be ideally plastic with $\sigma_y = 35{,}000$ psi, the behavior of the beam for different values of P will be discussed.

For this beam

$$W_{\text{elastic}} = \frac{3(6)^2}{6} = 18 \text{ in.}^3$$

$$W_{\text{plastic}} = \frac{3(6)^2}{4} = 27 \text{ in.}^3$$

[12] See Bibliography Sec. 5.11, Nos. 5 and 6 at the end of this chapter.

$M_{\text{elastic}} = \sigma_y W_{\text{elastic}} = 35(18)10^3 = 63(10^4)$ in.-lb $= 5.25(10^4)$ ft-lb

$M_{\text{plastic}} = \sigma_y W_{\text{plastic}} = 35(27)10^3 = 94.5(10)^4$ in.-lb $= 7.875(10^4)$ ft-lb

Plastic deformation starts at the fixed end of the beam under the load

$$P = \frac{M_{\text{elastic}}}{4} = \frac{5.25(10^4)}{4} = 13{,}125 \text{ lb}$$

The built-in end section of the beam will become completely plastic when

$$P = P_{\max} = \frac{M_{\text{plastic}}}{4} = \frac{7.875(10^4)}{4} = 19{,}687.5 \text{ lb}$$

In such conditions, the beam will collapse since the built-in end of the beam cannot offer any further resistance and will behave like a hinge (plastic hinge).

Example 5.11.2. A beam of the same material and of the same cross section as the beam of Example 5.11.1 is 10 ft long and simply supported at its ends. It is subjected to a uniformly distributed load p_0 over its entire length. The behavior of the beam for different values of p_0 will be discussed.

Plastic deformation will start at the center of the beam for a uniformly distributed load

$$p_0 = \frac{8M_{\text{elastic}}}{l^2} = \frac{8(5.25)10^4}{(10^2)} = 4{,}200 \text{ lb/ft}$$

The beam will collapse, the center section of the beam becoming a plastic hinge, when the uniformly distributed load reaches the collapse load

$$p_0 = p_{\max} = \frac{8M_{\text{plastic}}}{l^2} = \frac{8(7.875)10^4}{(10)^2} = 6300 \text{ lb/ft}$$

5.12 Limit or Plastic Design of Beams

In the first ten Sections of this chapter problems of bending of straight bars have been discussed considering the material of the beams to be linearly elastic. In Sec. 5.11 the bending of beams has been analyzed for cases where the stress exceeds the elastic limit and the beams are in elastic-plastic conditions. This exceeding of the elastic limit occurs more frequently than one would suppose, even in the most elaborately and accurately designed structures, as is evidenced by the appearance of permanent deflections or deformations. Particularly important is the behavior of certain hyperstatical structures in which, due to external loads, the stress in certain sections reaches the elastic limit of the material. As a consequence, a plastic deformation occurs, the effect of which is in general the same as that produced by an elastic dislocation.[13] (The plastic deformation which occurred in Examples 5.11.1 and 5.11.2 when the elastic capacity of the beams was reached could be compared to the plastic hinge. This plastic hinge allows large rotations to occur in the sections of a beam where the plastic moment M_p has been reached.[14]) The stresses will decrease in such sections while they will increase in other less heavily loaded sections of the structure. This reaction of the material to the external forces which would be expected to destroy it may be compared to a natural defense set up by the structure itself.

The idea naturally occurs that one should imitate and supplement the automatic reaction of the structure, by artificially creating certain dislocations which tend to increase the resistance at the section in question to the external conditions under which it would fail. The problem in these cases, then, is to carry out, on the structure itself, when it is being set up, a very delicate operation to create such dislocations. Since the dislocations to be produced and permanently impressed on the structure are usually very small, it is necessary to adjust them very accurately if one is to avoid an insufficient effect on the one hand, or a remedy worse than the original defect on the other. It is not our intent here to give even an outline of how it is possible to deal with these problems or to show how their solution by application of the theory of elastic dislocations constitutes a new contribution to modern engineering. There is a considerable amount of literature dealing with these subjects. Those who wish to study these subjects further should read some of the available material.[15] We shall limit ourselves instead to a few examples of limit or plastic design of statically indeterminate beams. It will be shown how, by inserting a sufficient number of plastic hinges at the points of maximum moments in a statically indeterminate beam, a kinematically admissible collapse mechanism is created. This enables one to determine the maximum or limit load of the beam.

Consider the beam of length l with both ends built-in and subjected to the uniformly distributed load p_0 as shown in Fig. 5.12.1 (a). According to the elastic analysis, the maximum bending moments occur at the built-in ends of the beam and are

$$M_A = M_B = -\frac{p_0 l^2}{12}$$

while at the center of the beam the bending moment is

$$M_c = \frac{p_0 l^2}{24}$$

---

[13] See Bibliography Sec. 5.12, Nos. 9, 15, 46, and 47 at the end of this chapter.

[14] It should be pointed out that a plastic hinge is different from a frictionless hinge since in the hinges of the first type large rotations occur at the constant moment M_p while in the second type of hinges rotations occur at zero moment.

[15] See for instance Bibliography Sec. 5.12, No. 47 at the end of this chapter.

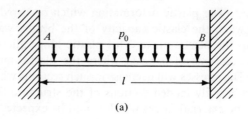

(a)

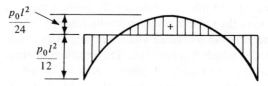

(b) Elastic bending moment

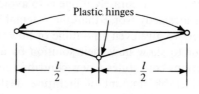

(c) Collapse mechanism

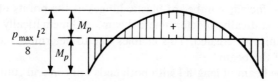

(d) Distribution of bending moments at collapse

Figure 5.12.1

[see Fig. 5.12.1 (b)]. Upon increasing the load, plastic hinges develop at the ends. This occurs when

$$M_A = M_B = \sigma_y W_{\text{plastic}} = \frac{pl^2}{12}$$

or under a load

$$p_0 = \frac{12\sigma_y W_{\text{plastic}}}{l^2}$$

The collapse mechanism occurs when, upon further increasing the distributed load, a third plastic hinge is formed at the center of the beam [see Fig. 5.12.1 (c)]. Since the maximum bending moment in a simply supported, uniformly loaded beam is

$$M_{max} = \frac{p_0 l^2}{8}$$

then at the instant of collapse, the following condition must be satisfied [see Fig. 5.12.1 (d)].

$$\frac{p_{max} l^2}{8} = 2M_p = 2\sigma_y W_{plastic}$$

From the above equation the collapse load

$$p_{max} = \frac{16\sigma_y W_{plastic}}{l^2}$$

immediately follows.

APPENDIX 5.A

Fourier Series Expansions

The representation of periodic functions by means of trigonometric series was developed by Fourier and the series representation is commonly called the Fourier expansion of the function. In this appendix the conditions under which a function $f(x)$ may be represented by a Fourier series are discussed.

For practical applications, the conditions under which a function $f(x)$ of period $2T$[16] may be represented by a Fourier series are contained in the theorem of Dirichlet which may be stated as follows:

---

[16] A function $f(x)$ is periodic with period $2T$, if it satisfies the condition $f(x) = f(x + 2T)$.

Fourier, Jean-Baptiste-Joseph. French mathematician (b. Auxerre 1768; d. Paris 1830). He was professor at the Ecole Normale and later at the Ecole Polytechnique in Paris. In 1798, Fourier accompanied Napoleon to Egypt and, upon his return to France in 1801, wrote an introduction to the history of the Egyptian expedition of Napoleon. He was elected a member of the French Academy of Sciences in 1817 and became its permanent secretary in 1822. His most famous scientific contributions were on heat flow (his *Théorie analytique de la chaleur* was published in Paris in 1822) and on trigonometric series. [Reproduction from the Vito Volterra collection, Villa Volterra, Ariccia (Rome).]

If $f(x)$ is a bounded periodic function which in any one period has a finite number of maxima and minima and a finite number of discontinuities, then the function may be represented by the series

$$\frac{a_0}{2} + \sum_{n=1}^{\infty} \left(a_n \cos \frac{n\pi x}{T} + b_n \sin \frac{n\pi x}{T} \right), \quad n = 1, 2, \ldots \quad (5.\text{A}.1)$$

which converges at every point x_0 to the value

$$\frac{f(x_0^+) + f(x_0^-)}{2}$$

From the theorem it follows that the series converges to $f(x_0)$ at every point x_0 where the function is continuous. If the function is discontinuous at x_0 with two different values $f(x_0^+)$ and $f(x_0^-)$ (see Fig. 5.A.1) then the series converges to the arithmetic mean of these values. Further, it can be shown that if the values of the function are not equal at the end points of a period, i.e.,

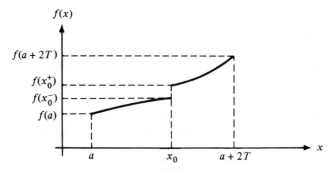

Figure 5.A.1

$f(a) \neq f(a + 2T)$, then the Fourier series converges to $\frac{1}{2}[f(a + 2T) + f(a)]$ at the end points.

If a function satisfies the Dirichlet conditions, then the integral of the function can be obtained by term-by-term integration of its Fourier series. The series which results from a term-by-term differentiation, however, will converge more slowly than the original series and may diverge in some cases. Caution must be exercised in termwise differentiation of Fourier series.[17]

To determine the coefficients in the expansion

$$f(x) = \frac{a_0}{2} + \sum_{n=1}^{\infty} \left(a_n \cos \frac{n\pi x}{T} + b_n \sin \frac{n\pi x}{T} \right) \quad (5.A.2)$$

we shall make use of the relations

$$\left. \begin{array}{l} \int_a^{a+2T} \cos \frac{n\pi x}{T} \cos \frac{m\pi x}{T} dx = \begin{cases} 0 & m \neq n \\ T & m = n \neq 0 \\ 2T & m = n = 0 \end{cases} \\ \int_a^{a+2T} \sin \frac{n\pi x}{T} \cos \frac{m\pi x}{T} dx = 0 \quad \text{for all } m, n \\ \int_a^{a+2T} \sin \frac{m\pi x}{T} \sin \frac{n\pi x}{T} dx = \begin{cases} 0 & m \neq n \\ T & m = n \neq 0 \\ 0 & m = n = 0 \end{cases} \end{array} \right\} \quad (5.A.3)$$

where m and n are integers. The above equations, known as the orthogonality conditons for the trigonometric functions, are easily verified.

To determine coefficient a_n in Eq. (5.A.2) we multiply both sides of that equation by $\cos m\pi x/T$ and integrate over any interval of length $2T$. Then according to Eqs. (5.A.3) the only non-zero term on the right side will be the one containing a_n and we have

$$\int_a^{a+2T} f(x) \cos \frac{n\pi x}{T} dx = a_n \int_a^{a+2T} \cos^2 \frac{n\pi x}{T} dx = a_n T$$

or

$$a_n = \frac{1}{T} \int_a^{a+2T} f(x) \cos \frac{n\pi x}{T} dx, \quad n = 0, 1, 2, \ldots \quad (5.A.4)$$

In a similar manner we multiply both sides of Eq. (5.A.2) by $\sin m\pi x/T$ and integrate over an interval of length $2T$ and obtain

---

[17] The conditions for convergence of the series which results from termwise differentiation of a Fourier series are discussed in Reference No. 8, Bibliography Sec. Appendix V at the end of this chapter.

$$b_n = \frac{1}{T}\int_a^{a+2T} f(x) \sin \frac{n\pi x}{T} dx, \quad n = 1, 2, \ldots \tag{5.A.5}$$

The integrals in Eqs. (5.A.4) and (5.A.5) can always be integrated since the function $f(x)$ is known. In some instances, of course, it may be necessary to use approximate methods of integration.

Example 5.A.1. Obtain the Fourier expansion of the function $f(x)$ defined as follows:

$$f(x) = \begin{cases} 1 & \text{for } 0 < x < l \\ 0 & \text{for } l < x < 2l \end{cases} \tag{5.A.6}$$

with $f(x + 2l) = f(x)$ for all x. The function is shown in Fig. 5.A.2.

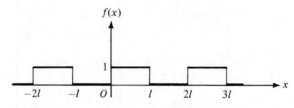

Figure 5.A.2

To evaluate coefficients a_n and b_n we let $a = 0$ and $T = l$ in Eqs. (5.A.4) and (5.A.5); then

$$a_0 = \frac{1}{l}\int_0^{2l} f(x)\, dx = \frac{1}{l}\int_0^{l} (1)\, dx = 1$$

$$a_n = \frac{1}{l}\int_0^{2l} f(x) \cos \frac{n\pi x}{l} dx = \frac{1}{l}\int_0^{l} (1) \cos \frac{n\pi x}{l} dx = 0, \quad n = 1, 2, \ldots$$

$$b_n = \frac{1}{l}\int_0^{2l} f(x) \sin \frac{n\pi x}{l} dx = \frac{1}{l}\int_0^{l} (1) \sin \frac{n\pi x}{l} dx = \frac{1 - \cos n\pi}{n}$$

$$= \frac{1 - (-1)^n}{n\pi}$$

or

$$b_n = \begin{cases} 0 & \text{if } n = 2, 4, 6, \ldots \\ \dfrac{2}{n\pi} & \text{if } n = 1, 3, 5, \ldots \end{cases}$$

Hence, the Fourier series expansion for $f(x)$ is

$$f(x) = \frac{1}{2} + \frac{2}{\pi} \sum_{n=1,3,\ldots}^{\infty} \frac{1}{n} \sin \frac{n\pi x}{l} \tag{5.A.7}$$

The above relation converges to $f(x)$ at points where $f(x)$ is continuous. At points of discontinuity, i.e., at $x = 0, \pm l, \pm 2l, \ldots$, the series converges to an arithmetic mean value of $\frac{1}{2}$. This is easily verified by inspection of Eq. (5.A.7). In this example, we defined the function so that it is periodic. However, if the function were defined only for $0 < x < 2l$, we could imagine the function to be periodic and proceed exactly as we have done above. In this way we would obtain the Fourier expansion of the function for $0 < x < 2l$.

Example 5.A.2. Obtain the Fourier expansion of the function defined as follows:

$$f(x) = \begin{cases} -x & \text{for } -2 < x < 0 \\ x & \text{for } 0 < x < 2 \end{cases} \tag{5.A.8}$$

with $f(x) = f(x + 4)$ for all x. This function is shown in Fig. 5.A.3.

To evaluate the coefficients a_n and b_n we set $a = -2$ and $T = 2$ in Eqs. (5.A.4) and (5.A.5); then we find that

$$a_0 = \frac{1}{2}\int_{-2}^{2} f(x)dx = \frac{1}{2}\int_{-2}^{0}(-x)dx + \frac{1}{2}\int_{0}^{2} x\,dx = 2$$

$$a_n = \frac{1}{2}\int_{-2}^{2} f(x)\cos\frac{n\pi x}{2}dx$$

$$= \frac{1}{2}\int_{-2}^{0}(-x)\cos\frac{n\pi x}{2}dx + \frac{1}{2}\int_{0}^{2}(x)\cos\frac{n\pi x}{2}dx$$

$$= \frac{4}{n^2\pi^2}(\cos n\pi - 1)$$

$$= -\frac{8}{n^2\pi^2} \quad n = 1, 3, 5, \ldots$$

$$b_n = \frac{1}{2}\int_{-2}^{2} f(x)\sin\frac{n\pi x}{2}dx$$

$$= \frac{1}{2}\int_{-2}^{0}(-x)\sin\frac{n\pi x}{2}dx + \frac{1}{2}\int_{0}^{2}(x)\sin\frac{n\pi x}{2}dx$$

$$= 0$$

Substituting the above coefficients into the Fourier series [Eq. (5.A.2)] gives

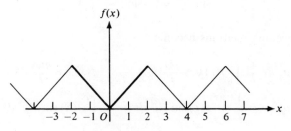

Figure 5.A.3

$$f(x) = 1 - \frac{8}{\pi^2} \sum_{n=1,3,\ldots}^{\infty} \frac{1}{n^2} \cos \frac{n\pi x}{2} \tag{5.A.9}$$

which converges to $f(x)$ for all values of x.

In many applications the evaluation of the coefficients in the Fourier series can be simplified. For instance, in the case of an even function, i.e., when $f(x) = f(-x)$ for all values of x, coefficients b_n are all zero and coefficients a_n are obtained by integration over one half the period. To demonstrate the validity of these statements we shall take $a = -T$ in Eqs. (5.A.4) and (5.A.5). Then

$$a_n = \frac{1}{T} \int_{-T}^{T} f(x) \cos \frac{n\pi x}{T} dx$$

$$= \frac{1}{T} \int_{-T}^{0} f(x) \cos \frac{n\pi x}{T} dx + \frac{1}{T} \int_{0}^{T} f(x) \cos \frac{n\pi x}{T} dx$$

$$b_n = \frac{1}{T} \int_{-T}^{T} f(x) \sin \frac{n\pi x}{T} dx$$

$$= \frac{1}{T} \int_{-T}^{0} f(x) \sin \frac{n\pi x}{T} dx + \frac{1}{T} \int_{0}^{T} f(x) \sin \frac{n\pi x}{T} dx$$

If, in the integrals taken from $-T$ to 0, we let $x = -t$, then $dx = -dt$. Also, $x = -T$ implies that $t = T$, and $x = 0$ implies that $t = 0$; hence the above relations become

$$a_n = \frac{1}{T} \int_{T}^{0} f(-t) \cos \left(-\frac{n\pi t}{T} \right)(-dt) + \frac{1}{T} \int_{0}^{T} f(x) \cos \frac{n\pi x}{T} dx$$

$$b_n = \frac{1}{T} \int_{T}^{0} f(-t) \sin \left(-\frac{n\pi x}{T} \right)(-dt) + \frac{1}{T} \int_{0}^{T} f(x) \cos \frac{n\pi x}{T} dx$$

If we note that $f(t) = f(-t)$ and that

$$\cos \frac{n\pi t}{T} = \cos \left(-\frac{n\pi t}{T} \right)$$

$$\sin \frac{n\pi t}{T} = -\sin \left(-\frac{n\pi t}{T} \right)$$

then the preceding relations become

$$a_n = \frac{1}{T} \int_{0}^{T} f(t) \cos \frac{n\pi t}{T} dt + \frac{1}{T} \int_{0}^{T} f(x) \cos \frac{n\pi x}{T} dx$$

$$b_n = -\frac{1}{T} \int_{0}^{T} f(t) \sin \frac{n\pi t}{T} dt + \frac{1}{T} \int_{0}^{T} f(x) \sin \frac{n\pi x}{T} dx$$

The two integrals in each of the above relations are identical, except for the dummy variable of integration which is immaterial; therefore, we obtain the following relations for the case where $f(x)$ is an even function:

$$\left. \begin{array}{l} a_n = \dfrac{2}{T} \displaystyle\int_0^T f(x) \cos \dfrac{n\pi x}{T} dx \\ b_n = 0 \end{array} \right\} \qquad (5.A.10)$$

For the case of an odd function, i.e., a function such that $f(x) = -f(-x)$ for all x, the coefficients can again be simplified. Following a procedure similar to that used in the case of the even function, we find that

$$\left. \begin{array}{l} a_n = 0 \\ b_n = \dfrac{2}{T} \displaystyle\int_0^T f(x) \sin \dfrac{n\pi x}{T} dx \end{array} \right\} \qquad (5.A.11)$$

are the Fourier coefficients for odd functions.

The choice of axes will determine whether a function is odd or even. For instance, the sawtooth function shown in Fig. 5.A.4 is an odd function, if

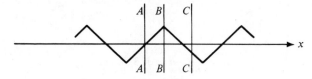

Figure 5.A.4

line A-A is chosen as the y-axis. If line B-B is chosen as the y-axis, then the function in an even function. If line C-C is chosen as the y-axis, then the function is neither odd nor even and the Fourier expansion would contain both sines and cosines.

Example 5.A.3. Obtain the sine series expansion for the function defined as follows:

$$f(x) = 1 \quad \text{for} \quad 0 < x < l \qquad (5.A.12)$$

The function is shown as the solid line in Fig. 5.A.5 (a).

Since we wish to represent the function by a sine series, we shall treat it as an odd function of period $2l$. This extended function is indicated by the dashed lines in Fig. 5.A.5 (a). Since the function is an odd function, the coefficients are given by Eqs. (5.A.11); therefore,

$$a_n = 0$$

$$b_n = \frac{2}{l} \int_0^l f(x) \sin \frac{n\pi x}{l} dx$$

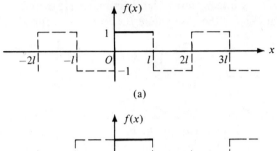

Figure 5.A.5

$$= \frac{2}{l} \int_0^l (1) \sin \frac{n\pi x}{l} \, dx$$

$$= \frac{2}{n\pi}(1 - \cos n\pi)$$

or finally

$$b_n = \frac{4}{n\pi}, \quad n = 1, 3, 5, \ldots$$

By substituting these coefficients into the Fourier series, we obtain

$$f(x) = \frac{4}{\pi} \sum_{n=1,3,5,\ldots}^{\infty} \frac{1}{n} \sin \frac{n\pi x}{l} \tag{5.A.13}$$

This expansion represents the extended function shown in Fig. 5.A.5 (a); however, we are concerned only with the interval $0 < x < l$. The series expansion obtained in Example 5.A.1 [see Eq. (5.A.7)] also represents the function defined by Eq. (5.A.12); in that case, however, the extended function is neither odd nor even and the expansion is more complicated. The function defined by Eq. (5.A.12) could also be expanded in a cosine series by letting the extended function be an even function as indicated in Fig. 5.A.5 (b). For this choice of extended function the period is $4l$.

BIBLIOGRAPHY

Sections 5.1, 5.2, 5.3. DIFFERENTIAL EQUATIONS OF EQUILIBRIUM, NAVIER'S FLEXURE FORMULA, JOURAWSKI'S SHEARING STRESS FORMULA; DIFFERENTIAL EQUATIONS FOR DEFLECTIONS OF ELASTIC BEAMS ACCORDING TO THE BERNOULLI-EULER THEORY; SOLUTION OF BEAM DEFLECTION PROBLEMS BY DIRECT INTEGRATION

1. Abhandlungen uber das Gleichgewicht und die Schwingungen der Ebenen Elastischen Kurven Von Jakob Bernoulli (1691, 1694, 1695) und Leohard Euler (1744)," Verlag von Wilhelm Engelmann, Leipzig, 1910.
2. Bernoulli, Jacob, "Collected Works," Vol. 2, Geneva, Switzerland, 1744.
3. Euler, L., "Methodus Inveniendi Lineas Curvas," apud Marcum Michaelem Bousquet et Socios, Lausanne and Geneva, Switzerland, 1744.
4. Föppl, A., "Vorlesungen uber Technische Mechanik," B. G. Teubner, Berlin, Germany, 1922.
5. Guidi, C., "Lezioni di Scienza delle Costruzioni," V. Bona, Torino, Italy, 1938.
6. Jourawski, D. J., "Sur la résistance d'un corps prismatique," *Ann. Ponts Chaussées*, 2nd Semestre, 1856, pp. 328.
7. Lamé, G., "Leçons sur la théorie mathématique de l'élasticité des corps solides," Bachelier, Paris, France, 1852.
8. Magnel, G., "Cours de Stabilité des Constructions," Gand, Belgium, 1942.
9. Navier, L. M. H., "De la Résistance des Corps Solides," avec des notes et des appendices par Barré de Saint-Venant, Dunod, Paris, France, 1864.
10. Popov, E. P., "Introduction to Mechanics of Solids," Prentice-Hall Inc., Englewood Cliffs, N.J., 1968.
11. Straub H., "A History of Civil Engineering," Massachusetts Institute of Technology Press, Cambridge, Mass., 1964.
12. Stüssi, E., "Statique Appliquée et Résistance des Matériaux," Dunod, Paris, France, 1949.
13. Timoshenko, S. P., "History of Strength of Materials," McGraw-Hill Book Company, New York, N.Y., 1953.
14. Timoshenko, S. P., and D. H. Young, "Elements of Strength of Materials," D. Van Nostrand Co., Inc., Princeton, N.P., 1962.
15. Timoshenko, S. P., and J. N. Goodier, "Theory of Elasticity," McGraw-Hill Book Company, New York, N.Y., 1951, p. 321.

Sections 5.4, 5.5, 5.6, and 5.7. MACAULAY'S USE OF SINGULARITY FUNCTIONS FOR STUDYING DEFLECTIONS OF BEAMS; USE OF SINGULARITY FUNCTIONS IN THE CASE OF BEAMS OF VARIABLE CROSS SECTIONS; USE OF TAYLOR'S AND MACLAURIN'S SERIES FOR STUDYING DEFLECTIONS OF BEAMS; GENERAL DEFLECTION EQUATION FOR BEAMS OF UNIFORM CROSS SECTION

1. Brungraber, R. S., "Singularity Functions in the Solution of Beam Deflection Problems," *J. Eng. Educ.*, **55**, No. 9, 1965, pp. 278–280.
2. Carslaw, H. S., and J. C. Jaeger, "Operational Methods in Applied Mathematics," Oxford University Press, London, England, 1941.
3. Case, J., "The Strength of Materials," Edward Arnold (Publishers) Ltd., London, England. 1943.
4. Crandall, S. H., and N. C. Dahl, "An Introduction to the Mechanics of Solids," McGraw-Hill Book Company, New York, N.Y., 1959.
5. Hetenyi, M., "Deflection of Beams of Varying Cross Section," *J. Appl. Mech.*, **Vol. 4**, 1937, pp. A 49–52.

6. Guillemin, E. A., "Introductory Circuit Analysis," John Wiley & Sons Inc., New York, N.Y., 1953.

7. Macaulay, W. H., "Note on the Deflection of Beams," *Messenger Math.*, **Vol. 48**, 1919, pp. 129–130.

8. Popov, E. P., "Introduction to Mechanics of Solids," Prentice-Hall, Inc., Englewood Cliffs, N.J., 1968.

9. Thomson, W. T., "Deflection of Beams by the Operational Method," *J. Franklin Inst.*, **Vol. 247**, No. 6, 1949, pp. 557–568.

10. ——, "Laplace Transformation, Theory and Engineering Applications," Prentice-Hall, Inc., Englewood Cliffs, N.J., 1950.

Sections 5.8, 5.9. MOHR'S CONJUGATE BEAM METHOD; CLAPEYRON'S THREE-MOMENT EQUATION FOR SOLVING CONTINUOUS BEAMS

1. Anger, G., "Lignes d'influence dans les poutres continues," Ernst, Berlin, Germany, 1948–49.

2. Belluzzi, Odone, "Scienza delle Costruzioni," **Vol. 1**, Zanichelli, Bologna, Italy, 1946.

3. Benscoter, S. U., "Matrix Analysis of Continuous Beams," American Society of Civil Engineers, 1946, New York, p. 1091.

4. ——, "Numerical Transformation Procedures in Continuous Beam Analysis," *J. Franklin Inst.* 1947, **Vol. 244**, pp. 15–26.

5. Bollinger, O. E., "Tables pour Poutres Continues," Dunod, Paris, France, 1951.

6. Borg, S. F., "Additional Interpretation of the Solution of the Straight-Beam Differential Equation," *J. Franklin Inst.*, 1950, **Vol. 250**, pp. 249–256.

7. Bouasse, H., "Théorie de l'élasticité-résistance des matériaux," Delagrave, Paris, France, 1920.

8. Ceradini, C., "Meccanica Applicata alle Costruzioni," Vallardi, Milan, Italy, 1921.

9. Cesari, L., F. Conforto, and C. Minelli, "Travi Continue Inflesse e Sollecitate Assialmente," *Pubbl. Ist. Appl. Calcolo*, Consiglio Nazionale delle Ricerche, Rome, Italy, 1941.

10. Clapeyron, B. P. E., "Calcul d'une poutre élastique reposant librement sur des appuis inégalement espacés," *Compt. Rend.*, **Vol. 45**, pp. 1076.

11. Flamard, E., "Calcul des systèmes élastiques de la construction," Gauthier-Villars, Paris, France, 1918.

12. Grashof, F., "Theorie der Elastizität und Festigkeit," Gaertner, Berlin, Germany, 1878.

13. Griot, G., "Kontinuierliche Balken," Selbstverlag, Zurich, Switzerland, 1934.

14. Kleinlogel, A., "Belastungsglieder," Ernst, Berlin, Germany, 1931.

15. ——, "Der Durchlaufende Träger," Ernst, Berlin, Germany, 1934.

16. L'Hermite, R., "Résistance des Matériaux Théorique et Expérimentale," Dunod, Paris, France, 1954.

17. Mohr, O., "Abhandlungen aus dem Gebrete der Technischen Mechanik," Ernst, Berlin, Germany, 1928.

18. Morsch, E., "Der Durchlaufende Träger," Wittwer, Stuttgart, Germany, 1923.
19. Muller-Breslau, H., "Die Graphische Statik der Baukonstruktionen," Kroner, Leipzig, Germany, 1912.
20. Ritter, W., "Anwendungen der Graphischen Statik Der Kontinuierliche Balken," Raustein, Zurich, Switzerland, 1900.
21. Staack, J., "Rahmen und Balken," Julius Springer, Berlin, Germany, 1921.
22. Strassner, A., "Berechnung Statisch Unbestimmter Systems Der Einfache und Durchlaufende Balken," Ernst, Berlin, Germany, 1929.
23. Suter, E., "Die Methode der Festpuntke," Julius Springer, Berlin, Germany, 1923.
24. Timoshenko, S. P., "History of Strength of Materials," McGraw-Hill Book Company, New York, N.Y., 1953.
25. Timoshenko, S. P., and D. H. Young, "Elements of Strength of Materials," D. Van Nostrand Co., Inc., Princeton, N.J., 1962.
26. Vallette, R., "Poutres Continues," *Travaux*, January 1943.

Section 5.10. APPLICATION OF TRIGONOMETRIC SERIES TO THE STUDY OF DEFLECTIONS OF BEAMS

1. Inglis, C. E., "A Mathematical Treatise on Vibrations in Railway Bridges," Chapter I, Cambridge University Press, London, England, 1934.
2. Timoshenko, S. P., "Application of Normal Coordinates in Analyzing Bending of Beams and Plates," *Bull. Kiev Polytech. Inst.*, 1909.
3. ———, "Strength of Materials," Part I, 3rd ed., Chapter III, D. Van Nostrand Co., Inc., Princeton, N.J., 1955.
4. ———, "Strength of Materials," Part II, 3rd ed., Chapter II, D. Van Nostrand Co., Inc., Princeton, N.J., 1956.
5. Volterra, E., and E. C. Zachmanoglou, "Dynamics of Vibrations," Chapter 1.6, Charles E. Merrill Books, Inc., New York, N.Y., 1965.
6. Westergaard, H. M., "Buckling of Elastic Structures," *Proc. Am. Soc. Civil Engrs.*, **Vol. 47,** 1921, pp. 455–533.

Section 5.11. ELASTIC-PLASTIC BENDING OF BEAMS

1. Melan, E., "Theory of Statically Indeterminate Systems," *Preliminary Publication, Second Cong. Intern. Assoc. Bridges and Structural Eng.*, Berlin-Munich, Germany, October 1–11, 1936.
2. Prager, W., "Mécanique des solides isotropes au delà du domain élastique," *Mem. Sci. Math.*, **Fasc. 88,** 1937.
3. Volterra, E., "Results of Experiments on Metallic Beams Bent Beyond the Elastic Limit," *J. Inst. Civil Engrs., London,* **Vol. 20, No. 5,** 1943, pp. 1–20. See correspondence on the above paper by Sir John Baker, Dr. J. W. Roderick, and Professor G. Winter in the *J. Inst. Civil Engrs., London,* **Vol. 20, No. 8,** 1943, pp. 349–356.
4. ———, "Travi Metalliche Inflesse in Regime Elasto-Plastico," *Giorn. Genio Civile,* **Fasc. 5,** 1946.

5. ———, "Risultati di Ricerche Sperimentali sul Beton e sul Beton Armato," *Ric. Ing.*, **Vol. 2, Nos. 2 and 3**, 1934, pp. 1–76.

6. ———, "On the Deformation of Reinforced Concrete Structures and on the Calculation of Pre-Stressed Reinforced Beams," *J. Inst. Struct. Engrs.*, London, **Vol. 21, No. 4**, 1943, pp. 123–138. See discussion on the above paper by Dr. K. Hajnal-Konyi, R. H. Squire, Dr. K. W. Mautner, K. J. Sommerfeld, Dr. P. Abeles, and E. A. Scott in *J. Inst. Struct. Engrs.*, London, **Vol. 21, No. 10**, 1943, pp. 409–426.

Section 5.12. LIMIT OR PLASTIC DESIGN OF BEAMS

1. ASCE "Manual of Engineering Practice No. 41," Commentary on Plastic Design in Steel, New York, N.Y., 1963.

2. Baker, Sir John, M. R. Horne, and J. Heyman, "Plastic Behavior and Design," *The Steel Skeleton*, **Vol. 2**, Cambridge University Press, London, England, 1956.

3. Baker, Sir John and J. F. Roderick, "Investigation into the Behaviour of Welded Rigid Framed Structure," *First Interim Rept. Trans. Inst. Welding*, **Vol. 1, No. 4**, 1938; *Second Interim Rept. ibid*, **Vol. 3, No. 2**, 1940.

4. Beedle, L. S., "Plastic Design of Steel Frames," John Wiley & Sons, Inc., New York, N.Y., 1958.

5. Beedle, L. S., B. Thurliman, and R. L. Ketter, "Plastic Design in Structural Steel," American Institute of Steel Construction, Inc., New York, N.Y., 1955.

6. Bleich, F., "La ductilité de l'acier. Son application au dimensionnement des systemes hyperstatiques," *L'Ossature Métallique*, 1934, p. 93.

7. ———, "La théorie et la recherche expérimentale en construction métallique," *L'Ossature Métallique*, 1934, p. 627.

8. Bleich, H., "Ueber die Bemessung statisch unbestimmter Stahltragwerke unter Berüchksichtigung des elastisch—plastischen Verhaltens des Baustoffes," *Bauingenieur*, **No. 13**, 1937, p. 261.

9. Colonnetti, G., "Scienza delle costruzioni," Einaudi, Turin, Italy, 1955.

10. ———, "Il secondo principio di reciprocità e le sue applicazioni al calcolo delle deformazioni permanenti," *Atti Accad. Nazl. Lincei, Rend.*, Serie 6, **Vol. 27**, 1 Semestre, 1938.

11. ———, "La statica dei corpi elasto-plastici," *Pontificia Accad. Sci. Comment.* **Vol. 2**, 1938.

12. ———, "Su la resistenza alla flessione in regime elasto plastico," *Pontificia Accad. Sci.*, **Vol. 3**, 1938.

13. ———, "Alla ricerca dei fondamenti sperimentali della teoria dell equilibrio elastoplastico," *Atti Accad. Nazl. Lincei, Rend.*, Serie 6, **Vol. 28**, 2 Semestre, 1938.

14. ———, "Risoluzione grafica del problema della flessione in regime elasto-plastico," *Pontificia Accad. Sci. Comment.* **Vol. 3**, 1939.

15. Danusso, A., "Le Autotensioni (Spunti teorici ed applicazioni pratiche)," *Rend. Sem. Milano*, 1934.

16. ———, "Le ragioni ed i fondamenti della ricerca sperimentale sulle costruzioni. Nuove ricerche sulle Costruzioni civili a cura dell Istituto Nazionale per gli Studi e la Sperimentazione nell Industria Edilizia," Edizione della Bussola, Rome, Italy, 1946.

17. ———, "La Plasticitá nella Scienza delle Costruzioni," (Symposium in onore di Arturo Danusso) Nicola Zanichelli Bologna, Italy, 1956. See papers by A. Berio, M. Brozzu, F. Campus, W. S. Dorn, and H. J. Greenberg, L. Finzi, H. Geiringer, G. Grandori, P. G. Hodge, Jr., B. Venkatraman, G. Macchi, F. K. G. Odquist, W. Olszak, W. Prager, and F. Stussi.
18. Drucker, D. C., H. J. Greenberg, and W. Prager, "The Safety Factor of an Elastic-Plastic Body in Plane Strain," *J. Appl. Mech.*, Vol. **18**, 1951, pp. 371–378.
19. Drucker, D. C., W. Prager, and H. J. Greenberg, "Extended Limit Design Theorems for Continuous Media," *Quart. Appl. Math.*, Vol. **9**, 1952, pp. 381–389.
20. Freudenthal, A. M., "The Inelastic Behavior of Engineering Materials and Structures," John Wiley & Sons, Inc., New York, N.Y., 1950.
21. Fritsche, J., "Grundlagen des Plastizitätstheorie," *Preliminary Publication, Second Congr. Intern. Assoc. Bridges and Structural Eng.*, Berlin-Munich, Germany, October 1–11, 1936.
22. Greenberg, H. J., and W. Prager, "Limit Design of Beams and Frames," *Proc. Am. Soc. Civil Engrs.*, Vol. **77**, 1951.
23. Heyman, J., "Plastic Design of Portal Frames," Cambridge University Press, London, England, 1957.
24. Hill, R., "The Mathematical Theory of Plasticity," Oxford University Press, London, England, 1950.
25. Hodge, P. G., "Plastic Analysis of Structures," McGraw-Hill Book Company, New York, N.Y., 1959.
26. Hoffman, O., and G. Sachs, "Introduction to the Theory of Plasticity for Engineers," McGraw-Hill Book Company, New York, N.Y., 1953.
27. Iliouchine, A. A., "Plasticité," Eyrolles, Paris, France, 1956.
28. Levi, F., and E. Giacchero, "Conferme sperimentali della teoria di Colonnetti sul l'equilibrio elasto-plastico," *Acta Pontificia Accad. Sci.*, 1939.
29. Levi, F., and G. Pizzetti, "Fluage-plasticité-precontrainte," Dunod, Paris, France, 1951.
30. Meier-Leibnitz, H., "Contributions to the Problem of Ultimate Carrying Capacity of Simple and Continuous Beams of Structural Steel and Timber," *Bautechnik* Vol. **1**, No. 6, 1927.
31. ———, "Versuch, Ausdentung and Auswenaung der Ergebnisse," *Preliminary Publication Second Congr. Intern. Assoc. Bridges and Structural Eng.*, Berlin-Munich, October 1–2, 1936, pp. 121–136.
32. ———, "Die Beziehungen M (P) in Mittelfeld," *Final Rept. Intern. Congr. Bridges and Structural Eng.*, Berlin-Munich, Germany, 1936, pp. 70–73.
33. Melan, E., "Theory of Statically Indeterminate Systems," *Preliminary Publication Second Congr. Intern. Assoc. Bridges and Structural Eng.*, Berlin-Munich, Germany, October 1–2, 1936.
34. Nadai, A., "Theory of Flow and Fracture of Solids," McGraw-Hill Book Company, New York, N.Y., 1950.
35. Neal, B. G., "The Plastic Methods of Structural Analysis," John Wiley & Sons, Inc., New York, N.Y., 1956.

36. Pippard, A. J. S., and Sir John Baker, "Analysis of Engineering Structures," Edward Arnold (Publishers) Ltd., London, England, 1943.
37. Prager, W., "Mécanique des Solides Isotropes au delà du Domain Elastique," *Mem. Sci. Math.*, **Fasc. 88,** 1937.
38. ———, "Problèmes de plasticité théorique," Dunod, Paris, France, 1958.
39. ———, "Introduction to Plasticity," Addison-Wesley Publishing Co., Reading, Mass., 1959.
40. Prager, W., and P. Hodge, "Theory of Perfectly Plastic Solids," John Wiley & Sons, Inc., New York, N.Y., 1951.
41. Rjanitsyn, A. R., "Calcul a la rupture et plasticité des constructions," Eyrolles, Paris, France, 1959.
42. Sokolowsky, W. W., "The Theory of Plasticity: An Outline of Work Done in Russia," *J. Appl. Mech.*, 1946, **Vol. 13, No. 1,** pp. A.1–A.10.
43. Stassi D'Alia, F., "Teoria della Plasticità e sue Applicazioni," Denaro, Palermo, Italy, 1958.
44. Stüssi, F., "Zur Auswertung von Versuchen über das Traglastverfahren," *Final Rept. Intern. Congr. Bridges and Structural Eng.*, Berlin-Munich, Germany, 1936, pp. 7–76.
45. Van Den Broek, J. A., "Theory of Limit Design," John Wiley & Sons, Inc., New York, N.Y., 1952.
46. Volterra, E., "Some Recent Applications of the Theory of Elastic Dislocations in Civil Engineering," *Rept. Brit. Assoc. Advan. Sci.*, **Vol. 11, No. 5,** 1942, pp. 78–79.
47. Volterra, V., and E. Volterra, "Sur les Distorsions des Corps Elastiques, Théorie et Applications," *Mem. Sci. Math.*, **Fas. 147,** 1960.
48. Westergaard, H. M., "Theory of Elasticity and Plasticity," Harvard University Press, Cambridge, Mass., 1952.
49. Winter, G., "Trends in Steel Design and Research," *Buildings Res. Congr.*, **Div. I, Part II,** 1951, pp. 81–88.
50. Zhudin, N. H., "Calcul des portiques en acier en tenant compte des déformations plastiques," *L'Ossature Métallique*, 1937, p. 79.

Appendix V. FOURIER SERIES EXPANSIONS

1. Carslaw, H. S., "Introduction to the Theory of Fourier's Series and Integrals," 3rd ed., Dover Publications, Inc., New York, N.Y., 1930.
2. Eagle, A., "A Practical Treatise on Fourier's Theorem and Harmonic Analysis for Physicists and Engineers," Longmans, Green & Co., London, England, 1925.
3. Pipes, L. A., "Applied Mathematics for Engineers and Physicists," McGraw-Hill Book Company, New York, N.Y., 1958, Chap. III.
4. Sneddon, I. N., "Fourier Series," Routledge & Kegan Paul, Ltd., London, England, 1961.
5. Sokolnikoff, I. S., and E. S., Sokolnikoff, "Higher Mathematics for Engineers and Physicists," 2nd ed., McGraw-Hill Book Company, 1941, Chap. II.

6. Volterra, E., and E. C. Zachmanoglou, "Dynamics of Vibrations," Charles E. Merrill Books, Inc., New York, N.Y., 1965, Chap. 1.5.
7. Whittaker, E. T., and G. Robinson, "The Calculus of Observations," 4th ed., Blackie & Son, Ltd., Glasgow, 1952, Chapter X.
8. Whittaker, E. T., and G. N. Watson, "Modern Analysis," The Macmillan Company, New York, N.Y., 1943.

PROBLEMS

5-1. For the uniform beam shown in Fig. P.5.1, determine the maximum bending stress and the maximum shearing stress.

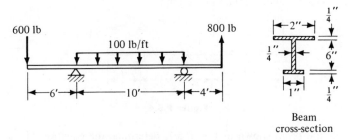

Figure P.5.1

5-2. Determine the maximum bending stress and the maximum shearing stress for the uniform beam shown in Fig. P.5.2.

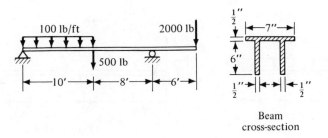

Figure P.5.2

5-3. For the uniform beam shown in Fig. P.5.3, determine the maximum bending stress and the maximum shearing stress.

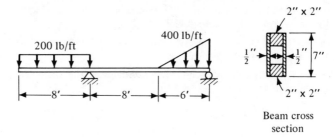

Figure P.5.3

Note: In problems 5-4 through 5-10 use the second order differential equation of the elastic curve and singularity functions to obtain solutions.

5-4. Determine the equation of the deflection curve for the beam shown in Fig. P.5.4. What is the maximum deflection?

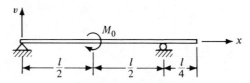

Figure P.5.4

5-5. For the beam shown in Fig. P.5.1, determine the equation of the elastic curve. What is the maximum deflection?

5-6. For the beam shown in Fig. P.5.2, determine the equation of the elastic curve. What is the maximum deflection?

5-7. For the beam shown in Fig. P.5.5, determine the support reactions and the equation of the elastic curve.

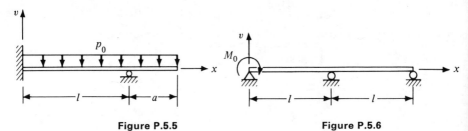

Figure P.5.5   Figure P.5.6

5-8. Determine the support reactions and the equation of the elastic curve for the beam shown in Fig. P.5.6.

5-9. Determine the equation of the elastic curve and the maximum bending moment for the beam shown in Fig. P.5.7.

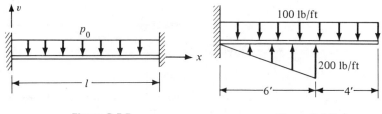

Figure P.5.7   Figure P.5.8

5-10. Determine the equation of the elastic curve for the beam shown in Fig. P.5.8.

Note: In Problems 5-11 through 5-21 use the fourth order differential equation of the elastic curve and singularity functions to obtain solutions.

5-11. Determine the equation of the elastic curve for the beam shown in Fig. P.5.1.

5-12. Determine the equation of the elastic curve and the maximum deflection for the beam shown in Fig. P.5.2.

5-13. Determine the equation of the elastic curve for the beam shown in Fig. P.5.3.

5-14. Determine the equation of the elastic curve and the maximum deflection for the beam shown in Fig. P.5.4.

5-15. Determine the support reactions and the equation of the elastic curve for the beam shown in Fig. P.5.6.

5-16. Determine the support reactions and the equation of the elastic curve for the beam shown in Fig. P.5.9.

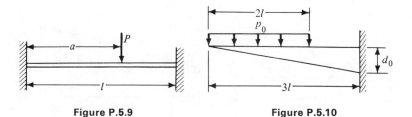

Figure P.5.9   Figure P.5.10

5-17. The beam shown in Fig. P.5.10 has rectangular cross sections with constant thickness and linearly varying depth. Determine the equation of the elastic curve.

5-18. The beam shown in Fig. P.5.11 has rectangular cross sections with constant thickness and linearly varying depth. Determine the equation of the elastic curve.

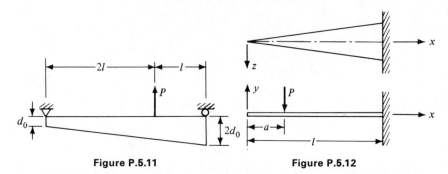

Figure P.5.11   Figure P.5.12

5-19. The beam shown in Fig. P.5.12 has constant depth and linearly varying thickness. Determine the equation of the elastic curve.

5-20. Determine the equation of the deflection curve for the beam shown in Fig. P.5.13.

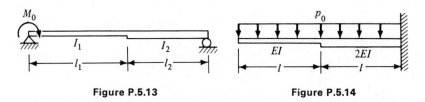

Figure P.5.13   Figure P.5.14

5-21. Determine the equation of the elastic curve for the beam shown in Fig. P.5.14.

Note: Use the Taylor's Series method discussed in Secs. 5.6 and 5.7 to obtain solutions for Problems 5-22 through 5-35.

5-22. Determine the equation of the elastic curve for the beam shown in Fig. P.5.15.

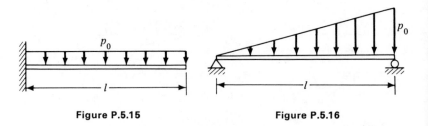

Figure P.5.15   Figure P.5.16

5-23. Determine the equation of the elastic curve for the beam shown in Fig. P.5.16.

5-24. Determine the equation of the elastic curve for the beam shown in Fig. P.5.17.

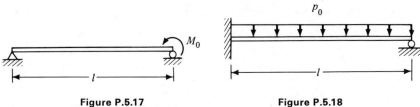

Figure P.5.17   Figure P.5.18

5-25. For the beam shown in Fig. P.5.18 determine the support reactions and the equation of the elastic curve. Draw the shearing force and bending moment diagrams for the beam.

5-26. Determine the support reactions and the equation of the elastic curve for the beam shown in Fig. P.5.19.

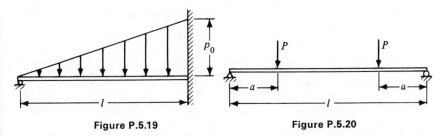

Figure P.5.19   Figure P.5.20

5-27. Determine the equation of the elastic curve for the beam shown in Fig. P.5.20.

5-28. Determine the equation of the elastic curve for the beam shown in Fig. P.5.21.

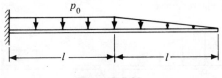

Figure P.5.21

5-29. Determine the equation of the elastic curve for the beam shown in Fig. P.5.22.

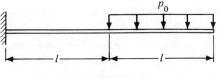

Figure P.5.22

5-30. Determine the equation of the elastic curve for the beam shown in Fig. P.5.23.

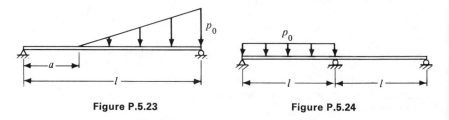

Figure P.5.23   Figure P.5.24

5-31. Determine the support reactions and the equation of the elastic curve for the beam shown in Fig. P.5.24.

5-32. Determine the support reactions and the equation of the elastic curve for the beam shown in Fig. P.5.25.

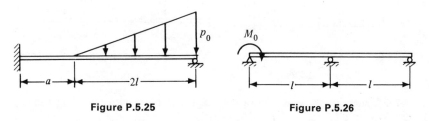

Figure P.5.25   Figure P.5.26

5-33. Determine the support reactions and the equation of the elastic curve for the beam shown in Fig. P.5.26.

5-34. For the beam shown in Fig. P.5.27, determine the support reactions and the equation of the elastic curve.

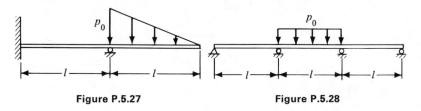

Figure P.5.27   Figure P.5.28

5-35. For the beam shown in Fig. P.5.28, determine the support reactions and the equation of the elastic curve. Draw the shearing force and bending moment diagrams for the beam.

Note: Use the conjugate beam method to obtain solutions to Problems 5-36 through 5-45.

5-36. For the beam shown in Fig. P.5.29, determine the slope of the elastic curve at the left-hand support and the deflection at midspan.

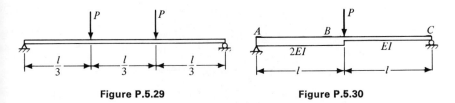

Figure P.5.29   Figure P.5.30

5-37. Determine the slope of the elastic curve at A and the deflection at B for the beam shown in Fig. P.5.30.

5-38. Determine the equation of the elastic curve for the beam shown in Fig. P.5.31.

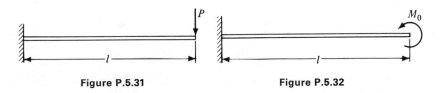

Figure P.5.31   Figure P.5.32

5-39. Determine the equation of the elastic curve for the beam shown in Fig. P.5.32.

5-40. Determine slope and deflection at the end of the beam shown in Fig. P.5.33.

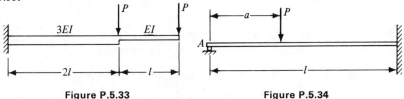

Figure P.5.33   Figure P.5.34

5-41. Determine the support reaction at the left end of the beam shown in Fig. P.5.34.

5-42. Determine the support reactions for the beam shown in Fig. P.5.35.

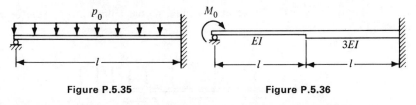

Figure P.5.35   Figure P.5.36

5-43. Determine the support reaction at the left end of the beam shown in Fig. P.5.36.

5-44. Determine the support reactions for the beam shown in Fig. P.5.37.

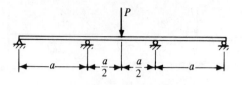

Figure P.5.37

5-45. Determine the support reactions for the beam shown in Fig. P.5.38.

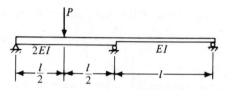

Figure P.5.38

Note: Solve Problems 5-46 through 5-51 by use of the three-moment equation discussed in Sec. 5.9.

5-46. Determine the bending moments at the supports and the support reactions for the beam shown in Fig. P.5.28. Draw the shearing force and bending moment diagrams for the beam.

5-47. Determine the bending moments at the supports and the support reactions for the beam shown in Fig. P.5.37.

5-48. For the beam shown in Fig. P.5.39, determine the support reactions. Draw the shearing force and bending moment diagrams for the beam.

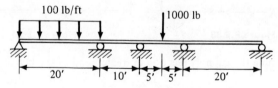

Figure P.5.39

5-49. Draw the shearing force and bending moment diagrams for the beam shown in Fig. P.5.40.

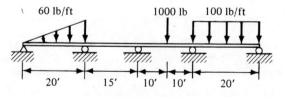

Figure P.5.40

5-50. Draw the shearing force and bending moment diagrams for the beam shown in Fig. P.5.24.

5-51. Draw the shearing force and bending moment diagrams for the beam shown in Fig. P.5.41.

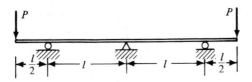

Figure P.5.41

5-52. Show that the deflection of a horizontal beam which is simply supported at both ends and which carries a distributed load, whose intensity varies uniformly from zero at the end $x = 0$ to p_0 at the end $x = l$, can be expressed by

$$v(x) = \frac{2p_0 l^4}{\pi^5 EI} \sum_{n=1}^{\infty} \frac{(-1)^{n+1}}{n^5} \sin \frac{n\pi x}{l}$$

5-53. Find the Fourier series expansion for the deflection of a horizontal, simply supported beam of length l which is subjected to a uniform load p_0 from $x = 0$ to $x = l/2$.

5-54. Express in a sine series, the deflection of a horizontal beam of length l which is simply supported at its ends and subjected to a distributed load whose intensity varies linearly from zero at both ends to the value p_0 in the middle.

5-55. Express in a sine series, the deflection of a horizontal beam which is simply supported at both ends and which carries a distributed load whose intensity varies linearly from the value p_0 at the end $x = 0$ to the value $2p_0$ at the end $x = l$.

5-56. Express in a sine series, the deflection of a horizontal beam of length l which is simply supported at both ends and carries a distributed load whose intensity varies linearly from the value 0 at the end $x = 0$ to the value p_0 at the middle.

5-57. Express in a sine series, the deflection of a horizontal beam of length l which is simply supported at both ends and which carries a distributed load whose intensity $p(x)$ varies according to the law

$$p(x) = \frac{4p_0 x(l - x)}{l^2}$$

in the interval $0 \leq x \leq l$.

5-58 to 62. Determine the maximum load P_{max} at which the beams shown in Figs. P.5.42 through P.5.46 will collapse in accordance with the plastic theory.

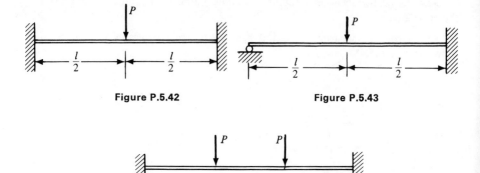

Figure P.5.42   **Figure P.5.43**

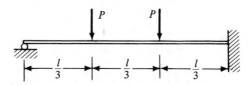

Figure P.5.44

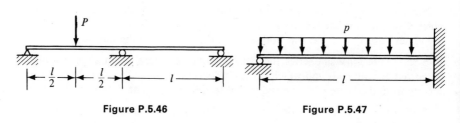

Figure P.5.45

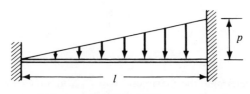

Figure P.5.46   **Figure P.5.47**

Note: Use the method discussed in the Appendix to solve Problems 5-67 through 5-72.

5-63 to 66. Determine the maximum load p_{max} at which the beams shown in Figs. P.5.47 through P.5.50 will collapse in accordance with the plastic theory.

Figure P.5.48

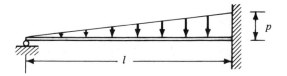

Figure P.5.49

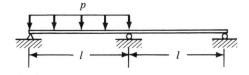

Figure P.5.50

5-67. Show that the Fourier series expansion of function $f(t)$ defined in the finite interval $-\pi < t < \pi$ by

$$f(t) = 2t + t^2$$

is

$$f(t) = \frac{\pi^2}{3} + 4 \sum_{n=1}^{\infty} \cos n\pi \left(\frac{1}{n^2} \cos nt - \frac{1}{n} \sin nt \right)$$

Find the value toward which this Fourier series converges at $t = \pi$.

5-68. Find the Fourier series expansion of the function

$$f(t) = e, \quad -\pi < t < \pi$$

5-69. Show that the Fourier cosine expansion for the function

$$f(t) = \cos \frac{t}{2}$$

in the interval $0 < t < \pi$ is

$$f(t) = \frac{2}{\pi} - \frac{4}{\pi} \sum_{n=1}^{\infty} \frac{\cos n\pi}{4n^2 - 1} \cos nt,$$

while its Fourier sine expansion is

$$f(t) = \frac{8}{\pi} \sum_{n=1}^{\infty} \frac{n}{4n^2 - 1} \sin nt$$

Draw diagrams of the periodic functions represented by the above two series for all values of t.

5-70. Find the Fourier sine and cosine series expansions for the function $f(t) = \cos t$ in the interval $0 \leq t \leq \pi/2$.

5-71. Show that the Fourier series expansion for the function $f(t) = t \sin t$ in the interval $0 \leq t \leq \pi$ is

$$f(t) = \frac{\pi}{2} \sin t - \frac{16}{\pi} \sum_{n=1}^{\infty} \frac{n}{(4n-1)^2} \sin 2nt$$

5-72. Show that the Fourier series expansion for function $f(t)$ defined in the finite interval $-\pi \leq t \leq \pi$ by

$$f(t) = \begin{cases} 0 & \text{for } -\pi \leq t \leq 0 \\ \sin t & \text{for } 0 \leq t \leq \pi \end{cases}$$

is

$$f(t) = \frac{1}{\pi} - \frac{2}{\pi} \sum_{n=1}^{\infty} \frac{\cos 2nt}{4n^2 - 1} + \frac{1}{2} \sin t$$

Draw a diagram of the periodic function represented by this series for all values of t.

CHAPTER

6

BENDING OF A CURVED BEAM OUT OF ITS INITIAL PLANE

6.0 Introduction

Bending of curved beams is an important topic in the strength of material because such beams have many practical applications. Many well-known elasticians including Barré de Saint-Venant, B. Biezeno, H. Marcus, and A. J. S. Pippard, have contributed to this field of research.[1]

In this chapter, Saint-Venant's two equations for a curved beam bent out of its plane of initial curvature will be derived first. Then, by expressing the conditions of equilibrium of the beam, two subsidiary equations will be obtained. By use of these four equations the following problems have been solved in closed form:[2]

[1] For references see Bibliography at the end of this chapter.
[2] See bibliography number 27 at the end of this chapter.

354 BENDING OF A CURVED BEAM Sec. 6.0

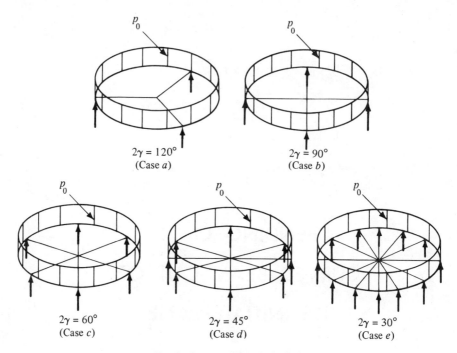

Circular beams uniformly loaded

Figure 6.0.1

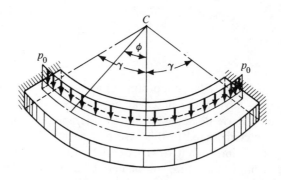

Uniform loading

Figure 6.0.2

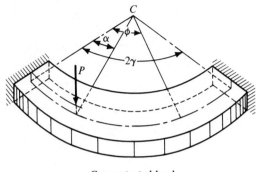

Concentrated load

Figure 6.0.3

1. Bending of a circular beam subjected to a uniformly distributed load and symmetrically supported (see Fig. 6.0.1).

2. Bending of a circular arc bow girder subjected to a uniformly distributed load (see Fig. 6.0.2).

3. Bending of a circular arc bow girder subjected to a concentrated load (see Fig. 6.0.3).

Numerical results for deflections, angles of twist, bending moments, and twisting moments are presented in tabular form for the above cases.

6.1 Bending of a Curved Beam out of Its Plane of Initial Curvature: Saint-Venant's Equations and Equations of Equilibrium

Consider the curved beam built-in at one end and having constant inertia characteristics as shown in Fig. 6.1.1. The beam is referred to a system of cartesian coordinates $Oxyz$ with the origin O at the centroid of the cross section of the beam, the x- and y-axes coinciding with the principal axes of inertia of the cross section and the z-axis coinciding with the tangent to the center line at O. The xz-plane coincides with the horizontal plane of initial curvature of the beam, with x taken positive toward the center of curvature of the center line of the beam, y taken positive downward, z taken positive away from the built-in end, and the arc s of the center line measured from the fixed end. If M_x and M_z are, respectively, moments acting on the cross section at O about the x- and z-axes (M_x being the bending moment and M_z the twisting moment), V the shear force in the direction of the y-axis, v the displacement of centroid O in the direction of the y-axis, β the angle of twist of the cross section about the z-axis (considered as positive, if the rotation is counterclockwise with respect to the z-axis), EI_x the flexural rigidity, K the torsional rigidity, θ the angle of twist per unit length at the same cross section, R the initial radius of curvature of the center line of the beam, and R_1 the radius of curvature of the center line at O after the beam is deformed in the principal plane yz, then the following two equations can be derived:

$$\left. \begin{array}{l} \dfrac{EI_x}{R_1} = M_x \\ K\theta = M_z \end{array} \right\} \tag{6.1.1}$$

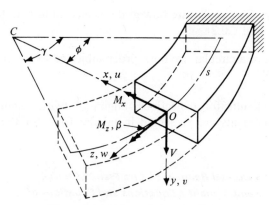

Diagram showing symbols and orientation

Figure 6.1.1

In order to obtain expressions for the radius of curvature R_1 and the twist θ as functions of displacement v and the angle β, v and β will be assumed to be small quantities. The final values of the radius of curvature and of the twist will be obtained by adding the separate effects produced on the beam by the linear displacement v and by the angular displacement β.

Suppose first that an element ds of the curved bar (see Fig. 6.1.2) is subjected to a displacement dv in the y-direction. As a consequence, the element of the beam will rotate with respect to the principal x-axis (CO) through an angle dv/ds. Due to this rotation, the axis $x_1(CO_1)$ of the adjacent cross section O_1 will displace into the new position $x_2(CO_2)$, the angle

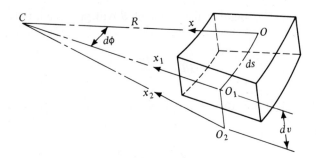

Figure 6.1.2

O_1CO_2 between the x_1- and x_2-axes being equal to dv/R. The twist per unit length of the beam will be

$$(\theta)_1 = \frac{1}{R}\frac{dv}{ds} \tag{6.1.2}$$

In addition, a displacement v in the y-direction will produce a curvature $1/R_1$ of the center line of the beam in the principal yz-plane, analogous to that in a straight bar, and which will be given by

$$\left(\frac{1}{R_1}\right)_1 = -\frac{d^2v}{ds^2} \tag{6.1.3}$$

Suppose now that the same beam is subjected to a small angular displacement β (see Fig. 6.1.3). The corresponding curvature will be given by

$$\left(\frac{1}{R_1}\right)_2 = \frac{\sin\beta}{R} \doteq \frac{\beta}{R} \tag{6.1.4}$$

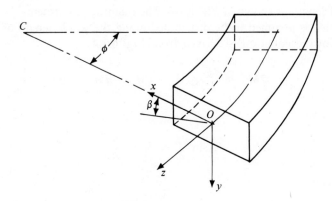

Figure 6.1.3

At the same time, the angular displacement β will produce a twist per unit length

$$(\theta)_2 = \frac{d\beta}{ds} \tag{6.1.5}$$

Adding Eqs. (6.1.2) and (6.1.5) and Eqs. (6.1.3) and (6.1.4) gives the two equations

$$\left.\begin{aligned}\frac{1}{R_1} &= \frac{\beta}{R} - \frac{d^2v}{ds^2} \\ \theta &= \frac{d\beta}{ds} + \frac{1}{R}\frac{dv}{ds}\end{aligned}\right\} \tag{6.1.6}$$

By substituting Eqs. (6.1.6) into Eqs. (6.1.1), Saint-Venant's equations for a curved beam bent out of its plane of initial curvature are obtained; thus,

$$\left.\begin{aligned}M_x &= EI_x\left(\frac{\beta}{R} - \frac{d^2v}{ds^2}\right) \\ M_z &= K\left(\frac{d\beta}{ds} + \frac{1}{R}\frac{dv}{ds}\right)\end{aligned}\right\} \tag{6.1.7}$$

The two subsidiary equations

$$\left.\begin{aligned}\frac{d}{ds}\left[\frac{dM_x}{ds} + \frac{M_z}{R}\right] &= p \\ \frac{dM_z}{ds} - \frac{M_x}{R} &= 0\end{aligned}\right\} \tag{6.1.8}$$

are obtained by expressing the conditions for equilibrium of the element of beam of length ds shown in Fig. 6.1.4. The quantity p represents the external load which is assumed to be distributed along the whole length of the bar. The following equation is derived from Fig. 6.1.4:

$$2M_x \sin\frac{d\phi}{2} + M_z - (M_z + dM_z) = 0$$

However, since

$$2\sin\frac{d\phi}{2} \doteq d\phi = \frac{ds}{R}$$

one obtains

$$M_x \frac{ds}{R} - dM_z = 0$$

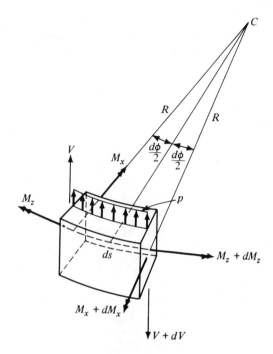

Figure 6.1.4

from which the second of Eqs. (6.1.8) is obtained. Also, from Fig. 6.1.4, it follows that

$$V - (V + dV) + p\,ds = 0$$

from which

$$\frac{dV}{ds} = p$$

In addition,

$$M_x - (M_x + dM_x) + V\,ds - 2M_z \sin\frac{d\phi}{2} = 0$$

or

$$-dM_x + V\,ds - M_z\,d\phi = 0$$

from which

$$V = \frac{dM_x}{ds} + M_z\frac{d\phi}{ds} = \frac{dM_x}{ds} + \frac{M_z}{R}$$

and finally

$$p = \frac{dV}{ds} = \frac{d}{ds}\left[\frac{dM_x}{ds} + \frac{M_z}{R}\right]$$

which is the first of Eqs. (6.1.8).

6.2 Bending of a Circular Beam Subjected to a Uniformly Distributed Load and Supported Symmetrically[3]

If p_0 is the uniformly distributed load per unit length of the beam and, if 2γ is the angular distance between the points of support (see Fig. 6.0.1), then by using the notation

$$\left. \begin{array}{l} Y = \dfrac{v}{R} \\[4pt] a = \dfrac{EI_x}{R} \\[4pt] \mu = \dfrac{K}{EI_x} \\[4pt] v = \dfrac{p_0 R^3}{EI_x} \end{array} \right\} \quad (6.2.1)$$

Eqs. (6.1.7) become

$$\left. \begin{array}{l} a\left(\beta - \dfrac{d^2 Y}{d\phi^2}\right) = M_x \\[6pt] \mu a \left(\dfrac{d\beta}{d\phi} + \dfrac{dY}{d\phi}\right) = M_z \end{array} \right\} \quad (6.2.2)$$

where the angle ϕ is measured from the bisector of the angle between two points of support. Equations (6.1.8) may be written as

$$\left. \begin{array}{l} \dfrac{dM_z}{d\phi} - M_x = 0 \\[6pt] \dfrac{d^2 M_x}{d\phi^2} + \dfrac{dM_z}{d\phi} - av = 0 \end{array} \right\} \quad (6.2.3)$$

while the boundary conditions in this case are

$$\left. \begin{array}{l} \beta(\gamma) = \beta(-\gamma) \\[4pt] \dfrac{d\beta(\gamma)}{d\phi} = \dfrac{d\beta(-\gamma)}{d\phi} \end{array} \right\} \quad (6.2.4a)$$

$$\left. \begin{array}{l} Y(\gamma) = Y(-\gamma) \\[4pt] \dfrac{dY(\gamma)}{d\phi} = \dfrac{dY(-\gamma)}{d\phi} \end{array} \right\} \quad (6.2.4b)$$

[3] Part of Section 6.2 was first discussed in the paper entitled "Deflections of a Circular Beam Out of Its Initial Plane" by Enrico Volterra (see bibliography number 27 at the end of this chapter) and is presented here with the permission of the American Society of Civil Engineers.

BENDING OF A CURVED BEAM

$$\left.\begin{array}{l}\dfrac{d^2Y(\gamma)}{d\phi^2} = \dfrac{d^2Y(-\gamma)}{d\phi^2} \\ \dfrac{dM_x(\gamma^-)}{d\phi} - \dfrac{dM_x(\gamma^+)}{d\phi} = 2a\gamma v\end{array}\right\} \quad (6.2.4c)$$

From Eqs. (6.2.2), (6.2.3), and (6.2.4) one obtains

$$\frac{\beta}{v} = \frac{\mu+1}{2\mu}\left[2 - \frac{\gamma}{\sin\gamma}\phi\sin\phi - \frac{\gamma}{\sin\gamma}(1 + \gamma\cot\gamma)\cos\phi\right] \quad (6.2.5a)$$

$$\frac{Y}{v} = \frac{1}{2\mu}\left\{\phi^2 - 2 + \frac{(\mu+1)\gamma}{\sin\gamma}\phi\sin\phi\right\}$$

$$+ \frac{\gamma}{\sin\gamma}[\mu + 3 + (\mu+1)\gamma\cot\gamma]\cos\phi + A' \quad (6.2.5b)$$

$$\frac{M_x}{av} = 1 - \frac{\gamma}{\sin\gamma}\cos\phi \quad (6.2.5c)$$

and

$$\frac{M_z}{av} = \phi - \frac{\gamma}{\sin\gamma}\sin\phi \quad (6.2.5d)$$

In the case of rigid supports, $Y = 0$ for $\phi = \gamma$, and the constant of Eq. (6.2.5b) assumes the value,

$$A' = -\frac{1}{2\mu}\left\{\gamma^2 - 2 + (\mu+1)\gamma^2 + \gamma\cot\gamma[\mu + 3 + (\mu+1)\gamma\cot\gamma]\right\} \quad (6.2.6)$$

Values of the functions β/v, Y/v, M_x/av, and M_z/av are given in Tables 6.2.1, 6.2.2, 6.2.3, 6.2.4, and 6.2.5 for the following values of the parameters: $\mu = 0.7$ and 1.00; $2\gamma = 120°, 90°, 60°, 45°,$ and $30°$. In the computation of numerical values, the constant A' was assumed to be equal to zero.

TABLE 6.2.1

$2\gamma = 120°$	$\mu = 0.7$		$\mu = 1.00$		$\mu = 0.7; 1.00$	
ϕ	$\dfrac{\beta}{v}$	$\dfrac{Y}{v}$	$\dfrac{\beta}{v}$	$\dfrac{Y}{v}$	$\dfrac{M_x}{av}$	$\dfrac{M_z}{av}$
0	0.07252	2.6549	0.05972	2.1495	−0.2092	0
10°	0.06381	2.6591	0.05255	2.1535	−0.1908	−0.03544
20°	0.03930	2.6710	0.03237	2.1648	−0.1363	−0.06450
30°	0.0²376	2.6881	0.0²310	2.1812	−0.0472	−0.08100
40°	−0.03518	2.7066	−0.02897	2.1990	0.0737	−0.07913
50°	−0.06744	2.7218	−0.05554	2.2136	0.2227	−0.05364
60°	−0.08107	2.7281	−0.06676	2.2197	0.3954	0

TABLE 6.2.2

$2\gamma = 90°$	$\mu = 0.7$		$\mu = 1.00$		$\mu = 0.7; 1.00$	
ϕ	$\dfrac{\beta}{v}$	$\dfrac{Y}{v}$	$\dfrac{\beta}{v}$	$\dfrac{Y}{v}$	$\dfrac{M_x}{av}$	$\dfrac{M_z}{av}$
0	0.02055	2.5662	0.01692	2.0938	−0.11072	0
10°	0.01625	2.5681	0.01339	2.0957	−0.09385	−0.01834
20°	0.0^2475	2.5733	0.0^2391	2.1008	−0.04374	−0.03082
30°	$−0.0^2994$	2.5799	$−0.0^2818$	2.1072	0.03809	−0.03176
40°	−0.02133	2.5850	−0.01756	2.1121	0.14914	−0.01583
45°	−0.02319	2.5858	−0.01910	2.1129	0.21460	0

TABLE 6.2.3

$2\gamma = 60°$	$\mu = 0.7$		$\mu = 1.00$		$\mu = 0.7; 1.00$	
ϕ	$\dfrac{\beta}{v}$	$\dfrac{Y}{v}$	$\dfrac{\beta}{v}$	$\dfrac{Y}{v}$	$\dfrac{M_x}{av}$	$\dfrac{M_z}{av}$
0	0.0^2376	2.4922	0.0^2310	2.0441	−0.0472	0
5°	0.0^2332	2.4924	0.0^2273	2.0443	−0.0432	$−0.0^2400$
10°	0.0^2206	2.4930	0.0^2170	2.0448	−0.0313	$−0.0^2731$
15°	0.0^3225	2.4938	0.0^3185	2.0456	−0.0115	$−0.0^2924$
20°	$−0.0^2182$	2.4946	$−0.0^2150$	2.0465	0.0160	$−0.0^2910$
25°	$−0.0^2354$	2.4954	$−0.0^2291$	2.0472	0.0509	$−0.0^2623$
30°	$−0.0^2428$	2.4957	$−0.0^2352$	2.0475	0.0931	0

TABLE 6.2.4

$2\gamma = 45°$	$\mu = 0.7$		$\mu = 1.00$		$\mu = 0.7; 1.00$	
ϕ	$\dfrac{\beta}{v}$	$\dfrac{Y}{v}$	$\dfrac{\beta}{v}$	$\dfrac{Y}{v}$	$\dfrac{M_x}{av}$	$\dfrac{M_z}{av}$
0	0.0^2116	2.4648	0.0^3956	2.0252	−0.0262	0
5°	0.0^3920	2.4649	0.0^3758	2.0253	−0.0223	$−0.0^2217$
10°	0.0^3273	2.4652	0.0^3225	2.0256	−0.0106	$−0.0^2366$
15°	$−0.0^3559$	2.4655	$−0.0^3461$	2.0259	0.0^2879	$−0.0^2379$
20°	$−0.0^2121$	2.4658	$−0.0^3100$	2.0262	0.03571	$−0.0^2191$
22.5°	$−0.0^2132$	2.4658	$−0.0^2109$	2.0263	0.05194	0

TABLE 6.2.5

$2\gamma = 30°$	$\mu = 0.7$		$\mu = 1.00$		$\mu = 0.7; 1.00$	
ϕ	$\dfrac{\beta}{v}$	$\dfrac{Y}{v}$	$\dfrac{\beta}{v}$	$\dfrac{Y}{v}$	$\dfrac{M_x}{av}$	$\dfrac{M_z}{av}$
0	0.0^3225	2.4448	0.0^3185	2.01133	−0.0115	0
5°	0.0^3123	2.4448	0.0^3102	2.00753	$−0.0^2767$	$−0.0^3893$
10°	0.0^3109	2.4449	$−0.0^4896$	1.9961	0.0^2385	$−0.0^2111$
15°	$−0.0^3257$	2.4450	$−0.0^3212$	1.9773	0.02295	0

Example 6.2.1. A reinforced concrete circular beam, which has a radius $R = 20$ ft, a modulus of elasticity $E = 3 \times 10^6$ psi, and a square cross section (16 × 16 in.), rests on eight rigid supports and carries a load p_0 of 0.5 tons per ft (Fig. 6.0.1, case d).

The following values are computed from the given data: $I_x = 5{,}451$ in.^{4}, $a = 5{,}688.54$ ft-tons, $v = 0.0239$, $av = 133.11$ ft-tons and $\mu = 0.7$. Table 6.2.4 is applicable in solving this problem and with the given data deflections, angles of twist, bending moments, and twisting moments for various positions along the beam may be computed. These values are presented in Table 6.2.6.

TABLE 6.2.6

RESULTS FOR EXAMPLE 6.2.1

ϕ, degrees	β, radians	v, inches	M_x, foot-tons	M_z, foot-tons
0	−0.000812	0.005616	3.4875	0
5	−0.000644	0.005064	2.9683	0.2888
10	−0.000191	0.003360	1.4110	0.4872
15	0.000391	0.001680	−1.1700	0.5045
20	0.000847	0.000820	−4.7533	0.2542
22.5	0.000924	0	−6.9137	0

6.3 Bending of a Circular Arc Bow Girder Subjected to a Uniformly Distributed Load (see Fig. 6.0.2)[4]

Equations (6.2.1), (6.2.2), and (6.2.3) are also valid in this case provided that the following boundary conditions are assumed:

$$\beta(\gamma) = \beta(-\gamma) = Y(\gamma) = Y(-\gamma) = \frac{dY(\gamma)}{d\phi} = \frac{dY(-\gamma)}{d\phi} = 0 \quad (6.3.1)$$

From Eqs. (6.2.2), (6.2.3), and (6.3.1) one obtains

$$\frac{\beta}{v} = \frac{\mu + 1}{2\mu}\left[2 + D_1\phi \sin\phi - \left(\frac{2}{\cos\gamma} + D_1\gamma \tan\gamma\right)\cos\phi\right] \quad (6.3.2a)$$

$$\frac{Y}{v} = \frac{1}{2\mu}\left[\phi^2 - \gamma^2 - 2(\mu + 1)\left(1 - \frac{\cos\phi}{\cos\gamma}\right)\right]$$
$$+ \frac{D_1}{2\mu}\{-(\mu + 1)\phi \sin\phi + [(\mu + 1)\gamma \tan\gamma - 2]\cos\phi + 2\cos\gamma\} \quad (6.3.2b)$$

$$\frac{M_x}{av} = 1 + D_1 \cos\phi \quad (6.3.2c)$$

---

[4]Part of Section 6.3 was first discussed in the paper entitled "Deflections of a Circular Beam Out of Its Initial Plane" by Enrico Volterra (see bibliography number 27 at the end of this chapter) and is presented here with the permission of the American Society of Civil Engineers.

TABLE 6.3.1

$2\gamma = 180°$

	$\mu = 0.50$		$\mu = 1.00$		$\mu = 1.33$		$\mu = 0.50; 1.00; 1.33$	
ϕ	$\dfrac{\beta}{v}$	$\dfrac{Y}{v}$	$\dfrac{\beta}{v}$	$\dfrac{Y}{v}$	$\dfrac{\beta}{v}$	$\dfrac{Y}{v}$	$\dfrac{M_x}{av}$	$\dfrac{M_z}{av}$
0	0.49503	−0.41595	0.42920	−0.38966	0.41287	−0.38314	−0.27324	0
10°	0.47520	−0.40435	0.41448	−0.37905	0.39941	−0.37278	−0.25390	−0.04656
20°	0.41808	−0.37073	0.37193	−0.34825	0.36047	−0.34267	−0.19645	−0.08641
30°	0.33063	−0.31856	0.30632	−0.30028	0.30028	−0.29575	−0.10266	−0.11302
40°	0.22403	−0.25333	0.22533	−0.23998	0.22566	−0.23667	0.02464	−0.12029
50°	0.11309	−0.18211	0.13915	−0.17366	0.14562	−0.17156	0.18158	−0.10269
60°	0.01546	−0.11300	0.05990	−0.10867	0.07093	−0.10760	0.36338	−0.05546
70°	−0.04937	−0.05446	0.0²101	−0.05293	0.01351	−0.05255	0.56453	0.02528
80°	−0.06114	−0.01452	−0.02354	−0.01599	−0.01421	−0.01424	0.77890	0.14237
90°	0	0	0	0	0	0	1	0.29756

BENDING OF A CURVED BEAM

$$\frac{M_z}{av} = \phi + D_1 \sin \phi \qquad (6.3.2d)$$

where

$$D_1 = 2\frac{\gamma \cos \gamma - (1 + \mu) \sin \gamma}{(1 + \mu)\gamma + (\mu - 1) \sin \gamma \cos \gamma} \qquad (6.3.3)$$

In the particular case in which $\gamma = 90°$,

$$D_1 \gamma \tan \gamma + \frac{2}{\cos \gamma} = \frac{\pi^2 + 4(\mu - 1)}{\pi(1 + \mu)} \qquad (6.3.4)$$

Values of the functions β/v, Y/v, M_x/av, and M_z/av are given in Tables 6.3.1, 6.3.2, 6.3.3, and 6.3.4 for the following values of the parameters: $\mu = 0.5$, 1.0, and 1.33; and $2\gamma = 180°$, 150°, 120°, and 90°.

TABLE 6.3.2

$2\gamma = 150°$

ϕ	$\dfrac{\beta}{v}$	$\dfrac{Y}{v}$	$\dfrac{M_x}{av}$	$\dfrac{M_z}{av}$	
0	0.25811	−0.18147	−0.20761	0	$\mu = 0.50$
10°	0.24487	−0.17445	−0.18927	−0.03517	
20°	0.20721	−0.15437	−0.13478	−0.06396	
30°	0.15123	−0.12400	−0.04582	−0.08021	
40°	0.08672	−0.08774	0.07491	−0.07811	
50°	0.02662	−0.05117	0.22376	−0.05242	
60°	−0.01372	−0.02062	0.39619	0.0²137	
70°	−0.01738	−0.0²252	0.58697	0.08695	
75°	0	0	0.68744	0.14253	
0	0.21797	−0.17268	−0.21701	0	$\mu = 1.00$
10°	0.20816	−0.16613	−0.19852	−0.03680	
20°	0.18014	−0.14733	−0.14361	−0.06718	
30°	0.13810	−0.11879	−0.05396	−0.08490	
40°	0.08875	−0.08450	0.06772	−0.08415	
50°	0.04096	−0.04963	0.21772	−0.05962	
60°	0.0²528	−0.02019	0.39150	−0.0²676	
70°	−0.0²668	−0.0²249	0.58376	0.07812	
75°	0	0	0.68502	0.13346	
0	0.20650	−0.17042	−0.22065	0	$\mu = 1.33$
10°	0.19757	−0.16398	−0.20210	−0.03743	
20°	0.17204	−0.14551	−0.14703	−0.06842	
30°	0.13362	−0.11744	−0.05711	−0.08672	
40°	0.08834	−0.08365	0.06493	−0.08648	
50°	0.04376	−0.04923	0.21538	−0.06240	
60°	0.0²952	−0.02007	0.38968	−0.0²991	
70°	−0.0²418	−0.0²248	0.58251	0.07470	
75°	0	0	0.68407	0.12994	

TABLE 6.3.3

$2\gamma = 120°$

ϕ	$\dfrac{\beta}{v}$	$\dfrac{Y}{v}$	$\dfrac{M_x}{av}$	$\dfrac{M_z}{av}$	
0	0.1156	−0.06710	−0.14516	0	$\mu = 0.5$
10°	0.1074	−0.06319	−0.12776	−0.02432	
20°	0.0845	−0.05225	−0.07610	−0.04260	
30°	0.0524	−0.03652	0.0²826	−0.04898	
40°	0.0196	−0.01952	0.12276	−0.03796	
50°	0.0²234	−0.0²571	0.26391	−0.0²458	
60°	0	0	0.42742	0.05546	
0	0.09310	−0.06442	−0.15399	0	$\mu = 1.00$
10°	0.08710	−0.06072	−0.13646	−0.02585	
20°	0.07033	−0.05032	−0.08439	−0.04562	
30°	0.04646	−0.03531	−0.0³618	−0.05339	
40°	0.02138	−0.01898	0.11599	−0.04364	
50°	0.0²283	−0.0²559	0.25823	−0.01134	
60°	0	0	0.42301	0.04782	
0	0.08640	−0.06369	−0.15704	0	$\mu = 1.33$
10°	0.08099	−0.06004	−0.13946	−0.02639	
20°	0.06585	−0.04979	−0.08726	−0.04667	
30°	0.05046	−0.03497	−0.0²203	−0.05492	
40°	0.02124	−0.01883	0.11366	−0.04560	
50°	0.0²381	−0.0²556	0.25627	−0.01368	
60°	0	0	0.42148	0.04517	

Example 6.3.1. A circular bow girder which is made of reinforced concrete has a radius $R = 10$ ft, a modulus of elasticity $E = 3 \times 10^6$ psi, a rectangular cross section (20 in. wide × 12 in. deep), and carries a uniformly distributed load of 0.25 tons per ft. The angular distance between the built-in ends is $2\gamma = 120°$.

The following values are computed from the given data: $I_x = 8,000$ in.[4], $a = 8,333.33$ ft-tons, $v = 3 \times 10^{-3}$, $av = 25$ ft-tons, and $\mu = 0.5$. The results for this example are given in Table 6.3.5 and are based on Table 6.3.3.

TABLE 6.3.4

$2\gamma = 90°$

ϕ	$\dfrac{\beta}{v}$	$\dfrac{Y}{v}$	$\dfrac{M_x}{av}$	$\dfrac{M_z}{av}$	
0	0.04018	−0.01917	−0.08889	0	$\mu = 0.50$
10°	0.03565	−0.01726	−0.07235	−0.01455	
20°	0.02368	−0.01216	−0.02323	−0.02336	
30°	0.0²912	−0.0²572	0.05699	−0.02085	
40°	−0.0³314	−0.0³795	0.16586	−0.0²180	
45°	0	0	0.23004	0.01543	
0	0.03043	−0.01857	−0.0935	0	$\mu = 1.00$
10°	0.02721	−0.01673	−0.0769	−0.01536	
20°	0.01865	−0.01182	−0.0276	−0.02494	
30°	0.0²801	−0.0²558	0.0530	−0.02316	
40°	0.0³499	−0.0³780	0.1623	−0.0²477	
45°	0	0	0.2268	0.01216	
0	0.0276	−0.0184	−0.09495	0	$\mu = 1.33$
10°	0.0248	−0.0166	−0.07831	−0.0156	
20°	0.0171	−0.0117	−0.02891	−0.0254	
30°	0.0²754	−0.0²554	0.05175	−0.0239	
40°	0.0³627	−0.0³776	0.16122	−0.0²569	
45°	0	0	0.22575	0.01115	

TABLE 6.3.5

RESULTS FOR EXAMPLE 6.3.1

ϕ, degrees	β, radians	v, inches	M_x, foot-tons	M_z, foot-tons
0	0.0003468	0.02415	3.6290	0
10	0.0003222	0.02275	3.1940	0.6080
20	0.0002535	0.01881	1.9025	1.0650
30	0.0001572	0.01315	− 0.2065	1.2245
40	0.0000588	0.007030	− 3.0690	0.9490
50	0.00000702	0.002052	− 6.5977	0.1145
60	0	0	−10.6855	−1.3865

6.4 Bending Of a Circular Arc Bow Girder Subjected to a Concentrated Load[5]

If P is the concentrated load, 2γ the angular distance between the extremities of the beam, and α the angular distance between the end of the beam ($\phi = 0$) and the point where the concentrated force is applied (Fig. 6.0.3), it follows that Eqs. (6.2.1) and (6.2.2) are still valid with the exception that $v = PR^2/EI_x$.

For equilibrium the following equations must be satisfied:

$$\left.\begin{array}{r}\dfrac{dM_z}{d\phi} - M_x = 0 \\[6pt] \dfrac{d^2 M_x}{d\phi^2} + \dfrac{dM_z}{d\phi} = 0\end{array}\right\} \quad (6.4.1)$$

The boundary conditions and the conditions of continuity are, in this case,

$$\beta(0) = \beta(2\gamma) = Y(0) = Y(2\gamma) = \frac{dY(0)}{d\phi} = \frac{dY(2\gamma)}{d\phi} = 0 \quad (6.4.2a)$$

$$\frac{dM_x(\alpha^+)}{d\phi} - \frac{dM_x(\alpha^-)}{d\phi} = -av \quad (6.4.2b)$$

and β, $d\beta/d\phi$, Y and $dY/d\phi$ are all continuous for $\phi = \alpha$. Because of the loading discontinuity at $\phi = \alpha$, the analytical expressions for the bending and twisting moments M_x and M_z, for displacement v, and for the angle of twist β are different in the intervals $(0, \alpha)$ and $(\alpha, 2\gamma)$. If $\beta = \bar{\beta}$, $Y = \bar{Y}$, $M_x = \bar{M}_x$, and $M_z = \bar{M}_z$ for $0 \leq \phi \leq \alpha$ and if $\beta = \bar{\beta} + \beta^*$, $Y = \bar{Y} + Y^*$, $M_x = \bar{M}_x + M_x^*$, and $M_z = \bar{M}_z + M_z^*$ for $\alpha \leq \phi \leq 2\gamma$, then the solution of Eqs. (6.2.2), (6.4.1), and (6.4.2) gives

$$\beta^* = \frac{\mu + 1}{2\mu} v[(\phi - \alpha)\cos(\phi - \alpha) - \sin(\phi - \alpha)] \quad (6.4.3a)$$

$$Y^* = -\frac{v}{2\mu}[2(\phi - \alpha)$$
$$+ (\mu + 1)(\phi - \alpha)\cos(\phi - \alpha) - (\mu + 3)\sin(\phi - \alpha)] \quad (6.4.3b)$$

[5] Part of Section 6.4 was first discussed in the paper entitled "Deflections of a Circular Beam Out of its Initial Plane" by Enrico Volterra (see bibliography number 27 at the end of this chapter) and is presented here with the permission of the American Society of Civil Engineers.

$$M_x^* = -av\sin(\phi - \alpha) \qquad (6.4.3c)$$
$$M_z^* = -av[1 - \cos(\phi - \alpha)] \qquad (6.4.3d)$$

and

$$\bar{\beta} = vA(\mu + 1)\phi \sin\phi$$
$$+ vB[-(\mu + 1)\phi \cos\phi + (\mu - 1)\sin\phi] + 2vC\sin\phi \qquad (6.4.4a)$$
$$\bar{Y} = vA[2 - (\mu + 1)\phi \sin\phi - 2\cos\phi]$$
$$+ vB(\mu + 1)(\phi \cos\phi - \sin\phi) + 2vC(\phi - \sin\phi) \qquad (6.4.4b)$$
$$\bar{M}_x = 2\mu va(A\cos\phi + B\sin\phi) \qquad (6.4.4c)$$
$$\bar{M}_z = 2\mu va(A\sin\phi - B\cos\phi + C) \qquad (6.4.4d)$$

The constants A, B, and C are determined from the following equations:

$$A = [(\mu + 1)\gamma^2 - 2 + 2\cos\gamma]\frac{L_1}{D_2}\sin\gamma + [-(\mu + 1)\gamma^2 \cos\gamma$$
$$+ (\mu - 1)\gamma \sin\gamma + 2\sin^2\gamma]\frac{L_2}{D_2} - (1 - \cos\gamma)\frac{L_3}{D_2} \qquad (6.4.5a)$$

$$B = -[(\mu + 1)\gamma^2 \cos\gamma + (\mu - 1)\gamma \sin\gamma + 2(1 - \cos\gamma)^2]\frac{L_1}{D_2}$$
$$+ [-(\mu + 1)\gamma^2 + 2 - 2\cos\gamma]\frac{L_2}{D_2}\sin\gamma + \frac{L_3}{D_2}\sin\gamma \qquad (6.4.5b)$$

$$C = -\frac{L_1}{D_2}\sin\gamma - (1 - \cos\gamma)\frac{L_2}{D_2} + [(\mu + 1)\gamma - (\mu - 1)\sin\gamma]\frac{L_3}{D_2} \qquad (6.4.5c)$$

where

$$D_2 = [(\mu + 1)\gamma + (\mu - 1)\sin\gamma]\Delta \qquad (6.4.6)$$
$$\Delta = 4(1 - \cos\gamma) - \gamma[(\mu + 1)\gamma - (\mu - 1)\sin\gamma] \qquad (6.4.7)$$
$$L_1 = \frac{\mu + 1}{2\mu}[(\gamma - \alpha)\cos(\gamma - \alpha) - \sin(\gamma - \alpha)] \qquad (6.4.8a)$$
$$L_2 = \frac{1}{2\mu}[(\mu + 1)(\gamma - \alpha)\sin(\gamma - \alpha) + 2\cos(\gamma - \alpha) - 2] \qquad (6.4.8b)$$
$$L_3 = \frac{1}{\mu}[\sin(\gamma - \alpha) - (\gamma - \alpha)] \qquad (6.4.8c)$$

Values of the constants A, B, and C are given in Tables 6.4.1, 6.4.2, 6.4.3, and 6.4.4 for the following values of the parameters: $\mu = 0.564, 0.7, 0.85, 1.00, 1.15,$ and 1.33; $2\gamma = 180°, 150°, 120°$ and $90°$ and for the angle α varying in increments of $5°$ through the interval $(0, \gamma)$.

TABLE 6.4.1(a)

$2\gamma = 180°$

α	μ	0.564	0.7	0.85	1	1.15	1.33
5°	A	−0.075608	−0.060858	−0.050079	−0.042542	−0.036976	−0.031958
	B	0.88321	0.71162	0.58604	0.49813	0.43316	0.37454
	C	0.88544	0.71337	0.58746	0.49932	0.43418	0.37541
10°	A	−0.14715	−0.11834	−0.097316	−0.082628	−0.071790	−0.062024
	B	0.87356	0.70384	0.57963	0.49269	0.42842	0.37044
	C	0.88195	0.71046	0.58500	0.49717	0.43230	0.37376
15°	A	−0.21386	−0.17188	−0.14126	−0.11989	−0.10413	−0.089935
	B	0.85799	0.68130	0.56930	0.48391	0.42079	0.36384
	C	0.87576	0.70534	0.58068	0.49346	0.42902	0.37089
20°	A	−0.27509	−0.22096	−0.18151	−0.15400	−0.13372	−0.11546
	B	0.83701	0.67439	0.55538	0.47208	0.41050	0.35494
	C	0.86663	0.69782	0.57439	0.48805	0.42427	0.36675
25°	A	−0.33029	−0.26517	−0.21776	−0.18470	−0.16034	−0.13842
	B	0.81113	0.65354	0.53821	0.45748	0.39781	0.34397
	C	0.85440	0.68780	0.56603	0.48088	0.41798	0.36128
30°	A	−0.37901	−0.30419	−0.24973	−0.21177	−0.18381	−0.15867
	B	0.78089	0.62917	0.51814	0.44042	0.38298	0.33114
	C	0.83896	0.67521	0.55556	0.47191	0.41014	0.35446
35°	A	−0.42093	−0.33776	−0.27724	−0.23507	−0.20401	−0.17608
	B	0.74685	0.60175	0.49556	0.42122	0.36628	0.31671
	C	0.82026	0.66001	0.54295	0.46114	0.40074	0.34630
40°	A	−0.45584	−0.36572	−0.30016	−0.25448	−0.22085	−0.19060
	B	0.70959	0.57173	0.47083	0.40021	0.34801	0.30091
	C	0.79833	0.64223	0.52825	0.44859	0.38980	0.33682
45°	A	−0.48361	−0.38800	−0.31843	−0.26997	−0.23429	−0.20220
	B	0.66969	0.53958	0.44436	0.37770	0.32844	0.28399
	C	0.77323	0.62195	0.51150	0.43433	0.37738	0.32607

TABLE 6.4.1(b)

$2\gamma = 180°$

α	μ	0.564	0.7	0.85	1	1.15	1.33
50°	A	−0.50426	−0.40459	−0.33207	−0.28155	−0.24435	−0.21089
	B	0.62773	0.50577	0.41651	0.35404	0.30786	0.26619
	C	0.74512	0.59927	0.49281	0.41843	0.36355	0.31410
55°	A	−0.51787	−0.41559	−0.34115	−0.28928	−0.25107	−0.21671
	B	0.58427	0.47076	0.38768	0.32953	0.28655	0.24777
	C	0.71417	0.57436	0.47230	0.40101	0.34840	0.30101
60°	A	−0.52465	−0.42114	−0.34578	−0.29325	−0.25456	−0.21974
	B	0.53989	0.43500	0.35823	0.30450	0.26478	0.22895
	C	0.68064	0.54739	0.45013	0.38219	0.33206	0.28689
65°	A	−0.52488	−0.42148	−0.34616	−0.29364	−0.25494	−0.22011
	B	0.49512	0.39892	0.32853	0.27925	0.24282	0.20996
	C	0.64479	0.51861	0.42650	0.36214	0.31465	0.27186
70°	A	−0.51894	−0.41690	−0.34253	−0.29065	−0.25239	−0
	B	0.45047	0.36295	0.29890	0.25406	0.22092	0.19102
	C	0.60696	0.48826	0.40159	0.34103	0.29633	0.25605
75°	A	−0.50726	−0.40775	−0.33517	−0.28449	−0.24712	−0.21345
	B	0.40642	0.32746	0.26967	0.22922	0.19932	0.17235
	C	0.56750	0.45663	0.37565	0.31905	0.27726	0.23960
80°	A	−0.49037	−0.39444	−0.32440	−0.27546	−0.23935	−0.20680
	B	0.36343	0.29282	0.24114	0.20497	0.17824	0.15411
	C	0.52679	0.42402	0.34892	0.29641	0.25763	0.22266
85°	A	−0.46883	−0.37741	−0.31059	−0.26386	−0.22935	−0.19823
	B	0.32189	0.25936	0.21359	0.18155	0.15787	0.13650
	C	0.48524	0.39075	0.32165	0.27332	0.23761	0.20540
90°	A	−0.44326	−0.35714	−0.29412	−0.25000	−0.21739	−0.18797
	B	0.28219	0.22736	0.18724	0.15915	0.13840	0.11967
	C	0.44326	0.35714	0.29412	0.25000	0.21739	0.18797

TABLE 6.4.2(a)

$2\gamma = 150°$

α	μ	0.564	0.7	0.85	1	1.15	1.33
5°	A	−0.074712	−0.060127	−0.049471	−0.042022	−0.036522	−0.031563
	B	0.88281	0.71130	0.58578	0.49792	0.43298	0.37438
	C	0.88460	0.71269	0.58689	0.49884	0.43376	0.37504
10°	A	−0.14355	−0.11541	−0.094888	−0.080554	−0.069979	−0.060454
	B	0.87195	0.70256	0.57861	0.49183	0.42770	0.36983
	C	0.87860	0.70773	0.58273	0.49525	0.43061	0.37230
15°	A	−0.20576	−0.16530	−0.13582	−0.11525	−0.10008	−0.086430
	B	0.85441	0.68848	0.56704	0.48203	0.41919	0.36249
	C	0.86827	0.69925	0.57564	0.48916	0.42527	0.36764
20°	A	−0.26076	−0.20936	−0.17193	−0.14583	−0.12661	−0.10931
	B	0.83075	0.66947	0.55143	0.46880	0.40772	0.35260
	C	0.85344	0.68712	0.56554	0.48051	0.41769	0.36106
25°	A	−0.30810	−0.24725	−0.20297	−0.17211	−0.14939	−0.12895
	B	0.80153	0.64601	0.53218	0.45248	0.39356	0.34039
	C	0.83407	0.67135	0.55245	0.46932	0.40792	0.35257
30°	A	−0.34748	−0.27876	−0.22878	−0.19396	−0.16832	−0.14527
	B	0.76741	0.61861	0.50970	0.43343	0.37704	0.32614
	C	0.81020	0.65198	0.53641	0.45563	0.39597	0.34222
35°	A	−0.37877	−0.30381	−0.24930	−0.21133	−0.18338	−0.15825
	B	0.72903	0.58781	0.48442	0.41201	0.35846	0.31012
	C	0.78198	0.62915	0.51755	0.43956	0.38199	0.33010

TABLE 6.4.2(b)

$2\gamma = 150°$

α	μ	0.564	0.7	0.85	1	1.15	1.33
40°	A	−0.40196	−0.32240	−0.26453	−0.22424	−0.19457	−0.16790
	B	0.68709	0.55415	0.45680	0.38861	0.33816	0.29261
	C	0.74964	0.60306	0.49603	0.42125	0.36605	0.31631
45°	A	−0.41720	−0.33465	−0.27460	−0.23278	−0.20199	−0.17430
	B	0.64229	0.51820	0.42730	0.36360	0.31648	0.27391
	C	0.71352	0.57395	0.47207	0.40088	0.34834	0.30100
50°	A	−0.42476	−0.34078	−0.27967	−0.23710	−0.20575	−0.17756
	B	0.59534	0.48052	0.39638	0.33740	0.29374	0.25430
	C	0.67399	0.54217	0.44593	0.37869	0.32906	0.28434
55°	A	−0.42502	−0.34110	−0.28001	−0.23742	−0.20606	−0.17785
	B	0.54694	0.44168	0.36450	0.31037	0.27029	0.23406
	C	0.63152	0.50806	0.41791	0.35491	0.30842	0.26651
60°	A	−0.41848	−0.33600	−0.27592	−0.23402	−0.20314	−0.17536
	B	0.49779	0.40222	0.33210	0.28290	0.24645	0.21349
	C	0.58664	0.47205	0.38835	0.32985	0.28666	0.24773
65°	A	−0.40575	−0.32597	−0.26780	−0.22721	−0.19728	−0.17033
	B	0.44855	0.36268	0.29963	0.25536	0.22255	0.19286
	C	0.53990	0.43457	0.35760	0.30379	0.26405	0.22822
70°	A	−0.38750	−0.31153	−0.25608	−0.21735	−0.18877	−0.16304
	B	0.39986	0.32356	0.26749	0.22810	0.19887	0.17242
	C	0.49190	0.39611	0.32606	0.27706	0.24087	0.20822
75°	A	−0.36450	−0.29328	−0.24123	−0.20484	−0.17798	−0.15376
	B	0.35231	0.28535	0.23609	0.20144	0.17572	0.15242
	C	0.44326	0.35714	0.29412	0.25000	0.21739	0.18797

TABLE 6.4.3

$2\gamma = 120°$

α	μ	0.564	0.7	0.85	1	1.15	1.33
5°	A	−0.073269	−0.058951	−0.048496	−0.041188	−0.035793	−0.030931
	B	0.88171	0.71041	0.58505	0.49729	0.43243	0.37391
	C	0.88302	0.71140	0.58583	0.49793	0.43297	0.37436
10°	A	−0.13785	−0.11078	−0.091050	−0.077276	−0.067119	−0.057972
	B	0.86761	0.69907	0.57573	0.48940	0.42558	0.36800
	C	0.87237	0.70269	0.57857	0.49171	0.42752	0.36962
15°	A	−0.19315	−0.15509	−0.12736	−0.10803	−0.093793	−0.080977
	B	0.84488	0.68081	0.56074	0.47668	0.41455	0.35848
	C	0.85454	0.68816	0.56649	0.48138	0.41850	0.36179
20°	A	−0.23882	−0.19161	−0.15726	−0.13334	−0.11572	−0.099874
	B	0.81426	0.65623	0.54056	0.45958	0.39972	0.34569
	C	0.82965	0.66795	0.54975	0.46709	0.40604	0.35098
25°	A	−0.27472	−0.22029	−0.18072	−0.15317	−0.13289	−0.11466
	B	0.77660	0.62601	0.51577	0.43858	0.38150	0.32998
	C	0.79803	0.64235	0.52859	0.44906	0.39032	0.33737
30°	A	−0.30090	−0.24119	−0.19781	−0.16761	−0.14539	−0.12542
	B	0.73282	0.59090	0.48697	0.41418	0.36034	0.31173
	C	0.76014	0.61174	0.50334	0.42757	0.37162	0.32118
35°	A	−0.31765	−0.25456	−0.20873	−0.17684	−0.15337	−0.13229
	B	0.68393	0.55169	0.45482	0.38693	0.33671	0.29135
	C	0.71658	0.57664	0.47443	0.40299	0.35024	0.30270
40°	A	−0.32540	−0.26075	−0.21380	−0.18112	−0.15708	−0.13548
	B	0.63097	0.50922	0.41998	0.35742	0.31111	0.26927
	C	0.66812	0.53764	0.44234	0.37574	0.32656	0.28223
45°	A	−0.32477	−0.26028	−0.21342	−0.18080	−0.15680	−0.13524
	B	0.57500	0.46434	0.38317	0.32622	0.28404	0.24592
	C	0.61558	0.49542	0.40764	0.34628	0.30097	0.26013
50°	A	−0.31654	−0.25375	−0.20810	−0.17631	−0.15292	−0.13190
	B	0.51712	0.41791	0.34507	0.29392	0.25602	0.22174
	C	0.55991	0.45073	0.37094	0.31514	0.27393	0.23678
55°	A	−0.30159	−0.24187	−0.19841	−0.16814	−0.14585	−0.12582
	B	0.45841	0.37079	0.30638	0.26111	0.22754	0.19716
	C	0.50213	0.40436	0.33287	0.28286	0.24591	0.21259
60°	A	−0.28094	−0.22544	−0.18501	−0.15682	−0.13606	−0.11739
	B	0.39992	0.32382	0.26779	0.22837	0.19911	0.17261
	C	0.44326	0.35714	0.29412	0.25000	0.21739	0.18797

TABLE 6.4.4

$2\gamma = 90°$

α	μ	0.564	0.7	0.85	1	1.15	1.33
5°	A	−0.070846	−0.056993	−0.046879	−0.039812	−0.034596	−0.029895
	B	0.87876	0.70802	0.58307	0.49561	0.43097	0.37264
	C	0.87958	0.70863	0.58354	0.49599	0.43128	0.37290
10°	A	−0.12847	−0.10322	−0.084816	−0.071976	−0.062510	−0.053987
	B	0.85626	0.68989	0.56815	0.48293	0.41994	0.36311
	C	0.85912	0.69201	0.56977	0.48423	0.42103	0.36401
15°	A	−0.17287	−0.13874	−0.11391	−0.096605	−0.083859	−0.072393
	B	0.82041	0.66103	0.54441	0.46277	0.40243	0.34798
	C	0.82596	0.66516	0.54757	0.46531	0.40454	0.34973
20°	A	−0.20435	−0.16387	−0.13445	−0.11397	−0.098895	−0.085343
	B	0.77281	0.62277	0.51295	0.43607	0.37924	0.32795
	C	0.78125	0.62904	0.51777	0.43994	0.38245	0.33061
25°	A	−0.22354	−0.17915	−0.14692	−0.12449	−0.10799	−0.093166
	B	0.71531	0.57656	0.47499	0.40386	0.35126	0.30380
	C	0.72646	0.58485	0.48136	0.40898	0.35552	0.30732
30°	A	−0.23137	−0.18535	−0.15195	−0.12872	−0.11163	−0.096288
	B	0.64988	0.52402	0.43182	0.36724	0.31947	0.27634
	C	0.66327	0.53398	0.43949	0.37340	0.32459	0.28058
35°	A	−0.22899	−0.18341	−0.15033	−0.12732	−0.11041	−0.095219
	B	0.57865	0.46681	0.38483	0.32737	0.28485	0.24645
	C	0.59361	0.47796	0.39341	0.33428	0.29060	0.25121
40°	A	−0.21780	−0.17444	−0.14296	−0.12108	−0.10499	−0.090540
	B	0.50379	0.40668	0.33543	0.28545	0.24845	0.21501
	C	0.51955	0.41844	0.34448	0.29275	0.25452	0.22004
45°	A	−0.19934	−0.15968	−0.13088	−0.11085	−0.096124	−0.082894
	B	0.42753	0.34540	0.28506	0.24270	0.21131	0.18294
	C	0.44326	0.35714	0.29412	0.25000	0.21739	0.18797

Example 6.4.1. A circular bow girder which is made of reinforced concrete has a radius $R = 20$ ft, a modulus of elasticity $E = 3 \times 10^6$ psi, a square cross section (24 × 24 in.), and is subjected to a concentrated load $P = 10$ tons at the center. The angle between the built-in ends is $2\gamma = 90°$ so that P is at $\alpha = 45°$.

From this information, the following data are computed: $I_x = 27,648$ in.^{4}, $\mu = 0.7$, $v = 0.013889$, $a = 14,400$ ft-tons, and $2\mu v a = 280$ ft-tons. From Table 6.4.4 the values $A = -0.16$, $B = 0.345$, and $C = 0.357$ are obtained. The results for this problem, obtained by applying Eqs. 6.4.4, are given in Table 6.4.5.

TABLE 6.4.5

RESULTS FOR EXAMPLE 6.4.1

ϕ, degrees	β, radians	v, inches	M_x, foot-tons	M_z, foot-tons
0	0	0	−44.7104	3.2872
5	−0.00002012	0.002544	−36.1116	−0.2408
10	−0.00004321	0.009888	−27.2356	−3.0072
15	−0.0001223	0.020520	−18.1552	−4.9840
20	−0.0002243	0.033468	− 8.9376	−6.1712
25	−0.0003358	0.046584	0.3500	−6.5464
30	−0.0004443	0.059328	9.6348	−6.1124
35	−0.0005371	0.070032	18.8468	−4.8664
40	−0.0006015	0.077376	27.9160	−2.8252
45	−0.0006257	0.080352	36.7696	0.0000

BIBLIOGRAPHY

1. Biezeno, C. B., "Über die quasi-statische Berechunung geschlossener Kreisförmiger Ringe Konstanten Querschnittes," *Z. Angew. Math. Mech.*, **Vol. 7**, 1928, p. 237.

2. Biezeno, C. B., and R. Grammel, "Engineering Dynamics," **Vol. II**, Blackie & Son, Ltd., Glasgow, Scotland, 1956.

3. Dusterbehn, F., "Ringsförmige Träger," *Eisenbau*, 1920, pp. 73–80.

4. Gibson, A. H., and E. G. Ritchie, "A Study of the Circular Arc Bow Girder," Constable & Co., Ltd., London, England, 1914.

5. Hogan, B., "The Solution of Closed Rings with Equally Spaced Supports," *Bull. Univ. Utah*, **Vol. 34**, 1943.

6. ——, "The Solution of Closed Rings with Equally Spaced Supports and Moments-Loadings," *Ibid.*, **Vol. 34**, 1944.

7. ——, "Moment Distribution Applied to Combined Flexure and Torsion," *Ibid.*, **Vol. 35**, 1945.

8. ——, "The Deflection of Circular Arcs Subjected to Combined Flexure and Torsion," *Ibid.*, **Vol. 35**, 1947.

9. ——, "The Derivation of Two-Five Theorems for Continuous Plane Curved Beams," *Ibid.*, **Vol. 36**, 1947.

10. Kannenberg, B. G., "Zur Theorie Torsionsfester Ringe," *Eisenbau*, 1913, pp. 329–334.

11. Love, A. E. H., "A Treatise on the Mathematical Theory of Elasticity," 4th ed., Dover Publications, Inc., New York, N.Y., 1944, Chapter XXI.

12. Marcus, H., "Abriss einer allgemeinen Theorie des eigenspannten Trägers mit raumlich gefundener Mittelinie," *Z. Bauwesen*, 1914, pp. 198–223.

13. ——, "Die Elastische Linie des Doppelt gegrümmten Trägers," *Ibid.*, 1919, pp. 163–180.

14. Mayer, R., "Ueber Elastizität und Stabilität des geschlossenen und offenen Kreisbogen," *Z. Math. Physik*, 1913, pp. 246–320.

15. Moorman, R. B., "Stresses in Curved Beams under Loads Normal to the Plane of Its Axis," *Bull. Iowa State College*, No. 145, 1940.

16. Moorman, R. B., and M. B. Tate, "Influence Lines for Horizontally Curved Fixed-End Beams of Circular Arch Plan," *Miss. Univ. Bull.*, Vol. 48, No. 26 (Eng. Series No. 35), pp. 1–39.

17. Pippard, A. J. S., "An Application of the Principle of Superposition to Certain Structural Problems," *J. Inst. Civil Engrs.*, London, 1938, pp. 1–27.

18. ———, "Studies in Elastic Structures," Edward Arnold & Co., London, England, 1952.

19. Pippard, A. J. S., and F. L. Barrow, "The Stress Analysis of Bow Girders," Building Res. Tech. Paper No. 1, Department of Scientific and Industrial Research, His Majesty's Stationery Office, London, England, 1926, pp. 1–27.

20. ———, "Analysis of Engineering Structures," 2nd ed., Edward Arnold & Co., London, England, 1945.

21. Résal, H., "De la Déformation qu'éprouve une pièce à simple ou à double courbure sous l'action de forces qui lui font subir en meme temps une flexion et une torsion," *J. Math. Pures Appl.*, Vol. 3, 1877, pp. 307–322.

22. Saint-Venant, Barré de, "Mémoire sur le calcul de la résistance et de la flexion des pièces solides à simple ou à double courbure, en prenant simultanément en considération les divers efforts auxquels elles peuvent etre soumises dans tous les sens," *Compt. Rend.*, Vol. 17, 1843, pp. 942, 1020–1031.

23. Schleicher, F., "Die elastische Verschiebungen gekrümmter Stabe als Drehungen beziehungweise Schraubungen um Achsen von Momentenflächen," *Z. Angew. Math. Mech.*, Vol. 4, 1924 p. 475.

24. Timoshenko, S. P., "Strength of Materials—Advanced Theory and Problems," Pt. II, The Macmillan Company, New York, N.Y., 1936, pp. 467–473.

25. Timoshenko, S. P., and J. M. Gere, "Theory of Elastic Stability," 2nd ed., McGraw-Hill Book Company, New York, N.Y., 1961, Chapter VII.

26. Velutini, B., "Analysis of Continuous Circular Curved Beams," *J. Am. Concrete Inst.*, Vol. 22, 1950, pp. 217–228.

27. Volterra, E., "Deflections of a Circular Beam out of Its Initial Plane," Paper No. 2727, *Trans. Am. Soc. Civil Engrs.*, Vol. 120, 1955, pp. 65–91.

PROBLEMS

6-1. A reinforced concrete circular beam has a radius $R = 15$ ft, modulus of elasticity $E = 3 \times 10^6$ psi, a square cross section (20×20 in.), rests on twelve rigid supports, and carries a load $p_0 = 0.30$ tons per ft. Compute the values of the bending and twisting moments and the deflection and angle of twist for the beam.

6-2. Solve Problem 1 for the case where the beam rests on eight rigid supports.

6-3. Solve Problem 1 for the case where the beam rests on six rigid supports.

6-4. Solve Problem 1 for the case where the beam rests on four rigid supports.

6-5. A circular bow girder which is made of reinforced concrete has a radius $R = 12$ ft, modulus of elasticity $E = 3 \times 10^6$ psi, a square cross section (15 × 15 in.), and is subjected to a uniformly distributed load of 0.40 tons per ft. The angular distance between the built-in ends is $2\gamma = 180°$. Compute the values of M_x, M_z, β, and v for the beam.

6-6. Solve Problem 5 for the case in which $2\gamma = 150°$.

6-7. Solve Problem 5 for the case in which $2\gamma = 120°$.

6-8. Solve Problem 5 for the case in which $2\gamma = 90°$.

6-9. Solve Problem 5 for the case in which $2\gamma = 60°$.

6-10. Solve Problem 5 for the case in which $2\gamma = 45°$.

6-11. Solve Problem 5 for the case in which $2\gamma = 30°$.

6-12. A reinforced concrete circular bow girder has a radius $R = 20$ ft, modulus of elasticity $E = 3 \times 10^6$ psi, a square cross section (20 × 20 in.), and is subjected to a uniformly distributed load of 0.30 tons per ft. Two concentrated loads of 2 tons each are applied at angular distances of 60° from the ends. The angle between the built-in ends is $2\gamma = 180°$. Compute the values of M_x, M_z, β, and v for the beam.

CHAPTER

7

BEAMS ON ELASTIC FOUNDATIONS

7.0 Introduction

In this chapter the problem of beams resting on elastic foundations will be discussed. Foundations are usually assumed to follow the hypothesis proposed in 1867 by E. Winkler.[1] According to this hypothesis, the deflection at every point of the foundation is proportional to the pressure applied at that point and independent of pressures acting at nearby points of the foundation. This assumption is equivalent to considering the foundation to be composed of independent elastic springs. The foundation is characterized by the foundation modulus which is generally denoted by the letter k. The foundation modulus represents the reaction expressed in lb/sq in. when the deflection is 1 in.; therefore, k has units of lb/cu in. For sand, k may vary from 25 to 100

---

[1] According to some recently published work (see bibliography number 14 at the end of this chapter) Leonard Euler seems to have been the first to formulate the hypothesis, although it is generally attributed to Winkler.

lb/cu in. For gravel, the value of k is even more uncertain and it can vary from 200 to 1200 lb/cu in.

The hypothesis that a foundation has such an elastic behavior may seem strange at first, but experiments done on railway tracks show that the foundation behaves elastically. If one observes a train moving on a railroad track, one notes that everytime a wheel passes over a crosstie, the crosstie moves down and afterwards comes up again. Another indication of the elasticity of soils is that in general soils are good conductors of sound and of seismic waves. However, Winkler's hypothesis that the effect produced by a concentrated force on the foundation applies only at the point of application is not exact since near-by points of the foundation are also affected (see Fig. 7.0.1). However, this influence reduces very rapidly, usually by an exponential law, so that the effects of the deflections of near-by points of the foundation reduce very rapidly.

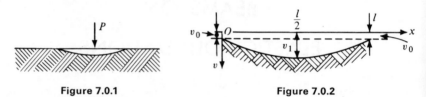

Figure 7.0.1   **Figure 7.0.2**

To take into account the effect of nearby points of the foundation, C. Jodi proposed in 1936 the influence function

$$w(x - \xi) = Ae^{-B|x-\xi|} \tag{7.0.1}$$

where ξ is the point under the concentrated load, while x is a generic point of the foundation, and A and B are two constants of the foundation which can be easily determined experimentally. In order to determine these constants, assume that a foundation of unit width is subjected to a uniformly distributed load of intensity p_0 acting over a length l. The deflection v at the generic point under load will be

$$v = Ap_0 \int_0^x e^{-B|x-\xi|} \, d\xi + Ap_0 \int_x^l e^{-B|x-\xi|} \, d\xi$$

or

$$v = \frac{Ap_0}{B}[2 - e^{-Bx} - e^{-B(l-x)}] \tag{7.0.2}$$

By measuring v at the two ends of the foundation, $x = 0$ and $x = l$, and at the midpoint $x = l/2$ (see Fig. 7.0.2) and letting

$$v(0) = v(l) = v_0$$

and

$$v\left(\frac{l}{2}\right) = v_1$$

one obtains

$$\left.\begin{array}{l} B = -\dfrac{2}{l} \log\left(\dfrac{2v_0}{v_1} - 1\right) \\[6pt] A = \dfrac{B}{4p_0} \dfrac{v_1^2}{v_1 - v_0} \end{array}\right\} \qquad (7.0.3)$$

The use of Jodi's hypothesis makes computations much more complicated, however, and the approximation obtained is not worth the greater effort (see Sec. 7.4).

After deriving the equation of the elastic line for the straight bar, according to the Winkler hypothesis, deflections of the beam of infinite length and of the beam of finite length resting on a Winkler soil will be discussed. A general discussion of the deflection of a beam resting on the most general type of elastic soil follows. Next, deflections of circular beams resting on elastic foundations and loaded by symmetrical, concentrated forces acting perpendicular to the plane of original curvature of the beam will be analyzed. This is a problem which is important in practice, its most common application being the case of circular, reinforced concrete beams used as foundations of water tanks (see Fig. 7.0.3). It will first be assumed that the beam is free to rotate at the sections where the loads are applied. However, very frequently the vertical concentrated loads are transmitted to the circular

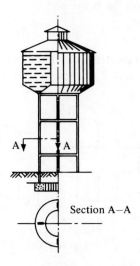

Figure 7.0.3

beam through structural members which are rigidly connected to the beam. These structural members will restrain the beam from rotating at the points of application of the concentrated loads. Deflections of circular beams under such conditions are analyzed. The chapter ends with a discussion of the use of the method of harmonic analysis in determining deflections of circular beams resting on elastic foundations. This method allows one to consider not only the case of symmetrical loading but also the case of antisymmetrical loading. Antisymmetrical loading is induced, in the case of circular, reinforced concrete beams used as foundations of water tanks, by the wind action on the water tank.

7.1 The Equation of the Elastic Line for the Straight Bar According to the Winkler Hypothesis

Let b be the breadth of the beam resting on a Winkler foundation. The reaction of the foundation for a 1-in. length of the beam, if the deflection of the beam is 1 in., will be given by

$$\beta = kb$$

which has units of lb/sq in. Therefore, if the deflection of the beam is v, the reaction of the elastic soil on which the beam rests will be

$$q = \beta v$$

Consider now a beam resting on a Winkler soil and subjected to a distributed load of intensity p directed downward (see Fig. 7.1.1). The total distributed load acting on the beam will be

$$p - q = p - \beta v \tag{7.1.1}$$

If the beam has constant inertia characteristics (EI = constant), the differential equation for the deflection of the beam[2] will be

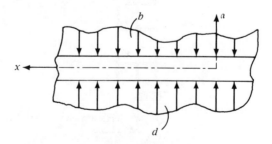

Figure 7.1.1

---

[2]In this chapter the deflection and loading are assumed to be positive downward.

$$EI\frac{d^4v}{dx^4} = p - \beta v$$

or

$$EI\frac{d^4v}{dx^4} + \beta v = p \tag{7.1.2}$$

If $p = 0$, the above equation becomes

$$EI\frac{d^4v}{dx^4} + \beta v = 0$$

or

$$\frac{d^4v}{dx^4} + 4\alpha^4 v = 0 \tag{7.1.3}$$

where

$$\alpha^4 = \frac{\beta}{4EI}$$

The general integral of Eq. (7.1.3) is

$$v = C_1 e^{\alpha x} \sin \alpha x + C_2 e^{\alpha x} \cos \alpha x + C_3 e^{-\alpha x} \sin \alpha x + C_4 e^{-\alpha x} \cos \alpha x \tag{7.1.4}$$

where C_1, C_2, C_3, and C_4 are constants of integration which are determined from the boundary conditions of the beam.[3]

For the case in which

$$p = p_0 = \text{constant}$$

Eq. (7.1.3) can be written as

$$\frac{d^4v}{dx^4} + 4\alpha^4 v = \frac{p_0}{EI} \tag{7.1.5}$$

or

$$\frac{d^4v}{dx^4} + 4\alpha^4 \left[v - \frac{p_0}{\beta} \right] = 0 \tag{7.1.6}$$

The integral of Eq. (7.1.6) is

$$v = \frac{p_0}{\beta} + e^{\alpha x}[C_1 \sin \alpha x + C_2 \cos \alpha x] + e^{-\alpha x}[C_3 \sin \alpha x + C_4 \cos \alpha x] \tag{7.1.7}$$

---

[3] It can be proved that Eq. (7.1.4) is the general solution of the differential equation [Eq. (7.1.3)] by differentiating Eq. (7.1.4) four times and substituting into Eq. (7.1.3).

Once the deflection is determined [i.e., constants C_1, C_2, C_3, and C_4 of Eq. (7.1.4) or Eq. (7.1.7) are known], the slope θ, the bending moment M, the shear force V, and the soil reaction q can be determined by use of the following equations[4]

$$\theta = \frac{dv}{dx}$$

$$M = -EI\frac{d^2v}{dx^2}$$

$$V = -EI\frac{d^3v}{dx^3}$$

$$q = \beta v$$

The successive derivatives of Eq. (7.1.4) are

$$\frac{dv}{dx} = \alpha e^{\alpha x}[(C_1 - C_2)\sin \alpha x + (C_1 + C_2)\cos \alpha x]$$
$$+ \alpha e^{-\alpha x}[-(C_3 + C_4)\sin \alpha x + (C_3 - C_4)\cos \alpha x]$$

$$\frac{d^2v}{dx^2} = 2\alpha^2 e^{\alpha x}[-C_2 \sin \alpha x + C_1 \cos \alpha x] + 2\alpha^2 e^{-\alpha x}[C_4 \sin \alpha x - C_3 \cos \alpha x]$$

$$\frac{d^3v}{dx^3} = 2\alpha^3 e^{\alpha x}[-(C_1 + C_2)\sin \alpha x + (C_1 - C_2)\cos \alpha x]$$
$$+ 2\alpha^3 e^{-\alpha x}[(C_3 - C_4)\sin \alpha x + (C_3 + C_4)\cos \alpha x]$$

7.2 The Beam of Infinite Length

Consider the beam of infinite length resting on a Winkler foundation and subjected to a concentrated load P at the section $x = 0$ as shown in Fig. 7.2.1. Since for $x = \infty$, $v = M = 0$, it follows from Eq. (7.1.4) that $C_1 =$

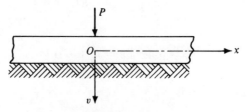

Figure 7.2.1

---

[4] The minus sign appears in the moment and shear equations because the deflection is assumed to be positive downward in this chapter.

$C_2 = 0$. Under the load P, i.e., for $x = 0$, $dv/dx = 0$ by symmetry, from which it follows that $C_3 = C_4 = C$ and Eq. (7.1.4) reduces to

$$v = Ce^{-\alpha x}[\sin \alpha x + \cos \alpha x] \qquad (7.2.1)$$

The constant C is determined from the condition that

$$V = -\frac{P}{2} = -EI\frac{d^3v}{dx^3} \text{ for } x = 0 \qquad (7.2.2)$$

By use of Eq. (7.2.1) and Eq. (7.2.2) one finds that

$$C = \frac{P}{8\alpha^3 EI} = \frac{P\alpha}{2\beta}$$

and the deflection expression becomes

$$v = \frac{P\alpha}{2\beta}e^{-\alpha x}[\sin \alpha x + \cos \alpha x] = \frac{P\alpha}{\sqrt{2}\beta}e^{-\alpha x}\sin\left(\alpha x + \frac{\pi}{4}\right) \qquad (7.2.3)$$

Equation (7.2.3) shows that the deflection curve is a *damped* sinusoid due to the presence of term $e^{-\alpha x}$. The wave length Λ is defined by the formula

$$\Lambda = \frac{2\pi}{\alpha} = 2\pi\sqrt[4]{\frac{4EI}{\beta}} \qquad (7.2.4)$$

The bending moment is given by

$$M = \frac{P}{4\alpha}e^{-\alpha x}[\cos \alpha x - \sin \alpha x] = \frac{P}{2\sqrt{2}\alpha}e^{-\alpha x}\cos\left(\alpha x + \frac{\pi}{4}\right) \qquad (7.2.5)$$

From Eqs. (7.2.3) and (7.2.5), one obtains the following values under the load (i.e., at $x = 0$):

$$v_{max} = \frac{P\alpha}{2\beta}$$

$$M_{max} = \frac{P}{4\alpha}$$

$$q_{max} = \frac{P\alpha}{2}$$

The above very simple results are rigorous for a beam of infinite length resting on a Winkler soil. However, they can also be applied in the case of a beam of finite length, if the length of the beam is not small as compared with the wave length Λ. Moreover, by using the principle of superposition, the above results can also be utilized when one has to deal with combinations of concentrated loads acting on the beam.

The functions which appear in Eqs. (7.2.3) and (7.2.5) are denoted, respectively, by $F_1(\alpha x)$ and $F_2(\alpha x)$, i.e.,

$$F_1(\alpha x) = e^{-\alpha x}[\sin \alpha x + \cos \alpha x] = \sqrt{2}\, e^{-\alpha x} \sin\left(\alpha x + \frac{\pi}{4}\right)$$

$$F_2(\alpha x) = e^{-\alpha x}[\cos \alpha x - \sin \alpha x] = \sqrt{2}\, e^{-\alpha x} \cos\left(\alpha x + \frac{\pi}{4}\right)$$

Values of these functions are given in Table 7.2.1 for several values of αx varying from 0 to 10.

Example 7.2.1. A steel beam has a square cross section with 2-in. sides, is 60 ft long, is supported by an elastic Winkler foundation of modulus $k = 400$ lb/cu in.[3], and is subjected at its center to three equal loads of 2 tons each (as shown in Fig. 7.2.2). Compute the deflection and the bending moment of the beam under the central load. Since

$$\beta = kb = 400(2) = 800 \text{ lb/in.}^2$$
$$I = \tfrac{16}{12} = 1.33 \text{ in.}^4$$
$$E = 30(10^6) \text{ psi}$$

it follows that

$$\alpha = \sqrt[4]{\frac{\beta}{4EI}} = \sqrt[4]{\frac{800}{4(30)10^6(1.33)}} = 0.048 \text{ in.}^{-1}$$

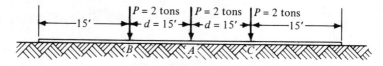

Figure 7.2.2

The wave length $\Lambda = 2\pi/\alpha = 2\pi/0.048 = 131$ in. is much smaller than the length of the beam and the results for the bar of infinite length can be applied to this case. The deflection at A is given by

$$v = \frac{P\alpha}{2\beta}[1 + 2e^{-\alpha d}(\sin \alpha d + \cos \alpha d)]$$
$$= \frac{(2)(2000)(0.048)}{(2)(800)}[1 + 2e^{-1.728}(\sin 1.728 + \cos 1.728)]$$
$$= 0.1496 \text{ in.}$$

TABLE 7.2.1

αx	$F_1(\alpha x)$	$F_2(\alpha x)$	αx	$F_1(\alpha x)$	$F_2(\alpha x)$	αx	$F_1(\alpha x)$	$F_2(\alpha x)$
0.0	1.000000	1.000000	1.2	0.389865	−0.171585	5.0	−0.004550	0.008372
0.1	0.990650	0.809984	1.4	0.284922	−0.201096	5.5	0.000013	0.005780
0.2	0.965067	0.639754	$\frac{\pi}{2}$	0.207880	−0.207880	6.0	0.001687	0.003073
0.3	0.926657	0.488804	1.8	0.123420	−0.198532	2π	0.001867	0.001867
0.4	0.87844	0.356371	2.0	0.066741	−0.179379	6.5	0.001792	0.001145
0.5	0.823067	0.241494	2.5	−0.016636	−0.114887	7.0	0.001287	0.000088
0.6	0.762836	0.143071	3.0	−0.042263	−0.056315	7.5	0.000711	−0.000327
0.7	0.699718	0.059900	π	−0.043214	−0.043214	8.0	0.000283	−0.000381
$\frac{\pi}{4}$	0.644794	0.000000	3.5	−0.038821	−0.017686	8.5	0.000040	−0.000285
0.9	0.571205	−0.065749	4.0	−0.025833	0.001889	9.0	−0.000062	−0.000163
1.0	0.508326	−0.110794	4.5	−0.013201	0.008518	10.0	−0.000063	−0.000013

The bending moment at A is given by

$$M = \frac{P}{4\alpha}[1 + 2e^{-\alpha d}(\cos \alpha d - \sin \alpha d)]$$
$$= \frac{(2)(2000)}{(4)(0.048)}[1 + 2e^{-1.728}(\cos 1.728 - \sin 1.728)$$
$$= 12{,}500 \text{ in.-lb}$$

Example 7.2.2. Consider a beam of infinite length loaded with a uniformly distributed load of intensity p_0 over a central region BC of length c as shown in Fig. 7.2.3. The deflection and the bending moment are to be computed at point A which is at distances a and b from the extremities B and C.

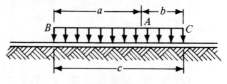

Figure 7.2.3

An elementary load $p_0\, dx$ at a distance x from point A produces at A a deflection

$$dv = \frac{p_0\, dx\, \alpha}{2\beta} e^{-\alpha x}[\sin \alpha x + \cos \alpha x]$$

and a bending moment

$$dM = \frac{p_0\, dx}{4} e^{-\alpha x}[\cos \alpha x - \sin \alpha x]$$

Therefore, the total deflection of point A will be

$$v = \int_0^a \frac{p_0 \alpha}{2\beta} e^{-\alpha x}[\sin \alpha x + \cos \alpha x]\, dx + \int_0^b \frac{p_0 \alpha}{2\beta} e^{-\alpha x}[\sin \alpha x + \cos \alpha x]\, dx$$

or

$$v = \frac{p_0}{2\beta}[2 - e^{-\alpha a}\cos \alpha a - e^{-\alpha b}\cos \alpha b]$$

while the total bending moment at point A will be

$$M = \int_0^a \frac{p_0}{4\alpha} e^{-\alpha x}[\cos \alpha x - \sin \alpha x]\, dx + \int_0^b \frac{p_0}{4\alpha} e^{-\alpha x}[\cos \alpha x - \sin \alpha x]\, dx$$

or

$$M = \frac{p_0}{4\alpha^2}[e^{-\alpha a}\sin \alpha a + e^{-\alpha b}\sin \alpha b]$$

If A is at the midpoint of BC, then $a = b = c/2$ and one finds that

$$v_0 = \frac{p_0}{\beta}\left[1 - e^{-\alpha c/2} \cos \frac{\alpha c}{2}\right]$$

and

$$M_0 = \frac{p_0}{2\alpha^2} e^{-\alpha c/2} \sin \frac{\alpha c}{2}$$

If c is large then $e^{-\alpha c/2}$ is a small quantity and the above formulas reduce to

$$v_0 = \frac{p_0}{\beta} \quad \text{and} \quad M_0 = \frac{p_0}{2\alpha^2}$$

Example 7.2.3. The long beam shown in Fig. 7.2.4 is subjected at the extremity $x = 0$ to a vertical force P and to a couple M. Determine the deflection and the rotation of the beam at $x = 0$. Since, for $x = \infty$, $v = M = 0$, then $C_1 = C_2 = 0$, and Eq. (7.1.4) becomes in this case

$$v = e^{-\alpha x}[C_3 \sin \alpha x + C_4 \cos \alpha x] \tag{7.2.6}$$

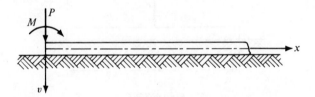

Figure 7.2.4

The two constants C_3 and C_4 of Eq. (7.2.6) are determined from the boundary conditions at $x = 0$:

$$EI\frac{d^2v}{dx^2} = -M$$

$$EI\frac{d^3v}{dx^3} = -V = P$$

It follows that

$$C_3 = \frac{M}{2\alpha^2 EI}$$

$$C_4 = \frac{P - \alpha M}{2\alpha^3 EI}$$

and the deflection curve becomes

$$v = \frac{e^{-\alpha x}}{2\alpha^3 EI}[\alpha M \sin \alpha x + (P - \alpha M) \cos \alpha x]$$

while the slope equation is

$$\theta = \frac{dv}{dx} = -\frac{e^{-\alpha x}}{2\alpha^2 EI}[P \sin \alpha x + (P - 2\alpha M) \cos \alpha x]$$

At the extremity $x = 0$ of the beam, one obtains

$$v = \frac{P - \alpha M}{2\alpha^3 EI}$$

$$\theta = -\frac{P - 2\alpha M}{2\alpha^2 EI}$$

7.3 The Beam of Finite Length

Consider the beam of length l resting on an elastic foundation and subjected to a concentrated load P acting at its center as shown in Fig. 7.3.1. The foun-

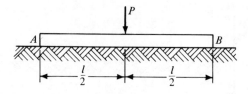

Figure 7.3.1

dation is assumed to react in both senses (i.e., it reacts up if the beam deflects down, or it reacts down if the beam deflects up, as in the case when the beam is deeply immersed in the foundation). The deflection line of the beam is given by Eq. (7.1.4), i.e.,

$$v = C_1 e^{\alpha x} \sin \alpha x + C_2 e^{\alpha x} \cos \alpha x + C_3 e^{-\alpha x} \sin \alpha x + C_4 e^{-\alpha x} \cos \alpha x \quad (7.1.4)$$

By assuming the origin O of the x-y coordinate system at the center of the beam, the following boundary conditions must be satisfied:

$$\left. \begin{array}{l} \text{at } x = 0: \quad \dfrac{dv}{dx} = 0, \quad -EI\dfrac{d^3v}{dx^3} = V = -\dfrac{P}{2} \\[6pt] \text{at } x = \dfrac{l}{2}: \quad M = -EI\dfrac{d^2v}{dx^2} = 0, \quad V = -EI\dfrac{d^3v}{dx^3} = 0 \end{array} \right\} \quad (7.3.1)$$

From Eqs. (7.1.4) and (7.3.1) one obtains

$$C_1 = \frac{P\alpha}{4\beta} K_1$$
$$C_2 = \frac{P\alpha}{4\beta} K_2$$
$$C_3 = \frac{P\alpha}{4\beta} K_3 \quad (7.3.2)$$
$$C_4 = \frac{P\alpha}{4\beta} K_4$$

where

$$K_1 = \frac{\sin \alpha l + \cos \alpha l - e^{-\alpha l}}{\sin \alpha l + \sinh \alpha l}$$
$$K_2 = \frac{2 - \sin \alpha l + \cos \alpha l + e^{-\alpha l}}{\sin \alpha l + \sinh \alpha l} \quad (7.3.3)$$
$$K_3 = 2 - K_1$$
$$K_4 = 2 + K_2$$

Once constants C_i and K_i ($i = 1, 2, 3, 4$) are known, one is able to compute the values of deflection v, slope θ, bending moment M, shear force V, and reaction q of the foundation at each point of the beam. In particular, under the load P (at $x = 0$) one has

$$v(0) = v_{\max} = \frac{P\alpha}{2\beta}(1 + K_2)$$
$$M(0) = M_{\max} = \frac{P}{4\alpha}(1 - K_1) \quad (7.3.4)$$

At the extremities of the beam ($x = \pm l/2$), one obtains

$$v_A = v_B = \frac{P\alpha}{2\beta} K$$

where

$$K = \frac{4 \cos \frac{\alpha l}{2} \cosh \frac{\alpha l}{2}}{\sin \alpha l + \sinh \alpha l} \quad (7.3.5)$$

In Table 7.3.1, values of the coefficients K_1, K_2, K_3, K_4, and K are given for different values of αl. If the beam is long, the results given in this section are very close to those given in Sec. 7.2. for the beam of infinite length; in fact, the expressions for $v_{\max}$ and $M_{\max}$ differ, respectively, by factors $(1 + K_2)$

TABLE 7.3.1

αl	K_1	K_2	K_3	K_4	K
0.00	1.000000	∞	1.000000	∞	∞
0.01	0.995000	199.000000	1.005000	201.000000	200.000000
0.02	0.990000	99.000000	1.010000	101.000000	100.000000
0.03	0.985000	65.666667	1.015000	67.666667	66.666666
0.04	0.980000	49.000002	1.020000	51.000002	49.999998
0.05	0.975000	39.000003	1.025000	41.000003	39.999995
0.10	0.950000	19.000025	1.050000	21.000025	19.999963
0.20	0.900001	9.000200	1.099999	11.000200	9.999700
0.30	0.850007	5.667342	1.149993	7.667342	6.665654
0.40	0.800028	4.001600	1.199972	6.001600	4.997601
0.50	0.750087	3.003124	1.249931	5.003124	3.995315
1.00	0.502756	1.024812	1.497244	3.024812	1.962808
$\frac{\pi}{2}$	0.239942	0.365880	1.760058	2.365880	1.134875
2.00	−0.078881	0.178541	1.921119	2.178541	0.735186
3.00	−0.088459	0.090430	2.088459	2.090430	0.065519
π	−0.090331	0.090331	2.090331	2.090331	−0.000000
4.00	−0.053848	0.079956	2.053848	2.079956	−0.236026
5.00	−0.009311	0.044363	2.009311	2.044363	−0.268299
6.00	0.003367	0.016095	1.996633	2.016095	−0.197919
2π	0.003728	0.011212	1.996272	2.011212	−0.173179
7.00	0.002568	0.003821	1.997432	2.003821	−0.113082

and $(1 - K_1)$. For $\alpha l = \pi$ (i.e., for $l = \Lambda/2$) $-K_1 = K_2 = 0.09031$ and it follows that the formulas

$$v_{\max} = \frac{P\alpha}{2\beta}$$

$$M_{\max} = \frac{P}{4\alpha}$$

found in the case of a beam of infinite length give values which are too small by 8.25 per cent. For $\alpha l = 2\pi$ (i.e., for $l = \Lambda$) $1 + K_2 = 1.011$ and $1 - K_1 = 0.996$ and it follows that for the beam of infinite length $v_{\max}$ is too small by 1.1 per cent and $M_{\max}$ is too large by 0.4 per cent. If the beam is sufficiently long, the central part of the beam goes down into the foundation while the extremities of the beam go up. The length of the zone of the beam which goes down varies with the length l of the beam from a minimum of π/α for a beam of length $l = \pi/\alpha$ to a maximum $3\pi/2\alpha$ for a beam of infinite length.

Example 7.3.1. Consider a beam of length l subjected to two equal loads P at its extremities as shown in Fig. 7.3.2. By taking the origin O in the middle of the beam and using Eq. (7.1.4), i.e.,

$$v = C_1 e^{\alpha x} \sin \alpha x + C_2 e^{\alpha x} \cos \alpha x + C_3 e^{-\alpha x} \sin \alpha x + C_4 e^{-\alpha x} \cos \alpha x \quad (7.1.4)$$

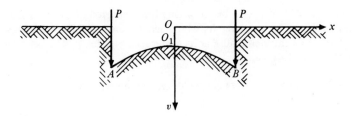

Figure 7.3.2

then the following boundary conditions must be satisfied:

$$\left.\begin{array}{l} \text{at } x = 0: \quad \dfrac{dv}{dx} = 0, \quad V = -EI\dfrac{d^3v}{dx^3} = 0 \\[6pt] \text{at } x = \pm\dfrac{l}{2}: \quad M = -EI\dfrac{d^2v}{dx^2} = 0, \quad V = -EI\dfrac{d^3v}{dx^3} = P \end{array}\right\} \quad (7.3.6)$$

From Eqs. (7.1.4) and (7.3.6) one obtains

$$\left.\begin{array}{l} C_1 = -C_3 = \dfrac{2P\alpha}{\beta} \dfrac{\sin\frac{\alpha l}{2} \sinh\frac{\alpha l}{2}}{\sin \alpha l + \sinh \alpha l} \\[10pt] C_4 = C_1 \cotg\dfrac{\alpha l}{2} \cotgh\dfrac{\alpha l}{2} = \dfrac{2P\alpha}{\beta} \dfrac{\cos\frac{\alpha l}{2} \cosh\frac{\alpha l}{2}}{\sin \alpha l + \sinh \alpha l} \end{array}\right\} \quad (7.3.7)$$

Once the constants $C_i (i = 1, 2, 3, 4)$ are known, one is able to compute values of deflection v, slope θ, bending moment M, shear force V, and foundation reaction q at each point of the beam by the use of Eq. (7.1.4). In particular, at the midpoint of the beam one has

$$v(0) = 2C_2 = \frac{4P\alpha}{\beta} \frac{\cos\frac{\alpha l}{2} \cosh\frac{\alpha l}{2}}{\sin \alpha l + \sinh \alpha l}$$

$$M(0) = -4\alpha^2 EI C_1 = -\frac{2P}{\alpha} \frac{\sin\frac{\alpha l}{2} \sinh\frac{\alpha l}{2}}{\sin \alpha l + \sinh \alpha l}$$

The deflection at $x = 0$ becomes zero if $\alpha l = \pi$; therefore, if $l > \pi/\alpha$ and the foundation reacts only upward there is a part of the beam which does not rest on the foundation. The deflection at either end of the beam is given by the expression

$$v_A = v_B = \frac{4P\alpha}{\beta} \frac{\sin^2\frac{\alpha l}{2} \sinh^2\frac{\alpha l}{2} + \cos^2\frac{\alpha l}{2} \cosh^2\frac{\alpha l}{2}}{\sin \alpha l + \sinh \alpha l}$$

Example 7.3.2. Consider a beam of length l supported at the ends A and B by two rigid supports, resting over its whole length on a Winkler foundation and subjected to a concentrated load P at its center (see Fig. 7.3.3). The deflection curve is given by Eq. (7.1.4), i.e.,

$$v = C_1 e^{\alpha x} \sin \alpha x + C_2 e^{\alpha x} \cos \alpha x + C_3 e^{-\alpha x} \sin \alpha x + C_4 e^{-\alpha x} \cos \alpha x \quad (7.1.4)$$

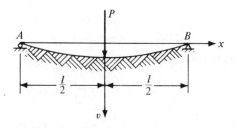

Figure 7.3.3

Besides the four constants $C_i (i = 1, 2, 3, 4)$, the reaction R at A and B is also unknown. The boundary conditions in this case are

$$\text{for } x = 0: \quad \frac{dv}{dx} = 0$$

$$\text{for } x = \frac{l}{2}: \quad v = 0, \quad M = 0, \quad V = -R$$

The remaining condition needed to evaluate the unknowns is obtained from the equation of equilibrium

$$2 \int_0^{l/2} \beta y \, dx + 2R = P$$

One finds for the reaction R at supports A and B .

$$R_A = R_B = R = P \frac{\cos \frac{\alpha l}{2} \cosh \frac{\alpha l}{2}}{\cos \alpha l + \cosh \alpha l}$$

At the center of the beam one has

$$\left.\begin{aligned} v(0) &= \frac{P\alpha}{2\beta} \frac{\sinh \alpha l - \sin \alpha l}{\cosh \alpha l + \cos \alpha l} \\ M(0) &= \frac{P}{4\alpha} \frac{\sinh \alpha l + \sin \alpha l}{\cosh \alpha l + \cos \alpha l} \end{aligned}\right\} \quad (7.3.8)$$

For $\alpha l = 1$, Eqs. (7.3.8) become

$$v_1 = 0.1802 \frac{P\alpha}{2\beta}$$

$$M_1 = 0.9680 \frac{P}{4\alpha}$$

while in the absence of supports A and B, one obtains [See Eqs. (7.3.4)]

$$v_{\max} = \frac{P\alpha}{2\beta}(1 + K_2) = 2.0248 \frac{P\alpha}{2\beta} = 11.2 v_1$$

$$M_{\max} = \frac{P}{4\alpha}(1 - K_1) = 0.4972 \frac{P}{4\alpha} = 0.51 M_1$$

The reactions R become zero for $\alpha l = \pi, 3\pi, \ldots$, i.e., for $l = \Lambda/2, 3\Lambda/2, \ldots$. If $l > \Lambda/2$, R becomes negative.

Example 7.3.3. The beam shown in Fig. 7.3.4 has the rigid central part of length d (being, for instance, part of a column) while the two lateral parts of lengths a

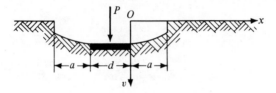

Figure 7.3.4

are elastic. The beam is supported by a Winkler foundation and is subjected to a concentrated load of intensity P acting at its middle point. The deflection curve is given by Eq. (7.1.4), i.e.,

$$v = C_1 e^{\alpha x} \sin \alpha x + C_2 e^{\alpha x} \cos \alpha x + C_3 e^{-\alpha x} \sin \alpha x + C_4 e^{-\alpha x} \cos \alpha x \quad (7.1.4)$$

Besides the four constants $C_i (i = 1, 2, 3, 4)$ the displacement v_0 of the rigid base is also unknown. The four boundary conditions are

$$\text{at } x = 0: \quad v = v_0, \quad \frac{dv}{dx} = 0$$

$$\text{at } x = a: \quad M = 0, \quad V = 0$$

and the equilibrium equation is

$$2\int_0^a \beta v \, dx + \beta v_0 d = P$$

From the above relations one obtains

$$\left.\begin{aligned}C_1 &= \frac{P\alpha}{4\beta} K_1 \\ C_2 &= \frac{P\alpha}{4\beta} K_2 \\ C_3 &= \frac{P\alpha}{4\beta} K_3 \\ C_4 &= \frac{P\alpha}{4\beta} K_4\end{aligned}\right\} \quad (7.3.9)$$

where

$$\left.\begin{aligned} K_1 &= \frac{\sin 2\alpha a + \cos 2\alpha a - e^{-2\alpha a}}{\sin 2\alpha a + \sinh 2\alpha a + \alpha d(\cos^2 \alpha a + \cosh^2 \alpha a)}, \quad K_3 = 2 - K_1 \\ K_2 &= \frac{2 - \sin 2\alpha a + \cos 2\alpha a + e^{-2\alpha a}}{\sin 2\alpha a + \sin 2\alpha a + \alpha d(\cos^2 \alpha a + \cosh^2 \alpha a)}, \quad K_4 = 2 + K_2 \end{aligned}\right\} \quad (7.3.10)$$

One finds that

$$v_0 = \frac{P\alpha\,(\cos^2 \alpha a + \cosh^2 \alpha a)}{\beta[\sin 2\alpha a + \sinh 2\alpha a + \alpha d(\cos^2 \alpha a + \cosh^2 \alpha a)]}$$

$$M_{\max} = M(0) = \frac{P(\cosh \alpha a - \cos \alpha a)}{4\alpha\,[\sin 2\alpha a + \sinh 2\alpha a + ad(\cos^2 \alpha a + \cosh^2 \alpha a)]}$$

$$v(a) = \frac{2P\alpha \cos \alpha a \cosh \alpha a}{\beta[\sin 2\alpha a + \sinh 2\alpha a + \alpha d(\cos^2 \alpha a + \cosh^2 \alpha a)]}$$

The above relations reduce to those for a beam of length $2a$ subjected to a concentrated load P at its center if $d = 0$.

7.4 The General Problem of the Straight Beam on an Elastic Foundation[5]

So far the problem of the beam resting on an elastic foundation has been discussed assuming that the foundation followed Winkler's hypothesis. In this section, the more general hypothesis that the behavior of the foundation is expressed by a Green's function of the type

$$w(x - \xi) \quad (7.4.1)$$

[5] Part of Section 7.4 was first discussed in the two papers entitled "Sul Problema Generale della Trave Poggiata su Suolo Elastico" by Enrico Volterra (see bibliography number 32 at the end of this chapter) and is presented here with the permission of the Accademia Nazionale dei Lincei.

will be assumed. This function is the most general expression for representing a displacement at point x due to a unit force at point ξ. In the particular case of a Winkler's foundation expression (7.4.1) becomes

$$w(x - \xi) = \begin{cases} \frac{1}{\beta} & \text{for } x = \xi \\ 0 & \text{for } x \neq \xi \end{cases} \tag{7.4.2}$$

while in the case of Jodi's foundation

$$w(x - \xi) = Ae^{-B|x-\xi|} \tag{7.4.3}$$

As before, let v be the deflection of the beam (which is the same as the deflection of the foundation), p the intensity of the distributed load acting on the beam, q the reaction of the foundation, and EI the inertia characteristics of the bar. Then the deflection of the beam is given by:

$$EI \frac{d^4v}{dx^4} = p - q \tag{7.4.4}$$

The deformation of the foundation for a load q is given by

$$v = \int_{-\infty}^{\infty} q(\xi)w(x - \xi)\,d\xi \tag{7.4.5}$$

Under the assumption that p, q, w, and v remain finite for $|x| \to \infty$, the following Fourier's transforms are defined:

$$V(u) = \frac{1}{\sqrt{2\pi}} \int_{-\infty}^{\infty} v(x)e^{ixu}\,dx \tag{7.4.6a}$$

$$P(u) = \frac{1}{\sqrt{2\pi}} \int_{-\infty}^{\infty} p(x)e^{ixu}\,dx \tag{7.4.6b}$$

$$Q(u) = \frac{1}{\sqrt{2\pi}} \int_{-\infty}^{\infty} q(x)e^{ixu}\,dx \tag{7.4.6c}$$

$$W(u) = \frac{1}{\sqrt{2\pi}} \int_{-\infty}^{\infty} w(x)e^{ixu}\,dx \tag{7.4.6d}$$

Since p and w are known functions, then $P(u)$ and $W(u)$ will be known functions.

From Eqs. (7.4.4), (7.4.6c) and (7.4.6b), one finds that

$$P(u) - Q(u) = \frac{EI}{\sqrt{2\pi}} \int_{-\infty}^{\infty} \frac{d^4v}{dx^4} e^{ixu}\,dx \tag{7.4.7}$$

If the integral of Eq. (7.4.7) can be successively integrated by parts, then, and under the assumption that the function v and its three first derivatives go to zero as $|x| \to \infty$, it follows that

$$\frac{1}{\sqrt{2\pi}} \int_{-\infty}^{\infty} \frac{d^4v(x)}{dx^4} e^{iux} dx = -\frac{1}{\sqrt{2\pi}} \int_{-\infty}^{\infty} \frac{d^3v}{dx^3} (iu) e^{iux} dx$$

$$= \ldots = \frac{1}{\sqrt{2\pi}} \int_{-\infty}^{\infty} v(iu)^4 e^{iux} dx = u^4 V(u) \qquad (7.4.8)$$

In view of Eq. (7.4.8), Eq. (7.4.7) can be written as

$$P(u) - Q(u) = EI\, u^4\, V(u) \qquad (7.4.9)$$

From Eqs. (7.4.6a) and (7.4.5) one obtains

$$\left.\begin{aligned}
V(u) &= \frac{1}{\sqrt{2\pi}} \int_{-\infty}^{\infty} \left\{ \int_{-\infty}^{\infty} q(\xi)\, w(x - \xi)\, d\xi \right\} e^{iux}\, dx \\
&= \int_{-\infty}^{\infty} q(\xi) \left\{ \frac{1}{\sqrt{2\pi}} \int_{-\infty}^{\infty} w(x - \xi)\, e^{iux}\, dx \right\} d\xi \\
&= \int_{-\infty}^{\infty} q(\xi) \left\{ \frac{1}{\sqrt{2\pi}} \int_{-\infty}^{\infty} w(t)\, e^{i(\xi+t)u}\, dt \right\} d\xi \\
&= \int_{-\infty}^{\infty} q(\xi) e^{i\xi u}\, W(u)\, d\xi = \sqrt{2\pi}\, Q(u)\, W(u)
\end{aligned}\right\} \qquad (7.4.10)$$

By use of Eqs. (7.4.9) and (7.4.10), the unknown functions $V(u)$ and $Q(u)$ can be determined. For the function $V(u)$, for instance, one finds that

$$V(u) = \frac{P(u)}{EI u^4 + \dfrac{1}{\sqrt{2\pi}\, W(u)}} \qquad (7.4.11)$$

The actual displacement v is finally given by the following equation:

$$v = \frac{1}{\sqrt{2\pi}} \int_{-\infty}^{\infty} V(u) e^{-iux}\, du \qquad (7.4.12)$$

Example 7.4.1. Suppose that function p represents a total load F uniformly distributed over a length $2l$ between $x = a - l$ and $x = a + l$. The case of a concentrated load is immediately derived by letting $l \to 0$ and the case of a general distribution of load is also immediately derived by using the principle of superposition. One, therefore, obtains

$$p = \begin{cases} 0 & \text{for } x < a - l \\ \dfrac{F}{2l} & \text{for } a - l < x < a + l \\ 0 & \text{for } x > a + l \end{cases}$$

From the above equation it follows that

$$P(u) = \frac{1}{\sqrt{2\pi}} \frac{F}{2l} \int_{a-l}^{a+l} e^{ixu} \, dx$$

$$= \frac{1}{\sqrt{2\pi}} \frac{F}{2l} \frac{1}{iu} \{ e^{i(a+l)u} - e^{i(a-l)u} \}$$

$$= \frac{1}{\sqrt{2\pi}} F e^{iau} \frac{\sin lu}{lu} \qquad (7.4.13)$$

In the limiting case of a concentrated load at the point $x = a$, Eq. (7.4.13) becomes

$$P(u) = \frac{1}{\sqrt{2\pi}} F e^{iau}$$

(a) *Solution according to Winkler's hypothesis.* One has in this particular case

$$q = \beta v$$

and it follows that

$$W(u) = \frac{1}{\sqrt{2\pi}} \frac{1}{\beta}$$

From Eq. (7.4.12) the deflection equation is obtained; thus,

$$v = \frac{F}{2\pi} \int_{-\infty}^{\infty} \left\{ \frac{e^{i(a-x)u}}{EIu^4 + \beta} \frac{\sin lu}{lu} \right\} du$$

or

$$v = \frac{F}{\pi} \int_{0}^{\infty} \frac{\cos(a-x)u \, \dfrac{\sin lu}{lu}}{EIu^4 + \beta} \, du \qquad (7.4.14)$$

The integral of Eq. (7.4.14) can be computed in closed form;[6] hence, one obtains the solution

---

[6] See bibliography number 32 at the end of this chapter.

$$v = \frac{F}{4l\beta}\left\{e^{-(k/\sqrt{2})(a-x+l)}\cos\frac{k}{\sqrt{2}}(a-x+l)\right.$$
$$\left. - e^{-(k/\sqrt{2})(a-x+l)}\cos\frac{k}{\sqrt{2}}(a-x+l)\right\}, \quad x \leq a-l$$

$$v = \frac{F}{4l\beta}\left\{2 - e^{(k/\sqrt{2})(a-x+l)}\cos\frac{k}{\sqrt{2}}(a-x-l)\right.$$
$$\left. - e^{-(k/\sqrt{2})(a-x+l)}\cos\frac{k}{\sqrt{2}}(a-x+l)\right\}, \quad a-l \leq x \leq a+l \quad (7.4.15)$$

$$v = \frac{F}{4l\beta}\left\{-e^{-(k/\sqrt{2})(a-x+l)}\cos\frac{k}{\sqrt{2}}(a-x-l)\right.$$
$$\left. + e^{(k/\sqrt{2})(a-x+l)}\cos\frac{k}{\sqrt{2}}(a-x+l)\right\}, \quad x \geq a+l$$

where $k = \sqrt[4]{\beta/EI}$

(b) *Solution according to Jodi's hypothesis.* One has in this particular case that

$$w(x) = Ae^{-B|x|}$$

It follows that

$$W(u) = \frac{1}{\sqrt{2\pi}}\int_{-\infty}^{\infty} Ae^{-B|x|}e^{ixu}\,dx$$
$$= \frac{1}{\sqrt{2\pi}}\int_{-\infty}^{0} Ae^{-B|x|}e^{ixu}\,dx + \frac{1}{\sqrt{2\pi}}\int_{0}^{\infty} Ae^{-B|x|}e^{ixu}\,dx$$
$$= \frac{A}{\sqrt{2\pi}}\left\{\frac{1}{B+iu} + \frac{1}{B-iu}\right\}$$

or finally

$$W(u) = \frac{2AB}{\sqrt{2\pi}(B^2+u^2)}$$

The solution is given by

$$v = \frac{F}{2\pi}\int_{-\infty}^{\infty}\left\{\frac{e^{i(a-x)u}\frac{\sin lu}{lu}}{EIu^4 + \frac{B^2+u^2}{2AB}}\right\}du$$

or

$$v = \frac{F}{\pi}\int_{0}^{\infty}\left\{\frac{\cos(a-x)u\,\frac{\sin lu}{lu}}{EIu^4 + \frac{B^2+u^2}{2AB}}\right\}du \quad (7.4.16)$$

The integral of Eq. (7.4.16) can be computed in closed form[7] and one obtains the solution

[7] See bibliography number 32 at the end of this chapter.

$$v = \frac{FA}{2lB} \left\{ e^{-p(a-x-l)} \left[\cos q(a-x-l) + \frac{m}{n} \sin q(a-x-l) \right] \right.$$
$$\left. - e^{-p(a-x+l)} \left[\cos q(a-x+l) + \frac{m}{n} \sin q(a-x+l) \right] \right\} \text{ for } x \le a-l$$

$$v = \frac{FA}{2lB} \left\{ 2 - e^{p(a-x-l)} \left[\cos q(a-x-l) + \frac{m}{n} \sin q(a-x+l) \right] \right.$$
$$\left. - e^{-p(a-x+l)} \left[\cos q(a-x+l) + \frac{m}{n} \sin q(a-x+l) \right] \right\} \text{ for } a-l \le x \le a+l$$

$$v = \frac{FA}{2lB} \left\{ -e^{p(a-x-l)} \left[\cos q(a-x-l) - \frac{m}{n} \sin q(a-x-l) \right] \right.$$
$$\left. + e^{p(a-x+l)} \left[\cos q(a-x+l) - \frac{m}{n} \sin q(a-x+l) \right] \right\} \text{ for } x \ge a+l$$

(7.4.17)

In Eqs. (7.4.17) $m = 1/4ABEI$, $n = m\sqrt{8AB^3EI - 1}$ (one assumes that $8AB^3EI > 1$), $p = \sqrt{\frac{1}{2}(\sqrt{m^2 + n^2} + m)}$, and $q = \sqrt{\frac{1}{2}(\sqrt{m^2 + n^2} - m)}$.

7.5 Bending of a Circular Beam Resting on an Elastic Foundation[8]

Deflections of circular beams resting on elastic foundations and loaded by symmetrical, concentrated forces acting perpendicular to the plane of original curvature of the beam and at an angular distance 2γ (see Fig. 7.5.1) will be considered in this section. The curved beam will be referred to the system of rectangular coordinates $Oxyz$ as shown in Fig. 7.5.2. Saint-Venant's equations reduce in this case to the two equations[9]

$$\left. \begin{array}{r} EI_x \left(\dfrac{\psi}{R} - \dfrac{d^2v}{ds^2} \right) = M_x \\ C\theta = C \left(\dfrac{d\psi}{ds} + \dfrac{1}{R} \dfrac{dv}{ds} \right) = M_z \end{array} \right\} \quad (7.5.1)$$

If β is the elastic constant of the foundation, P the external concentrated force, 2γ the angular distance between the points of application of the external

---

[8] Part of Section 7.5 was first discussed in the paper entitled "Bending of a Circular Beam Resting on an Elastic Foundation" by Enrico Volterra (see bibliography number 33 at the end of this chapter) and is presented here with the permission of the American Society of Mechanical Engineers.

[9] In this chapter, torsional rigidity and angle of twist of curved beams are represented by C and ψ, respectively, instead of K and β as in Chapter 6.

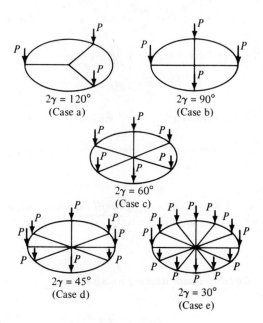

Figure 7.5.1

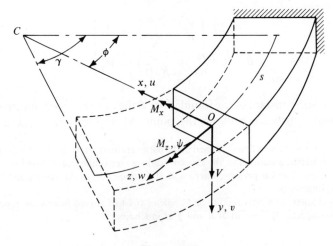

Figure 7.5.2

Sec. 7.5   BEAMS ON ELASTIC FOUNDATIONS   403

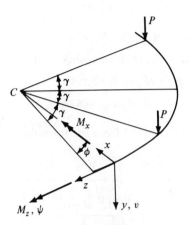

Figure 7.5.3

forces (see Figs. 7.5.1 and 7.5.3) and if

$$\eta = \frac{v}{R}, \quad a = \frac{EI_x}{R}, \quad \mu = \frac{C}{EI_x}, \quad v = \frac{PR^2}{EI_x}, \quad \lambda = \frac{\beta R^4}{EI_x}$$

then Eqs. (7.5.1) become

$$\left.\begin{array}{r} a\left(\psi - \dfrac{d^2\eta}{d\phi^2}\right) = M_x \\[6pt] \mu a\left(\dfrac{d\psi}{d\phi} + \dfrac{d\eta}{d\phi}\right) = M_z \end{array}\right\} \quad (7.5.2)$$

where the angle ϕ is measured from the bisector of the angle between two points of loading.

For equilibrium the following equations must be satisfied:

$$\left.\begin{array}{r} \dfrac{dM_z}{d\phi} - M_x = 0 \\[6pt] \dfrac{d^2 M_x}{d\phi^2} - \dfrac{dM_z}{d\phi} - a\gamma\eta = 0 \end{array}\right\} \quad (7.5.3)$$

The boundary conditions are

$$\left.\begin{array}{r} \psi(\gamma) = \psi(-\gamma), \; \eta(\gamma) = \eta(-\gamma), \; \dfrac{d^2\eta(\gamma)}{d\phi^2} = \dfrac{d^2\eta(-\gamma)}{d\phi^2} \\[6pt] \dfrac{d\psi(\gamma)}{d\phi} = \dfrac{d\psi(-\gamma)}{d\phi}, \; \dfrac{d\eta(\gamma)}{d\phi} = \dfrac{d\eta(-\gamma)}{d\phi}, \; \dfrac{dM_x(\gamma^+)}{d\phi} - \dfrac{dM_x(\gamma^-)}{d\phi} = av \end{array}\right\} \quad (7.5.4)$$

The solution of Eqs. (7.5.2) and (7.5.3) with the boundary conditions expressed by Eqs. (7.5.4) gives the following expressions for the functions η, ψ, M_x, and M_z:

$$\eta = A \cosh m\phi + B_1 \cosh r\phi \cos s\phi + B_2 \sinh r\phi \sin s\phi \quad (7.5.5)$$
$$\psi = C \cosh m\phi + D_1 \cosh r\phi \cos s\phi + D_2 \sinh r\phi \sin s\phi \quad (7.5.6)$$
$$M_x = aE \cosh m\phi + aF_1 \cosh r\phi \cos s\phi + aF_2 \sinh r\phi \sin s\phi \quad (7.5.7)$$
$$M_z = aG \sinh m\phi + aH_1 \sinh r\phi \cos s\phi + aH_2 \cosh r\phi \sin s\phi \quad (7.5.8)$$

where m, r, and s are determined from the formulas

$$m = \sqrt{-\tfrac{2}{3} + \tfrac{1}{3}(u_1 - u_2)}$$

$$r = \frac{1}{2}\sqrt{\frac{2}{m}}\sqrt{\frac{\lambda}{\mu} - m^2 - 2}$$

$$s = \frac{1}{2}\sqrt{\frac{2}{m}}\sqrt{\frac{\lambda}{\mu} + m^2 + 2}$$

with

$$u_1^3 = 9\lambda + 1 + \frac{27}{2}\frac{\lambda}{\mu} + \sqrt{27\lambda\left[(\lambda+1)^2 + \frac{9\lambda+1}{\mu} + \frac{27}{4}\frac{\lambda}{\mu^2}\right]}$$

$$u_2^3 = -9\lambda - 1 - \frac{27}{2}\frac{\lambda}{\mu} + \sqrt{27\lambda\left[(\lambda+1)^2 + \frac{9\lambda+1}{\mu} + \frac{27}{4}\frac{\lambda}{\mu^2}\right]}$$

The constants $A, \ldots, H_2$ are calculated from the following formulas:

$$A = \frac{v}{2mT \sinh(m\gamma)}\left(-m^2 + \frac{1}{\mu}\right)$$

$$B_1 = v\left\{\left[m^2(m^2+2) + \frac{2}{m^2}\frac{\lambda}{\mu} + \frac{1}{\mu}(3m^2+2)\right]R + \left(-m^2 + \frac{1}{\mu}\right)4rsS\right\}$$

$$B_2 = v\left\{\left[m^2(m^2+2) + \frac{2}{m^2}\frac{\lambda}{\mu} + \frac{1}{\mu}(3m^2+2)\right]S + \left(m^2 - \frac{1}{\mu}\right)4rsR\right\}$$

$$C = \frac{A}{\mu+1}\left(m^2 + \frac{\lambda}{\mu^2} - \mu\right)$$

$$D_1 = -\frac{\mu}{\mu+1}\left\{\left[\frac{1}{2}\left(m^2 + \frac{1}{\mu}\right)(m^2+2) + 1\right]B_1 + \left(m^2 - \frac{1}{\mu}\right)2rsB_2\right\}$$

$$D_2 = \frac{\mu}{\mu+1}\left\{-\left[\frac{1}{2}\left(m^2 + \frac{1}{\mu}\right)(m^2+2) + 1\right]B_2 + \left(m^2 - \frac{1}{\mu}\right)2rsB_1\right\}$$

$$E = \frac{A\mu}{\mu+1}\left(\frac{\lambda}{\mu m^2} - m^2 - 1\right)$$

TABLE 7.5.1

$2\gamma = 120°$ (CASE a)

$\lambda = 60$			$\mu = 0.7$			$\lambda = 100$			$\mu = 1$	
ϕ	$\dfrac{\psi}{v}$	$\dfrac{\eta}{v}$	$\dfrac{M_x}{av}$	$\dfrac{M_z}{av}$		$\dfrac{\psi}{v}$	$\dfrac{\eta}{v}$	$\dfrac{M_x}{av}$	$\dfrac{M_z}{av}$	
0°	0.01584	0.0^3326	-0.03896	0		0.0^2969	-0.0^3660	-0.02610	0	
10°	0.01417	0.0^2115	-0.03794	-0.0^2674		0.0^2875	-0.0^3116	-0.02657	-0.0^2459	
20°	0.0^2929	0.0^2355	-0.03343	-0.01305		0.0^2593	0.0^2150	-0.02631	-0.0^2924	
30°	0.0^2171	0.0^2723	-0.02123	-0.01797		0.0^2136	0.0^2408	-0.02045	-0.01345	
40°	-0.00741	0.01156	0.0^2489	-0.01965		-0.0^2443	0.0^2728	-0.0^2143	-0.01561	
50°	-0.01581	0.01547	0.05188	-0.01505		-0.01006	0.01034	0.03973	-0.01266	
60°	-0.01974	0.01726	0.12517	0		-0.01282	0.01181	0.11070	0	

TABLE 7.5.2
$2\gamma = 90°$ (CASE b)

	$\lambda = 60$		$\mu = 0.7$			$\lambda = 100$		$\mu = 1$	
ϕ	$\dfrac{\psi}{v}$	$\dfrac{\eta}{v}$	$\dfrac{M_x}{av}$	$\dfrac{M_z}{av}$	$\dfrac{\psi}{v}$	$\dfrac{\eta}{v}$	$\dfrac{M_x}{av}$	$\dfrac{M_z}{av}$
0°	0.01000	0.0^2611	-0.05167	0	0.0^2716	0.0^2255	-0.04373	0
10°	0.0^2798	0.0^2703	-0.04517	-0.0^2864	0.0^2574	0.0^2331	-0.03890	-0.0^2736
20°	0.0^2247	0.0^2952	-0.02425	-0.01493	0.0^2184	0.0^2540	-0.02267	-0.01294
30°	-0.0^2479	0.01277	0.01488	-0.01605	-0.0^2340	0.0^2818	0.01093	-0.01427
40°	-0.01066	0.01537	0.07674	-0.00841	-0.0^2776	0.01047	0.06895	-0.0^2770
45°	-0.01166	0.01381	0.11717	0	-0.0^2851	0.01087	0.10883	0

TABLE 7.5.3
$2\gamma = 60°$ (CASE c)

	$\lambda = 60$		$\mu = 0.7$			$\lambda = 100$		$\mu = 1$	
ϕ	$\dfrac{\psi}{v}$	$\dfrac{\eta}{v}$	$\dfrac{M_x}{av}$	$\dfrac{M_z}{av}$	$\dfrac{\psi}{v}$	$\dfrac{\eta}{v}$	$\dfrac{M_x}{av}$	$\dfrac{M_z}{av}$
0°	0.0^2341	0.01445	-0.04246	0	0.0^2272	0.0^2815	-0.04091	0
5°	0.0^2301	0.01463	-0.03899	-0.0^2360	0.0^2240	0.0^2831	-0.03764	-0.0^2347
10°	0.0^2188	0.01511	-0.02854	-0.0^2660	0.0^2150	0.0^2878	-0.02774	-0.0^2638
15°	0.0^3214	0.01583	-0.01096	-0.0^2838	0.0^3176	0.0^2946	-0.01093	-0.0^2812
20°	-0.0^2164	0.01662	0.01394	-0.0^2830	-0.0^2131	0.01022	0.01316	-0.0^2807
25°	-0.0^2322	0.01729	0.04631	-0.0^2573	-0.0^2257	0.01087	0.04493	-0.0^2559
30°	-0.0^2390	0.01758	0.08622	0	-0.0^2312	0.01115	0.08461	0

TABLE 7.5.4
$2\gamma = 45°$ (CASE d)

	$\lambda = 60$		$\mu = 0.7$			$\lambda = 100$		$\mu = 1$	
ϕ	$\dfrac{\psi}{v}$	$\dfrac{\eta}{v}$	$\dfrac{M_x}{av}$	$\dfrac{M_z}{av}$	$\dfrac{\psi}{v}$	$\dfrac{\eta}{v}$	$\dfrac{M_x}{av}$	$\dfrac{M_z}{av}$	
0°	0.0^2145	0.02061	−0.03272	0	0.0^2118	0.01213	−0.03233	0	
5°	0.0^2115	0.02073	−0.02788	-0.0^2271	0.0^3940	0.01225	−0.02758	-0.0^2268	
10°	0.0^3344	0.02108	−0.01336	-0.0^2458	0.0^3281	0.01259	−0.01329	-0.0^2454	
15°	-0.0^3700	0.02152	0.01089	-0.0^2476	-0.0^3571	0.01302	0.01070	-0.0^2472	
20°	-0.0^2152	0.02186	0.04489	-0.0^2240	-0.0^2124	0.01336	0.04451	-0.0^2238	
22°30′	-0.0^2166	0.02192	0.06551	0	-0.0^2135	0.01342	0.06511	0	

TABLE 7.5.5
$2\gamma = 30°$ (CASE e)

	$\lambda = 60$		$\mu = 0.7$			$\lambda = 100$		$\mu = 1$	
ϕ	$\dfrac{\psi}{v}$	$\dfrac{\eta}{v}$	$\dfrac{M_x}{av}$	$\dfrac{M_z}{av}$	$\dfrac{\psi}{v}$	$\dfrac{\eta}{v}$	$\dfrac{M_x}{av}$	$\dfrac{M_z}{av}$	
0°	0.0^3429	0.03165	−0.02191	0	0.0^3352	0.01892	−0.02186	0	
5°	0.0^3236	0.03173	−0.01460	-0.0^2170	0.0^3194	0.01900	−0.01457	-0.0^2170	
10°	-0.0^3206	0.03192	0.0^2732	-0.0^2212	-0.0^3170	0.01918	0.0^2729	-0.0^2212	
15°	-0.0^3489	0.03203	0.04375	0	-0.0^3402	0.01930	0.04370	0	

$$F_1 = -\frac{\mu}{2(\mu+1)}(m^2+1)(m^2B_1 + 4rsB_2)$$

$$F_2 = \frac{\mu}{2(\mu+1)}(m^2+1)(-m^2B_2 + 4rsB_1)$$

$$G = \frac{E}{m}$$

$$H_1 = m\sqrt{\frac{\mu}{\lambda}}(rF_1 - sF_2)$$

$$H_2 = m\sqrt{\frac{\mu}{\lambda}}(rF_2 + sF_1)$$

where

$$T = 3m^4 + 4m^2 + 1 + \lambda$$

$$R = \frac{m}{4rsT}\sqrt{\frac{\mu}{\lambda}} \frac{s \sinh r\gamma \cos s\gamma + r \cosh r\gamma \sin s\gamma}{\cosh 2r\gamma - \cos 2s\gamma}$$

$$S = \frac{m}{4rsT}\sqrt{\frac{\mu}{\lambda}} \frac{s \cosh r\gamma \sin s\gamma - r \sinh r\gamma \cos s\gamma}{\cosh 2r\gamma - \cos 2s\gamma}$$

Values of the functions ψ/v, η/v, M_x/av and M_z/av are given in Tables 7.5.1, 7.5.2, 7.5.3, 7.5.4, and 7.5.5 for the following values of the parameters: $\lambda = 60$, $\mu = 0.7$; $\lambda = 100$ and $\mu = 1$.

Example 7.5.1. Consider a circular beam of radius $R = 25$ ft which is made of reinforced concrete ($E = 3 \times 10^6$ psi) of square cross section (depth 30 in., width 30 in.) and is subjected to eight concentrated loads $P = 75$ tons each ($2\gamma = 45°$). The constant of the foundation is assumed to be $\beta = 1500$ lb/in.². It follows that: $I_x = 6750$ in.⁴, $\lambda = 60$, $\mu = 0.7$; $v = 0.0666$, $a = 28,125$ ft-tons, and $av = 1873.12$ ft-tons. In Table 7.5.6, the results obtained in this particular case by use of Table 7.5.4 are presented.

TABLE 7.5.6

ϕ, deg-min	ψ, radians	v, in.	M_x, ft-tons	M_z, ft-tons
0°	0.0⁴966	0.4122	−61.288	0
5°	0.0⁴766	0.4146	−52.222	−5.076
10°	0.0⁴229	0.4216	−25.025	−8.579
15°	−0.0⁴466	0.4304	20.398	−8.916
20°	−0.0³1033	0.4372	84.084	−4.495
22°30′	−0.0³1106	0.4384	122.708	0

7.6 Bending of a Constrained Circular Beam Resting on an Elastic Foundation[10]

Suppose that the circular beam discussed in Sec. 7.5 is restrained at the points of application of the loads from rotating about the center line (see Fig. 7.6.1). In this case, while Eqs. (7.5.2) and (7.5.3) remain the same, the boundary conditions become

$$\left. \begin{array}{l} \psi(\gamma) = \psi(-\gamma) = 0, \quad \eta(\gamma) = \eta(-\gamma), \quad \dfrac{d\eta(\gamma)}{d\phi} = -\dfrac{d\eta(-\gamma)}{d\phi}, \\[6pt] \dfrac{d^2\eta(\gamma)}{d\phi^2} = \dfrac{d^2\eta(-\gamma)}{d\phi^2}, \quad \dfrac{dM_x(\gamma^+)}{d\phi} - \dfrac{dM_x(\gamma^-)}{d\phi} = av, \end{array} \right\} \quad (7.6.1)$$

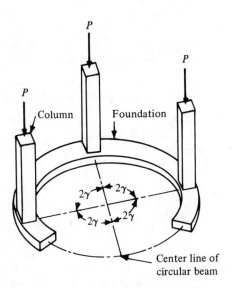

Figure 7.6.1

The solution of Eqs. (7.5.2) and (7.5.3), with the boundary conditions expressed by Eqs. (7.6.1), gives the following expressions for functions η, ψ, M_x, and M_z:

$$\eta = A \cosh m\phi + B_1 \cosh r\phi \cos s\phi + B_2 \sinh r\phi \sin s\phi$$
$$\psi = C \cosh m\phi + D_1 \cosh r\phi \cos s\phi + D_2 \sinh r\phi \sin s\phi$$

---

[10] Part of Section 7.6 was first discussed in the paper entitled "Bending of a Constrained Circular Beam Resting on an Elastic Foundation" by Enrico Volterra and R. Chung (see bibliography number 36 at the end of this chapter) and is presented here with the permission of the American Society of Civil Engineers.

$$M_x = (E \cosh m\phi + F_1 \cosh r\phi \cos s\phi + F_2 \sinh r\phi \sin s\phi)a$$
$$M_z = (G \sinh m\phi + H_1 \sinh r\phi \cos s\phi + H_2 \cosh r\phi \sin s\phi)a$$

where m, r, and s are determined from the formulas

$$m = \sqrt{u_1 + u_2 - \tfrac{2}{3}}$$
$$r = \frac{1}{2}\sqrt{\frac{2}{m}\sqrt{\frac{\lambda}{\mu}} - m^2 - 2}$$
$$s = \frac{1}{2}\sqrt{\frac{2}{m}\sqrt{\frac{\lambda}{\mu}} + m^2 + 2}$$

with

$$u_1 = \frac{1}{3}\sqrt[3]{9\lambda + 1 + \frac{27\lambda}{2\mu} + \sqrt{27\lambda(\lambda+1)^2 + \frac{9\lambda+1}{\mu} + \frac{27\lambda}{4\mu^2}}}$$

$$u_2 = \frac{1}{3}\sqrt[3]{9\lambda + 1 + \frac{27\lambda}{2\mu} - \sqrt{27\lambda(\lambda+1)^2 + \frac{9\lambda+1}{\mu} + \frac{27\lambda}{4\mu^2}}}$$

The constants $A, \ldots, H_2$ are calculated from the following formulas:

$$A = \frac{v(rP + sQ)}{2m\left[rR + sS - \dfrac{2rs}{1+\mu}\sqrt{\dfrac{\lambda}{\mu}}\,T\right]}$$

$$B_1 = \frac{v\left[\mu m U \sinh m\gamma + \left(m^2 + \dfrac{\lambda}{m^2} - \mu\right) X \cosh m\gamma\right]}{2\mu m\left[rR + sS - \dfrac{2rs}{1+\mu}\sqrt{\dfrac{\lambda}{\mu}}\,T\right]}$$

$$B_2 = -\frac{v\left[\mu m V \sinh m\gamma - \left(m^2 + \dfrac{\lambda}{m^2} - \mu\right) Y \cosh m\gamma\right]}{2\mu m\left[rR + sS - \dfrac{2rs}{1+\mu}\sqrt{\dfrac{\lambda}{\mu}}\,T\right]}$$

$$E = \frac{A\mu}{1+\mu}\left(\frac{\lambda}{\mu m^2} - m^2 - 1\right)$$

$$F_1 = -\frac{\mu}{2(1+\mu)}(1+m^2)(m^2 B_1 + 4rs B_2)$$

$$F_2 = -\frac{\mu}{2(1+\mu)}(1+m^2)(m^2 B_2 - 4rs B_1)$$

$$G = \frac{E}{m}$$

$$H_1 = m\sqrt{\frac{\mu}{\lambda}}(rF_1 - sF_2)$$

$$H_2 = m\sqrt{\frac{\mu}{\lambda}}(rF_2 + sF_1)$$

where

$$P = \left[-(1+m^2)^2 - \frac{1}{m}\left(m^2 - \frac{1}{\mu}\right)\sqrt{\frac{\lambda}{\mu}}\right]\sin sy \cos sy$$

$$Q = \left[-(1+m^2) + \frac{1}{m}\left(m^2 - \frac{1}{\mu}\right)\sqrt{\frac{\lambda}{\mu}}\right]\sinh ry \cosh ry$$

$$R = \left\{\lambda(1+m^2)\left(\frac{1}{\mu m^2} - 2\right) + \left[m(1+m^2)\right.\right.$$
$$\left.\left. - \frac{\lambda}{m^3}\left(m^2 - \frac{1}{\mu}\right)\sqrt{\frac{\lambda}{\mu}}\right]\right\}\sinh my \sin sy \cos sy$$

$$S = \left\{\lambda(1+m^2)\left(\frac{1}{\mu m^2} - 2\right) - \left[m(1+m^2)\right.\right.$$
$$\left.\left. - \frac{\lambda}{m^3}\left(m^2 - \frac{1}{\mu}\right)\sqrt{\frac{\lambda}{\mu}}\right]\right\}\sinh my \sinh ry \cosh ry$$

$$T = \frac{1}{m^2}(1+m^2)\left(m^2 + \frac{\lambda}{m^2} - \mu\right)\cosh my\,(\sinh^2 ry \cos^2 sy$$
$$+ \cosh^2 ry \sin^2 sy)$$

$$U = \left[\frac{1}{2}\left(m^2 + \frac{1}{\mu}\right)(m^2 + 2) + 1\right]\sinh ry \sin sy$$
$$+ 2rs\left(m^2 - \frac{1}{\mu}\right)\cosh ry \cos sy$$

$$V = -\left[\frac{1}{2}\left(m^2 + \frac{1}{\mu}\right)(m^2 + 2) + 1\right]\cosh ry \cos sy$$
$$+ 2rs\left(m^2 - \frac{1}{\mu}\right)\sinh ry \sin sy$$

$$X = r \cosh ry \sin sy + s \sinh ry \cos sy$$

$$Y = r \sinh ry \cos sy - s \cosh ry \sin sy$$

Values of the functions η/v, ψ/v, M_x/av, M_z/av are given in Tables 7.6.1, 7.6.2, 7.6.3, 7.6.4, and 7.6.5 for the following values of the parameters: $\lambda = 60$, $\mu = 0.7$; $\lambda = 60$, $\mu = 1.0$; $\lambda = 100$, $\mu = 0.7$; $\lambda = 100$, $\mu = 1.0$.

Example 7.6.1. Consider a circular beam having the same characteristics as the beam of Example 7.5.1, but which cannot rotate. In Table 7.6.6 the results, obtained for this particular case from Table 7.6.4, are given. If one compares the results given by Tables 7.5.6 and 7.6.6, one sees that the values in the ψ columns are completely different, the values for v are practically the same, while the values in the M_x and M_z columns are fairly close.

TABLE 7.6.1

$2\gamma = 120°$ (CASE a)

	ϕ	$\dfrac{\eta}{v}$	$\dfrac{\psi}{v}$	$\dfrac{M_x}{av}$	$\dfrac{M_z}{av}$
$\lambda = 60$ $\mu = 0.7$	0° 10° 20° 30° 40° 50° 60°	0.000765 0.0001570 0.003902 0.007469 0.011673 0.015458 0.017188	0.025509 0.024198 0.020025 0.013813 0.006654 0.000806 0.000000	−0.027786 −0.026544 −0.021398 −0.008300 0.018768 0.066487 0.140059	0 −0.004780 −0.009044 −0.011791 −0.011130 −0.004035 0.013592
$\lambda = 60$ $\mu = 1.0$	0° 10° 20° 30° 40° 50° 60°	0.000835 0.001629 0.003929 0.007454 0.011614 0.015370 0.017092	0.023236 0.021999 0.018414 0.012936 0.006579 0.001232 0.000000	−0.029301 −0.027973 −0.022594 −0.009176 0.018209 0.066157 0.139811	0 −0.005039 −0.009534 −0.012463 −0.011926 −0.004908 0.012671
$\lambda = 100$ $\mu = 0.7$	0° 10° 20° 30° 40° 50° 60°	−0.000520 0.000021 0.001627 0.004188 0.007367 0.010391 0.011843	0.019018 0.018112 0.015406 0.011042 0.005655 0.000981 0.000000	−0.016618 −0.017019 −0.016567 −0.010423 0.008925 0.050354 0.121438	0 −0.002927 −0.005897 −0.008384 −0.008770 −0.003982 0.010543
$\lambda = 100$ $\mu = 1.0$	0° 10° 20° 30° 40° 50° 60°	−0.000482 0.000054 0.001644 0.004182 0.007337 0.010344 0.011791	0.017425 0.016615 0.014201 0.010318 0.005515 0.001175 0.000000	−0.017854 −0.018180 −0.017521 −0.011092 0.008543 0.050182 0.121342	0 −0.003138 −0.006295 −0.008924 −0.009401 −0.004660 0.009844

TABLE 7.6.2
$2\gamma = 90°$ (CASE b)

	ϕ	$\dfrac{\eta}{v}$	$\dfrac{\psi}{v}$	$\dfrac{M_x}{av}$	$\dfrac{M_z}{av}$
$\lambda = 60$ $\mu = 0.7$	0° 10° 20° 30° 40° 45°	0.006335 0.007237 0.009686 0.012885 0.015440 0.015869	0.017598 0.015783 0.010887 0.004640 0.000163 0.000000	−0.043027 −0.036459 −0.015351 0.024001 0.086021 0.126474	0 −0.007130 −0.011885 −0.011429 −0.002185 0.007040
$\lambda = 60$ $\mu = 1.0$	0° 10° 20° 30° 40° 45°	0.006379 0.007267 0.009680 0.012837 0.015387 0.015702	0.015396 0.013850 0.009670 0.004292 0.000305 0.000000	−0.044266 −0.037641 −0.016389 0.023126 0.085252 0.125720	0 −0.007343 −0.012293 −0.012003 0.002901 0.006258
$\lambda = 100$ $\mu = 0.7$	0° 10° 20° 30° 40° 45°	0.002622 0.003385 0.005479 0.006269 0.010556 0.010948	0.015340 0.013815 0.009650 0.004207 0.000177 0.000000	−0.035777 −0.031054 −0.014760 0.018815 0.076825 0.116707	0 −0.005974 −0.010174 −0.010124 −0.002181 0.006209
$\lambda = 100$ $\mu = 1.0$	0° 10° 20° 30° 40° 45°	0.002659 0.003411 0.005478 0.008238 0.010504 0.010894	0.013451 0.012148 0.008579 0.003885 0.000296 0.000000	−0.036951 −0.032157 −0.015681 0.018099 0.076243 0.116143	0 −0.006174 −0.010552 −0.010645 −0.002813 0.005527

TABLE 7.6.3

$2\gamma = 60°$ (CASE c)

	ϕ	$\dfrac{\eta}{v}$	$\dfrac{\psi}{v}$	$\dfrac{M_x}{av}$	$\dfrac{M_z}{av}$
$\lambda = 60$ $\mu = 0.7$	0° 5° 10° 15° 20° 25° 30°	0.014517 0.014689 0.015174 0.015884 0.016675 0.017342 0.017631	0.006623 0.006242 0.005168 0.003601 0.001879 0.000476 0.000000	−0.039032 −0.035562 −0.025103 −0.007514 0.017388 0.049767 0.089673	0 −0.003305 −0.006004 −0.007479 −0.007102 −0.004227 0.001803
$\lambda = 60$ $\mu = 1.0$	0° 5° 10° 15° 20° 25° 30°	0.014523 0.014692 0.015172 0.015876 0.016659 0.017322 0.017608	0.005613 0.005295 0.004396 0.003083 0.001634 0.000439 0.000000	−0.039559 −0.036086 −0.025617 −0.008017 0.016897 0.049285 0.089193	0 −0.003351 −0.006095 −0.007615 −0.007281 −0.004448 0.001540
$\lambda = 100$ $\mu = 0.7$	0° 5° 10° 15° 20° 25° 30°	0.008175 0.008340 0.008807 0.009494 0.010261 0.010911 0.011193	0.006418 0.006052 0.005016 0.003503 0.001833 0.000466 0.000000	−0.037538 −0.034274 −0.024382 −0.007583 0.016495 0.048254 0.087927	0 −0.003181 −0.005789 −0.007236 −0.006902 −0.004133 0.001751
$\lambda = 100$ $\mu = 1.0$	0° 5° 10° 15° 20° 25° 30°	0.008183 0.008346 0.008809 0.009489 0.010249 0.010895 0.011175	0.005441 0.005135 0.004268 0.002999 0.001599 0.000430 0.000000	−0.038063 −0.034794 −0.024889 −0.008076 0.016026 0.047799 0.087476	0 −0.003227 −0.005880 −0.007370 −0.007077 −0.004349 0.001495

TABLE 7.6.4
$2\gamma = 45°$ (CASE d)

	ϕ	$\dfrac{\eta}{v}$	$\dfrac{\psi}{v}$	$\dfrac{M_x}{av}$	$\dfrac{M_z}{av}$
$\lambda = 60$ $\mu = 0.7$	0°	0.020635	0.002941	−0.031173	0
	5°	0.020762	0.002649	−0.026338	−0.002580
	10°	0.021102	0.001865	−0.011814	−0.004315
	15°	0.021541	0.000864	0.012440	−0.004359
	20°	0.021886	0.000101	0.046432	−0.001860
	22°30′	0.021943	0.000000	0.067053	0.000607
$\lambda = 60$ $\mu = 1.0$	0°	0.020634	0.002463	−0.031422	0
	5°	0.020760	0.002220	−0.026586	−0.002601
	10°	0.021099	0.001569	−0.012060	−0.004358
	15°	0.021535	0.000734	0.012196	−0.004423
	20°	0.021878	0.000091	0.046190	−0.001946
	22°30′	0.021935	0.000000	0.066811	0.000510
$\lambda = 100$ $\mu = 0.7$	0°	0.012143	0.002912	−0.030794	0
	5°	0.012268	0.002623	−0.026048	−0.002549
	10°	0.012605	0.001848	−0.011754	−0.004269
	15°	0.013039	0.000857	0.012230	−0.004319
	20°	0.013381	0.000100	0.046038	−0.001848
	22°30′	0.013438	0.000000	0.066633	0.000601
$\lambda = 100$ $\mu = 1.0$	0°	0.012144	0.002438	−0.031043	0
	5°	0.012268	0.002199	−0.026296	−0.002571
	10°	0.012603	0.001555	−0.011998	−0.004312
	15°	0.013035	0.000728	0.011989	−0.004383
	20°	0.013375	0.000091	0.045800	−0.001933
	22°30′	0.013432	0.000000	0.066395	0.000506

TABLE 7.6.5

$2\gamma = 30°$ (CASE e)

	ϕ	$\dfrac{\eta}{v}$	$\dfrac{\psi}{v}$	$\dfrac{M_x}{av}$	$\dfrac{M_z}{av}$
$\lambda = 60$ $\mu = 0.7$	0° 5° 10° 15°	0.031661 0.031741 0.031925 0.032042	0.000894 0.000704 0.000270 0.000000	−0.021441 −0.014627 −0.007792 0.044227	0 −0.001658 −0.002041 0.000124
$\lambda = 60$ $\mu = 1.0$	0° 5° 10° 15°	0.031660 0.031740 0.031923 0.032040	0.000742 0.000585 0.000225 0.000000	−0.021521 −0.014207 0.007712 0.044148	0 −0.001665 −0.002055 0.000103
$\lambda = 100$ $\mu = 0.7$	0° 5° 10° 15°	0.018925 0.019006 0.019189 0.019306	0.000893 0.000703 0.000269 0.000000	−0.021390 −0.014102 0.007765 0.044173	0 −0.001655 −0.002037 0.000124
$\lambda = 100$ $\mu = 1.0$	0° 5° 10° 15°	0.018925 0.019005 0.019188 0.019305	0.000741 0.000584 0.000225 0.000000	−0.021470 −0.014182 0.007686 0.044094	0 −0.001662 −0.002051 0.000103

TABLE 7.6.6

CONSTRAINED BEAM

ϕ, deg-min	ψ, radians	v, in.	M_x, ft-tons	M_z, ft-tons
0°	0.0³1959	0.4122	−58.391	0
5°	0.0³1764	0.4148	−49.334	−4.833
10°	0.0³1242	0.4216	−22.129	−8.082
15°	0.0⁴575	0.4304	23.301	−8.165
20°	0.0⁵673	0.4373	86.973	−3.494
22°30′	0	0.4384	125.598	1.137

7.7 Deflections of Circular Beams Resting on Elastic Foundations Obtained by the Method of Harmonic Analysis[9]

In this section, a solution in terms of trigonometric series is given for the problem of a circular beam resting on an elastic foundation. This solution allows one to study not only the case of symmetrical loading (see Fig. 7.7.1)

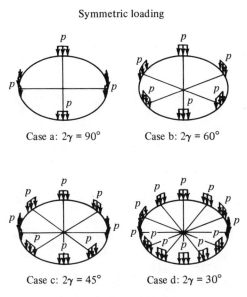

Figure 7.7.1

which was considered in Sec. 7.5 and 7.6 but also the case of antisymmetrical loading (see Fig. 7.7.2). As already pointed out, the problem of a circular beam resting on an elastic foundation has a practical application in civil engineering in the design of circular, reinforced concrete beams used as foundations of water tanks (see Fig. 7.0.3). In this instance, not only the case of symmetrical loading has to be taken into consideration, but also the case of antisymmetrical loading which is induced by the wind action on the water tank.

---

[9]Part of Section 7.7 was first discussed in the paper entitled "Deflections of Circular Beams Resting on Elastic Foundations Obtained by Methods of Harmonic Analysis" by Enrico Volterra (see bibliography number 35 at the end of this chapter) and is presented here with the permission of the American Society of Mechanical Engineers.

Antisymmetric loading

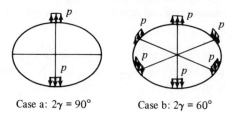

Case a: $2\gamma = 90°$ Case b: $2\gamma = 60°$

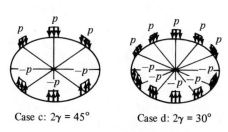

Case c: $2\gamma = 45°$ Case d: $2\gamma = 30°$

Figure 7.7.2

For this study one has, in addition to Eqs. (7.5.2), the equilibrium equations which are

$$\left.\begin{array}{r}\dfrac{dM_z}{d\phi} - M_x = 0 \\ \dfrac{d^2M_x}{d\phi^2} + \dfrac{dM_z}{d\phi} + R^2[p - R\beta\eta] = 0\end{array}\right\} \quad (7.7.1)$$

where p represents the externally distributed force and $\eta = \dfrac{v}{R}$ (see Fig. 7.7.3). From Eqs. (7.5.2) and (7.7.1) one concludes that

$$\left.\begin{array}{r}-\psi + \mu\dfrac{d^2\psi}{d\phi^2} + (1+\mu)\dfrac{d^2\eta}{d\phi^2} = 0 \\ \dfrac{d^2\psi}{d\phi^2}(1+\mu) + \mu\dfrac{d^2\eta}{d\phi^2} - \dfrac{d^4\eta}{d\phi^4} + \dfrac{R^2}{a}[p - \beta R\eta] = 0\end{array}\right\} \quad (7.7.2)$$

By combining Eqs. (7.7.2), it follows that

$$\left.\begin{array}{r}\psi = \dfrac{1+2\mu}{1+\mu}\dfrac{d^2\eta}{d\phi^2} + \dfrac{\mu}{1+\mu}\dfrac{d^4\eta}{d\phi^4} - \dfrac{\mu R^2}{a(1+\mu)}[p - \beta R\eta] \\ \dfrac{d^6\eta}{d\phi^6} + 2\dfrac{d^4\eta}{d\phi^4} + \dfrac{d^2\eta}{d\phi^2}\left(1 + \dfrac{\beta R^3}{a}\right) - \dfrac{R^3}{a\mu}\beta\eta = -\dfrac{R^2}{a\mu}p + \dfrac{R^2}{a}\dfrac{d^2p}{d\phi^2}\end{array}\right\} \quad (7.7.3)$$

The periodicity conditions are

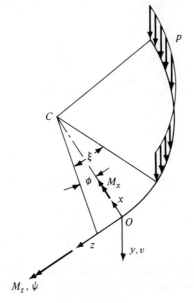

Figure 7.7.3

$$\left.\begin{array}{l}\psi(0) = \psi(2\pi), \quad \left(\dfrac{d\psi}{d\phi}\right)_0 = \left(\dfrac{d\psi}{d\phi}\right)_{2\pi}, \quad \eta(0) = \eta(2\pi), \\[6pt] \left(\dfrac{d\eta}{d\phi}\right)_0 = \left(\dfrac{d\eta}{d\phi}\right)_{2\pi}, \quad \left(\dfrac{d^2\eta}{d\phi^2}\right)_0 = \left(\dfrac{d^2\eta}{d\phi^2}\right)_{2\pi}, \quad \left(\dfrac{d^3\eta}{d\phi^3}\right)_0 = \left(\dfrac{d^3\eta}{d\phi^3}\right)_{2\pi}\end{array}\right\} \quad (7.7.4)$$

If a concentrated load P is acting on the section ξ, then[10]

$$p = \frac{P}{2\pi R}\left\{1 + 2\sum_{n=1}^{\infty}\cos n(\xi - \phi)\right\} \quad (7.7.5)$$

The solution of Eqs. (7.5.2) and (7.7.3) is expressed by the following relations:

$$\eta = \frac{P}{2\pi R^2 \beta} + \frac{PR}{a\mu\pi}\sum_{n=1}^{\infty} Y_n \cos n(\xi - \phi) \quad (7.7.6)$$

where

$$Y_n = \frac{\mu n^2 - 1}{n^6 - 2n^4 + n^2(1 + \lambda) + \dfrac{\lambda}{\mu}} \quad (7.7.7)$$

$$\psi = \frac{PR}{a\mu\pi}\sum_{n=1}^{\infty} B_n \cos n(\xi - \phi) \quad (7.7.8)$$

---

[10] It should be noted that the series in Eq. (7.7.5), which is used for representing concentrated loads, is not a convergent series (see bibliography number 16 at the end of this chapter), but this will not interfere with the legitimacy of the harmonic process which is to be used to determine deflections.

where
$$B_n = -\frac{(1+\mu)n^2}{1+\mu n^2} Y_n \qquad (7.7.9)$$

$$M_x = \frac{PR}{\mu\pi} \sum_{n=1}^{\infty} M_{xn} \cos n(\xi - \phi) \qquad (7.7.10)$$

where
$$M_{xn} = B_n + n^2 Y_n \qquad (7.7.11)$$

and
$$M_z = \frac{PR}{\mu\pi} \sum_{n=1}^{\infty} M_{zn} \cos n(\xi - \phi) \qquad (7.7.12)$$

where
$$M_{zn} = -\frac{M_{xn}}{n} \qquad (7.7.13)$$

Now, if p is the intensity of a distributed load at the section ξ, then by applying the principle of superposition, the following expressions are obtained:

$$\eta = \frac{1}{2\pi\beta R^2} \int_0^{2\pi} p(\xi)\, d\xi$$
$$+ \frac{R}{a\mu\pi} \sum_{n=1}^{\infty} Y_n \left[\cos n\phi \int_0^{2\pi} p(\xi) \cos n\xi\, d\xi + \sin n\phi \int_0^{2\pi} p(\xi) \sin n\xi\, d\xi \right] \quad (7.7.14)$$

$$\psi = \frac{R}{a\mu\pi} \sum_{n=1}^{\infty} B_n \left[\cos n\phi \int_0^{2\pi} p(\xi) \cos n\xi\, d\xi + \sin n\phi \int_0^{2\pi} p(\xi) \sin n\xi\, d\xi \right] \quad (7.7.15)$$

$$M_x = \frac{R}{\mu\pi} \sum_{n=1}^{\infty} M_{xn} \left[\cos n\phi \int_0^{2\pi} p(\xi) \cos n\xi\, d\xi + \sin n\phi \int_0^{2\pi} p(\xi) \sin n\xi\, d\xi \right] \quad (7.7.16)$$

$$M_z = \frac{R}{\mu\pi} \sum_{n=1}^{\infty} M_{zn} \left[\sin n\phi \int_0^{2\pi} p(\xi) \cos n\xi\, d\xi - \cos n\phi \int_0^{2\pi} p(\xi) \sin n\xi\, d\xi \right] \quad (7.7.17)$$

(a) Case of symmetrical loading. If the loads are symmetrical (see Fig. 7.7.1) so that p is an even function of ξ then through obvious reductions, the coefficients of $\sin n\phi$ in Eqs. (7.7.14), (7.7.15), and (7.7.16) and of $\cos n\phi$ in Eq. (7.7.17) vanish and one obtains

$$\eta = \frac{2}{\pi\beta R^2} \int_0^{\pi/2} p(\xi)\, d\xi + \frac{4R}{a\mu\pi} \sum_{n=2,4,6\ldots}^{\infty} Y_n \cos n\phi \int_0^{\pi/2} p(\xi) \cos n\xi\, d\xi \quad (7.7.18)$$

$$\psi = \frac{4R}{a\mu\pi} \sum_{n=2,4,6\ldots}^{\infty} B_n \cos n\phi \int_0^{\pi/2} p(\xi) \cos n\xi\, d\xi \quad (7.7.19)$$

$$M_x = \frac{4R}{\mu\pi} \sum_{n=2,4,6,\cdots}^{\infty} M_{xn} \cos n\phi \int_0^{\pi/2} p(\xi) \cos n\xi \, d\xi \quad (7.7.20)$$

$$M_z = \frac{4R}{\mu\pi} \sum_{n=2,4,6,\cdots}^{\infty} M_{zn} \sin n\phi \int_0^{\pi/2} p(\xi) \cos n\xi \, d\xi \quad (7.7.21)$$

Tables 7.7.1 and 7.7.2 contain the first 24 coefficients of the series expressing functions η, ψ, M_x, and M_z for $\mu = 1$, $\lambda = 100$ and for $\mu = 0.7$, $\lambda = 60$.

TABLE 7.7.1

COEFFICIENTS FOR SYMMETRICAL LOADING: $\mu = 1$; $\lambda = 100$

n	Y_n	B_n	M_{xn}	M_{zn}
2	0.0²559701	−0.0²895522	0.0134328	−0.0²671640
4	0.0²283019	−0.0²532742	0.0399556	−0.0²998890
6	0.0³732217	−0.0²142485	0.0249350	−0.0²415583
8	0.0³241828	−0.0³476215	0.0150000	−0.0²187500
10	0.0⁴981948	−0.0³194445	0.0²962504	−0.0³962504
12	0.0⁴483262	−0.0⁴959858	0.0²686299	−0.0³571916
14	0.0⁴260953	−0.0⁴519257	0.0²506275	−0.0³361625
16	0.0⁴152950	−0.0⁴304710	0.0²388505	−0.0³242816
18	0.0⁵954489	−0.0⁴190310	0.0²307351	−0.0³170750
20	0.0⁵626128	−0.0⁴124913	0.0²249202	−0.0³124601
22	0.0⁵427583	−0.0⁵853403	0.0²206097	−0.0⁴938804
24	0.0⁵301841	−0.0⁵602636	0.0²173257	−0.0⁴721904

TABLE 7.7.2

COEFFICIENTS FOR SYMMETRICAL LOADING: $\mu = 0.70$; $\lambda = 60$

n	Y_n	B_n	M_{xn}	M_{zn}
2	0.0²497630	−0.0²890496	0.0110002	−0.0²550010
4	0.0²219557	−0.0²489504	0.0302341	−0.0²755852
6	0.0³522163	−0.0²121971	0.0175782	−0.0²292970
8	0.0³169806	−0.0³403382	0.0104643	−0.0²130804
10	0.0⁴699665	−0.0³167525	0.0²672913	−0.0³672913
12	0.0⁴338250	−0.0⁴813395	0.0²478946	−0.0³399122
14	0.0⁴182458	−0.0⁴439906	0.0²353219	−0.0³252299
16	0.0⁴106951	−0.0⁴258297	0.0²271212	−0.0³169507
18	0.0⁵667610	−0.0⁴161422	0.0²214692	−0.0³119273
20	0.0⁵437960	−0.0⁴105983	0.0²174124	−0.0⁴870620
22	0.0⁵299094	−0.0⁵724233	0.0²144037	−0.0⁴654714
24	0.0⁵211157	−0.0⁵511541	0.0²121115	−0.0⁴504646

(*b*) *Antisymmetrical case of loading.* If the loads are antisymmetrical (see Fig. 7.7.2) so that p is an odd function of ξ, analogous reductions lead to

$$\eta = \frac{4R}{a\mu\pi} \sum_{n=1,3,5,\ldots}^{\infty} Y_n \sin n\phi \int_0^{\pi/2} p(\xi) \sin n\xi \, d\xi \quad (7.7.22)$$

$$\psi = \frac{4R}{a\mu\pi} \sum_{n=1,3,5,\ldots}^{\infty} B_n \sin n\phi \int_0^{\pi/2} p(\xi) \sin n\xi \, d\xi \quad (7.7.23)$$

$$M_x = \frac{4R}{\mu\pi} \sum_{n=1,3,5,\ldots}^{\infty} M_{xn} \sin n\phi \int_0^{\pi/2} p(\xi) \sin n\xi \, d\xi \quad (7.7.24)$$

$$M_z = \frac{-4R}{\mu\pi} \sum_{n=1,3,5,\ldots}^{\infty} M_{zn} \cos n\phi \int_0^{\pi/2} p(\xi) \sin n\xi \, d\xi \quad (7.7.25)$$

Tables (7.7.3) and (7.7.4) contain the first 24 coefficients of the series expressing functions η, ψ, M_x, and M_z for $\mu = 1$, $\lambda = 100$ and for $\mu = 0.7$, $\lambda = 60$.

TABLE 7.7.3

COEFFICIENTS FOR ANTISYMMETRICAL LOADING: $\mu = 1$; $\lambda = 100$

n	Y_n	B_n	M_{xn}	M_{zn}
1	0	0	0	0
3	0.0^{2}507614	−0.0^{2}913705	0.0365483	−0.0121828
5	0.0^{2}141176	−0.0^{2}271492	0.0365483	−0.0^{2}651582
7	0.0^{3}407138	−0.0^{3}797990	0.0191518	−0.0^{2}273597
9	0.0^{3}151918	−0.0^{3}300131	0.0120053	−0.0^{2}133392
11	0.0^{4}683916	−0.0^{3}135662	0.0^{2}813972	−0.0^{3}739974
13	0.0^{4}350961	−0.0^{4}697793	0.0^{2}586146	−0.0^{3}450882
15	0.0^{4}198018	−0.0^{4}394283	0.0^{2}441597	−0.0^{3}294398
17	0.0^{4}120001	−0.0^{4}239174	0.0^{2}344411	−0.0^{3}202595
19	0.0^{5}768873	−0.0^{4}153350	0.0^{2}276030	−0.0^{3}145279
21	0.0^{5}515091	−0.0^{4}102785	0.0^{2}226127	−0.0^{3}107679
23	0.0^{5}357894	−0.0^{5}714437	0.0^{2}188612	−0.0^{4}820052

TABLE 7.7.4

COEFFICIENTS FOR ANTISYMMETRICAL LOADING: $\mu = 0.70$; $\lambda = 60$

n	Y_n	B_n	M_{xn}	M_{zn}
1	−0.0^{2}205882	0.0^{2}205882	0	0
3	0.0^{2}441037	−0.0^{2}924365	0.0304497	−0.0101499
5	0.0^{2}103217	−0.0^{2}237120	0.0234330	−0.0^{2}468660
7	0.0^{3}287263	−0.0^{3}677875	0.0133980	−0.0^{2}191400
9	0.0^{3}106431	−0.0^{3}253996	0.0^{2}836692	−0.0^{3}929668
11	0.0^{4}478355	−0.0^{3}114816	0.0^{2}566327	−0.0^{3}514843
13	0.0^{4}245393	−0.0^{4}590959	0.0^{2}408805	−0.0^{3}314465
15	0.0^{4}137393	−0.0^{4}331564	0.0^{2}305818	−0.0^{3}203879
17	0.0^{5}839161	−0.0^{4}202794	0.0^{2}240489	−0.0^{3}141464
19	0.0^{5}537736	−0.0^{4}130078	0.0^{2}192822	−0.0^{3}101485
21	0.0^{5}380291	−0.0^{5}872167	0.0^{2}158016	−0.0^{4}752957
23	0.0^{5}250358	−0.0^{5}606375	0.0^{2}131833	−0.0^{4}573187

Example 7.7.1. Consider a circular beam of radius $R = 25$ ft made of reinforced concrete ($E = 3 \times 10^6$ psi) of square cross section (depth 30 in., width 30 in.) with eight concentrated loads $P = 50$ tons each ($2\gamma = 45°$) (see Fig. 7.7.4). The constant of the foundation is assumed to be $\beta = 1500$ lb/sq in. It follows that

$$I_x = 6.75(10^4) \text{ in.}^4$$

$$\lambda = 60$$

$$\mu = 0.7$$

$$v = \frac{R^2 P}{EI_x} = 0.04444$$

$$a = \frac{EI_x}{R} = 28,125 \text{ ft-tons}$$

$$av = 1250 \text{ ft-tons}$$

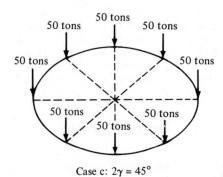

Case c: $2\gamma = 45°$

Figure 7.7.4

From Eqs. (7.7.18), (7.7.19), (7.7.20), and (7.7.21) one obtains in this particular case:

$$\eta = \frac{4P}{\pi \beta R} + \frac{8R^2 P}{a\mu\pi} \sum_{n=8,16,24,\ldots}^{\infty} Y_n \cos n\phi \cos n(22°30') \qquad (7.7.26)$$

$$\psi = \frac{8RP}{a\mu\pi} \sum_{n=8,16,24,\ldots}^{\infty} B_n \cos n\phi \cos n(22°30') \qquad (7.7.27)$$

$$M_x = \frac{8RP}{\mu\pi} \sum_{n=8,16,24,\ldots}^{\infty} M_{xn} \cos n\phi \cos n(22°30') \qquad (7.7.28)$$

$$M_z = \frac{8RP}{\mu\pi} \sum_{n=8,16,24,\ldots}^{\infty} M_{zn} \sin n\phi \cos n(22°30') \qquad (7.7.29)$$

In Table 7.7.5, results obtained by using the first three terms of the series given by the approximate Eqs. (7.7.26), (7.7.27), (7.7.28), and (7.7.29) are presented. In Table 7.7.6, results for the same type of beam are given using the solution in closed form which was presented in Sec. 7.5.

TABLE 7.7.5

ϕ	v, inches	ψ, radians	M_x, ft-tons	M_z, ft-tons
0°	0.2751	0.0^{4}6187	−40.7586	0
5°	0.2766	0.0^{4}4964	−31.7226	−3.2630
10°	0.2809	0.0^{4}1483	−16.8680	−5.3951
15°	0.2867	−0.0^{4}4585	12.1185	−5.8185
20°	0.2910	−0.0^{4}6489	57.1428	−2.7282
22°30′	0.2917	−0.0^{4}7022	65.4239	0

TABLE 7.7.6

ϕ	v, inches	ψ, radians	M_x, ft-tons	M_z, ft-tons
0°	0.2748	0.0^{4}6444	−40.9000	0
5°	0.2764	0.0^{4}5111	−34.850	−3.3875
10°	0.2811	0.0^{4}1529	−16.700	−5.7250
15°	0.2869	−0.0^{4}3111	13.6125	−5.9500
20°	0.2915	−0.0^{4}6755	56.1125	−3.0000
22°30′	0.2923	−0.0^{4}7377	81.8875	0

BIBLIOGRAPHY

1. Barden, L., "Influence Coefficients for Beams Resting on Soil," *Civil Eng.*, London, Vol. **58**, No. **682**, 1963, pp. 601–605.
2. Biot, M. A., "Bending of an Infinite Beam on an Elastic Foundation," *J. Appl. Mech.*, Vol. **4**, 1937, p. 17.
3. Boussinesq, J. V., "Application des Potentiels à l'Etude de l'Equilibre et du Mouvement des Solides Elastiques," Gauthier-Villars, Paris, France, 1885, p. 90.
4. Essenburg, F., "Shear Deformation in Beams on Elastic Foundations," *J. Appl. Mech.*, Vol. **2**, 1962, pp. 313–317.
5. Euler, L., "De Pressione Ponderis in Planum cui Incumbit," *Novi. Comm. Acad. Petropolit.*, T. **18**, St. Petersburg, 1774, pp. 289–329.
6. ———, "Von dem Druck eines mit einem Gewicht Beschwerten Tisches auf eine Fläche," *Hindenburg's Archiv Reinen Angew. Math.* Vol. **1**, 1795, pp. 74–80.
7. Franciosi, V., "Contributo allo Studio delle Travi su Mezzo Elastico," *L'Ingegnere*, 1957.
8. Freudenthal, A. M., and H. G. Lorsch, "The Infinite Elastic Beam on a Linear Viscoelastic Foundation," *Proc. Am. Soc. Civil Engrs.*, Vol. **83**, E.M. 1 1957, pp. 1–22.
9. Fung., Y. C., "Foundations of Solid Mechanics," Prentice-Hall, Inc., Englewood Cliffs, N.J., 1965, pp. 195–197.
10. Hendry, A. W., "New Method for the Analysis of Beams on Elastic Foundations,"

I and II, *Civil Eng., London,* **Vol. 53, No. 621,** 1958, pp. 297–299; **No. 622,** 1958, pp. 444–446.

11. Hetenyi, M., "Beams on Elastic Foundation," The University of Michigan Press, Ann Arbor, Michigan, 1946.

12. ——, "A General Solution for the Bending of Beams on an Elastic Foundation of Arbitrary Continuity," *J. Appl. Phys.,* **Vol. 21, No. 1,** 1950, pp. 55–58.

13. ——, "A Comparison of Various Solutions for Beams on Elastic Foundations," *Acta Tech. Acad. Sci. Hung.,* **Vol. 26, Nos. 1** and **2,** 1959, pp. 21–28.

14. ——, "Beams and Plates on Elastic Foundations and Related Problems," *Appl. Mech. Rev.,* **Vol. 19, No. 2,** 1966, pp. 95–102.

15. ——, "Beams on Elastic Foundation," in "Handbook of Engineering Mechanics," W. Flugge, Ed., McGraw-Hill Book Company, New York, N.Y., 1962.

16. Inglis, Sir Charles, "A Mathematical Treatise on Vibrations in Railway Bridges," Cambridge University Press, London, England, 1934.

17. Jodi, C., "Procedimenti per Ricerche Sperimentali su Suoli Elastico," *Atti Accad. Nazl. Lincei, Rend., Serie 6,* 1935.

18. Kármán, T. von, and M. Biot, "Mathematical Methods in Engineering," McGraw-Hill Book Company, New York, N.Y., 1940, Chapter VIII, Sec. 6.

19. Kerr, A. D., "Elastic and Viscoelastic Foundation Models," *J. Appl. Mech.,* 1964, pp. 491–498.

20. Krylov, A. N., "Analysis of Beams on Elastic Foundation" (in Russian) 2nd ed., Russian Academy of Sciences, Russia, 1931.

21. Love, A. E. H., "A Treatise on the Mathematical Theory of Elasticity," 4th ed., Dover Publications, Inc., New York, N.Y., 1944.

22. Malter, H., "Numerical Solutions for Beams on Elastic Foundations," *Proc. Am. Soc. Civil Engrs.,* **Vol. 84, St. 2,** 1958, Paper 1562.

23. Raymondi, C., "Sul Problema delle Travi con Appoggio Elastico Continuo," *Atti Ist. Sci. Costruzioni Univ. Pisa,* 1957.

24. ——, "Contributo allo Studio della Trave su Suolo Elastico," *Atti Ist. Sci. Costruzioni Univ. Pisa,* **No. 60,** 1958.

25. ——, "Complementi allo Studio della Trave su Suolo Elastico," *Atti Ist. Sci. Costruzioni Univ. Pisa,* **No. 65,** 1959.

26. ——, "Sull'Impostazione Euristica per Via Differenziale del Problema della Trave su Suolo Elastico," *Atti Ist. Sci. Costruzioni Univ. Pisa,* **No. 82,** 1961.

27. Rodriguez, D. A., "Three-Dimensional Bending of a Ring on an Elastic Foundation," *J. Appl. Mech.,* **Vol. 3,** 1961, pp. 461–463.

28. Saint-Venant, Barré de, "Memoire sur le calcul de la résistance et de la flexion des pièces solides a simple ou a double courbure, en prenant simultanément en considération les divers efforts auxquels elles peuvent etre soumises dans tous les sens," *Compt. Rend.,* **Vol. 17,** 1843, pp. 942–954, 1020–1031.

29. Seide, P., "Elasto-Plastic Bending of Beams on Elastic Foundations," *J. Aerospace Sci.,* **Vol. 23, No. 6,** 1956, pp. 563–570.

30. Timoshenko, S. P., "Strength of Materials, Part II, Advanced Theory and Problems," The Macmillan Company, New York, N.Y., 1936, pp. 467–473.

31. ——, "History of Strength of Materials," McGraw-Hill Book Company, New York, N.Y., 1953.

32. Volterra, E., "Sul Problema Generale della Trave Poggiata su Suolo Elastico," Nota 1 and 2, *Atti Accad. Nazl. Lincei, Rend., Serie 7,* **Vol. 2, Fasc. 3** and **Fasc. 4.,** 1947.

33. ——, "Bending of a Circular Beam Resting on an Elastic Foundation," *J. Appl. Mech., Trans. ASME,* **Vol. 74,** 1952, pp. 1–4.

34. ——, "Deflections of Circular Beams Resting on Elastic Foundations under Symmetric or Antisymmetric Loadings," *Proc. Eighth Intern. Congr. Theoret. Appl. Mech.,* Istanbul, Turkey, 1952.

35. ——, "Deflections of Circular Beams Resting on Elastic Foundations Obtained by Methods of Harmonic Analysis," *J. Appl. Mech., Trans. ASME,* **Vol. 73,** 1953, pp. 227–232.

36. Volterra, E., and R. Chung, "Bending of a Constrained Circular Beam Resting on an Elastic Foundation," *Trans. Am. Soc. Civil Engrs.,* **Vol. 120,** 1955, pp. 301–310.

37. Winkler, E., "Die Lehre von der Elastizität und Festigkeit," Dominicus, Prague, Czechoslovakia, 1867.

38. Zanaboni, O., "Soluzione Abbreviata della Trave su Suolo Elastico," *Ing. Ferroviaria,* 1955.

39. Zimmermann, H., "Die Berechnung des Eisenbahnoberbaues," Berlin, Germany, 1888.

PROBLEMS

7-1. A steel bar of square cross section with 2-in. sides is 24 ft long, is supported by a foundation of modulus $k = 400$ lb/in.^{3}, and is subjected in its central part to three equal loads P of 2 tons each at a distance of 3 ft (see Fig. P.7.1). Compute the deflection and the bending moment under the central load.

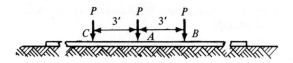

Figure P.7.1

7-2. If the portion CB of the beam in Problem 7.1 is rigid, determine the bending moment at A.

7-3. Solve Problem 7.1 for the case where the total length of the beam is 12 ft.

7-4. Solve Problem 7.1 for the case where the beam is a wooden beam of rectangular cross section (breadth 10 in., and height 12 in.) and of total length 20 ft. Assume that for wood $E = 1.6 \times 10^6$ psi.

7-5. Compute the maximum bending moment, the maximum deflection, and the maximum bending stress for a railroad rail subjected to a single wheel load of 12.5 tons. The rail is supported by ties and ballast having a spring constant $\beta =$

2000 lb/sq in. Assume that $I = 88.6$ in.^{4}, $E = 30 \times 10^6$ psi, the depth of the rail is $7\frac{1}{8}$ in. and that the distance of the centroidal axis of the cross section of the rail from the top surface is 3.90 in.

7-6. If the rail of Problem 7.5 is subjected to two concentrated loads of 20 tons which are 4 ft apart, compute the maximum deflection, the maximum bending moment, and the maximum normal stress in the rail.

7-7. Solve Problem 7.6 for the case where there are three concentrated loads of 20 tons which are 4 ft apart.

7-8. Compute the maximum deflection, the maximum bending moment, and the maximum normal stress in the rail of Problem 7.6, if the rail is 4 ft long and carries a single concentrated load of 20 tons at its center.

7-9. A 10 I 35 American Standard beam 12 ft long is subjected to a load P of 10 tons at its center, rests on an elastic foundation having $\beta = 1500$ lb/sq in. and is supported at its ends by two immovable supports. Compute

1. The maximum deflection.
2. The maximum bending moment.
3. The maximum normal stress.
4. The reactions at the supports.

For such a beam $I = 146$ in.^{4}

7-10. Show that the deflection and the bending moment at the middle of the uniformly loaded beam of length l with hinged ends and resting on an elastic foundation as shown in Fig. P.7.2 are, respectively,

$$v_c = \frac{p_0}{\beta}\left[1 - \frac{2\cosh\frac{\alpha l}{2}\cos\frac{\alpha l}{2}}{\cosh \alpha l + \cos \alpha l}\right]$$

$$M_c = \frac{p_0}{\alpha^2}\frac{\cosh \alpha l \sin \alpha l}{\cosh \alpha l + \cos \alpha l}$$

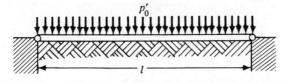

Figure P.7.2

7-11. The beam of length l shown in Fig. P.7.3 has built in ends and rests on an elastic foundation. If the beam is subjected to a uniformly distributed load p_0 and a concentrated force P at the center then show that the bending moments at the ends are

$$M = -\frac{P}{\alpha}\frac{\sinh\frac{\alpha l}{2}\sin\frac{\alpha l}{2}}{\alpha \sinh \alpha l + \sin \alpha l} - \frac{p_0}{2\alpha^2}\frac{\sinh \alpha l - \sin \alpha l}{\sinh \alpha l + \sin \alpha l}$$

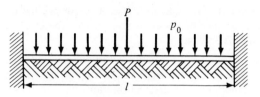

Figure P.7.3

7-12. Show that the deflection curve for the beam of length l resting on an elastic foundation with a load P applied at one end (see Fig. P.7.4) is

$$v(x) = \frac{2P\alpha}{\beta(\sinh^2 \alpha l - \sin^2 \alpha l)}$$
$$\times [\sinh \alpha l \cos \alpha x \cosh \alpha (l - x) - \sin \alpha l \cosh \alpha x \cos \alpha (l - x)]$$

Figure P.7.4

7-13. Show that the deflection curve for the beam of length l hinged at both ends and resting on an elastic foundation (see Fig. P.7.5) and bent by a couple M_0 applied at the end is

$$v(x) = \frac{2M_0 \alpha^2}{\beta(\cosh^2 \alpha l - \cos^2 \alpha l)}$$
$$\times [\cosh \alpha l \sin \alpha x \sinh \alpha (l - x) - \cos \alpha l \sinh \alpha x \sin \alpha (l - x)]$$

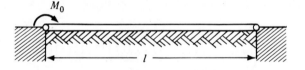

Figure P.7.5

7-14. A circular beam of radius $R = 30$ ft is made of reinforced concrete ($E = 3 \times 10^6$ psi) of rectangular cross section (depth 30 in., width 20 in.) and is subjected to six concentrated loads $P = 100$ tons each ($2\gamma = 60°$). The constant of the foundation is assumed to be $\beta = 1000$ psi. Compute deflections, angle of twist, and bending and twisting moments for such a beam.

7-15. Solve Problem 7.14 for the case where the beam cannot rotate at the sections where the loads are applied.

7-16. Solve Problem 7.14 by using the harmonic analysis method. Compare the results with those of Problem 7.14.

CHAPTER

8

BENDING OF PLATES

8.0 Introduction

In this chapter, problems of bending of thin plates with various shapes, boundary conditions, and loading characteristics are discussed. First the classical Lagrange equation of equilibrium is derived following the simple but rigorous method usually employed in elementary courses of the strength of materials. Then problems of circular plates with various loads and boundary conditions are considered, followed by the solution of the problem of an elliptic plate which is built-in at the edge and subjected to a uniformly distributed load. Navier's and Maurice Levy's solutions for rectangular plates under various loads and boundary conditions are then discussed. Next Nadai's and Woinowsky-Krieger's solutions for triangular plates are considered.

The principle of virtual work is then used to obtain rigorous and approximate solutions to the problem of bending of rectangular plates. Next the Ritz and Marcus methods for obtaining approximate solutions are presented. The approximate method proposed by Grashof and very often used by practical engineers is then discussed. The chapter ends with a general discussion of the bending of plates resting on elastic foundations, the foundation being assumed to follow Winkler's simple law as well as the most generalized law.

Lagrange, Giuseppe Luigi. Italian mathematician (b. Turin 1736; d. Paris 1813). At the age of nineteen, Lagrange was appointed Professor of Mathematics at the Royal School of Artillery in Turin and later was one of the founders of the Turin Academy of Sciences. From 1766 to 1787 he was director of the Berlin Academy of Sciences succeeding Euler. From 1787 until his death, he was a professor at the Ecole Normale and the Ecole Polytechnique in Paris. Napoleon honored Lagrange by making him a count and a senator. He is buried in the Pantheon in Paris. Lagrange is often said to be second only to Newton among modern mathematicians. In almost every branch of mathematics he left the mark of his genius. He was the discoverer of the calculus of variations, and he is considered to be the father of analytic mechanics. His *Mécanique Analytique* was first published in Paris in 1788. His complete works in 14 volumes were published in Paris in 1867 to 1892. (From *Dynamics of Vibrations* by Enrico Volterra and E. C. Zachmanoglou. Reprinted with permission of Charles E. Merrill Books, Inc., Columbus, Ohio.) Reproduction from the Vito Volterra collection, Villa Volterra, Ariccia (Rome).

8.1 Derivation of the Lagrange Equilibrium Equation for a Thin Plate[1]

The equilibrium equation for a thin, flat plate, first obtained by Lagrange in 1811, is

$$\frac{\partial^4 w}{\partial x^4} + 2\frac{\partial^4 w}{\partial x^2 \partial y^2} + \frac{\partial^4 w}{\partial y^4} = \frac{1-\nu^2}{EI}p \qquad (8.1.1)$$

[1] Part of Section 8.1 is from *Dynamics of Vibrations* by Enrico Volterra and E. C. Zachmanoglou, and is reproduced with the permission of Charles E. Merrill Books, Inc.

where w represents the displacement of a generic point of the middle surface of the plate, $I = h^3/12$ the moment of inertia, h the thickness of the plate, E and v Young's modulus and Poisson's ratio, respectively, for the material of the plate, and p the external load per unit area. In general, the load p may vary over the surface of the plate and is, therefore, considered to be a function of x and y. The plate is referred to a system of orthogonal coordinates $Oxyz$ with the horizontal xy plane coincident with the middle surface of the undeformed plate and the vertical z-axis directed positively downward (see Fig. 8.1.1). Components of the elastic displacement in the x-, y-, and z-directions will be indicated, respectively, by the letters u, v, and w.

To derive the Lagrange equation of equilibrium [Eq. (8.1.1)] the following hypotheses, attributed to Kirchhoff, are made:

1. It is assumed that a straight line initially normal to the middle surface of the plate remains straight and normal to the deformed middle surface of the plate and unchanged in length. This assumption corresponds to the Bernoulli-Navier hypothesis for the deflection of bars. In this hypothesis transverse shear and transverse normal strains are neglected.

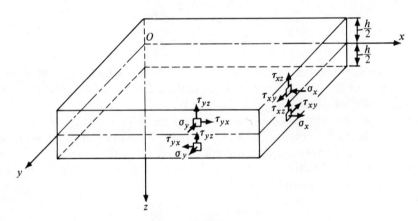

Figure 8.1.1

2. The normal stresses σ_x and σ_y (see Figs. 8.1.1 and 8.1.2) are assumed to be zero at the middle surface of the plate. This assumption is valid, if the deflections w are small as compared to the plate thickness h.
3. The displacement w is assumed to be small; consequently, the curvatures $1/\rho_x$ and $1/\rho_y$ of the middle surface can be expressed as

$$\frac{1}{\rho_x} = \frac{\partial^2 w}{\partial x^2}$$

$$\frac{1}{\rho_y} = \frac{\partial^2 w}{\partial y^2}$$

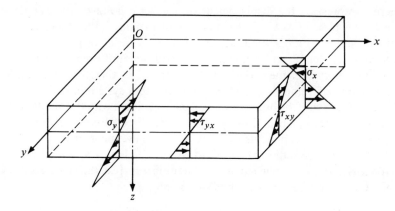

Figure 8.1.2

and the angles ϕ_x and ϕ_y that the tangent plane to the middle surface makes with the x- and y-axes can be expressed as

$$\phi_x \simeq \tan \phi_x = \frac{\partial w}{\partial x}$$

$$\phi_y \simeq \tan \phi_y = \frac{\partial w}{\partial y}$$

As a consequence of the first two assumptions, stresses σ_x and σ_y at a point within the plate are proportional to the distance z of the point from the middle surface of the plate (see Fig. 8.1.2).

Under the third assumption, points on the middle surface of the plate can be assumed to be displaced only in the z-direction. For other points of the plate, the corresponding u and v displacements in the x- and y-directions are proportional to their distances from the middle surface. Thus, a point is displaced by the amount

$$u = -z\phi_x = -z\frac{\partial w}{\partial x}$$

in the x-direction, and by the amount

$$v = -z\frac{\partial w}{\partial y}$$

in the y-direction. The components of strain are

$$\left.\begin{aligned}\epsilon_x &= \frac{\partial u}{\partial x} = -z\frac{\partial^2 w}{\partial x^2} \\ \epsilon_y &= \frac{\partial v}{\partial y} = -z\frac{\partial^2 w}{\partial y^2} \\ \gamma_{xy} &= \frac{\partial u}{\partial y} + \frac{\partial v}{\partial x} = -2z\frac{\partial^2 w}{\partial x \partial y}\end{aligned}\right\} \quad (8.1.2)$$

and the corresponding stresses are

$$\begin{aligned}
\sigma_x &= \frac{E}{1-v^2}(\epsilon_x + v\epsilon_y) = -\frac{Ez}{1-v^2}\left(\frac{\partial^2 w}{\partial x^2} + v\frac{\partial^2 w}{\partial y^2}\right) \\
\sigma_y &= \frac{E}{1-v^2}(\epsilon_y + v\epsilon_x) = -\frac{Ez}{1-v^2}\left(\frac{\partial^2 w}{\partial y^2} + v\frac{\partial^2 w}{\partial x^2}\right) \\
\tau_{xy} &= G\gamma_{xy} = -2Gz\frac{\partial^2 w}{\partial x \partial y} = -\frac{Ez}{(1+v)}\frac{\partial^2 w}{\partial x \partial y}
\end{aligned} \quad (8.1.3)$$

The bending and twisting moments on surfaces normal to the x-axis are denoted by M_x and M_{xy}, respectively (see Fig. 8.1.3). Similarly, M_y and M_{yx} denote bending and twisting moments on surfaces normal to the y-axis. The bending and twisting moments acting on areas of unit width and height h may be expressed as follows:

$$\begin{aligned}
M_x &= \int_{-h/2}^{h/2} \sigma_x(1)z\, dz \\
M_y &= \int_{-h/2}^{h/2} \sigma_y(1)z\, dz \\
M_{xy} &= M_{yx} = \int_{-h/2}^{h/2} \tau_{xy}(1)z\, dz = \int_{-h/2}^{h/2} \tau_{yx}(1)z\, dz
\end{aligned} \quad (8.1.4)$$

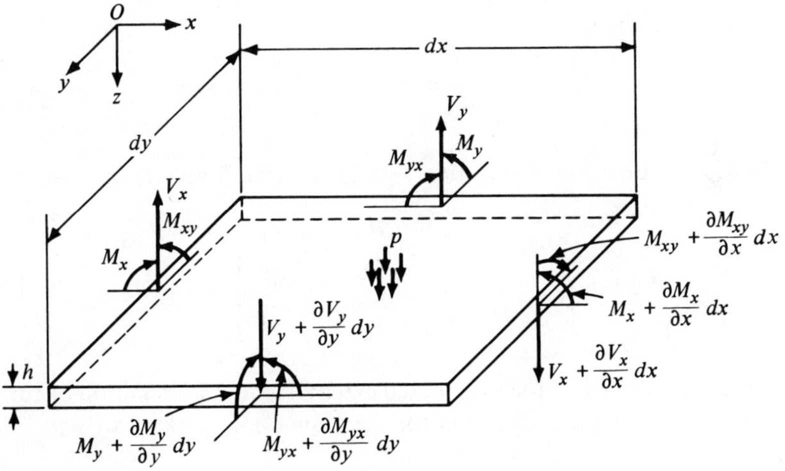

Figure 8.1.3

By substituting Eq. (8.1.3) into Eq. (8.1.4), and noting that

$$I = \int_{-h/2}^{h/2} z^2(1)\, dz = \frac{h^3}{12}$$

and letting

$$D = \frac{EI}{(1-v^2)} = \frac{Eh^3}{12(1-v^2)}$$

we obtain the relations

$$\left.\begin{aligned} M_x &= -D\left(\frac{\partial^2 w}{\partial x^2} + v\frac{\partial^2 w}{\partial y^2}\right) \\ M_y &= -D\left(\frac{\partial^2 w}{\partial y^2} + v\frac{\partial^2 w}{\partial x^2}\right) \\ M_{xy} &= -(1-v)D\frac{\partial^2 w}{\partial x \partial y} \end{aligned}\right\} \quad (8.1.5)$$

Now consider the equilibrium conditions for the plate element shown in Fig. 8.1.3. The forces per unit length on surfaces normal to the x- and y-axes are denoted by V_x and V_y, respectively. The condition of moment equilibrium about the y-axis is expressed by the equation

$$V_x\,dx\,dy - \frac{\partial M_x}{\partial x}dx\,dy - \frac{\partial M_{xy}}{\partial y}dx\,dy = 0$$

or

$$V_x = \frac{\partial M_x}{\partial x} + \frac{\partial M_{xy}}{\partial y} \quad (8.1.6)$$

In the same way, the condition of moment equilibrium about the x-axis is expessed by the equation

$$V_y = \frac{\partial M_y}{\partial x} + \frac{\partial M_{xy}}{\partial y} \quad (8.1.7)$$

Substitution of Eqs. (8.1.5) into Eqs. (8.1.6) and (8.1.7) gives

$$\left.\begin{aligned} V_x &= -D\frac{\partial}{\partial x}\left(\frac{\partial^2 w}{\partial x^2} + \frac{\partial^2 w}{\partial y^2}\right) \\ V_y &= -D\frac{\partial}{\partial y}\left(\frac{\partial^2 w}{\partial x^2} + \frac{\partial^2 w}{\partial y^2}\right) \end{aligned}\right\} \quad (8.1.8)$$

If p is the intensity of the distributed external load, the condition of equilibrium of forces in the z-direction for the plate element of area $dx\,dy$ and thickness h (see Fig. 8.1.3) is expressed by

$$p\,dx\,dy + \frac{\partial V_x}{\partial x}dx\,dy + \frac{\partial V_y}{\partial y}dx\,dy = 0$$

or

$$\frac{\partial V_x}{\partial x} + \frac{\partial V_y}{\partial y} = -p \quad (8.1.9)$$

Substitution of Eqs. (8.1.8) into Eq. (8.1.9) gives the equilibrium equation

$$\left(\frac{\partial^4 w}{\partial x^4} + 2\frac{\partial^4 w}{\partial x^2 \partial y^2} + \frac{\partial^4 w}{\partial y^4}\right) = \frac{(1-v^2)p}{EI} = \frac{p}{D} \qquad (8.1.1)$$

which was to be derived.

8.2 Deflection of a Circular Plate

Lagrange's equation [Eq. (8.1.1)] transforms, in the case of a circular plate symmetrically bent with respect to its center O, to the equation

$$\left(\frac{d^2}{dr^2} + \frac{1}{r}\frac{d}{dr}\right)\left(\frac{d^2 w}{dr^2} + \frac{1}{r}\frac{dw}{dr}\right) = \frac{p}{D} \qquad (8.2.1)$$

In Eq. (8.2.1) w represents the deflection of the circular plate and r the distance from its center O. The expressions for the stresses for this case are as follows[2]:

$$\left.\begin{array}{l} \sigma_r = -\dfrac{Ez}{1-v^2}\left(\dfrac{d^2 w}{dr^2} + \dfrac{v}{r}\dfrac{dw}{dr}\right) \\[2mm] \sigma_\theta = -\dfrac{Ez}{1-v^2}\left(\dfrac{1}{r}\dfrac{dw}{dr} + v\dfrac{d^2 w}{dr^2}\right) \\[2mm] \tau_{r\theta} = 0 \end{array}\right\} \qquad (8.2.2)$$

If $p = $ constant, the general solution of Eq. (8.2.1) is[3]

$$w = C_1 \log r + C_2 r^2 \log r + C_3 r^2 + C_4 + \frac{pr^4}{64D} \qquad (8.2.3)$$

where C_1, C_2, C_3, and C_4 are constants to be determined from the boundary conditions.

(a) Plate built-in at the edge and subjected to a uniformly distributed load[4]. The two constants C_1 and C_2 must, in this case, be zero in order for w and

---

[2] Equations (8.2.2) are easily derived as follows: replace the second derivatives in Eqs. (8.1.3) by the equivalent expressions in polar coordinates, (see Appendix 3.A); then, by setting θ equal to zero and noting that w is independent of θ because of symmetry, one obtains Eqs. (8.2.2).
[3] See Appendix 3.B.
[4] See Fig. 8.2.1.

d^2w/dr^2 to remain finite for $r = 0$. The remaining two constants C_3 and C_4 are determined from the boundary conditions

$$(w)_{r=a} = 0 \quad \text{or} \quad C_3 a^2 + C_4 + \frac{pa^4}{64D} = 0$$

$$\left(\frac{dw}{dr}\right)_{r=a} = 0 \quad \text{or} \quad 2C_3 a + \frac{pa^3}{16D} = 0$$

where a is the radius of the plate. It follows that

$$C_3 = -\frac{pa^2}{32D}$$

$$C_4 = \frac{pa^4}{64D}$$

and that

$$w = \frac{pa^4}{64D}\left(\frac{r^4}{a^4} - 2\frac{r^2}{a^2} + 1\right) \tag{8.2.4}$$

Now, in view of Eqs. (8.2.2), the stresses are

$$\left.\begin{aligned} \sigma_r &= -\frac{pa^2}{16I}z\left[(3+v)\frac{r^2}{a^2} - 1 - v\right] \\ \sigma_\theta &= -\frac{pa^2}{16I}z\left[(1+3v)\frac{r^2}{a^2} - 1 - v\right] \\ \tau_{r\theta} &= 0 \end{aligned}\right\} \tag{8.2.5}$$

The maximum deflection is at the center of the plate ($r = 0$) and is equal to

$$w_{\max} = \frac{pa^4}{64D} \tag{8.2.6}$$

The deflection surface [Eq. (8.2.4)] is a surface of revolution whose meridian section is an algebraic curve of the fourth degree, symmetrical with respect to the z-axis, with two points of inflection ($d^2w/dr^2 = 0$ at $r = \pm a/\sqrt{3}$), and with three points having zero slope (at the center, $r = 0$, and at the supports, $r = \pm a$). This curve is shown in Fig. 8.2.1. At the upper and lower free surface of the plate ($z = \pm h/2$) the stresses are

$$\left.\begin{aligned} (\sigma_r)_l^u &= \pm\frac{pa^2 h}{32I}\left[(3+v)\frac{r^2}{a^2} - 1 - v\right] \\ (\sigma_\theta)_l^u &= \pm\frac{pa^2 h}{32I}\left[(1+3v)\frac{r^2}{a^2} - 1 - v\right] \end{aligned}\right\} \tag{8.2.7}$$

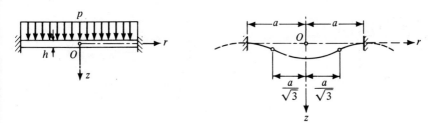

Figure 8.2.1

The positive sign indicates tension, while the minus sign indicates compression. At $r = 0$

$$(\sigma_r)_i^u = (\sigma_\theta)_i^u = \mp(1+\nu)\frac{pa^2h}{32I}$$

At $r = a$

$$\left.\begin{aligned}(\sigma_r)_i^u &= \pm\frac{pa^2h}{16I} \\ (\sigma_\theta)_i^u &= \pm\nu\frac{pa^2h}{16I}\end{aligned}\right\} \quad (8.2.8)$$

Since Poisson's ratio ν is smaller than unity it follows from Eqs. (8.2.8) that the extreme values for σ_r and σ_θ are given by σ_r at the built-in edge where

$$[\sigma_r(a)]_i^u = \sigma_{\substack{\max \\ \min}} = \pm\frac{pa^2h}{16I} \quad (8.2.9)$$

From Eqs. (8.2.7), one sees that $\sigma_r = 0$ at $r = a\sqrt{(1+\nu)/(3+\nu)}$ while $\sigma_\theta = 0$ at $r = a\sqrt{(1+\nu)/(1+3\nu)}$. These two values of r are greater than the value $a/\sqrt{3}$ where the meridian section of the deflection surface has a point of inflection.

(b) Plate simply supported at the edge and subjected to a uniformly distributed load.[5] In this case, as in the preceding one, $C_1 = C_2 = 0$. The boundary conditions which determine the remaining constants C_3 and C_4 are, in this case,

$$(w)_{r=a} = 0 \quad \text{or} \quad C_3 a^2 + C_4 + \frac{pa^4}{64D} = 0$$

$$(\sigma_r)_{r=a} = 0 \quad \text{or} \quad C_3 = -\frac{pa^2}{32D}\frac{3+\nu}{1+\nu}$$

[5] See Fig. 8.2.2.

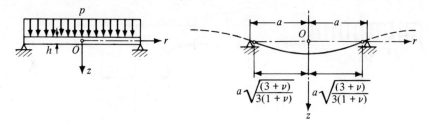

Figure 8.2.2

from which it follows that

$$C_4 = \frac{pa^4}{64D}\frac{5+\nu}{1+\nu}$$

Thus, the deflection equation is

$$w = \frac{pa^4}{64D}\left[\frac{r^4}{a^4} - 2\frac{(3+\nu)}{(1+\nu)}\frac{r^2}{a^2} + \frac{(5+\nu)}{(1+\nu)}\right] \quad (8.2.10)$$

and the stresses are given by

$$\left.\begin{array}{l}\sigma_r = \dfrac{pa^2}{16I}z(3+\nu)\left(1-\dfrac{r^2}{a^2}\right) \\[6pt] \sigma_\theta = \dfrac{pa^2}{16I}z\left[3+\nu - (1+3\nu)\dfrac{r^2}{a^2}\right] \\[6pt] \tau_{r\theta} = 0\end{array}\right\} \quad (8.2.11)$$

The maximum deflection, which occurs at the center of the plate ($r = 0$), is

$$w_{\max} = \frac{pa^4}{64D}\frac{(5+\nu)}{(1+\nu)} \quad (8.2.12)$$

If one compares Eq. (8.2.12) with Eq. (8.2.6) one finds the ratio

$$\frac{\text{maximum deflection for simply supported plate}}{\text{maximum deflection for built-in plate}} = \frac{5+\nu}{1+\nu}$$

which depends on Poisson's ratio. In the case of $\nu = 0.3$, this ratio is close to 4. In the case of beams

$$w_{\max} = \frac{5p_0 l^4}{384 EI} \quad \text{for a simply supported beam}$$

and

$$w_{\max} = \frac{p_0 l^4}{384 EI} \quad \text{for a built in beam}$$

and the corresponding ratio between the maximum deflections is equal to 5. The deflection curve [Eq. (8.2.10)] is a surface of revolution whose meridian section is an algebraic curve of the fourth degree, symmetrical with respect to the z-axis, with two points of inflection ($d^2w/dr^2 = 0$) at $r = \pm a\sqrt{[1 + (v/3)]/(1 + v)}$. These two points of inflection are nearer to the border than in the case of the built-in plate since $0 \leq v \leq 0.5$ (see Fig. 8.2.2).

At the upper and lower free surface of the plate ($z = \pm h/2$) the stresses are

$$(\sigma_r)_i^u = \mp \frac{pa^2h}{32I}(3 + v)\left(1 - \frac{r^2}{a^2}\right)$$
$$(\sigma_\theta)_i^u = \mp \frac{pa^2h}{32I}\left[3 + v - (1 + 3v)\frac{r^2}{a^2}\right]$$
(8.2.13)

Equations (8.2.13) show that the stress σ_r is negative (compression) at every point of the upper surface of the plate and positive (tension) at every point of the lower surface with the exception of the border ($r = a$) where $\sigma_r = 0$. The same applies to the stress σ_θ. It does not, however, become zero at the border ($r = a$). Instead, it reaches the value

$$[(\sigma_\theta)_i^u]_{r=a} = \mp \frac{pa^2h}{16I}(1 - v)$$

The stresses σ_r and σ_θ reach their maximum values at the center of the plate where they have the same value:

$$\sigma_{\substack{\min \\ \max}} = [(\sigma_r)_i^u]_{r=0} = [(\sigma_\theta)_i^u]_{r=0} = \mp(3 + v)\frac{pa^2h}{32I} \quad (8.2.14)$$

By comparing Eqs. (8.2.14) and (8.2.9) one sees that the ratio of the absolute values of the extreme values of the stresses for the simply supported and the built-in plates depends on Poisson's ratio and is equal to $(3 + v)/2$. The value of this ratio is between 1.5 and 1.75 since $0 \leq v \leq 0.5$. The corresponding ratio in the case of a simply supported beam and a built-in beam is equal to 1.5 and is independent of Poisson's ratio.

(c) *Influence of shear on the deflection of the plate.* In the elementary theory of deflection of thin plates, which has been used so far, the effects of the shear stress τ_{rz} and the normal stress σ_z have not been taken into consideration. The results which have been obtained are therefore approximate and their accuracy depends on the ratio h/a between the thickness h of the plate and the radius a of the plate.

If q denotes the shear force per unit length, it is possible to extend to bent plates the theory used for bent beams by assuming that the distribution of the shear stress across the thickness of the plate is parabolic with the maximum value occuring at the neutral axis.[6] Therefore, by letting

$$(\tau_{rz})_{max} = -1.5\frac{q}{h}$$

it follows that the corresponding shear strain is

$$\gamma_{rz} = \frac{\tau_{rz}}{G} = -1.5\frac{q}{hG}$$

Calling w_1 the additional deflection due to shear, then

$$\frac{dw_1}{dr} = \frac{\tau_{rz}}{G} = -1.5\frac{q}{hG}$$

Considering a simply supported circular plate of radius a (see Fig. 8.2.2), the additional deflection due to shear will be given by

$$w_1 = -\frac{1.5}{hG}\int_a^r q\, dr \qquad (8.2.15)$$

For a uniformly distributed load of intensity p_0

$$q = \frac{\pi p_0 r^2}{2\pi r} = \frac{p_0 r}{2}$$

and Eq. (8.2.15) becomes

$$w_1 = -\frac{1.5 p_0}{2hG}\int_a^r r\, dr = \frac{3p_0}{8hG}(a^2 - r^2) = \frac{p_0 h^2}{16D}\frac{(a^2 - r^2)}{(1 - v)}$$

The total deflection of a simply supported circular plate, taking into account the effect of shear, is given by

$$w = \frac{pa^4}{64D}\left[\frac{r^4}{a^4} - 2\frac{(3+v)}{(1+v)}\frac{r^2}{a^2} + \frac{5+v}{1+v} + \frac{4h^2}{a^2(1-v)}\left(1 - \frac{r^2}{a^2}\right)\right]$$

The maximum deflection occurs at the center of the plate ($r = 0$) and is given by

$$w_{max} = \frac{pa^4}{64D}\left[\frac{5+v}{1+v} + \frac{4h^2}{a^2(1-v)}\right]$$

(d) Plate built-in at the edge and subjected to a concentrated load at its center[7]. Since in this case $p = 0$, Eq. (8.2.3) becomes

$$w = C_1 \log r + C_2 r^2 \log r + C_3 r^2 + C_4 \qquad (8.2.16)$$

[6] See Section 5.1.
[7] See Fig. 8.2.3.

The constant $C_1 = 0$ in order that w be finite at $r = 0$. The two boundary conditions for the built-in edge are

$$(w)_{r=a} = 0 \quad \text{or} \quad C_2 a^2 \log a + C_3 a^2 + C_4 = 0 \qquad (8.2.17)$$

$$\left(\frac{dw}{dr}\right)_{r=a} = 0 \quad \text{or} \quad C_2 a(2 \log a + 1) + 2C_3 = 0 \qquad (8.2.18)$$

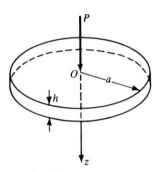

Figure 8.2.3

The third equation necessary to determine the three constants C_2, C_3, and C_4 of Eq. (8.2.16) is obtained by taking into account the shear force. Consider the forces acting on an element of the plate (see Fig. 8.2.4). These forces are:

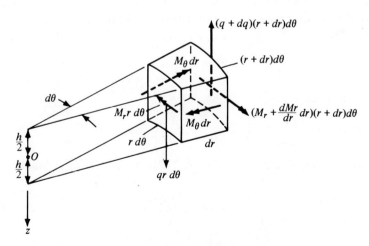

Figure 8.2.4

1. Forces $qr\, d\theta$ and $(q + dq)(r + dr)\, d\theta$ which are due to the presence of shear forces q and $q + dq$ per unit length.

2. The couples $M_r r\, d\theta$ and $[M_r + (dM_r/dr)\, dr](r + dr)\, d\theta$ which are due to normal stresses σ_r acting on the cylindrical surfaces $hr\, d\theta$ and $h(r + dr)\, d\theta$.
3. The two couples $M_\theta\, dr$ due to stresses σ_θ.

The above couples are represented in Fig. 8.2.4 by their corresponding vectors. By expressing the condition of equilibrium of the above forces and moments one obtains

$$(qr\, d\theta)\, dr + \left(M_r + \frac{dM_r}{dr}\, dr\right)(r + dr)\, d\theta - M_r r\, d\theta - 2(M_\theta\, dr)\frac{d\theta}{2} = 0$$

By simplifying and neglecting terms of higher order one has

$$q = \frac{M_\theta - M_r}{r} - \frac{dM_r}{dr} \qquad (8.2.19)$$

Now, by using Eqs. (8.2.2) for σ_r and σ_θ, one finds that

$$\left.\begin{aligned} M_r &= \int_{-h/2}^{h/2} z\sigma_r\, D\, dz = -D\left(\frac{d^2w}{dr^2} + \frac{\nu}{r}\frac{dw}{dr}\right) \\ M_\theta &= \int_{-h/2}^{h/2} z\sigma_\theta\, D\, dz = -D\left(\frac{1}{r}\frac{dw}{dr} + \nu\frac{d^2w}{dr^2}\right) \end{aligned}\right\} \qquad (8.2.20)$$

By introducing Eqs. (8.2.20) into Eq. (8.2.19) one obtains[8]

$$q = D\left[\frac{d^3w}{dr^3} + \frac{1}{r}\frac{d^2w}{dr^2} - \frac{1}{r^2}\frac{dw}{dr}\right] = D\frac{d}{dr}\left[\frac{1}{r}\frac{d}{dr}\left(r\frac{dw}{dr}\right)\right] \qquad (8.2.21)$$

The third condition is obtained by substituting the first three derivatives of Eq. (8.2.16) into Eq. (8.2.21) and noting that the shear force q must be equal to $P/2\pi r$; thus,

$$D\left[\frac{d^3w}{dr^3} + \frac{1}{r}\frac{d^2w}{dr^2} - \frac{1}{r^2}\frac{dw}{dr}\right] = \frac{4D}{r}C_2 = \frac{P}{2\pi r}$$

from which it follows that

$$C_2 = \frac{P}{8\pi D} \qquad (8.2.22)$$

By substituting Eq. (8.2.22) into Eqs. (8.2.17) and (8.2.18) one obtains

$$C_3 = -\frac{P}{16\pi D}(2\log a + 1)$$

$$C_4 = \frac{Pa^2}{16\pi D}$$

---

[8] Equation (8.2.21) applies also in the case in which there is a distributed external force p acting on the plate.

and finally

$$w = \frac{P}{16\pi D}\left(2r^2 \log \frac{r}{a} + a^2 - r^2\right) \qquad (8.2.23)$$

The maximum deflection is at the center of the plate. Its value is

$$w_{max} = \frac{Pa^2}{16\pi D} \qquad (8.2.24)$$

The equation for the maximum deflection of a circular plate which is built-in at the edge and subjected to a uniformly distributed load [Eq. (8.2.6)] may be written as

$$\frac{pa^4}{64D} = \frac{Pa^2}{64\pi D} \qquad (8.2.6)$$

where $P = p\pi a^2$ is the total distributed load. By comparing Eqs. (8.2.24) and (8.2.6) one finds that

$$\frac{\text{maximum deflection due to concentrated force}}{\text{maximum deflection due to distributed force}} = 4$$

The above ratio is independent of Poisson's ratio v. The stresses are derived by substituting Eq. (8.2.23) into Eqs. (8.2.2); thus,

$$\left.\begin{array}{l}\sigma_r = \dfrac{Pz}{4\pi I}\left[(1+v)\log \dfrac{a}{r} - 1\right] \\[2mm] \sigma_\theta = \dfrac{Pz}{4\pi I}\left[(1+v)\log \dfrac{a}{r} - v\right]\end{array}\right\} \qquad (8.2.25)$$

Equations (8.2.25) show that stresses σ_r and σ_θ become infinite at the center of the plate ($r = 0$) where the concentrated load is applied. Therefore, the theory which has been presented is not valid near the point of application of the concentrated force.

(e) *Plate simply supported at the edge and subjected to a concentrated load at its center.* Using Eq. (8.2.16) with $C_1 = 0$ the three equations necessary to determine the constants C_2, C_3, and C_4 are, in this case,

$$(w)_{r=a} = 0$$
$$(\sigma_r)_{r=a} = 0$$
$$q = \frac{P}{2\pi r}$$

It follows that the elastic surface w is given by

$$w = \frac{P}{16\pi D}\left[2r^2 \log \frac{r}{a} + \frac{3+\nu}{1+\nu}(a^2 - r^2)\right] \qquad (8.2.26)$$

The maximum deflection is at the center of the plate and its value is

$$w_{\max} = \frac{Pa^2}{16\pi D}\frac{3+\nu}{1-\nu} \qquad (8.2.27)$$

By letting $P = \pi a^2 p$ the above result can be compared with Eq. (8.2.12) which gives the maximum deflection for the circular plate which is simply supported at the edge and subjected to a uniformly distributed load p; thus,

$$\frac{w_{\max} \text{ due to concentrated load}}{w_{\max} \text{ due to distributed load}} = 4\left(\frac{3+\nu}{5+\nu}\right)$$

For $\nu = 0.30$, the above ratio is 2.49. By comparing Eqs. (8.2.27) and (8.2.24) one sees that the ratio of the maximum deflections of the plate under concentrated loads is

$$\frac{w_{\max} \text{ for simply supported plate}}{w_{\max} \text{ for built-in plate}} = \frac{3+\nu}{1+\nu}$$

For $\nu = 0.30$, the above ratio is 2.54.

The stresses are given by the equations

$$\left.\begin{array}{l}\sigma_r = \dfrac{Pz}{4\pi I}(1+\nu)\log\dfrac{a}{r} \\[1em] \sigma_\theta = \dfrac{Pz}{4\pi I}\left[(1+\nu)\log\dfrac{a}{r} + 1 - \nu\right]\end{array}\right\} \qquad (8.2.28)$$

Also, in this case, the stresses become infinite near the concentrated load ($r = 0$).

(f) Circular plate with a concentric circular hole bent by the couples M_1 and M_2 uniformly distributed along the edges and supported at the outer edge.[9] Since $q = 0$, Eq. (8.2.21) in this case becomes

$$\frac{d}{dr}\left[\frac{1}{r}\frac{d}{dr}\left(r\frac{dw}{dr}\right)\right] = 0 \qquad (8.2.29)$$

By integrating Eq. (8.2.29) one obtains

---

[9] See Fig. 8.2.5.

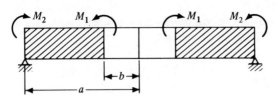

Figure 8.2.5

$$\frac{1}{r}\frac{d}{dr}\left(r\frac{dw}{dr}\right) = C_1 \quad \text{or} \quad \frac{d}{dr}\left(r\frac{dw}{dr}\right) = C_1 r$$

$$r\frac{dw}{dr} = C_1\frac{r^2}{2} + C_2 \quad \text{or} \quad \frac{dw}{dr} = C_1\frac{r}{2} + \frac{C_2}{r}$$

and finally

$$w = C_1\frac{r^2}{4} + C_2 \log r + C_3 \tag{8.2.30}$$

where C_1, C_2, and C_3 are constants which are determined by the three conditions

$$(M_r)_{r=a} = M_2$$
$$(M_r)_{r=b} = M_1$$
$$(w)_{r=a} = 0$$

Finally one obtains for the deflection w and for the radial moment M_r the expressions

$$w = -k_1\frac{r^2}{4} - k_2 \log \frac{r}{a} + k_3$$

$$M_r = D\left[-\frac{k_1}{2} - \frac{k_2}{r^2} + v\left(\frac{k_1}{2} + \frac{k_2}{r^2}\right)\right]$$

where

$$k_1 = \frac{2(a^2 M_2 - b^2 M_1)}{(1+v)D(a^2 - b^2)}$$

$$k_2 = \frac{a^2 b^2 (M_2 - M_1)}{(1-v)D(a^2 - b^2)}$$

$$k_3 = \frac{k_1 a^2}{4}$$

(g) Circular plate with a concentric circular hole bent by a uniformly distributed force applied at the inner edge and simply supported at the outer edge.[10] Equation (8.2.21) is in this case[11]

$$\frac{d}{dr}\left[\frac{1}{r}\frac{d}{dr}\left(r\frac{dw}{dr}\right)\right] = \frac{q}{D} = \frac{q_0 b}{Dr} \qquad (8.2.31)$$

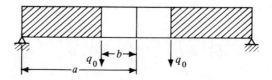

Figure 8.2.6

By integrating Eq. (8.2.31) one obtains

$$w = \frac{q_0 b}{8D} r^2 [2 \log r - 1] - \frac{q_0 b}{8D} r^2 + C_1 \frac{r^2}{4} + C_2 \log r + C_3$$

where C_1, C_2, and C_3 are constants which are determined by the three conditions

$$(M_r)_{r=a} = (M_r)_{r=b} = (w)_{r=a} = 0$$

One finally obtains for the deflection w, the expression

$$w = \frac{q_0 b}{4D} r^2 \left(\log \frac{r}{a} - 1\right) - \frac{k_1 r^2}{4} - k_2 \log \frac{r}{a} + k_3 \qquad (8.2.32)$$

where

$$k_1 = \frac{q_0 b}{2D}\left[\frac{1-v}{1+v} - \frac{2b^2}{(a^2-b^2)} \log \frac{b}{a}\right]$$

$$k_2 = -\frac{q_0 b}{2D}\frac{(1+v)}{(1-v)}\frac{a^2 b^2}{a^2 - b^2} \log \frac{b}{a}$$

$$k_3 = \frac{q_0 b a^2}{4D}\left[1 + \frac{1}{2}\frac{(1-v)}{(1+v)} - \frac{b^2}{a^2 - b^2} \log \frac{b}{a}\right]$$

(h) Other cases of bent circular plates. Once the results for case (f) and (g) are known it is possible by combining these results with the previously obtained results for the complete plate to discuss deflections of various types of circular plates with concentric holes. For instance, the case shown in Fig.

---

[10] See Fig. 8.2.6.
[11] The shear force at the inner edge is q_0, while the shear force q per unit length of circumference of radius r is $q = q_0 b/r$.

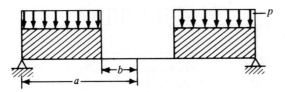

Figure 8.2.7

8.2.7 of a circular plate which has a concentric circular hole and is subjected to a uniformly distributed force of intensity p applied on the whole surface of the plate (the plate being simply supported at the outer edge) can be studied by superimposing results for the cases shown in Figs. 8.2.8 and 8.2.9, for

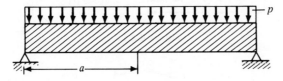

Figure 8.2.8

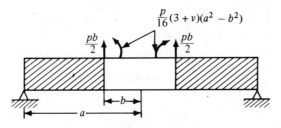

Figure 8.2.9

which solutions are known. The case shown in Fig. 8.2.8 is that of a circular plate subjected to a uniformly distributed load of intensity p over the whole area, while the case shown in Fig. 8.2.9 is that of an annular plate subjected on the inner border of radius b to a shear force per unit length

$$q = \frac{\pi b^2 p}{2\pi b} = \frac{pb}{2}$$

and to a bending moment

$$M_r = \frac{p}{16}(3+v)(a^2 - b^2)$$

In Figs. 8.2.10, 8.2.11, 8.2.12, 8.2.13, and 8.2.14 various cases of practical importance are presented. Results for these cases are obtained by superposition of the cases discussed in the previous sections. For cases I, II, and III let

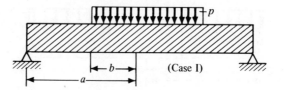

Figure 8.2.10

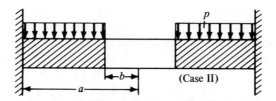

Figure 8.2.11

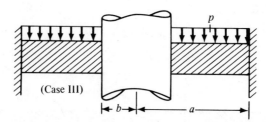

Figure 8.2.12

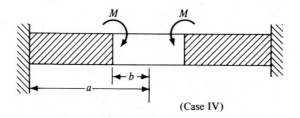

Figure 8.2.13

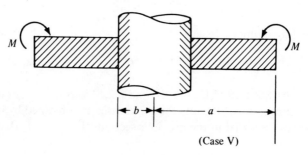

Figure 8.2.14

$$M_{max} = \alpha \frac{pa^4}{h} \quad \text{and} \quad w_{max} = \beta \frac{pa^4}{Eh^3}$$

represent, respectively, the maximum bending moment and maximum deflection where values of α and β are given in Table 8.2.1 for various values of the ratio a/b. For cases IV and V let

$$M_{max} = \alpha \frac{Ma^2}{h^2} \quad \text{and} \quad w_{max} = \beta \frac{Ma}{Eh^3}$$

TABLE 8.2.1

$\frac{a}{b}$	Case I[1]		Case II[2]		Case III[3]	
	α	β	α	β	α	β
1.25	1.13	0.62	0.10	0.003	0.07	0.001
1.50	0.96	0.51	0.26	0.01	0.18	0.005
2.00	0.70	0.34	0.48	0.05	0.36	0.02
3.00	0.40	0.17	0.65	0.13	0.55	0.06
4.00	0.26	0.10	0.71	0.16	0.63	0.09
5.00	0.185	0.06	0.73	0.18	0.67	0.11

[1] Fig. 8.2.10.
[2] Fig. 8.2.11.
[3] Fig. 8.2.12.

denote the maximum bending moment and maximum deflection, respectively, with values of α and β given in Table 8.2.2 for various values of the ratio a/b.

TABLE 8.2.2

$\frac{a}{b}$	Case IV[1]		Case V[2]	
	α	β	α	β
1.25	6	0.20	6.86	0.23
1.50	6	0.48	7.45	0.16
2.00	6	0.85	8.13	1.50
3.00	6	0.94	8.71	1.55
4.00	6	0.80	8.93	3.10
5.00	6	0.65	9.03	3.41

[1] Fig. 8.2.13.
[2] Fig. 8.2.14.

8.3 Bending of an Elliptic Plate Built-in at the Edge and Subjected to a Uniformly Distributed Load

Consider an elliptic plate (see Fig. 8.3.1) which has semi-axes a and b ($b > a$) and is built-in at the border and subjected to the uniformly distributed load

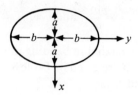

Figure 8.3.1

of intensity p. Let

$$w = C\left(1 - \frac{x^2}{a^2} - \frac{y^2}{b^2}\right)^2 \tag{8.3.1}$$

where C is a constant to be determined. From Eq. (8.3.1) it follows that

$$\left.\begin{array}{c} \dfrac{\partial^4 w}{\partial x^4} = \dfrac{24C}{a^4} \\[6pt] \dfrac{\partial^4 w}{\partial y^4} = \dfrac{24C}{b^4} \\[6pt] \dfrac{\partial^4 w}{\partial x^2 \partial y^2} = \dfrac{8C}{a^2 b^2} \end{array}\right\} \tag{8.3.2}$$

By substituting Eqs. (8.3.2) into Eq. (8.1.1), i.e,

$$\frac{\partial^4 w}{\partial x^4} + 2\frac{\partial^4 w}{\partial x^2 \partial y^2} + \frac{\partial^4 w}{\partial y^4} = \frac{p}{D} \tag{8.1.1}$$

one sees that Eq. (8.1.1) is satisfied if

$$C = \frac{p}{8D} \frac{a^4 b^4}{3a^4 + 2a^2 b^2 + 3b^4}$$

Thus, Eq. (8.3.1) becomes

$$w = \frac{p}{8D} \frac{a^4 b^4}{3a^4 + 2a^2 b^2 + 3b^4}\left(1 - \frac{x^2}{a^2} - \frac{y^2}{b^2}\right)^2 \tag{8.3.4}$$

Equation (8.3.4) satisfies the boundary conditions for an elliptic plate with a clamped edge. These boundary conditions are

$$\left.\begin{array}{c} w = 0 \\[4pt] \dfrac{\partial w}{\partial n} = 0 \end{array}\right\} \quad \text{for} \quad \frac{x^2}{a^2} + \frac{y^2}{b^2} = 1$$

where n is the normal to the plate boundary. The maximum deflection is at the center of the plate and its value is

$$w_{max} = \frac{pa^4b^4}{8D(3a^4 + 2a^2b^2 + 3b^4)} \tag{8.3.5}$$

The bending moments M_x and M_y and the twisting moment M_{xy} on sections perpendicular to the x- and y-axes are

$$\left. \begin{aligned} M_x &= -D\left[\frac{\partial^2 w}{\partial x^2} + v\frac{\partial^2 w}{\partial y^2}\right] = 4DC\left\{\left[\frac{1}{a^2} - \frac{3x^2}{a^4} - \frac{y^2}{a^2b^2}\right] + v\left[\frac{1}{b^2} - \frac{3y^2}{b^4} - \frac{x^2}{a^2b^2}\right]\right\} \\ M_y &= -D\left[\frac{\partial^2 w}{\partial y^2} + v\frac{\partial^2 w}{\partial x^2}\right] = 4DC\left\{\left[\frac{1}{b^2} - \frac{3y^2}{b^4} - \frac{x^2}{a^2b^2}\right] + v\left[\frac{1}{a^2} - \frac{3x^2}{a^4} - \frac{y^2}{a^2b^2}\right]\right\} \end{aligned} \right\} \tag{8.3.6}$$

$$M_{xy} = -(1-v)D\frac{\partial^2 w}{\partial x \partial y} = -(1-v)\frac{8DC}{a^2b^2}xy \tag{8.3.7}$$

The bending moments at the center of the plate ($x = y = 0$) are

$$\left. \begin{aligned} (M_x)_{\substack{x=0 \\ y=0}} &= 4DC\left(\frac{1}{a^2} + \frac{v}{b^2}\right) \\ (M_y)_{\substack{x=0 \\ y=0}} &= 4DC\left(\frac{1}{b^2} + \frac{v}{a^2}\right) \end{aligned} \right\} \tag{8.3.8}$$

The bending moments at the ends of the smaller axis ($x = \pm a$, $y = 0$) are

$$\left. \begin{aligned} (M_x)_{\substack{x=\pm a \\ y=0}} &= -\frac{8DC}{a^2} \\ (M_y)_{\substack{x=\pm a \\ y=0}} &= -v\frac{8DC}{a^2} \end{aligned} \right\} \tag{8.3.9}$$

The value of M_x given by the first of Eqs. (8.3.9) is the maximum (in absolute value) of the bending moments of the plate. The bending moments at the ends of the greater axis are obtained from Eq. (8.3.9) by substituting b in the place of a. In the case in which $a = b = R$ the results obtained in this section coincide with the results which were obtained in the case of a circular plate built-in at the border [Sec. 8.2. (a)]. In the case in which $b \gg a$, the results coincide with those of a uniformly loaded strip of span $2a$ which is built-in at both ends and subjected to a uniformly distributed load p. In this case, Eqs. (8.3.5), (8.3.8), and (8.3.9) become

$$w_{max} = \frac{p(2a)^4}{384D} \tag{8.3.5'}$$

$$M_x = \frac{p(2a)^2}{24} \text{ at the center of the strip} \qquad (8.3.8')$$

$$M_x = -\frac{p(2a)^2}{12} \text{ at the built-in ends} \qquad (8.3.9')$$

Example 8.3.1. Compare the results for an elliptic plate for which $b = 2a$ with those of a circular plate of radius $R = a$, both built-in at the border and subjected to a uniformly distributed load p.

For the elliptic plate one finds that

$$w_{max} = \frac{p}{8D} \frac{16a^8}{3a^4 + 8a^4 + 48a^4} = \frac{16}{59} \frac{pa^4}{8D} = 2.17 \frac{pa^4}{64D}$$

The bending moment M_x at the end of the minor axis of the ellipse is

$$M_x = -\frac{8D}{a^2} \frac{2}{59} \frac{pa^4}{D} = -2.17 \frac{pa^2}{8}$$

The bending moment at the center of the ellipse is

$$M_x = 4D \frac{2}{59} \frac{pa^4}{D} \left(\frac{1}{a^2} + \frac{\nu}{4a^2} \right) = 2.17 \left(1 + \frac{\nu}{4} \right) \frac{pa^2}{16}$$

From the above results, it follows that the maximum deflection and the maximum negative bending moment for the elliptic plate are 2.17 times greater than those for the circular plate which has its radius equal to the minor axis of the ellipse.

If $\nu = 0$, the maximum positive bending moment is also 2.17 times greater; if, instead $\nu = 0.30$, it is 1.80 times greater for the elliptic plate than it is for the circular plate.

8.4 Navier's Solution for Simply Supported Rectangular Plates

(a) Introduction. Consider the rectangular plate of sides a and b which is simply supported on the four sides and subjected to a distributed load p. The plate is referred to the system of cartesian coordinates $Oxyz$ shown in Fig. 8.4.1. Navier in 1820 gave a rigorous solution for the deflection of such a plate by means of a double trigonometric series. Suppose that p, which is a function of the two variables x and y, can be expressed in the domain $0 \le x \le a$, $0 \le y \le b$ by the double trigonometric series[12]

$$p = \sum_{m=1}^{\infty} \sum_{n=1}^{\infty} p_{mn} \left(\sin \frac{m\pi x}{a} \right) \left(\sin \frac{n\pi y}{b} \right) \qquad (8.4.1)$$

---
[12] See Appendix of Chapter 5.

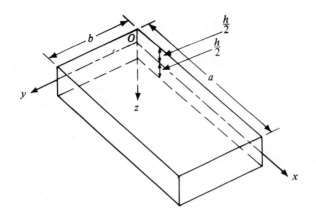

Figure 8.4.1

where

$$p_{mn} = \frac{4}{ab} \int_0^a \int_0^b p \sin \frac{m\pi x}{a} \sin \frac{n\pi y}{b} \, dx \, dy \tag{8.4.2}$$

Then Lagrange's equation [Eq. (8.1.1)] becomes, in this case,

$$\frac{\partial^4 w}{\partial x^4} + 2\frac{\partial^4 w}{\partial x^2 \partial y^2} + \frac{\partial^4 w}{\partial y^4} = \frac{1}{D} \sum_{m=1}^{\infty} \sum_{n=1}^{\infty} p_{mn} \sin \frac{m\pi x}{a} \sin \frac{n\pi y}{b} \tag{8.4.3}$$

Let

$$w = \sum_{m=1}^{\infty} \sum_{n=1}^{\infty} A_{mn} \sin \frac{m\pi x}{a} \sin \frac{n\pi y}{b} \tag{8.4.4}$$

By substituting Eq. (8.4.4) into Eq. (8.4.3), one finds that

$$A_{mn} = \frac{p_{mn}}{D\pi^4 \left(\frac{m^2}{a^2} + \frac{n^2}{b^2}\right)^2} \tag{8.4.5}$$

and that the solution of Lagrange's equation [Eq. (8.4.3)] is given by

$$w = \frac{1}{D\pi^4} \sum_{m=1}^{\infty} \sum_{n=1}^{\infty} \frac{p_{mn}}{\left(\frac{m^2}{a^2} + \frac{n^2}{b^2}\right)^2} \sin \frac{m\pi x}{a} \sin \frac{n\pi y}{b} \tag{8.4.6}$$

where the coefficients p_{mn} are computed from Eq. (8.4.2).

The solution expressed by Eq. (8.4.6) satisfies three of the four conditions required for a simply supported plate; these are

1. $\qquad w = 0 \text{ for } \begin{cases} x = 0 \text{ and } x = a \\ y = 0 \text{ and } y = b \end{cases}$

2. $\sigma_x = 0$ for $x = 0$ and $x = a$
3. $\sigma_y = 0$ for $y = 0$ and $y = b$

However, the fourth condition that

$$\tau_{xy} = \tau_{yx} = -\frac{Ez}{1+v}\frac{\partial^2 w}{\partial x \partial y}$$

be zero on the borders of the plate is not satisfied except when $z = 0$. It can be shown, however, that the presence of stresses τ_{xy} causes only a modification of the distribution of the reaction on the supports without changing the distribution of internal stresses and that, therefore, the ensuing solution is still valid.

(b) Case of uniformly distributed load. In the case of a uniformly distributed load $p = p_0 =$ constant and Eq. (8.4.2) becomes

$$p_{mn} = \frac{4p_0}{ab}\int_0^a \int_0^b \sin\frac{m\pi x}{a}\sin\frac{n\pi y}{b}\,dy\,dx = \frac{4p_0}{\pi^2 mn}(1-\cos m\pi)(1-\cos n\pi)$$

or

$$p_{mn} = \begin{cases} \dfrac{16p_0}{\pi^2 mn} & \text{for } \begin{cases} m = 1, 3, 5, \ldots, \infty \\ n = 1, 3, 5, \ldots, \infty \end{cases} \\ 0 & \text{for } \begin{cases} m = 2, 4, 6, \ldots, \infty \\ n = 2, 4, 6, \ldots, \infty \end{cases} \end{cases}$$

Equation (8.4.6) becomes, in this case,

$$w = \frac{16p_0}{\pi^6 D}\sum_{m=1,3,5\ldots}^{\infty}\sum_{n=1,3,5\ldots}^{\infty}\frac{1}{mn\left(\dfrac{m^2}{a^2}+\dfrac{n^2}{b^2}\right)^2}\sin\frac{m\pi x}{a}\sin\frac{n\pi y}{b} \qquad (8.4.7)$$

By using the notation

$$\sum = \sum_{m=1,3,5\ldots}\sum_{n=1,3,5\ldots} \quad \text{and} \quad \rho = \frac{a}{b}$$

Eq. (8.4.7) can be written in either of the following two forms:

$$w = \frac{16p_0 a^4}{\pi^6 D}\sum\frac{1}{mn(m^2+n^2\rho^2)^2}\sin\frac{m\pi x}{a}\sin\frac{n\pi y}{b} \qquad (8.4.8)$$

or

$$w = \frac{16p_0 b^4}{\pi^6 D}\sum\frac{1}{mn\left(\dfrac{m^2}{\rho^2}+n^2\right)^2}\sin\frac{m\pi x}{a}\sin\frac{n\pi y}{b} \qquad (8.4.9)$$

It can be shown that the double series given by Eqs. (8.4.7), (8.4.8), and (8.4.9) is absolutely and uniformly convergent and that all its derivatives of an order less than the fourth can be obtained by differentiating term by term in the series. The maximum deflection occurs at the center of the plate ($x = a/2$, $y = b/2$) and its value is

$$w_{max} = \frac{16 p_0 a^4}{\pi^6 D} \sum \frac{(-1)^{(m+n-2)/2}}{mn(m^2 + n^2 \rho^2)^2} \qquad (8.4.10)$$

Equation (8.4.10) is obtained directly from Eq. (8.4.8) since

$$\left(\sin \frac{m\pi x}{a}\right)_{x=a/2} = \sin \frac{m\pi}{2} = (-1)^{(m-1)/2}$$

$$\left(\sin \frac{n\pi y}{b}\right)_{y=b/2} = \sin \frac{n\pi}{2} = (-1)^{(n-1)/2}$$

By letting

$$\sum \frac{(-1)^{(m+n-2)/2}}{mn(m^2 + n^2 \rho^2)^2} = \Phi(\rho) \qquad (8.4.11)$$

Eq. (8.4.10) becomes

$$w_{max} = \frac{16 p_0 a^4}{\pi^6 D} \Phi(\rho) \qquad (8.4.12)$$

The series in Eq. (8.4.11) is a very rapidly convergent series. Values of $\Phi(\rho)$ are given in Table 8.4.1 for various combinations of m and n. In the first group, $\Phi(\rho)$ has been computed for $m = 1$, $n = 1$. The other combinations are as follows:

In the second group $m = 1, 3; n = 1$.
In the third group $m = 1, 3; n = 1, 3$.
In the fourth group $m = 1, 3, 5; n = 1, 3$.
In the fifth group $m = 1, 3, 5; n = 1, 3, 5$.

TABLE 8.4.1

$\rho = \frac{a}{b}$	$\frac{1}{3}$	$\frac{1}{2}$	1	2	3
1st group	0.810000	0.640000	0.250000	0.040000	0.010000
2nd group	0.722651	0.604546	0.243333	0.037784	0.008921
3rd group	0.738093	0.609587	0.244268	0.038096	0.009112
4th group	0.734015	0.608338	0.244038	0.038018	0.009061
5th group	0.735659	0.608766	0.244116	0.038045	0.009079

Table 8.4.1 shows how rapidly the series $\Phi(\rho)$, whose terms are inversely proportional to the sixth power of m and n, converges.

The normal stresses are given by the first two of Eqs. (8.1.3), i.e.,

$$\sigma_x = -\frac{Ez}{(1-v^2)}\left(\frac{\partial^2 w}{\partial x^2} + v\frac{\partial^2 w}{\partial y^2}\right) \\ \sigma_y = -\frac{Ez}{(1-v^2)}\left(\frac{\partial^2 w}{\partial x^2} - v\frac{\partial^2 w}{\partial y^2}\right) \tag{8.1.3}$$

By substituting w from Eq. (8.4.8) into the first of Eqs. (8.1.3) and letting $z = h/2$ one finds on the lower surface of the plate that

$$(\sigma_x)_{z=h/2} = \frac{8p_0 h}{\pi^4 I}\left[a^2 \sum \frac{m}{n(m^2 + n^2\rho^2)^2}\sin\frac{m\pi x}{a}\sin\frac{n\pi y}{b} \right. \\ \left. + vb^2 \sum \frac{n}{m\left(\frac{m^2}{\rho^2} + n^2\right)^2}\sin\frac{m\pi x}{a}\sin\frac{n\pi y}{b}\right] \tag{8.4.13}$$

The above normal stress will be a maximum at the center of the plate ($x = a/2$, $y = b/2$) where

$$\left\{\left(\frac{\partial \sigma_x}{\partial x}\right)_{z=h/2}\right\}_{\substack{x=a/2 \\ y=b/2}} = \left\{\left(\frac{\partial \sigma_x}{\partial y}\right)_{z=h/2}\right\}_{\substack{x=a/2 \\ y=b/2}} = 0$$

One finds that

$$(\sigma_x)_{\max} = \frac{8p_0 h}{\pi^4 I}\left[a^2 \sum \frac{m(-1)^{(m+n-2)/2}}{n(m^2 + n^2\rho^2)^2} + vb^2 \sum \frac{n(-1)^{(m+n-2)/2}}{m\left(\frac{m^2}{\rho^2} + 2\right)^2}\right] \tag{8.4.14}$$

By letting

$$N(\rho) = \sum \frac{m(-1)^{(m+n-2)/2}}{n(m^2 + n^2\rho^2)^2} \tag{8.4.15}$$

one has

$$(\sigma_x)_{\max} = \frac{8}{\pi^4}\frac{p_0 h}{I}\left[a^2 N(\rho) + vb^2 N\left(\frac{1}{\rho}\right)\right] \tag{8.4.16}$$

The terms of the series in Eq. (8.4.14) are inversely proportional to the fourth powers of m and n; thus, the series for the maximum value of σ_x converges more slowly than the series for the maximum deflection [see Eq. (8.4.10). Table 8.4.2 furnishes values of $N(\rho)$ for different values of ρ.

TABLE 8.4.2

$\rho = \dfrac{a}{b}$	0	$\tfrac{1}{3}$	$\tfrac{1}{2}$	1	2	3	∞
$N(\rho)$	0.7609	0.7153	0.5879	0.2242	0.0271	0.0042	0.0000

By similar computations one finds that the maximum value of σ_y occurs at the center of the plate and is given by

$$\sigma_{y\text{max}} = \frac{8}{\pi^4}\frac{p_0 h}{I}\left[va^2 N(\rho) + b^2 N\!\left(\frac{1}{\rho}\right)\right] \tag{8.4.17}$$

Both stresses $(\sigma_x)_{\text{max}}$ and $(\sigma_y)_{\text{max}}$ are positive and are therefore tensile stresses. From Eqs. (8.4.16) and (8.4.17), one obtains

$$(\sigma_x)_{\text{max}} - (\sigma_y)_{\text{max}} = \frac{8}{\pi^4}\frac{p_0 h}{I}\left[a^2(1-v)N(\rho) - b^2(1-v)N\!\left(\frac{1}{\rho}\right)\right]$$

$$= \frac{8}{\pi^4}\frac{p_0 b^2 h(1-v)}{I}\left[\rho^2 N(\rho) - N\!\left(\frac{1}{\rho}\right)\right] \tag{8.4.18}$$

The expression in brackets in Eq. (8.4.18) is positive if $\rho > 1$ while it is negative if $\rho < 1$. For the square plate where $\rho = 1$, $(\sigma_x)_{\text{max}} = (\sigma_y)_{\text{max}}$.

The shear stress τ_{xy} is given by the third of Eqs. (8.1.3), i.e.,

$$\tau_{xy} = -\frac{Ez}{(1+v)}\frac{\partial^2 w}{\partial x \partial y} \tag{8.1.3}$$

By substituting $z = h/2$ and the expression for w [from Eq. (8.4.8)] into the above equation one finds that on the lower surface of the plate

$$(\tau_{xy})_{z=h/2} = -\frac{8}{\pi^4}\frac{a^3}{b}\frac{p_0 h(1-v)}{I}\sum\frac{1}{(m^2+n^2\rho^2)^2}\cos\frac{m\pi x}{a}\cos\frac{n\pi y}{b} \tag{8.4.19}$$

The shear stress becomes zero at the center of the plate and along the axes of symmetry of the plate for which $x = a/2$ and $y = b/2$. Its absolute value is the same at the four corners O, A, B, and C of the plate (see Fig. 8.4.2) where the shear stress has a maximum value. This maximum value is obtained from Eq. (8.4.19) by letting $x = y = 0$. One obtains in this case

$$(\tau_{xy})_{\text{max}} = \frac{8}{\pi^4}\frac{p_0 h(1-v)a^2\rho}{I}T(\rho) \tag{8.4.20}$$

where

$$T(\rho) = \sum\frac{1}{(m^2+n^2\rho^2)^2} \tag{8.4.21}$$

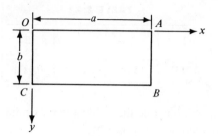

Figure 8.4.2

Table 8.4.3 furnishes values of $T(\rho)$ for various values of ρ.

TABLE 8.4.3

$\rho = \dfrac{a}{b}$	0	$\tfrac{1}{3}$	$\tfrac{1}{2}$	1	2	3	∞
$T(\rho)$	∞	1.2056	0.7928	0.2798	0.0495	0.0148	0.0000

Equations (8.4.17) and (8.4.20) may be written as

$$(\bar{\sigma}_y)_{\max} = \left[\nu N(\rho) + \frac{1}{\rho^2}N\left(\frac{1}{\rho}\right)\right]$$

$$(\bar{\tau}_{xy})_{\max} = (1-\nu)\rho T(\rho)$$

where

$$(\bar{\sigma}_y)_{\max} = \left(\frac{\pi^4 I}{8p_0 h a^2}\right)(\sigma_y)_{\max}$$

$$(\bar{\tau}_{xy})_{\max} = \left(\frac{\pi^4 I}{8p_0 h a^2}\right)(\tau_{xy})_{\max}$$

In Tables 8.4.4 and 8.4.5 values of $(\bar{\sigma}_y)_{\max}$ and $(\bar{\tau}_{xy})_{\max}$ are given for three different values of $\rho > 1$ and for $\nu = 0.25$ and $\nu = 0$. If one compares the figures of Tables 8.4.4 and 8.4.5, one sees that in the case $\nu = 0.25$ (case of a steel plate) the values for $(\bar{\sigma}_y)_{\max}$ are always greater than those for $(\bar{\tau}_{xy})_{\max}$. Instead, in the case $\nu = 0$ and $\rho = 1$ (case of a square plate) $(\bar{\tau}_{xy})_{\max} > (\bar{\sigma}_y)_{\max}$. It follows that in the case of plates of reinforced concrete which are nearly square and for which ν is small, one must employ reinforcement at the corners of the plate in order to avoid cracking due to shear stresses.

TABLE 8.4.4
($v = 0.25$)

$\rho = \dfrac{a}{b}$	1	2	3
$(\bar{\sigma}_y)_{\max}$	0.2802	0.1537	0.0806
$(\bar{\tau}_{xy})_{\max}$	0.2098	0.0792	0.0333

TABLE 8.4.5
($v = 0$)

$\rho = \dfrac{a}{b}$	1	2	3
$(\bar{\sigma}_y)_{\max}$	0.2242	0.1470	0.0796
$(\bar{\tau}_{xy})_{\max}$	0.2798	0.0990	0.0444

(c) Case of a concentrated load. In the case of a simply supported rectangular plate of sides a and b subjected to a concentrated load P acting on a generic point of coordinates $x = \xi$, $y = \eta$ (see Fig. 8.4.3) Navier's solution is

$$w = \frac{4P}{\pi^4 Dab} \sum_{m=1}^{\infty} \sum_{n=1}^{\infty} \frac{\sin \dfrac{m\pi\xi}{a} \sin \dfrac{n\pi\eta}{b}}{\left(\dfrac{m^2}{a^2} + \dfrac{n^2}{b^2}\right)^2} \sin \frac{m\pi x}{a} \sin \frac{n\pi y}{b} \quad (8.4.22)$$

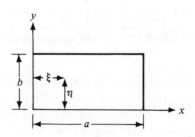

Figure 8.4.3

If load P is at the center of the plate ($\xi = a/2$, $\eta = b/2$) Eq. (8.4.22) becomes

$$w = \frac{4P}{\pi^4 Dab} \sum_{m=1,3,\ldots}^{\infty} \sum_{n=1,3,\ldots}^{\infty} \frac{\sin \dfrac{m\pi x}{a} \sin \dfrac{n\pi y}{b}}{\left(\dfrac{m^2}{a^2} + \dfrac{n^2}{b^2}\right)^2} \quad (8.4.23)$$

Equation (8.4.23) shows that $w = 0$ only when $x = 0$ or $x = a$ and $y = 0$ or $y = b$. For all other points x, y such that $0 < x < a$ and $0 < y < b, w > 0$. Suppose now that the plate is square ($a = b$) and that the load P is applied at its center. Then the maximum deflection is at the center of the plate and is equal to

$$w_{max} = \frac{4Pa^2}{\pi^4 D} \sum_{m=1,3,...}^{\infty} \sum_{n=1,3,...}^{\infty} \frac{1}{(m^2 + n^2)^2} \tag{8.4.24}$$

By computing the first nine terms of the series in Eq. (8.4.24) one gets

$$w_{max} = \frac{4Pa^2}{\pi^4 D} \left[\frac{1}{(2)^2} + \frac{2}{(10)^2} + \frac{1}{(18)^2} + \frac{2}{(25)^2} + \frac{2}{(34)^2} + \frac{1}{(50)^2} \right]$$

$$= \frac{4Pa^2}{\pi^4 D} 0.2782 = 0.01142 \frac{Pa^2}{D} \tag{8.4.25}$$

The exact value is $w_{max} = 0.01159 Pa^2/D$ and therefore the error is 1.47 per cent. The series in Eq. (8.4.25) is obtained by considering the following values of m and n:

$$m = 1, \text{ and } n = 1, 3, 5,$$
$$m = 3 \text{ and } n = 1, 3, 5,$$
$$m = 5 \text{ and } n = 1, 3, 5.$$

TABLE 8.4.6

$\rho = \frac{a}{b}$	1.0	1.1	1.2	1.4	1.5	1.6	1.8	2.0	3.0	4.0	∞
χ	0.01159	0.01266	0.01353	0.01485	0.01535	0.01570	0.01620	0.01651	0.01691	0.01693	0.01696

The exact value for the deflection is obtained by using simple series of rapid convergence, but whose terms are more difficult to compute. In Table 8.4.6, exact values are presented for the coefficient χ which gives the deflection at the center of a rectangular plate of sides a and b loaded by the concentrated load P applied at the center of the plate, i.e.,

$$w_{max} = \chi \frac{Pb^2}{D}$$

8.5 Maurice Levy's Solution for the Rectangular Plate

According to Maurice Levy, the solution of the problem of the rectangular plate of sides a and b (see Fig. 8.5.1), for particular boundary conditions on

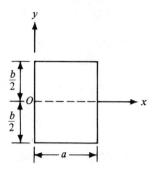

Figure 8.5.1

the sides $x = 0$ and $x = a$ and for general conditions of support on the sides $y = \pm b/2$, can be taken in the following form:

$$w = w_0 + \sum_{n=0}^{\infty} Y_n(y) \begin{cases} \sin \alpha_n x \\ \cos \alpha_n x \end{cases} \qquad (8.5.1)$$

In Eq. (8.5.1) w_0 is a particular solution of Lagrange's equation [Eq. (8.1.1)] which takes care of the loading while the function $Y_n(y)$ has four arbitrary constants and satisfies the boundary conditions on the sides $y = \pm b/2$. The principal advantage of Maurice Levy's solution as compared with the Navier solution is that instead of a double series one has to deal with a simple series.

In the following pages, Maurice Levy's method will be applied to the case of a simply supported rectangular plate subjected to a uniformly distributed load of intensity p_0.

To satisfy Lagrange's equation, i.e.,

$$\frac{\partial^4 w}{\partial x^4} + 2\frac{\partial^4 w}{\partial x^2 \partial y^2} + \frac{\partial^4 w}{\partial y^4} = \frac{p_0}{D} \qquad (8.1.1)$$

Eq. (8.5.1) will be taken in the form

$$w = w_0 + \sum_{n=1,3,\ldots}^{\infty} Y_n(y) \sin \frac{n\pi x}{a} \qquad (8.5.2)$$

with

$$w_0 = \frac{p_0 a^4}{24D}\left[\frac{x^4}{a^4} - 2\frac{x^3}{a^3} + \frac{x}{a}\right] \qquad (8.5.3)$$

Equation (8.5.2) satisfies the boundary conditions $w(0, y) = w(a, y) = 0$ and $M_x(0, y) = M_x(a, y) = 0$ for the plate simply supported at $x = 0$ and $x = a$. By substituting Eq. (8.5.2) into Eq. (8.1.1) one obtains

$$\sum_{n=1,3,\ldots}^{\infty} \left[\frac{d^4 Y_n}{dy^4} - \frac{2n^2\pi^2}{a^2}\frac{d^2 Y_n}{dy^2} + \frac{n^4\pi^4}{a^4} Y_n\right] \sin \frac{n\pi x}{a} = 0 \qquad (8.5.4)$$

Equation (8.5.4) is satisfied for any value of x if

$$\frac{d^4 Y_n}{dy^4} - \frac{2n^2\pi^2}{a^2}\frac{d^2 Y}{dy^2} + \frac{n^4\pi^4}{a^4}Y_n(y) = 0 \tag{8.5.5}$$

The general solution of Eq. (8.5.5) can be put in the form

$$Y_n = \frac{p_0 a^4}{D}\left[A_n \cosh\frac{n\pi y}{a} + B_n\frac{n\pi y}{a}\sinh\frac{n\pi y}{a}\right.$$
$$\left. + C_n \sinh\frac{n\pi y}{a} + D_n\frac{n\pi y}{a}\cosh\frac{n\pi y}{a}\right] \tag{8.5.6}$$

where A_n, B_n, C_n, and D_n are constants. Since Eq. (8.5.6) must be symmetrical with respect to the x-axis, i.e., it must have the same values if y is replaced by $-y$, it follows that

$$C_n = D_n = 0$$

and Eq. (8.5.2) becomes

$$w = \frac{p_0 a^4}{24D}\left[\left(\frac{x}{a}\right)^4 - 2\left(\frac{x}{a}\right)^3 + \left(\frac{x}{a}\right)\right]$$
$$+ \frac{p_0 a^4}{D}\sum_{n=1,3,\ldots}^{\infty}\left[A_n \cosh\frac{n\pi y}{a} + B_n\frac{n\pi y}{a}\sinh\frac{n\pi y}{a}\right]\sin\frac{n\pi x}{a} \tag{8.5.7}$$

By developing the first term of the above equation in a trigonometric series, i.e.,

$$\frac{p_0 a^4}{24D}\left[\left(\frac{x}{a}\right)^4 - 2\left(\frac{x}{a}\right)^3 + \left(\frac{x}{a}\right)\right] = \frac{4p_0 a^4}{\pi^5 D}\sum_{n=1,3,\ldots}^{\infty}\frac{1}{n}\sin\frac{n\pi x}{a}$$

Eq. (8.5.7) can be written as

$$w = \frac{p_0 a^4}{D}\sum_{n=1,3,5,\ldots}^{\infty}\left[\frac{4}{\pi^5 n^5} + A_n \cosh\frac{n\pi y}{a} + B_n\frac{n\pi y}{a}\sinh\frac{n\pi y}{a}\right]\sin\frac{n\pi x}{a} \tag{8.5.8}$$

Equation (8.5.8) satisfies Eq. (8.1.1) and the boundary conditions at $x = 0$ and $x = a$. The other conditions to be satisfied are

$$w = 0 \quad \text{and} \quad \frac{\partial^2 w}{\partial y^2} = 0 \quad \text{for} \quad y = \pm\frac{b}{2} \tag{8.5.9}$$

If Eqs. (8.5.9) are satisfied σ_y will be zero along the sides $y = \pm b/2$ since the condition $\partial^2 w/\partial x^2 = 0$ along the sides $y = \pm b/2$ is already satisfied. Equations (8.5.9) define the two constants A_n and B_n. In fact, the following two equations are obtained:

$$\left.\begin{array}{l}\dfrac{4}{\pi^5 n^5} + A_n \cosh \alpha_n + \alpha_n B_n \sinh \alpha_n = 0 \\ (A_n + 2B_n) \cosh \alpha_n + \alpha_n B_n \sinh \alpha_n = 0\end{array}\right\} \qquad (8.5.10)$$

where

$$\alpha_n = \frac{n\pi b}{2a}$$

From Eqs. (8.5.10) it follows that

$$A_n = -\frac{(\alpha_n \tanh \alpha_n + 2)}{\pi^5 n^5 \cosh \alpha_n}$$

$$B_n = \frac{2}{\pi^5 n^5 \cosh \alpha_n}$$

and Eq. (8.5.8) becomes

$$w = \frac{4p_0 a^4}{\pi^5 D} \sum_{n=1,3,5,\ldots}^{\infty} \frac{1}{n^5} \left[1 - \frac{\alpha_n \tanh \alpha_n + 2}{2 \cosh \alpha_n} \cosh \frac{2\alpha_n y}{b} \right.$$
$$\left. + \frac{\alpha_n}{2 \cosh \alpha_n} \frac{2y}{b} \sinh \frac{2\alpha_n y}{b} \right] \sin \frac{n\pi x}{a} \qquad (8.5.11)$$

The maximum deflection of the plate is obtained by letting $x = a/2$ and $y = 0$ in Eq. (8.5.11); thus, one gets

$$w_{\max} = \frac{4p_0 a^4}{\pi^5 D} \sum_{n=1,3,5,\ldots}^{\infty} \frac{(-1)^{(n-1)/2}}{n^5} \left[1 - \frac{\alpha_n \tanh \alpha_n + 2}{2 \cosh \alpha_n} \right]$$

Since

$$\sum_{n=1,3,5,\ldots}^{\infty} \frac{(-1)^{(n-1)/2}}{n^5} = \frac{5\pi^5}{2^9(3)}$$

one can write the maximum deflection equation for the plate, as follows:

$$w_{\max} = \frac{5p_0 a^4}{384 D} - \frac{4p_0 a^4}{\pi^5 D} \sum_{n=1,3,5,\ldots}^{\infty} \frac{(-1)^{(n-1)/2}}{n^5} \frac{\alpha_n \tanh \alpha_n + 2}{2 \cosh \alpha_n} \qquad (8.5.12)$$

The first term of Eq. (8.5.12) is the value which the deflection $w_{\max}$ would have in the case in which b is infinite. The series which appears in the second term converges very rapidly. In the case of the square plate, $a = b$ and $\alpha_n = n\pi/2$ and one has for the maximum deflection

$$w_{\max} = \frac{5p_0 a^4}{384 D} - \frac{4p_0 a^4}{\pi^5 D} [0.68562 - 0.00025 + \ldots]$$

From Eqs. (8.1.3) one can easily obtain the stresses σ_x, σ_y, and τ_{xy} by use of Eq. (8.5.11). The chief advantage of Levy's method over Navier's method is that instead of a double series one uses a simple series. It is in general simpler to perform numerical computations for simple series than for double series.

8.6 Nadai's Solution for a Plate in the Form of an Isosceles Right Triangle with Simply Supported Edges[13]

Navier's solution for the simply supported rectangular plate subjected to a concentrated load P acting at a generic point of coordinates ξ, η [see Eq. (8.4.22)] was used by Nadai to solve the problem of an isosceles right triangular plate with simply supported edges. This was accomplished by assuming a simply supported square plate (see Fig. 8.6.1) to be subjected to forces P

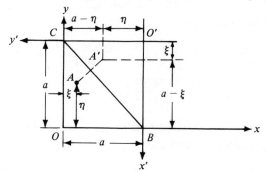

Figure 8.6.1

and $-P$ at points A and A', respectively. Point A has coordinates ξ and η with respect to the x-y coordinate system, and point A', which is the image of A with respect to the diagonal BC, has coordinates ξ and η with respect to the x'-y' coordinate system. The coordinates of A' with respect to the x-y coordinate system are $(a - \eta)$ and $(a - \xi)$. Due to the application of the two concentrated forces P and $-P$, the square plate is subjected to a deflection in which the diagonal line BC is a nodal line. Consequently the portion OBC of the plate is in the same condition as a simply supported triangular plate OBC. By use of Eq. (8.4.22), the deflection due to the force P at point A is found to be

[13] **Nadai, Arpad Ludwig.** American applied mechanicist of Hungarian origin (b. Budapest 1883; d. Pittsburgh, 1963). After studying at the Polytechnic Institutes of Zurich and Berlin, he taught at Berlin Polytechnic Institute and at Göttingen University. In 1929 Nadai came to the United States as a consulting mechanical engineer for Westinghouse Electric Corporation. He left important contributions in the mathematical theory of elasticity, in geomechanics, and especially in the theory of plasticity.

$$w_1 = \frac{4Pa^2}{\pi^2 D} \sum_{m=1}^{\infty} \sum_{n=1}^{\infty} \frac{\sin \frac{m\pi\xi}{a} \sin \frac{n\pi\xi}{a}}{(m^2 + n^2)^2} \sin \frac{m\pi x}{a} \sin \frac{n\pi x}{a} \quad (8.6.1)$$

By letting

$$-P \text{ replace } P$$
$$(a - \eta) \text{ replace } \xi$$
$$(a - \xi) \text{ replace } \eta$$

in Eq. (8.6.1), one finds that the deflection due to the force $-P$ at A' is

$$w_2 = -\frac{4Pa^2}{\pi^2 D} \sum_{m=1}^{\infty} \sum_{n=1}^{\infty} (-1)^{m+n} \frac{\sin \frac{m\pi\eta}{a} \sin \frac{n\pi\xi}{a}}{(m^2 + n^2)^2} \sin \frac{m\pi x}{a} \sin \frac{n\pi y}{a} \quad (8.6.2)$$

By adding Eqs. (8.6.1) and (8.6.2) one obtains the following expression for the deflection of the triangular plate:

$$w = w_1 + w_2$$

$$= \frac{4Pa^2}{\pi^2 D} \left[\sum_{m=1}^{\infty} \sum_{n=1}^{\infty} \frac{\sin \frac{m\pi\xi}{a} \sin \frac{n\pi\eta}{a}}{(m^2 + n^2)^2} \sin \frac{m\pi x}{a} \sin \frac{n\pi y}{a} \right.$$
$$\left. - \sum_{m=1}^{\infty} \sum_{n=1}^{\infty} (-1)^{m+n} \frac{\sin \frac{m\pi\eta}{a} \sin \frac{n\pi\xi}{a}}{(m^2 + n^2)^2} \sin \frac{m\pi x}{a} \sin \frac{n\pi y}{a} \right] \quad (8.6.3)$$

By use of Eq. (8.6.3) and the principle of superposition it is possible to compute the deflection of an isosceles right triangular plate for any conditions of loading.

Example 8.6.1. By use of Eq. (8.6.3), the deflection of a triangular plate subjected to a uniformly distributed load of intensity p_0 will be determined. By substituting $p_0 \, d\xi \, d\eta$ into Eq. (8.6.3) in the place of P and integrating over the area of the plate, one obtains

$$w = \frac{16 p_0 a^4}{\pi^6 D} \sum_{m=1,3,5,\ldots}^{\infty} \sum_{n=2,4,6,\ldots}^{\infty} \frac{n \sin \frac{m\pi x}{a} \sin \frac{n\pi y}{a}}{m(n^2 - m^2)(m^2 + n^2)^2}$$
$$+ \sum_{m=1,3,5,\ldots}^{\infty} \sum_{n=2,4,6,\ldots}^{\infty} \frac{m \sin \frac{m\pi x}{a} \sin \frac{n\pi y}{a}}{n(m^2 - n^2)(m^2 + n^2)^2} \quad (8.6.4)$$

The series in Eq. (8.6.4) is a very rapidly convergent series. The bending and twisting moments and shearing forces and the stresses acting on every point of the plate can be directly computed by use of Eqs. (8.1.5), (8.1.8), and (8.1.3).

8.7 Woinowsky-Krieger's Solution for the Plate in the Form of an Equilateral Triangle with Simply Supported Edges and Subjected to a Uniformly Distributed Load of Intensity p_0

It can be seen by substitution that the function

$$w = \frac{p_0}{64aD}[x^3 - 3y^2 - a(x^2 + y^2) + \tfrac{4}{27}a^3][\tfrac{4}{9}a^2 - x^2 - y^2] \quad (8.7.1)$$

satisfies Lagrange's equation [Eq. (8.1.1)] and the boundary conditions of simply supported edges for the triangular plate shown in Fig. 8.7.1 and is,

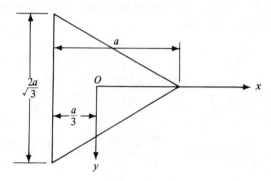

Figure 8.7.1

therefore, the solution of the problem. By using Eq. (8.7.1) and Eqs. (8.1.5), (8.1.8), and (8.1.3) the bending and twisting moments, the shearing forces, and the stresses at every point of the plate may be computed. The maximum bending moment occurs on the line bisecting an angle of the triangle (for example, the x-axis of Fig. 8.7.1). For the case in which $\nu = 0.30$, one finds that

$$(M_x)_{\max} = 0.0298 p_0 a^2 \quad \text{for} \quad x = -0.062a \quad y = 0$$
$$(M_y)_{\max} = 0.0259 p_0 a^2 \quad \text{for} \quad x = 0.129a \quad y = 0$$

At the center of the plate ($x = y = 0$) the bending moments are

$$M_x = M_y = (1 + \nu)\frac{p_0 a^2}{54}$$

8.8 Application of the Principle of Virtual Work to Obtain Rigorous and Approximate Solutions of the Problem of the Bending of Rectangular Plates

(a) *Rigorous solutions.* Consider a simply supported rectangular plate (Fig. 8.8.1) subjected to a force P perpendicular to the upper face and applied at the point of coordinates ξ, η. The deflection w of the plate can be developed in the double trigonometric series

$$w = \sum_{m=1}^{\infty} \sum_{n=1}^{\infty} B_{mn} \sin \frac{m\pi x}{a} \sin \frac{n\pi y}{b} \tag{8.8.1}$$

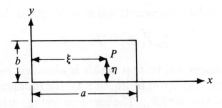

Figure 8.8.1

Each term of the series in Eq. (8.8.1) satisfies the boundary conditions for a simply supported plate; that is

$$w(0, y) = w(a, y) = w(x, 0) = w(x, b) = 0 \tag{8.8.2}$$

and

$$\sigma_x(0, y) = \sigma_x(a, y) = \sigma_y(x, 0) = \sigma_y(x, b) = 0 \tag{8.8.3}$$

Equations (8.8.3) are satisfied by Eq. (8.8.1) since [see Eqs. (8.1.3)]

$$\frac{\partial^2 w(0, y)}{\partial x^2} = \frac{\partial^2 w(a, y)}{\partial x^2} = \frac{\partial^2 w(x, 0)}{\partial x^2} = \frac{\partial^2 w(x, b)}{\partial x^2} = \frac{\partial^2 w(0, y)}{\partial y^2} = \frac{\partial^2 w(a, y)}{\partial y^2}$$

$$= \frac{\partial^2 w(x, 0)}{\partial y^2} = \frac{\partial^2 w(x, b)}{\partial y^2} = 0$$

The strain energy is given by the following equation:

$$U = G \iiint_V \left[\epsilon_x^2 + \epsilon_y^2 + \frac{\nu e^2}{1 - 2\nu} + \frac{1}{2} (\gamma_{xy}^2 + \gamma_{yz}^2 + \gamma_{zx}^2) \right] dx \, dy \, dz \tag{8.8.4}$$

By substituting Eqs. (8.1.2) into Eq. (8.8.4) and integrating with respect to the variable z, one obtains for the strain energy the expression

$$U = \frac{D}{2} \iint_\Omega \left[\left(\frac{\partial^2 w}{\partial x^2}\right)^2 + \left(\frac{\partial^2 w}{\partial y^2}\right)^2 + 2\nu \frac{\partial^2 w}{\partial x^2} \frac{\partial^2 w}{\partial y^2} + 2(1-\nu)\left(\frac{\partial^2 w}{\partial x \partial y}\right)^2 \right] dx\, dy$$

or

$$U = \frac{D}{2} \iint_\Omega \left[\left(\frac{\partial^2 w}{\partial x^2} + \frac{\partial^2 w}{\partial y^2}\right)^2 - 2(1-\nu)\left\{\frac{\partial^2 w}{\partial x^2} \frac{\partial^2 w}{\partial y^2} - \left(\frac{\partial^2 w}{\partial x \partial y}\right)^2\right\} \right] dx\, dy \quad (8.8.5)$$

where Ω is the area of the rectangular plate. By substituting Eq. (8.8.1) into Eq. (8.8.5) one obtains[14]

$$U = \frac{\pi^4 D a b}{8} \sum_{m=1}^{\infty} \sum_{n=1}^{\infty} B_{mn}^2 \left(\frac{m^2}{a^2} + \frac{n^2}{b^2}\right)^2 \quad (8.8.6)$$

The principle of virtual displacements will now be applied in the form

$$\sum \vec{F}_i \cdot \vec{\delta r}_i - \delta U = 0 \quad (8.8.7)$$

The virtual displacement will be that which is obtained by the variation of only one coefficient of the series in Eq. (8.8.1); for instance, the variation of coefficient B_{mn} by an amount δB_{mn}. The correspondsing variation of the strain energy, obtained from Eq. (8.8.6), will be

$$\delta U = \frac{\pi^4 D a b}{4} B_{mn} \left(\frac{m^2}{a^2} + \frac{n^2}{b^2}\right)^2 \delta B_{mn} \quad (8.8.8)$$

while the virtual displacement of the point of application of the force P is, in accordance with Eq. (8.8.1),

[14] In view of the orthogonality properties of the circular functions, i.e.,

$$\int_0^a \sin\frac{m\pi x}{a} \sin\frac{m'\pi x}{b} dx = \begin{cases} 0 & \text{for } m \neq m' \\ \frac{a}{2} & \text{for } m = m' \end{cases}$$

$$\int_0^b \sin\frac{n\pi y}{b} \sin\frac{n'\pi y}{b} dy = \begin{cases} 0 & \text{for } n \neq n' \\ \frac{b}{2} & \text{for } n = n' \end{cases}$$

one finds that

$$\iint_\Omega \left(\frac{\partial^2 w}{\partial x^2} + \frac{\partial^2 w}{\partial y^2}\right)^2 dx\, dy = \frac{\pi^4 ab}{4} \sum_{m=1}^{\infty} \sum_{n=1}^{\infty} B_{mn}^2 \left(\frac{m^2}{a^2} + \frac{n^2}{b^2}\right)^2 \quad \text{(a)}$$

$$\iint_\Omega \frac{\partial^2 w}{\partial x^2} \frac{\partial^2 w}{\partial y^2} dx\, dy = \frac{\pi^4}{4ab} \sum_{m=1}^{\infty} \sum_{n=1}^{\infty} m^2 n^2 B_{mn}^2 \quad \text{(b)}$$

$$\iint_\Omega \left(\frac{\partial^2 w}{\partial x \partial y}\right)^2 dx\, dy = \frac{\pi^4}{4ab} \sum_{m=1}^{\infty} \sum_{n=1}^{\infty} m^2 n^2 B_{mn}^2 \quad \text{(c)}$$

By adding Eqs. (a), (b), and (c) Eq. (8.8.6) is obtained.

$$(\delta w)_{\substack{x=\xi\\y=\eta}} = \delta B_{mn} \sin\frac{m\pi\xi}{a}\sin\frac{n\pi\eta}{b}$$

Since P is the only external force, the work done by the external forces will be

$$\sum \vec{F}_i \cdot \vec{\delta r}_i = P\sin\frac{m\pi\xi}{a}\sin\frac{n\pi\eta}{b}\delta B_{mn} \tag{8.8.9}$$

Equation (8.8.7) becomes, in view of Eqs. (8.8.8) and (8.8.9),

$$P\sin\frac{m\pi\xi}{a}\sin\frac{n\pi\eta}{b}\delta B_{mn} - \frac{\pi^4 Dab}{4}B_{mn}\left(\frac{m^2}{a^2}+\frac{n^2}{b^2}\right)^2\delta B_{mn} = 0$$

from which it follows that

$$B_{mn} = \frac{4P}{\pi^4 Dab}\frac{\sin\frac{m\pi\xi}{a}\sin\frac{n\pi\eta}{b}}{\left(\frac{m^2}{a^2}+\frac{n^2}{b^2}\right)^2} \tag{8.8.10}$$

By introducing Eq. (8.8.10) into Eq. (8.8.1) one obtains

$$w = \frac{4P}{\pi^4 Dab}\sum_{m=1}^{\infty}\sum_{n=1}^{\infty}\frac{\sin\frac{m\pi\xi}{a}\sin\frac{n\pi\eta}{b}}{\left(\frac{m^2}{a^2}+\frac{n^2}{b^2}\right)^2}\sin\frac{m\pi x}{a}\sin\frac{n\pi y}{b} \tag{8.8.11}$$

This result coincides with Eq. (8.4.22).

(b) Approximate solutions. The Principle of Virtual Work will now be applied to obtain an approximate solution for two different cases of bent plates.

Example 8.8.1. Consider a rectangular plate of sides a and b subjected to a uniformly distributed load of intensity p_0 and simply supported along its edges. The deflection curve of a simply supported bar of length l under a uniformly distributed load of intensity p_0 is given by

$$w = \frac{p_0 l^4}{24EI}\left[\frac{x^4}{l^4}-2\frac{x^3}{l^3}+\frac{x}{l}\right] = \frac{16}{5}w_0\left[\left(\frac{x}{l}\right)^4-2\left(\frac{x}{l}\right)^3+\left(\frac{x}{l}\right)\right]$$

where

$$w_0 = \frac{5p_0 l^4}{384EI}$$

represents the deflection at the center of the bar. By analogy with the bar, the deflection surface of a simply supported plate subjected to a uniformly distributed load p_0 can be written in the form

$$w = \frac{256 w_{00}}{25}\left[\left(\frac{x}{a}\right)^4 - 2\left(\frac{x}{a}\right)^3 + \left(\frac{x}{a}\right)\right]\left[\left(\frac{y}{b}\right)^4 - 2\left(\frac{y}{b}\right)^3 + \left(\frac{y}{b}\right)\right] \quad (8.8.12)$$

where w_{00} represents the deflection at the center of the plate, in other words at the point $x = a/2$, $y = b/2$.

Equation (8.8.12) satisfies the boundary conditions for the simply supported plate, i.e.,

$$w(0, y) = w(a, y) = 0, \quad w(x, 0) = w(x, b) = 0$$
$$\sigma_x(0, y) = \sigma_x(a, y) = 0, \quad \sigma_y(x, 0) = \sigma_y(x, b) = 0$$

The strain energy of the plate is obtained by substituting Eq. (8.8.12) into Eq. (8.8.5); thus[15]

$$U = 12.38 D \frac{(a^2 + b^2)}{a^3 b^3} w_{00}^2 \quad (8.8.13)$$

By choosing an infinitely small variation δw_{00} of w_{00} as the virtual displacement, the corresponding variations of w and U will be, respectively,

$$\delta w = \frac{256 \, \delta w_{00}}{25}\left(\frac{x^4}{a^4} - 2\frac{x^3}{a^3} + \frac{x}{a}\right)\left(\frac{y^4}{b^4} - 2\frac{y^3}{b^3} + \frac{y}{b}\right)$$

and

$$\delta U = 24.76 D \frac{(a^2 + b^2)^2}{a^3 b^3} w_{00} \, \delta w_{00}$$

The work done by the external forces will be

[15] By use of Eq. (8.8.12), one finds that

$$\iint_\Omega \left(\frac{\partial^2 w}{\partial x^2}\right)^2 dx \, dy = \frac{(3072)^2 w_{00}^2}{(25)^2 a^2} \iint_\Omega \left[\left(\frac{x^2}{a^2} - \frac{x}{a}\right)\left(\frac{y^4}{b^4} - \frac{2y^3}{b^3} + \frac{y}{b}\right)\right] dx \, dy$$
$$= 24.76 \frac{b}{a^3} w_{00}^2$$

In the same way:

$$\iint_\Omega \left(\frac{\partial^2 w}{\partial y^2}\right)^2 dx \, dy = 24.76 \frac{a}{b^3} w_{00}^2$$
$$\iint_\Omega \frac{\partial^2 w}{\partial x^2}\frac{\partial^2 w}{\partial y^2} dx \, dy = \iint_\Omega \left(\frac{\partial^2 w}{\partial x \partial y}\right)^2 dx \, dy = 24.76 \frac{w_{00}^2}{ab}$$

$$\sum \vec{F}_i \cdot \vec{\delta r}_i = \iint_\Omega \delta w \, p_0 \, dx \, dy$$

$$= \frac{256 p_0 \, \delta w_{00}}{25} \iint_\Omega \left(\frac{x^4}{a^4} - 2\frac{x^3}{a^3} + \frac{x}{a} \right) \left(\frac{y^4}{b^4} - 2\frac{y^3}{b^3} + \frac{y}{b} \right) dx \, dy$$

$$= 0.409 p_0 ab \, \delta w_{00} \tag{8.8.14}$$

where p_0 represents the load per unit area. By introducing the above expressions into the equation

$$\sum \vec{F}_i \cdot \vec{\delta r}_i - \delta U = 0$$

which expresses the Principle of Virtual Work, one gets

$$0.409 p_0 ab \, \delta w_{00} - 24.76 \, D \frac{a^2 + b^2}{a^3 b^3} w_{00} \, \delta w_{00} = 0$$

or

$$w_{00} = w_{\max} = 0.0165 \frac{1}{D} \frac{p_0 a^4 b^4}{(a^2 + b^2)^2} \tag{8.8.15}$$

Equation (8.8.15) becomes, in the case of the square plate ($a = b$),

$$w_{00} = 0.00412 \frac{p_0 a^4}{D} \tag{8.8.16}$$

The result given by Eq. (8.8.16) is 1.5 per cent greater than the exact value derived previously.

Example 8.8.2. In the case of a rectangular plate which has built-in edges of lengths a and b and is subjected to a uniformly distributed load of intensity p_0, one assumes for the deflection surface the expression[16]

$$w = 256 w_{00} \left(\frac{x^4}{a^4} - 2\frac{x^3}{a^3} + \frac{x^2}{a^2} \right) \left(\frac{y^4}{b^4} - 2\frac{y^3}{b^3} + \frac{y^2}{b^2} \right) \tag{8.8.17}$$

Equation (8.8.17) satisfies the boundary conditions for the built-in plate, i.e.,

$$w(0, y) = w(a, y) = 0, \quad w(x, 0) = w(x, b) = 0$$

$$\frac{\partial w(0, y)}{\partial x} = \frac{\partial w(a, y)}{\partial x} = 0, \quad \frac{\partial w(x, 0)}{\partial y} = \frac{\partial w(x, b)}{\partial y} = 0$$

---

[16] The choice of Eq. (8.8.17) is due to the fact that the deflection curve for a built-in beam of length l subjected to a uniformly distributed load of intensity p_0 is

$$w = \frac{p_0 l^4}{24 EI} \left(\frac{x^4}{l^4} - 2\frac{x^3}{l^3} + \frac{x^2}{l^2} \right) = 16 w_0 \left[\left(\frac{x}{l}\right)^4 - 2\left(\frac{x}{l}\right)^3 + \left(\frac{x}{l}\right)^2 \right]$$

where $w_0 = (p_0 l^4 / 384 EI)$ represents the deflection at the center of the bar.

The strain energy of the plate is given by[17]

$$U = \frac{D}{ab}\left(41.64\frac{a^4 + b^4}{a^2b^2} + 23.78\right)w_{00}^2 \qquad (8.8.18)$$

and it follows that

$$\delta U = \frac{2D}{ab}\left(41.61\frac{a^4 + b^4}{a^2b^2} + 23.78\right)w_{00}\,\delta w_{00} \qquad (8.8.19)$$

The variation of w is

$$\delta w = 256\,\delta w_{00}\left(\frac{x^4}{a^4} - 2\frac{x^3}{a^3} + \frac{x^2}{a^2}\right)\left(\frac{y^4}{b^4} - 2\frac{y^3}{b^3} + \frac{y^2}{b^2}\right) \qquad (8.8.20)$$

By introducing Eqs. (8.8.19) and (8.8.20) into the relation

$$\sum \vec{F_i}\cdot\delta\vec{r_i} - \delta U = 0$$

one obtains

$$0.2845 p_0 ab\,\delta w_{00} - \frac{2D}{ab}\left(41.61\frac{a^2 + b^2}{a^2b^2} + 23.78\right)w_{00}\,\delta w_{00} = 0$$

from which it is found that

$$w_{\max} = \frac{1}{D}\frac{p_0 a^2 b^2}{292.5\frac{a^4 + b^4}{a^2b^2} + 167} \qquad (8.8.21)$$

In the particular case of a square plate ($a = b$) Eq. (8.8.21) becomes

$$w_{00} = 0.00133\frac{p_0 a^4}{D}p_0 a^4 \qquad (8.8.22)$$

The value given by Eq. (8.8.22) is 5 per cent greater than the value which is obtained by the use of more accurate methods.

8.9 Ritz's Method Applied to the Bending of Plates

In the previous section, Eqs. (8.8.12) and (8.8.17) have been used to represent the elastic surface of the bent plate. These equations satisfy the boundary conditions of the plate but contain only one arbitrary constant. In order to obtain a better approximation, let

---

[17] By use of Eq. (8.8.17) one may show that

$$\iint_\Omega \left(\frac{\partial^2 w}{\partial x^2}\right)^2 dx\,dy = 83.22\frac{b}{a^3}w_{00}$$

$$\iint_\Omega \left(\frac{\partial^2 w}{\partial y^2}\right)^2 dx\,dy = 83.22\frac{a}{b^3}w_{00}$$

$$\iint_\Omega \frac{\partial^2 w}{\partial x^2}\frac{\partial^2 w}{\partial y^2}dx\,dy = \iint_\Omega \left(\frac{\partial^2 w}{\partial x\,\partial y}\right)^2 dx\,dy = 23.78\frac{w_{00}^2}{ab}$$

Sec. 8.9 BENDING OF PLATES

$$w = c_1\phi_1(x, y) + c_2\phi_2(x, y) + \ldots + c_k\phi_k(x, y) \qquad (8.9.1)$$

The functions $\phi_i(x, y)$ ($i = 1, 2, \ldots, k$) of Eq. (8.9.1) satisfy the boundary conditions of the plate while quantities c_i ($i = 1, 2, \ldots, k$) are parameters which are determined by applying the principle of virtual work. The use of Eq. (8.9.1) and the application of the principle of virtual work leads to an equation of the form

$$()\delta c_1 + ()\delta c_2 + \ldots + ()\delta c_k = 0 \qquad (8.9.2)$$

Since the variations δc_i ($i = 1, 2, \ldots, k$) are arbitrary, all terms in parentheses in Eq. (8.9.2) must be zero and a system of linear equations for determining parameters $c_1, c_2, \ldots, c_k$ is finally obtained.

The above method, known as Ritz' method, is often presented by introducing the total potential energy of the plate which is

$$E_{\text{total}} = U - \iint_\Omega wp \, dx \, dy$$

or

$$E_{\text{total}} = \iint_\Omega \left\{ \frac{D}{2}\left[\left(\frac{\partial^2 w}{\partial x^2} + \frac{\partial^2 w}{\partial y^2}\right)^2 - 2(1-\nu)\left(\frac{\partial^2 w}{\partial x^2}\frac{\partial^2 w}{\partial y^2} - \left[\frac{\partial^2 w}{\partial x \partial y}\right]^2\right)\right] - wp \right\} dx \, dy$$

Since in a position of stable equilibrium, the total potential energy must be a minimum, the following relations are obtained:

$$\frac{\partial E_{\text{total}}}{\partial c_i} = 0 \quad (i = 1, 2, \ldots, k) \qquad (8.9.3)$$

Equations (8.9.3) are identical to those obtained by equating to zero the coefficients of variations δc_i in Eq. (8.9.2).

Example 8.9.1. Consider the case of the rectangular plate of sides a and b which is built-in at the edges and subjected to a uniformly distribued load p_0 (see Fig. 8.9.1). The strain energy of plate, which is

$$U = \frac{D}{2} \iint_\Omega \left\{ \left(\frac{\partial^2 w}{\partial x^2} + \frac{\partial^2 w}{\partial y^2}\right)^2 - 2(1-\nu^2)\left(\frac{\partial^2 w}{\partial x^2}\frac{\partial^2 w}{\partial y^2} - \left[\frac{\partial^2 w}{\partial x \partial y}\right]^2\right) \right\} dx \, dy$$

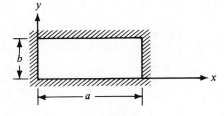

Figure 8.9.1

reduces, in the case of a plate with built-in edges, to[18]

$$U = \frac{D}{2} \iint_\Omega \left(\frac{\partial^2 w}{\partial x^2} - \frac{\partial^2 w}{\partial y^2}\right)^2 dx\, dy \qquad (8.9.4)$$

By assuming for w the expression

$$w = \sum_{m=1}^{k} \sum_{n=1}^{k} c_{mn}\left(1 - \cos\frac{2\pi mx}{a}\right)\left(1 - \cos\frac{2\pi ny}{b}\right) \qquad (8.9.5)$$

which satisfies the boundary conditions for a built-in plate, and substituting into Eq. (8.9.4) one obtains

$$U = \frac{D}{2}\int_0^a\int_0^b \left\{\sum_{m=1}^{k}\sum_{n=1}^{k} 4\pi^2 c_{mn}\left[\frac{m^2}{a^2}\cos\frac{2m\pi x}{a}\left(1 - \cos\frac{2n\pi y}{b}\right)\right.\right.$$
$$\left.\left. + \frac{n^2}{b^2}\cos\frac{2n\pi y}{b}\left(1 - \cos\frac{2m\pi x}{a}\right)\right]\right\}^2 dx\, dy$$

or

$$U = 2D\pi^4 ab\left\{\sum_{m=1}^{k}\sum_{n=1}^{k}\left[3\left(\frac{m}{a}\right)^4 + 3\left(\frac{n}{b}\right)^4 + 2\left(\frac{m}{a}\right)^2\left(\frac{n}{b}\right)^2\right]c_{mn}^2\right.$$
$$\left. + \sum_{m=1}^{k}\sum_{r=1}^{k}\sum_{s=1}^{k} 2\left(\frac{m}{a}\right)^4 c_{mr}c_{ms} + \sum_{r=1}^{k}\sum_{s=1}^{k}\sum_{n=1}^{k} 2\left(\frac{n}{b}\right)^4 c_{rn}c_{sn}\right\}, \quad r \neq s$$

The work done by the uniformly distributed load of intensity p_0 is

$$\int_0^a\int_0^b p_0 w\, dx\, dy = p_0 \int_0^a\int_0^b\left[\sum_{m=1}^{k}\sum_{n=1}^{k} c_{mn}\left(1 - \cos\frac{2m\pi x}{a}\right)\left(1 - \cos\frac{2n\pi y}{b}\right)\right] dx\, dy$$
$$= p_0 ab \sum_{m=1}^{k}\sum_{n=1}^{k} c_{mn}$$

The condition

[18] To prove that for a plate with built-in edges

$$\iint_\Omega \left(\frac{\partial^2 w}{\partial x^2}\frac{\partial^2 w}{\partial y^2}\right) dx\, dy = \iint_\Omega \left(\frac{\partial^2 w}{\partial x \partial y}\right)^2 dx\, dy$$

integrate the right-hand side of the above relation by parts to get

$$\iint_\Omega \frac{\partial^2 w}{\partial x \partial y}\frac{\partial^2 w}{\partial x \partial y} dx\, dy = \int_S \frac{\partial^2 w}{\partial x \partial y}\frac{\partial w}{\partial x} dx - \iint_\Omega \frac{\partial w}{\partial x}\frac{\partial^3 w}{\partial x \partial y^2} dx\, dy$$
$$= \int_S \frac{\partial^2 w}{\partial x \partial y}\frac{\partial w}{\partial x} dx - \int_S \frac{\partial w}{\partial x}\frac{\partial^2 w}{\partial y^2} dy + \iint_\Omega \frac{\partial^2 w}{\partial x^2}\frac{\partial^2 w}{\partial y^2} dx\, dy$$

For plates with built-in edges $\partial w/\partial x = \partial w/\partial y = 0$ along the edges, and consequently the first two integrals in the above expression become identically zero and one has that

$$\iint_\Omega \left[\frac{\partial^2 w}{\partial x^2}\frac{\partial^2 w}{\partial y^2} - \left(\frac{\partial^2 w}{\partial x \partial y}\right)^2\right] dx\, dy = 0.$$

$$\frac{\partial E_{\text{total}}}{\partial c_{mn}} = 0$$

gives the following equation:

$$4D\pi^4 ab\left\{\left[3\left(\frac{m}{a}\right)^4 + 3\left(\frac{n}{b}\right)^4 + 2\left(\frac{m}{a}\right)^2\left(\frac{n}{b}\right)^2\right]c_{mn} \\ + \sum_{r=1}^{k} 2\left(\frac{m}{a}\right)^4 c_{mr} + \sum_{r=1}^{k} 2\left(\frac{n}{b}\right)^4 c_{rn}\right\} - p_0 ab, \quad r \neq m \quad (8.9.6)$$

If only one parameter c_{11} is considered, Eq. (8.9.6) gives

$$c_{11} = \frac{p_0 a^4}{4\pi^4 D} \frac{1}{3 + 3\left(\frac{a}{b}\right)^4 + 2\left(\frac{a}{b}\right)^2}$$

In the case of a square plate ($a = b$) one obtains

$$c_{11} = \frac{p_0 a^4}{32\pi^4 D} \quad (8.9.7)$$

By substituting the above expression for c_{11} into Eq. (8.9.5) and letting $\nu = 0.3$, the maximum deflection, which occurs at the center of the plate, is found to be

$$w_{\max} = \frac{0.0140 p_0 a^4}{Eh^3}$$

which is only 1.5 per cent greater than the value obtained by using much longer and more elaborate methods. In the case in which one expresses the deflection w of Eq. (8.9.5) by the use of seven parameters, c_{11}, c_{12}, c_{21}, c_{22}, c_{13}, c_{31}, and c_{33}, the use of Eq. (8.9.6) leads to the following seven algebraic equations:

$$\left[3 + 3\left(\frac{a}{b}\right)^4 + 2\left(\frac{a}{b}\right)^2\right]c_{11} + 2c_{12} + 2\left(\frac{a}{b}\right)^4 c_{21} + 2c_{13} + 2\left(\frac{a}{b}\right)^4 c_{31} = \frac{p_0 a^4}{4\pi^4 D}$$

$$2c_{11} + \left[3 + 48\left(\frac{a}{b}\right)^4 + 8\left(\frac{a}{b}\right)^2\right]c_{12} + 2c_{13} + 32\left(\frac{a}{b}\right)^4 c_{22} = \frac{p_0 a^4}{4\pi^4 D}$$

$$a\left(\frac{a}{b}\right)^4 c_{11} + \left[48 + 3\left(\frac{a}{b}\right)^4 + 8\left(\frac{a}{b}\right)^2\right]c_{21} + 2\left(\frac{a}{b}\right)^4 c_{31} + 32c_{22} = \frac{p_0 a^4}{4\pi^4 D}$$

$$32c_{21} + 16\left[3 + 3\left(\frac{a}{b}\right)^4 + 2\left(\frac{a}{b}\right)^2\right]c_{22} + 32\left(\frac{a}{b}\right)^4 c_{12} = \frac{p_0 a^4}{4\pi^4 D}$$

$$2c_{11} + 2c_{12} + \left[3 + 243\left(\frac{a}{b}\right)^4 + 18\left(\frac{a}{b}\right)^2\right]c_{13} + 162\left(\frac{a}{b}\right)^4 c_{33} = \frac{p_0 a^4}{4\pi^4 D}$$

$$2\left(\frac{a}{b}\right)^4 c_{11} + 2\left(\frac{a}{b}\right)^4 c_{21} + \left[243 + 3\left(\frac{a}{b}\right)^4 + 18\left(\frac{a}{b}\right)^2\right]c_{31} + 162c_{33} = \frac{p_0 a^4}{4\pi^4 D}$$

$$162\left(\frac{a}{b}\right)^4 c_{13} + 162c_{31} + 81\left[3 + 3\left(\frac{a}{b}\right)^4 + 2\left(\frac{a}{b}\right)^2\right]c_{33} = \frac{p_0 a^4}{4\pi^4 D}$$

In the case of the square plate ($a = b$), the solution of the above equations gives

$$c_{11} = 0.11774 \frac{p_0 a^4}{4\pi^4 D}$$

$$c_{22} = 0.00189 \frac{p_0 a^4}{4\pi^4 D}$$

$$c_{33} = 0.00020 \frac{p_0 a^4}{4\pi^4 D}$$

$$c_{12} = c_{21} = 0.01184 \frac{p_0 a^4}{4\pi^4 D}$$

$$c_{13} = c_{31} = 0.00268 \frac{p_0 a^4}{4\pi^4 D}$$

By substituting the above values of parameter c_{mn} into Eq. (8.9.5) and letting $v = 0.3$, one finds that

$$w_{max} = \frac{0.0138 p_0 a^4}{E h^3}$$

8.10 Application of Finite Difference Equations to the Bending of Plates

By using the finite difference method, first proposed by Henri Marcus,[19] one avoids the integration of the fourth order differential equation

$$\frac{\partial^4 w}{\partial x^4} + 2 \frac{\partial^4 w}{\partial x^2 \partial y^2} + \frac{\partial^4 w}{\partial y^4} = \frac{p}{D} \tag{8.1.1}$$

In fact, two finite difference equations of the second order are substituted for Eq. (8.1.1). These equations are integrated by solving two systems of linear algebraic equations. Ususally the approximation which is obtained by the use of the method is a very good one and the results can be approximated to any degree of accuracy.

To explain the method, substitute for Eq. (8.1.1) a system of two second order equations by introducing M, a function of x and y, which is defined as follows:

---

[19] **Marcus, Henri.** American applied mechanicist and structural engineer of French origin (b. Smyrna, Turkey 1885; d. Washington, D.C. 1969). After receiving his degree in Civil Engineering at the Polytechnic Institute in Berlin, Marcus was the designer in Germany of many important structures in steel and reinforced concrete. He was one of the pioneers of the use of reinforced concrete in large structures and was later associated with the famous French engineer H. Freyssinet in the use of prestressed concrete for major bridges and maritime works in France. In 1942 be came to the United States as consultant on reinforced concrete and stress analysis for the U.S. Bureau of Yards and Docks and later for the Naval Research Laboratories in Washington, D.C. In addition to his work as a designer of structures, Henri Marcus left important contributions in the theories of elasticity and plasticity, stress analysis of plates, and elastic waves propagation.

$$M = -D\Delta w = -D\left(\frac{\partial^2 w}{\partial x^2} + \frac{\partial^2 w}{\partial y^2}\right) \qquad (8.10.1)$$

Equation (8.1.1) is thus replaced by the equations

$$\Delta M = \frac{\partial^2 M}{\partial x^2} + \frac{\partial^2 M}{\partial y^2} = -p \qquad (8.10.2)$$

and

$$\Delta w = \frac{\partial^2 w}{\partial x^2} + \frac{\partial^2 w}{\partial y^2} = -\frac{M}{D} \qquad (8.10.3)$$

From Eqs. (8.1.5), which are

$$\left. \begin{array}{l} M_x = -D\left(\dfrac{\partial^2 w}{\partial x^2} + v\dfrac{\partial^2 w}{\partial y^2}\right) \\[6pt] M_y = -D\left(\dfrac{\partial^2 w}{\partial y^2} + v\dfrac{\partial^2 w}{\partial x^2}\right) \\[6pt] M_{xy} = -(1-v)D\dfrac{\partial^2 w}{\partial x \partial y} \end{array} \right\} \qquad (8.1.5)$$

one obtains

$$M_x + M_y = -D(1+v)\left(\frac{\partial^2 w}{\partial x^2} + \frac{\partial^2 w}{\partial y^2}\right) \qquad (8.10.4)$$

By comparing Eqs. (8.10.4) and (8.10.3) one finds that

$$M = \frac{M_x + M_y}{(1+v)} \qquad (8.10.5)$$

i.e., that function M is the sum of moments M_x and M_y divided by $(1+v)$. It can be proved that M is an invariant at every point of the plate, i.e., it is independent of the directions of the x- and y-axes.[20]

[20] One proves that M, or in view of Eq. (8.10.5), that $(M_x + M_y)$ is an invariant in the following way. From the first two of Eqs. (8.1.5) one has that

$$M_x + M_y = -D(1+v)\left(\frac{\partial^2 w}{\partial x^2} + \frac{\partial^2 w}{\partial y^2}\right) = -D(1+v)\left(\frac{1}{\rho_x} + \frac{1}{\rho_y}\right)$$

A well known theorem by Euler states that, at any point of the deflection surface w, the sum of the curvatures in two perpendicular directions is a constant, i.e.,

$$\frac{1}{\rho_x} + \frac{1}{\rho_y} = \text{constant}$$

It therefore follows that

$$M = \frac{M_x + M_y}{1+v}$$

is a constant.

Consider now an elastic membrane, initially plane, subjected to a uniform tension of S pounds per unit length of its boundary and subjected to a distributed load p perpendicular to the plane of the membrane. If the deformations of the membrane are assumed to be small, the ordinates w of the deformed membrane satisfy the equation

$$\frac{\partial^2 w}{\partial x^2} + \frac{\partial^2 w}{\partial y^2} = -\frac{p}{S} \qquad (8.10.6)$$

By comparing Eqs. (8.10.2), (8.10.3), and (8.10.6), the following theorems given by Henri Marcus are derived:

1. A plane elastic membrane, having the same border as a plane plate, subjected in its plane to a uniform tension $S = 1$ and to the same load p as the plate, deforms in such a way that the ordinates w have the same values as M at every point of the corresponding plate.
2. If the above membrane is subjected to a distributed load $p = M/D$ its deformed surface is the same as the deformed elastic surface of the plate.

These two theorems for the deflected plate, given by Henri Marcus, correspond to the two theorems given by Otto Mohr for the deflected beam. Mohr's theorems are derived from the equations

$$\left.\begin{aligned}\frac{d^4 w}{dx^4} &= -\frac{p}{EI} \\ \frac{d^2 M}{dx^2} &= -p \\ \frac{d^2 w}{dx^2} &= \frac{M}{EI}\end{aligned}\right\} \qquad (8.10.7)$$

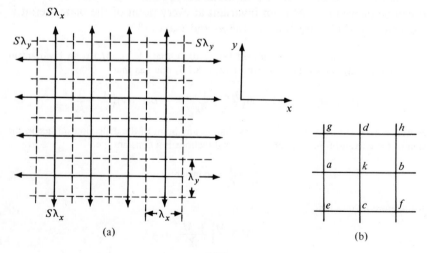

Figure 8.10.1

Assume now that a membrane having the same border as a plate is decomposed into two groups of strips, one group parallel to the x-axis and having widths $\Delta y = \lambda_y$ and the other parallel to the y-axis and having widths $\Delta x = \lambda_x$. Next replace these strips with two groups of wires placed in the middle of the width of these strips [see Fig. 8.10.1 (a)]. In this way, a set of orthogonal elastic wires is obtained. These wires are subjected to the same tensions $H_x = S\lambda_y$ and $H_y = S\lambda_x$ that are acting on the corresponding strips. At every joint k of the net there is a vertical concentrated load of intensity $p_k \lambda_x \lambda_y$ equal to the load acting on the rectangle $\lambda_x \lambda_y$ having that joint as center.

Now substitute for the membrane equation [Eq. (8.10.6)] the corresponding finite difference equation. One thus obtains for point k [see Fig. 8.10.1 (b)][21]

[21] Equations (8.10.8) are derived in the following way. Consider a function w of the variable x. In the neighborhood of a point k, at $x = 0$ say (see Fig. 8.10.2), the function w can be expanded in the Maclaurin series

$$w(x) = w(0) + x\frac{dw(0)}{dx} + \frac{x^2}{2}\frac{d^2w(0)}{dx^2} + \frac{x^3}{6}\frac{d^3w(0)}{dx^3} + \cdots \tag{a}$$

Now, if x is made equal to λ_x and to $-\lambda_x$, one will find that

$$\left.\begin{array}{l} w_b = w_k + \lambda_x \left(\dfrac{dw}{dx}\right)_k + \dfrac{\lambda_x^2}{2}\left(\dfrac{d^2w}{dx^2}\right)_k + \dfrac{\lambda_x^3}{6}\left(\dfrac{d^3w}{dx^3}\right)_k + \cdots \\[2mm] w_a = w_k - \lambda_x \left(\dfrac{dw}{dx}\right)_k + \dfrac{\lambda_x^2}{2}\left(\dfrac{d^2w}{dx^2}\right)_k - \dfrac{\lambda_x^3}{6}\left(\dfrac{d^3w}{dx^3}\right)_k + \cdots \end{array}\right\} \tag{b}$$

Adding these two series together one gets, by neglecting powers of λ_x greater than two, the following equation

$$w_a + w_b = 2w_k + \lambda_x^2 \left(\frac{d^2w}{dx^2}\right)_k$$

from which one obtains the finite difference approximation

$$\left(\frac{d^2w}{dx^2}\right)_k = \frac{w_a + w_b - 2w_k}{\lambda_x^2} \tag{c}$$

In the same way, by subtracting Eqs. (b) and neglecting terms in λ_x of order greater than two, one obtains a finite difference approximation to $(dw/dx)_k$; thus,

$$\left(\frac{dw}{dx}\right)_k = \frac{w_b - w_a}{2\lambda_x}$$

Similarly, one can show that

$$\left(\frac{d^2w}{dy^2}\right)_k = \frac{w_c + w_d - 2w_k}{\lambda_y^2}$$

and

$$\left(\frac{dw}{dy}\right)_k = \frac{w_d - w_c}{2\lambda_y}$$

$$\left. \begin{array}{l} \dfrac{\partial^2 w}{\partial x^2} = \dfrac{w_b + w_a - 2w_k}{\lambda_x^2} \\[2mm] \dfrac{\partial^2 w}{\partial y^2} = \dfrac{w_d + w_c - 2w_k}{\lambda_y^2} \end{array} \right\} \qquad (8.10.8)$$

If $S = 1$ the following equation can be substituted for Eq. (8.10.6):

$$\frac{2w_k - w_a - w_b}{\lambda_x^2} + \frac{2w_k - w_c - w_d}{\lambda_y^2} = p_k \qquad (8.10.9)$$

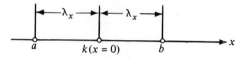

Figure 8.10.2

If, $\lambda_x = \lambda_y = \lambda$ Eq. (8.10.9) becomes[22]

$$4w_k - w_a - w_b - w_c - w_d = p_k \lambda^2 \qquad (8.10.10)$$

Similarly Eqs. (8.10.2) and (8.10.3) can be written in the following forms:

$$4M_k - M_a - M_b - M_c - M_d = p_k \lambda^2 \qquad (8.10.11)$$

$$4w_k - w_a - w_b - w_c - w_d = \frac{M_k}{D} \lambda^2 \qquad (8.10.12)$$

By writing Eqs. (8.10.11) and (8.10.12) for each joint of the net, one obtains two systems of linear algebraic equations having quantities M and w as unknowns. If the plate is simply supported on the border, the first system of equations can be solved independently of the second system. The second system is then solved after all the values of M have been determined. If the plate is constrained in any other way it is necessary to solve both systems of equations simultaneously.

Once values for M and w are obtained at all the joints k, the bending and twisting moments and the shear forces are obtained by the use of the Eqs. (8.1.5) and (8.1.8). The above equations become, in the present case

$$\left. \begin{array}{l} M_x = \dfrac{D}{\lambda^2}[(2w_k - w_a - w_b) + \nu(2w_k - w_c - w_D)] \\[2mm] M_y = \dfrac{D}{\lambda^2}[(2w_k - w_c - w_d) + \nu(2w_k - w_a - w_b)] \\[2mm] M_{xy} = \dfrac{D(1 - \nu)}{4\lambda^2}[(w_f + w_g) - (w_e + w_h)] \end{array} \right\} \qquad (8.10.13)$$

---

[22] It can be shown that Eq. (8.10.10) is the equilibrium equation for joint k subjected to the load $p_k \lambda^2$ and to unit tensions acting in the four wires which meet at joint k.

$$V_x = \frac{(M_b - M_a)}{2\lambda}$$
$$V_y = \frac{(M_d - M_c)}{2\lambda} \qquad (8.10.14)$$

By using the method of finite differences proposed by Marcus one avoids the integration of the differential equation [Eq. (8.1.1)]. Instead, one has to solve two systems of linear algebraic equations in the unknown quantities M and w. The method can be applied to any type of boundary condition of the plate.

In the case in which a concentrated load of intensity P_k acts on joint k Eq. (8.10.11) becomes:

$$4M_k - M_a - M_b - M_c - M_d = p_k \lambda^2 + P_k \qquad (8.10.15)$$

If the net representing the plate has few divisions, the moments M, although generally sufficiently accurate at a convenient distance from joint k, are not accurate near the joint. For this reason it is often necessary to use a better approximation to study the distribution of moments in the vicinity of joint k where the concentrated load is applied. For this purpose the area *efhg* is decomposed into smaller nets (see Fig. 8.10.3) and P_k is assumed to be uniformly distributed in the smaller square *mntq*.

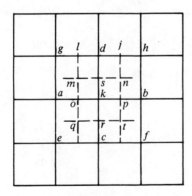

Figure 8.10.3

Example 8.10.1. In order to compute values of M and w for a square plate of sides a, which is simply supported at the border and subjected to a uniformly distributed load of intensity p_0, one proceeds in the following way:

First solution (see Fig. 8.10.4). Assume $\lambda = a/4$. Due to symmetry, values of M and w will be different only on the three joints 1, 2, and 3. At joints a, b, and c on the border, one has the following conditions:

$$M_a = M_b = M_c = 0$$
$$w_a = w_b = w_c = 0$$

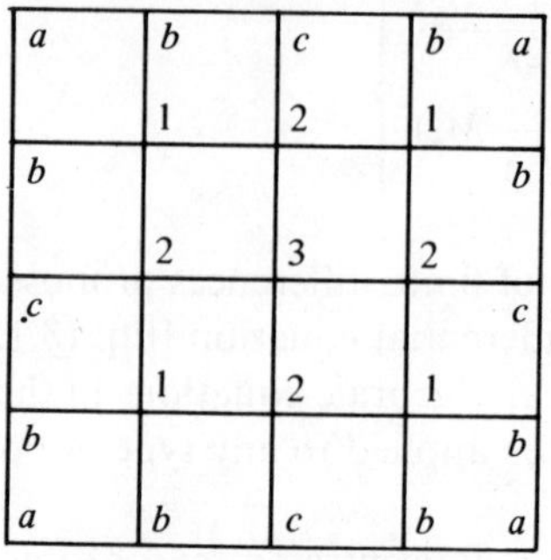

Figure 8.10.4

The use of Eq. (8.10.11) in this case gives

$$4M_1 - 2M_2 = p_0$$
$$4M_2 - M_1 - M_3 = p_0$$
$$4M_3 - 4M_2 = p_0$$

from which it follows that

$$M_1 = \frac{11p_0\lambda^2}{16} = \frac{11p_0 a^2}{256} = 0.04297 p_0 a^2$$

$$M_2 = \frac{14p_0\lambda^2}{16} = \frac{14p_0 a^2}{256} = 0.05469 p_0 a^2$$

$$M_3 = \frac{18p_0\lambda^2}{16} = \frac{18p_0 a^2}{256} = 0.07031 p_0 a^2$$

By applying Eq. (8.10.12) one finds

$$4w_1 - 2w_2 = \frac{11p_0\lambda^4}{16D}$$

$$4w_2 - w_1 - w_3 = \frac{14p_0\lambda^4}{16D}$$

$$4w_3 - 4w_2 = \frac{18p_0\lambda^4}{16D}$$

from which it follows that

$$w_1 = \frac{35p_0\lambda^4}{64D} = \frac{35p_0 a^4}{16384D} = 0.002136 \frac{p_0 a^4}{D}$$

$$w_2 = \frac{48p_0\lambda^4}{64D} = \frac{48p_0 a^4}{16384D} = 0.002930 \frac{p_0 a^4}{D}$$

$$w_3 = \frac{66p_0\lambda^4}{64D} = \frac{66p_0 a^4}{16384D} = 0.004028 \frac{p_0 a^4}{D}$$

At the center of the plate $M_x = M_y$ and Eq. (8.10.5) yields

$$M_x = M_y = (1 + v)\frac{M_3}{2} = 0.03515(1 + v)\frac{p_0 a^2}{D}$$

This value is 4.7 per cent less than the exact value which is $0.03684(1 + v)p_0a^2/D$. Deflection w_3 at the center of the plate is 0.79 per cent less than the exact value $0.00406 p_0 a^4/D$.

Second Solution (see Fig. 8.10.5). Assume $\lambda = a/8$. The values of w and M are unknown at the joints, 1, 2, 3, 4, 5, 6, 7, 8, 9, and 10. By applying Eq. (8.10.11) one obtains

$$4M_1 - 2M_2 = p_0 \lambda^2$$
$$4M_3 - M_2 - M_4 - M_6 = p_0 \lambda^2$$
$$4M_5 - 2M_2 - 2M_6 = p_0 \lambda^2$$
$$4M_7 - M_4 - 2M_6 - M_9 = p_0 \lambda^2$$
$$4M_9 - M_7 - 2M_8 - M_{10} = p_0 \lambda^2$$
$$4M_2 - M_1 - M_3 - M_5 = p_0 \lambda^2$$
$$4M_4 - 2M_3 - M_7 = p_0 \lambda^2$$
$$4M_6 - M_3 - M_5 - M_7 - M_8 = p_0 \lambda^2$$
$$4M_8 - 2M_6 - 2M_9 = p_0 \lambda^2$$
$$4M_{10} - 4M_9 = p_0 \lambda^2$$

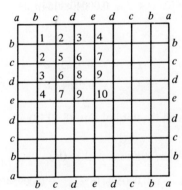

Figure 8.10.5

By solving these equations, one finds that

$$M_1 = 0.01778 p_0 a^2, \qquad M_6 = 0.05377 p_0 a^2$$
$$M_2 = 0.02774 p_0 a^2, \qquad M_7 = 0.05664 p_0 a^2$$
$$M_3 = 0.03291 p_0 a^2, \qquad M_8 = 0.06523 p_0 a^2$$
$$M_4 = 0.03452 p_0 a^2, \qquad M_9 = 0.06888 p_0 a^2$$
$$M_5 = 0.04466 p_0 a^2, \qquad M_{10} = 0.07278 p_0 a^2$$

By applying Eq. (8.10.12), one finds that

$$w_1 = 0.000663 \frac{p_0 a^4}{D}, \qquad w_6 = 0.002733 \frac{p_0 a^4}{D}$$

$$w_2 = 0.001186 \frac{p_0 a^4}{D}, \qquad w_7 = 0.002937 \frac{p_0 a^4}{D}$$

$$w_3 = 0.001515 \frac{p_0 a^4}{D}, \qquad w_8 = 0.003507 \frac{p_0 a^4}{D}$$

$$w_4 = 0.001627 \frac{p_0 a^4}{D}, \qquad w_9 = 0.003770 \frac{p_0 a^4}{D}$$

$$w_5 = 0.002134 \frac{p_0 a^4}{D}, \qquad w_{10} = 0.004055 \frac{p_0 a^4}{D}$$

At the center of the plate, one finds that $M_x = M_y = 0.03639(1 + v)p_0 a^2$ which is 1.22 per cent less than the exact value. The deflection w_{10} at the center is found to be 0.12 per cent less than the exact value. For $(M_x)_7$ and $(M_y)_7$ one obtains

$$(M_x)_7 = D[(2w_7 - 2w_6) + v(2w_7 - w_4 - w_9)]\Big/\left(\frac{a^2}{64}\right)$$
$$= [0.000408 + 0.000477v]64 p_0 a^2$$
$$(M_y)_7 = D[(2w_7 - w_4 - w_9) + v(2w_7 - 2w_6)]\Big/\left(\frac{a^2}{64}\right)$$
$$= [0.000477 + 0.000408v]64 p_0 a^2$$

If $v = 0.30$, one finds that

$$(M_x)_7 = 0.0353 p_0 a^2$$
$$(M_y)_7 = 0.0384 p_0 a^2$$

$(M_{xy})_6$ is obtained through the third of Eqs. (8.10.13) which, in this case, is

$$(M_{xy})_6 = D(1 - v)[(w_9 - w_2) - (w_6 - w_4)]\Big/\frac{a^2}{16}$$
$$= (1 - v)0.000596(16)p_0 a^2 = 0.00954(1 - v)p_0 a^2$$

The shear forces $(V_x)_6$ and $(V_y)_6$ are computed by use of Eqs. (8.10.14) which, in this case, are

$$(V_x)_6 = (M_7 - M_5)\Big/\left(\frac{2a}{8}\right) = 0.01198(4 p_0 a) = 0.0479 p_0 a$$
$$(V_y)_6 = (M_8 - M_3)\Big/\left(\frac{2a}{8}\right) = 0.03232(4 p_0 a) = 0.1293 p_0 a$$

Third Solution (see Fig. 8.10.6). Assume $\lambda = a/12$. Values of w and M are unknown at joints 1, 2, 3, 4, 5, 6, 7, 8, 9, 10, 11, 12, 13, 14, 15, 16, 17, 18, 19, 20, and 21. In this case the application of Eq. (8.10.11) gives the following system of equations:

$$4M_1 - 2M_2 = p_0\lambda^2$$
$$4M_2 - M_1 - M_3 - M_7 = p_0\lambda^2$$
$$4M_3 - M_2 - M_4 - M_8 = p_0\lambda^2$$
$$4M_4 - M_3 - M_5 - M_9 = p_0\lambda^2$$
$$4M_5 - M_4 - M_6 - M_{10} = p_0\lambda^2$$
$$4M_6 - 2M_5 - M_{11} = p_0\lambda^2$$
$$4M_7 - 2M_2 - 2M_8 = p_0\lambda^2$$
$$4M_8 - M_3 - M_7 - M_9 - M_{12} = p_0\lambda^2$$
$$4M_9 - M_4 - M_8 - M_{10} - M_{13} = p_0\lambda^2$$
$$4M_{10} - M_5 - M_9 - M_{11} - M_{14} = p_0\lambda^2$$
$$4M_{11} - M_6 - 2M_{10} - M_{15} = p_0\lambda^2$$
$$4M_{12} - 2M_8 - 2M_{13} = p_0\lambda^2$$
$$4M_{13} - M_9 - M_{12} - M_{14} - M_{16} = p_0\lambda^2$$
$$4M_{14} - M_{10} - M_{13} - M_{15} - M_{17} = p_0\lambda^2$$
$$4M_{15} - M_{11} - 2M_{14} - M_{18} = p_0\lambda^2$$
$$4M_{16} - 2M_{13} - 2M_{17} = p_0\lambda^2$$
$$4M_{17} - M_{14} - M_{16} - M_{18} - M_{19} = p_0\lambda^2$$
$$4M_{18} - M_{15} - 2M_{17} - M_{20} = p_0\lambda^2$$
$$4M_{19} - 2M_{17} - 2M_{20} = p_0\lambda^2$$
$$4M_{20} - M_{18} - 2M_{19} - M_{21} = p_0\lambda^2$$
$$4M_{21} - 4M_{20} = p_0\lambda^2$$

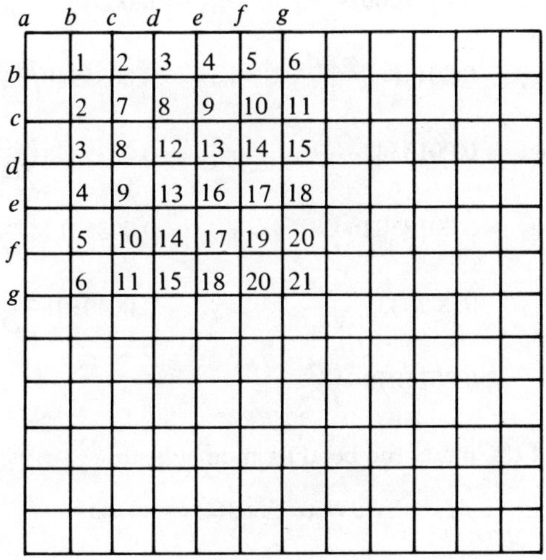

	a	b	c	d	e	f	g		
b		1	2	3	4	5	6		
c		2	7	8	9	10	11		
d		3	8	12	13	14	15		
e		4	9	13	16	17	18		
f		5	10	14	17	19	20		
g		6	11	15	18	20	21		

Figure 8.10.6

Solving these equations one finds that

$$M_1 = 0.009709 p_0 a^2, \quad M_{12} = 0.045005 p_0 a^2$$
$$M_2 = 0.015947 p_0 a^2, \quad M_{13} = 0.051847 p_0 a^2$$
$$M_3 = 0.020078 p_0 a^2, \quad M_{14} = 0.055754 p_0 a^2$$
$$M_4 = 0.022730 p_0 a^2, \quad M_{15} = 0.057024 p_0 a^2$$
$$M_5 = 0.024220 p_0 a^2, \quad M_{16} = 0.060007 p_0 a^2$$
$$M_6 = 0.024701 p_0 a^2, \quad M_{17} = 0.064695 p_0 a^2$$
$$M_7 = 0.027055 p_0 a^2, \quad M_{18} = 0.066224 p_0 a^2$$
$$M_8 = 0.034690 p_0 a^2, \quad M_{19} = 0.069852 p_0 a^2$$
$$M_9 = 0.039679 p_0 a^2, \quad M_{20} = 0.071536 p_0 a^2$$
$$M_{10} = 0.042504 p_0 a^2, \quad M_{21} = 0.073272 p_0 a^2$$
$$M_{11} = 0.043419 p_0 a^2,$$

By applying Eq. (8.10.12), one finds that

$$w_1 = 0.0003083 \frac{p_0 a^4}{D}, \quad w_{12} = 0.0021333 \frac{p_0 a^4}{D}$$
$$w_2 = 0.0005830 \frac{p_0 a^4}{D}, \quad w_{13} = 0.0025761 \frac{p_0 a^4}{D}$$
$$w_3 = 0.0008071 \frac{p_0 a^4}{D}, \quad w_{14} = 0.0028469 \frac{p_0 a^4}{D}$$
$$w_4 = 0.0009719 \frac{p_0 a^4}{D}, \quad w_{15} = 0.0029379 \frac{p_0 a^4}{D}$$
$$w_5 = 0.0010723 \frac{p_0 a^4}{D}, \quad w_{16} = 0.0031138 \frac{p_0 a^4}{D}$$
$$w_6 = 0.0011060 \frac{p_0 a^4}{D}, \quad w_{17} = 0.0034432 \frac{p_0 a^4}{D}$$
$$w_7 = 0.0011060 \frac{p_0 a^4}{D}, \quad w_{18} = 0.0035540 \frac{p_0 a^4}{D}$$
$$w_8 = 0.0015343 \frac{p_0 a^4}{D}, \quad w_{19} = 0.0038088 \frac{p_0 a^4}{D}$$
$$w_9 = 0.0018502 \frac{p_0 a^4}{D}, \quad w_{20} = 0.0039319 \frac{p_0 a^4}{D}$$
$$w_{10} = 0.0020431 \frac{p_0 a^4}{D}, \quad w_{21} = 0.0040591 \frac{p_0 a^4}{D}$$
$$w_{11} = 0.0021079 \frac{p_0 a^4}{D},$$

At the center of the plate, the bending moments are

$$M_x = M_y = 0.036636(1 + \nu) p_0 a^2$$

which are 0.55 per cent less than the exact values. The deflection w_{21} at the center is found to be 0.02 per cent less than the exact value.

Example 8.10.2. In order to compute values of M and w at various points of a square plate of sides a, which is built-in at the border and subjected to a uniformly distributed load of intensity p_0, the finite difference method will be used with $\lambda = a/4$ (see Fig. 8.10.7).

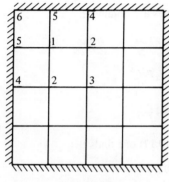

Figure 8.10.7

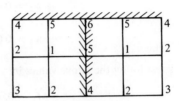

Figure 8.10.8

Values of M at joints 1, 2, 3, 4, 5, and 6 are unknown. Displacements at joints 4, 5, and 6 will be equal to zero. Let us assume that on the left of the plate there is another equal plate also built-in at the border and subjected to a uniformly distributed load of intensity p_0 (see Fig. 8.10.8). Then in this case, Eq. (8.10.11) gives

$$\left. \begin{array}{l} 4M_1 - 2M_2 - 2M_5 = p_0\lambda^2 \\ 4M_2 - 2M_1 - M_3 - M_4 = p_0\lambda^2 \\ 4M_3 - 4M_2 = p_0\lambda^2 \end{array} \right\} \quad (8.10.16)$$

while Eq. (8.10.12) gives

$$\left. \begin{array}{l} 4w_1 - 2w_2 = \dfrac{M_1}{D}\lambda^2 \\ \\ 4w_2 - 2w_1 - w_3 = \dfrac{M_2}{D}\lambda^2 \\ \\ 4w_3 - 4w_2 = \dfrac{M_3}{D}\lambda^2 \\ \\ -2w_2 = \dfrac{M_4}{D}\lambda^2 \\ \\ -2w_1 = \dfrac{M_5}{D}\lambda^2 \\ \\ 0 = M_6 \end{array} \right\} \quad (8.10.17)$$

By substituting Eqs. (8.10.17) into Eqs. (8.10.16), the following system of three equations is obtained:

$$\left.\begin{array}{c} 24w_1 - 16w_2 + 2w_3 = \dfrac{p_0\lambda^4}{D} \\ -16w_1 + 26w_2 - 8w_3 = \dfrac{p_0\lambda^4}{D} \\ 8w_1 - 32w_2 + 20w_3 = \dfrac{p_0\lambda^4}{D} \end{array}\right\} \quad (8.10.18)$$

Upon solving Eqs. (8.10.18) the following values are obtained:

$$w_1 = 0.0008174596 \frac{p_0 a^4}{D}$$

$$w_2 = 0.0012069874 \frac{p_0 a^4}{D}$$

$$w_3 = 0.0017995084 \frac{p_0 a^4}{D}$$

By substituting the above values into Eqs. (8.10.17) one finds that

$$M_1 = 0.0136938202 p_0 a^2$$
$$M_2 = 0.0222963483 p_0 a^2$$
$$M_3 = 0.0379213483 p_0 a^2$$
$$M_4 = -0.0386235955 p_0 a^2$$
$$M_5 = -0.0261587079 p_0 a^2$$
$$M_6 = 0$$

and consequently,

$$M_{3x} = M_{3y} = 0.01896(1 + \nu) p_0 a^2$$
$$M_{4x} = M_{4y} = -0.01931(1 + \nu) p_0 a^2$$

The exact values in this case are

$$w_3 = 0.00126 \frac{p_0 a^4}{D}$$

$$M_{3x} = M_{3y} = 0.0178(1 + \nu) p_0 a^2$$
$$M_{4x} = M_{4y} = -0.0256(1 + \nu) p_0 a^2$$

As in the previous example, the accuracy of the results can be improved by decreasing the net size λ.

8.11 Grashof's Method

Consider a rectangular plate which is either simply supported or built-in at the edges which have dimensions a and b [see Figs. 8.11.1 (a) and (b)].

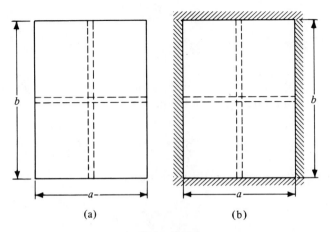

Figure 8.11.1

According to Grashof's method[23] the plate is represented by two strips perpendicular to one another and parallel to sides a and b. In the case of a plate subjected to a uniformly distributed load of intensity p_0 the plate is divided into beams of lengths a and b subjected, respectively, to the uniformly distributed loads

$$\left. \begin{array}{l} p_a = p_0 \dfrac{b^4}{a^4 + b^4} \\ p_b = p_0 \dfrac{a^4}{a^4 + b^4} \end{array} \right\} \quad (8.11.1)$$

In fact, in the case of the plate simply supported at the border, deflections of the two simply supported bars of lengths a and b under the uniformly distributed loads p_a and p_b will be

$$\left. \begin{array}{l} w_a = \dfrac{5 p_a a^4}{384 EI} \\ w_b = \dfrac{5 p_b b^4}{384 EI} \end{array} \right\} \quad (8.11.2)$$

The two conditions

$$w_a = w_b \quad \text{and} \quad p_0 = p_a + p_b \quad (8.11.3)$$

give the values shown in Eqs. (8.11.1) for p_a and p_b. In the case of a plate built-in at the border, Eqs. (8.11.2) become

[23] See Bibliography No. 16 at the end of this chapter.

$$w_a = \frac{p_a a^4}{384EI}$$
$$w_b = \frac{p_b b^4}{384EI} \quad \text{(8.11.4)}$$

and Eqs. (8.11.3) lead to the same values for p_a and p_b as those given by Eqs. (8.11.1). The central deflection of the plate is

$$w_{\max} = c_1 \frac{p_a b^4}{D} = c_1 p_0 \frac{a^4 b^4}{D(a^4 + b^4)}$$

The moments at the center of the plate are

$$M_x = c_2 p_a a^2 = c_2 p_0 a^2 \frac{b^4}{a^4 + b^4}$$

$$M_y = c_2 p_b b^2 = c_2 p_0 b^2 \frac{a^4}{a^4 + b^4}$$

and the moments at the border of the plate are

$$M_x = c_3 p_a a^2 = c_3 p_0 a^2 \frac{b^4}{a^4 + b^4}$$

$$M_y = c_3 p_b b^2 = c_3 p_0 b^2 \frac{a^4}{a^4 + b^4}$$

The coefficients c_i ($i = 1, 2$, and 3) have the following values:

$$\left. \begin{array}{l} c_1 = \frac{5}{384} \\ c_2 = \frac{1}{8} \\ c_3 = 0 \end{array} \right\} \text{ for the simply-supported plate}$$

$$\left. \begin{array}{l} c_1 = \frac{1}{384} \\ c_2 = \frac{1}{24} \\ c_3 = -\frac{1}{12} \end{array} \right\} \text{ for the built-in plate}$$

It is easy, by applying Grashof's method, to compute values of p_a and p_b for cases in which there are different boundary conditions.

For the case in which the plate is square $a = b$ and $p_a = p_b = p_0/2$ and the above formulas reduce to the forms shown in Table 8.11.1. The deflection of the built-in plate is given with a very good approximation by the formula

$$w_{\max} = 0.00130 \frac{p_0 a^4}{D}$$

TABLE 8.11.1

DEFLECTIONS AND BENDING MOMENTS FOR A SQUARE
PLATE BY THE GRASHOF METHOD

	Simply Supported Plate	Built-in Plate
Maximum deflection	$\dfrac{0.0065 p_0 a^4}{D}$	$\dfrac{0.0013 p_0 a^4}{D}$
Bending moment at the center	$0.0625 p_0 a^2$	$0.0208 p_0 a^2$
Bending moment at the border	0	$-0.0417 p_0 a^2$

The other results are not so good. For instance, the bending moment at the center of the simply supported plate is 30 per cent greater than the exact value for the case in which $v = 0.30$. However, due to the fact that this method is very simple and it is always on the safe side (i.e., it always gives results in excess of the actual values), it is used very often by practical engineers.

Example 8.11.1. Consider the case of a rectangular plate which is simply supported on three sides and built-in on one side of length b (see Fig. 8.11.2). By equat-

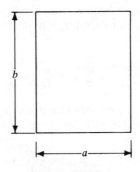

Figure 8.11.2

ing the deflections at the center, in accordance with Grashof's method, one obtains

$$\frac{p_a a^4}{192 EI} = \frac{5 p_b b^4}{384 EI}$$

from which

$$p_a = p_0 \frac{5 b^4}{2a^4 + 5b^4}$$

$$p_b = p_0 \frac{2 b^4}{2a^4 + 5b^4}$$

The maximum positive bending moments are

$$M_x = \frac{9}{128} \frac{5b^4}{2a^4 + 5b^4} p_0 a^2$$

$$M_y = \frac{1}{8} \frac{2a^4}{2a^4 + 5b^4} p_0 b^2$$

For the moment at the built-in side, one finds

$$M = -\frac{1}{8} \frac{5b^4}{2a^4 + 5b^4} p_0 a^2$$

For the case in which the plate is square ($a = b$), the above formulas reduce to

$$M_x = 0.0502 p_0 a^2$$
$$M_y = 0.0357 p_0 a^2$$
$$M = -0.0893 p_0 a^2$$

8.12 Bending of Plates on Elastic Foundations

(a) *Case of rectangular plates.* Assuming that the reaction q at any point of the surface of the plate in contact with the foundation is proportional to the deflection w at that point (Winkler's hypothesis), then Lagrange's equation [Eq. (8.1.1)] becomes

$$\frac{\partial^4 w}{\partial x^4} + 2\frac{\partial^4 w}{\partial x^2 \partial y^2} + \frac{\partial^4 w}{\partial y^4} = \frac{p}{D} - \frac{q}{D} = \frac{p}{D} - \frac{kw}{D} \quad (8.12.1)$$

In the case of a rectangular plate of sides a and b with the four edges simply supported, Eq. (8.12.1) can be easily integrated. Let p and w be represented by

$$p = \sum_{m=1}^{\infty} \sum_{n=1}^{\infty} p_{mn} \sin \frac{m\pi x}{a} \sin \frac{n\pi y}{b} \quad (8.12.2)$$

and

$$w = \sum_{m=1}^{\infty} \sum_{n=1}^{\infty} A_{mn} \sin \frac{m\pi x}{a} \sin \frac{n\pi y}{b} \quad (8.12.3)$$

where the coefficients p_{mn} of the series in Eq. (8.12.2) are determined from the known function p by use of the equation

$$p_{mn} = \frac{4}{ab} \int_0^a \int_0^b p \sin \frac{m\pi x}{a} \sin \frac{n\pi y}{b} \, dx \, dy \quad (8.12.4)$$

By substituting the series of Eqs. (8.12.2) and (8.12.3) into Eq. (8.12.1) and solving for A_{mn}, one obtains

$$A_{mn} = \frac{p_{mn}}{\pi^4 D\left(\dfrac{m^2}{a^2} + \dfrac{n^2}{b^2}\right)^2 + k} \qquad (8.12.5)$$

In the case of a concentrated force of intensity P applied at a point with coordinates ξ and η, one finds from Eq. (8.12.4) that

$$p_{mn} = \frac{4P}{ab} \sin \frac{m\pi\xi}{a} \sin \frac{n\pi\eta}{b} \qquad (8.12.6)$$

In view of Eqs. (8.12.5) and (8.12.6), Eq. (8.12.3) can be written as

$$w = \frac{4P}{ab} \sum_{m=1}^{\infty} \sum_{n=1}^{\infty} \frac{\sin \dfrac{m\pi\xi}{a} \sin \dfrac{n\pi\eta}{b}}{\pi^4 D\left(\dfrac{m^2}{a^2} + \dfrac{n^2}{b^2}\right)^2 + k} \sin \frac{m\pi x}{a} \sin \frac{n\pi y}{b} \qquad (8.12.7)$$

Equation (8.12.7) may be used with the superposition method to determine the deflection of rectangular plates under any kind of loading. In the case of a uniformly distributed load of intensity p_0 acting on the plate, $p_0 \, d\xi \, d\eta$ is substituted for P in Eq. (8.12.7). Then by integrating between 0 and a and between 0 and b, one finally obtains for the deflection surface of the plate the expression

$$w = \frac{16 p_0}{\pi^2} \sum_{m=1,3,5\ldots} \sum_{n=1,3,5\ldots} \frac{\sin \dfrac{m\pi x}{a} \sin \dfrac{n\pi y}{b}}{mn\left[\pi^4 D\left(\dfrac{m^2}{a^2} + \dfrac{n^2}{b^2}\right)^2 + k\right]} \qquad (8.12.8)$$

If in Eq. (8.12.8) $k = 0$, then Eq. (8.12.8) reduces to Eq. (8.4.7), i.e., to Navier's solution for the rectangular plate under a uniformly distributed load of intensity p_0.

(b) **Case of circular plates.** The problem of a circular plate resting on a continuous elastic foundation has been treated by S. P. Timoshenko and by R. De L'Hortet.[24] In this section, only an approximate strain energy method will be presented for calculating the deflection of a circular plate resting on an elastic foundation. In the case in which polar coordinates are used, instead of cartesian coordinates, Eq. (8.8.5) has the following form:

$$U_1 = \iint_{\Omega} \left\{ \frac{D}{2} \left[\frac{\partial^2 w}{\partial r^2} + \frac{1}{r} \frac{\partial w}{\partial r} + \frac{1}{r^2} \frac{\partial^2 w}{\partial \theta^2} \right]^2 - 2(1-v) \frac{\partial^2 w}{\partial r^2}\left(\frac{1}{r} \frac{\partial w}{\partial r} + \frac{1}{r^2} \frac{\partial^2 w}{\partial \theta^2} \right) \right.$$
$$\left. + 2(1-v)\left(\frac{1}{r} \frac{\partial^2 w}{\partial r \partial \theta} - \frac{1}{r^2} \frac{\partial w}{\partial \theta} \right)^2 \right] \right\} r \, dr \, d\theta$$

[24] See Bibliography Nos. 60 and 12 at the end of the chapter.

If there is symmetry with respect to the center of the plate of radius a, the above equation becomes

$$U_1 = 2\pi \int_0^a \left\{ \frac{D}{2}\left[\left(\frac{d^2w}{dr^2} + \frac{1}{r}\frac{dw}{dr}\right)^2 - \frac{2(1-v)}{r}\frac{dw}{dr}\frac{d^2w}{dr^2}\right]\right\} r\,dr \quad (8.12.9)$$

Assuming as a first approximation for the deflection w, the expression

$$w = c_1 + c_2 r^2 \quad (8.12.10)$$

where c_1 and c_2 are undetermined parameters, and substituting Eq. (8.12.10) into Eq. (8.12.9), one obtains for the strain energy of the plate the expression

$$U_1 = 4c_2^2 D\pi a^2(1+v)$$

The strain energy of deformation of the elastic foundation is

$$U_2 = \int_0^{2\pi}\int_0^a \frac{kw^2}{2} r\,dr\,d\theta = \frac{\pi k}{2}(c_1^2 a^2 + c_1 c_2 a^4 + \tfrac{1}{3}c_2^2 a^6)$$

If a concentrated load of intensity P is applied at the center of the plate ($r = 0$), the total energy of the system is expressed by

$$U = 4c_2^2 D\pi a^2(1+v) + \frac{\pi k}{2}(c_1^2 a^2 + c_1 c_2 a^4 + \tfrac{1}{3}c_2^2 a^6) - Pc_1$$

By noting that

$$\frac{\partial U}{\partial c_1} = \frac{\partial U}{\partial c_2} = 0$$

the two equations

$$c_1 + c_2 a^2 \left[\frac{2}{3} + 16D\frac{(1-v)}{ka^4}\right] = 0$$

$$c_1 + \frac{1}{2}c_2 a^2 = \frac{P}{\pi k a^2}$$

are obtained. It follows that

$$\left.\begin{aligned}c_1 &= \frac{P}{\pi k a^2}\left[1 + \frac{1}{2\left[\frac{1}{6} + 16D\frac{(1+v)}{ka^4}\right]}\right] \\ c_2 &= -\frac{P}{\pi k a^2\left[\frac{1}{6} + 16D\frac{(1+v)}{ka^4}\right]}\end{aligned}\right\} \quad (8.12.11)$$

By substituting Eqs. (8.12.11) into Eq. (8.12.10) and letting $r = 0$, one obtains for the maximum deflection

$$w_{max} = \frac{P}{\pi k a^2}\left[1 + \frac{1}{2\left[\frac{1}{6} + 16D\frac{(1+v)}{ka^4}\right]}\right] \quad (8.12.12)$$

In order to improve the accuracy of the method more terms must be added to the expression for the deflection.

(c) *The general problem of the plate resting on an elastic foundation.*[25] As in the case of the bar resting on an elastic foundation, it is possible to study the problem of the infinite plate of constant thickness resting on an elastic foundation for the general case in which the behavior of the foundation is represented by an influence function. In this case the influence function is

$$\psi(x - \xi, y - \eta) \quad (8.12.13)$$

Calling q the reaction of the foundation, the following two equations are derived:

$$D\left[\frac{\partial^4 w}{\partial x^4} + 2\frac{\partial^4 w}{\partial x^2 \partial y^2} + \frac{\partial^4 w}{\partial y^4}\right] = p - q \quad (8.12.14)$$

$$w = \int_{-\infty}^{\infty}\int_{-\infty}^{\infty} q(\xi, \eta)\psi(x - \xi, y - \eta)\,d\xi\,d\eta \quad (8.12.15)$$

The following Fourier's transforms are now defined:

$$P(u, v) = \frac{1}{2\pi}\int_{-\infty}^{\infty}\int_{-\infty}^{\infty} e^{iux+ivy} p(x, y)\,dx\,dy$$

$$Q(u, v) = \frac{1}{2\pi}\int_{-\infty}^{\infty}\int_{-\infty}^{\infty} e^{iux+ivy} q(x, y)\,dx\,dy$$

$$\Psi(u, v) = \frac{1}{2\pi}\int_{-\infty}^{\infty}\int_{-\infty}^{\infty} e^{iux+ivy} \psi(x, y)\,dx\,dy$$

$$W(u, v) = \frac{1}{2\pi}\int_{-\infty}^{\infty}\int_{-\infty}^{\infty} e^{iux+ivy} w(x, y)\,dx\,dy$$

From Eq. (8.12.14) one deduces that

$$\frac{D}{2\pi}\int_{-\infty}^{\infty}\int_{-\infty}^{\infty} e^{iux+ivy}\left[\frac{\partial^4 w}{\partial x^4} + 2\frac{\partial^4 w}{\partial x^2 \partial y^2} + \frac{\partial^4 w}{\partial y^4}\right]dx\,dy = P(u, v) - Q(u, v)$$

[25] Parts of Section 8.12(c) were first discussed in "Sul Calcolo della Deformazione della Piastra su Suolo Elastico" by C. Stechkewicz, and "Sul Problema Generale della Piastra Poggiata su Suolo Elastico" by Enrico Volterra (see number 55 and number 63 of the bibliography at the end of this chapter) and are presented here with the permission of the Accademia Nazionale dei Lincei.

By integrating by parts and assuming that w and its first three derivatives approach zero as $|x| \to \infty$ and $|y| \to \infty$, one obtains

$$D(u^2 + v^2)^2 \, W(u, v) = P(u, v) - Q(u, v) \qquad (8.12.16)$$

By use of Eq. (8.12.15), the equation for $W(u, v)$ may be written as follows:

$$W(u, v) = \frac{1}{2\pi} \int_{-\infty}^{\infty}\int_{-\infty}^{\infty} \left\{ e^{iux+ivy} \int_{-\infty}^{\infty}\int_{-\infty}^{\infty} q(\xi, \eta)\psi(x - \xi, y - \eta)\, d\xi\, d\eta \, dx\, dy \right\}$$

$$= \frac{1}{2\pi} \int_{-\infty}^{\infty}\int_{-\infty}^{\infty} \left\{ q(\xi, \eta) \int_{-\infty}^{\infty}\int_{-\infty}^{\infty} e^{iux+ivy} \psi(x - \xi, y - \eta)\, dx\, dy\, d\xi\, d\eta \right\}$$

$$= \frac{1}{2\pi} \int_{-\infty}^{\infty}\int_{-\infty}^{\infty} \left\{ q(\xi, \eta) \int_{-\infty}^{\infty}\int_{-\infty}^{\infty} e^{iu(\xi+s)+iv(\eta+t)} \psi(s, t)\, ds\, dt\, d\xi\, d\eta \right\}$$

$$= \frac{1}{2\pi} \int_{-\infty}^{\infty}\int_{-\infty}^{\infty} \left\{ e^{iu\xi+iv\eta} q(\xi, \eta) \int_{-\infty}^{\infty}\int_{-\infty}^{\infty} e^{ius+ivt} \psi(s, t)\, ds\, dt\, d\xi\, d\eta \right\}$$

$$= \Psi(u, v) \int_{-\infty}^{\infty}\int_{-\infty}^{\infty} e^{iu\xi+iv\eta} q(\xi, \eta)\, d\xi\, d\eta$$

$$= 2\pi \Psi(u, v) Q(u, v) \qquad (8.12.17)$$

From Eqs. (8.12.16) and (8.12.17) it follows that

$$W(u, v) = \frac{P(u, v)}{D(u^2 + v^2)^2 + \dfrac{1}{2\pi \Psi(u, v)}}$$

By applying the inversion formula for the double Fourier transform the final formula

$$w = \frac{1}{2\pi} \int_{-\infty}^{\infty}\int_{-\infty}^{\infty} e^{-iux-ivy} W(u, v)\, du\, dv \qquad (8.12.18)$$

is obtained for the deflection of the plate.

Example 8.12.1. Suppose that the function p represents a total load T uniformly distributed on an area $4l_1 l_2$ bounded by the lines

$$x = a - l_1, \quad x = a + l_1, \quad y = b - l_2, \quad y = b + l_2$$

The case of a concentrated load is immediately derived by letting $l_1 \to 0$ and $l_2 \to 0$, and the case of any other load distribution can be solved by applying the principle of superposition. It follows that

$$p(x, y) = \begin{cases} = \dfrac{T}{4l_1 l_2} & \text{for } \begin{cases} a - l_1 \leq x \leq a + l_1 \\ b - l_2 \leq y \leq b + l_2 \end{cases} \\ = 0 & \text{everywhere else} \end{cases}$$

and

$$P(u,v) = \frac{1}{2\pi} \frac{T}{4l_1 l_2} \int_{a-l_1}^{a+l_1} \int_{b-l_2}^{b+l_2} e^{iux+ivy} \, dx \, dy = \frac{T}{2\pi} e^{iau+ibv} \frac{\sin l_1 u}{l_1 u} \frac{\sin l_2 v}{l_2 v}$$

In the limiting case of a concentrated load T at point $x = a$, $y = b$, one obtains

$$P(u,v) = \frac{T}{2\pi} e^{iau} e^{ibv}$$

By assuming that Winkler's hypothesis is valid, one has

$$\psi(x - \xi, y - \eta) = \frac{1}{\beta} \quad \text{for} \quad x = \xi, y = \eta$$

$$\psi(x - \xi, y - \eta) = 0 \quad \text{for} \quad x \neq \xi, y \neq \eta$$

and the solution is given by

$$w = \frac{T}{4\pi^2} \int_{-\infty}^{\infty} \int_{-\infty}^{\infty} \frac{e^{i(a-x)u + i(b-y)v}}{D(u^2 + v^2)^2 + \beta} \frac{\sin l_1 u}{l_1 u} \frac{\sin l_2 v}{l_2 v} \, du \, dv$$

$$= \frac{T}{\pi^2} \int_{-\infty}^{\infty} \int_{-\infty}^{\infty} \frac{\cos(a-x)u \cos(b-y)v}{D(u^2 + v^2)^2 + \beta} \frac{\sin l_1 u}{l_1 u} \frac{\sin l_2 v}{l_2 v} \, du \, dv \qquad (8.12.19)$$

The function

$$F(x, y) = \int_0^\infty \int_0^\infty \frac{\sin xu \sin yv}{uv} \frac{du \, dv}{(u^2 + v^2)^2 + 1} \qquad (8.12.20)$$

has been computed by L. Stankiewicz. Numerical values of this function are presented in Table 8.12.1. In Table 8.12.1 only the values $y \leq x$ are given since the function $F(x, y)$ is symmetrical with respect to the variables x and y. In the last column (entitled $x = \infty$) of Table 8.12.1, the values of

$$\underset{x \to +\infty}{F}(x, y) = \frac{\pi^2}{4}\left(1 - e^{-y/\sqrt{2}} \cos \frac{y}{\sqrt{2}}\right)$$

are given. In order to use Table 8.12.1 for computing Eq. (8.12.19), one writes Eq. (8.12.19) in the following form:

$$w(x, y) = \frac{T}{4\pi^2 l_1 l_2 \beta}\{F[k(a - x + l_1), k(b - y + l_2)]$$

$$- F[k(a - x + l_1), k(b - y - l_2)] - F[k(a - x - l_1), y(b - y + l_2)]$$

$$+ F[k(a - x - l_1), k(b - y - l_2)]\}$$

where $\lambda = (\beta/D)^{1/4}$.

In the case in which the foundation reacts according to Jodi's hypothesis,[26] the influence function is

---

[26] See Bibliography No. 55 at the end of this chapter.

TABLE 8.12.1

FUNCTION $F(x, y)$

y \ x	0.0	0.5	1.0	1.5	2.0	2.5	3.0	3.5	4.0	4.5	5.0
0.0	0	0	0	0	0	0	0	0	0	0	0
0.5		0.277	0.499	0.656	0.756	0.817	0.846	0.858	0.861	0.858	0.854
1.0			0.906	1.199	1.383	1.498	1.549	1.576	1.580	1.574	1.566
1.5				1.591	1.842	1.998	2.074	2.103	2.107	2.097	2.084
2.0					2.138	2.320	2.409	2.442	2.444	2.431	2.414
2.5						2.517	2.610	2.642	2.640	2.623	2.602
3.0							2.704	2.734	2.728	2.706	2.681
3.5								2.759	2.749	2.722	2.694
4.0									2.734	2.705	2.674
4.5										2.672	2.639
5.0											2.605
5.5											
6.0											
6.5											
7.0											
7.5											
8.0											
9.0											
10.0											
11.0											
12.0											
∞											

y \ x	5.5	6.0	6.5	7.0	7.5	8.0	9.0	10.0	11.0	12.0	∞
0.0	0	0	0	0	0	0	0	0	0	0	0
0.5	0.850	0.846	0.844	0.842	0.842	0.842	0.842	0.842	0.842	0.843	0.842
1.0	1.557	1.551	1.546	1.543	1.542	1.541	1.541	1.542	1.542	1.543	1.542
1.5	2.072	2.062	2.056	2.051	2.049	2.049	2.049	2.050	2.050	2.051	2.050
2.0	2.401	2.387	2.378	2.373	2.370	2.369	2.370	2.371	2.371	2.372	2.372
2.5	2.582	2.567	2.557	2.551	2.548	2.547	2.547	2.549	2.550	2.550	2.550
3.0	2.658	2.641	2.630	2.623	2.620	2.619	2.619	2.621	2.622	2.623	2.622
3.5	2.670	2.651	2.638	2.631	2.628	2.626	2.628	2.629	2.630	2.630	2.631
4.0	2.647	2.627	2.614	2.606	2.602	2.601	2.603	2.605	2.606	2.606	2.606
4.5	2.611	2.590	2.577	2.569	2.565	2.564	2.566	2.568	2.569	2.570	2.570
5.0	2.575	2.554	2.540	2.532	2.530	2.528	2.539	2.532	2.533	2.534	2.534
5.5	2.546	2.524	2.510	2.502	2.499	2.498	2.500	2.502	2.504	2.505	2.504
6.0		2.502	2.488	2.481	2.477	2.477	2.479	2.482	2.483	2.484	2.483
6.5			2.475	2.467	2.463	2.464	2.466	2.468	2.470	2.471	2.470
7.0				2.460	2.457	2.456	2.459	2.461	2.463	2.464	2.463
7.5					2.454	2.453	2.456	2.459	2.460	2.461	2.461
8.0						2.453	2.456	2.459	2.460	2.461	2.460
9.0							2.459	2.462	2.463	2.463	2.463
10.0								2.464	2.466	2.466	2.466
11.0									2.467	2.468	2.467
12.0										2.468	2.468
∞											2.467

$$\psi(x - \xi, y - \eta) = Ae^{-B\sqrt{(x-\xi)^2+(y-\eta)^2}}$$

and

$$\Psi(u,v) = \frac{A}{2\pi} \int_{-\infty}^{\infty}\int_{-\infty}^{\infty} e^{iux+ivy}e^{-B\sqrt{x^2+y^2}}\,dx\,dy = \frac{AB}{(B^2 + u^2 + v^2)^{3/2}}$$

and the deflection of the plate is given by the formula

$$w = \frac{T}{4\pi^2} \int_{-\infty}^{\infty}\int_{-\infty}^{\infty} \frac{e^{i(a-x)u+i(b-y)v}}{D(u^2+v^2)^2 + \frac{1}{2\pi AB}(B^2+u^2+v^2)^{3/2}} \frac{\sin l_1 u}{l_1 u}\frac{\sin l_2 v}{l_2 v}\,du\,dv$$

$$= \frac{T}{\pi^2} \int_0^{\infty}\int_0^{\infty} \frac{\cos(a-x)u \cos(b-y)v}{D(u^2+v^2)^2 + \frac{1}{2\pi AB}(B^2+u^2+v^2)^{3/2}} \frac{\sin l_1 u}{l_1 u}\frac{\sin l_2 v}{l_2 v}\,du\,dv$$

BIBLIOGRAPHY

1. Almansi, E., "Sulle Deformazioni delle Piastre Elastiche," *Atti Accad. Nazl. Lincei, Rend.*, Serie 6, **Vol. 16**, 1932: Nota I, pp. 473–477; Nota II pp. 597–603; Also **Vol. 18**, 1933: Nota III, pp. 12–19; Nota IV, pp. 197–203; Nota V, pp. 425–431; Nota VI, pp. 776–783; Nota VII, pp. 1031–1038, Vol. 17, 1933: Nota VIII, pp. 3–9; Nota IX, pp. 77–82.

2. Bassali, W. A., and F. R. Barsoum, "The Transverse Flexure of Uniformly Loaded Curvilinear and Rectilinear Polygonal Plates," *Proc. Cambridge Phil. Soc.*, **Vol. 62**, 1966, p. 523.

3. Biezeno, C. B., and R. Grammel, "Engineering Dynamics," **Vol. I**, "Theory of Elasticity," Blackie & Son, Ltd., Glasgow, Scotland, 1955.

4. Boggio, T., "Sull' Equilibrio delle Piastre Elastiche Piane," *Rend. Ist. Lombardo*, Serie 2, **Vol. 34**, 1901, pp. 793–808.

5. ——, "Sull' Equilibrio delle Piastre Elastiche Incastrate," *Atti Accad. Nazl. Lincei, Rend.*, Serie 5, **Vol. 10**, 1901, pp. 197–205.

6. ——, "Sulla Deformazione delle Piastre Elastiche Soggette al Calore," *Atti Accad. Torino*, **Vol. 40**, 1904–1905, pp. 219–240.

7. ——, "Sulla Deformazione delle Piastre Elastiche Cilindriche di Grossezza Qualunque," *Atti Accad. Nazl. Lincei, Rend.*, Serie 5, **Vol. 13**, 1904, pp. 419–427.

8. Bolle, L., "Contribution au Problème Lineaire de Flexion d'une Plaque Elastique," *Bull. Tech. Suisse Romande*, **No. 21**, p. 281; **No. 22**, p. 293, 1947.

9. Boussinesq, J. V., "Applications des Potentiels," Paris, France, 1885.

10. Cerruti, V., "Sulla Deformazione di uno Strato Isotropo Indefinito Limitato da Due Piani Paralleli," *Atti Accad. Nazl. Lincei, Rend.*, Serie 4, **No. 1**, 1884–85, pp. 521–522.

11. De L'Hortet, "Plaque Circulaire Reposant sur un Sol Compressible," *Ann. Tech. Aviation Civ.*, **No. 3**, 1948.

12. Donnell, L. H., "The Effect of Transverse Shear Deformation on the Bending of Elastic Plates," *J. Appl. Mech.*, 1946, p. 249.

13. Föppl, A., and L. Föppl, "Drang und Zwang," Oldenbourg, Munich, Germany, 1924.
14. Fung, Y. C., "Foundations of Solid Mechanics," Prentice-Hall, Inc., Englewood Cliffs, N.J., 1965.
15. Galerkin, B. G., "Berechnung der frei gelagerten elliptischen Platte auf Biegung," *Z. Angew. Math. Mech.*, 1923.
16. Grashof, H., "Theorie der Elastizität," Berlin, Germany, 1878.
17. Green, A. E., "On Reissner's Theory of Bending of Elastic Plates," *Quart. Appl. Math.*, Vol. 7, No. 2, 1949.
18. Hadamard, J., "Mémoire sur le problème d'analyse relative à l'equilibre de la plaque élastique encatrée," *Mem. Acad. Sci.*, T. 33, No. 4, 1908.
19. Hencky, H., "Über die Berucksichtigung der Schubverzenung in ebenen Platten," *Ing. Arch.*, B. 16, 1947, pp. 72–76.
20. ———, "Neuere Verfahren in der Festigkeitslehre," Oldenburg, Munich, Germany, 1951.
21. Hertz, H., "On the Equilibrium of Floating Elastic Plates," *Wiedermann's Ann.*, Vol. 22, 1884, p. 449.
22. Hetenyi, M., "Beams on Elastic Foundations," University of Michigan Press, Ann Arbor, Mich., 1946.
23. ———, "Beams and Plates on Elastic Foundations and Related Problems," *Appl. Mech. Rev.*, Vol. 19, No. 2, pp. 95–102.
24. Holl, D. L., "Analysis of Plates, Examples by Different Methods," *J. Appl. Mech.*, 1936, p. 81.
25. Jodi, C., "Procedimenti per Ricerche Sperimentali su Suoli Elastici," *Atti Accad. Nazl. Lincei, Rend., Serie VI*, 1936.
26. Kambo, L., "Le Lastre Piane," Bardi, Rome, Italy, 1944.
27. Kirchhoff, G., "Über das Gleichgewicht und die Bewegung einer elastischen Scheibe," *J. Math. (Crelle)*, 1850, p. 51.
28. ———, "Vorlesungen uber Mathematische Physik," *Mechanik*, 1877.
29. Kromm, A., "Verallgemeinerte Theorie der Plattenstatik," *Ing. Arch.*, B. 21, 1953, pp. 266–286.
30. ———, "Über die Randquerkrafte bei gestutzten Platten," *Z. Angew. Math. Mech.*, B. 35, H. 6, 1955, pp. 231–242.
31. Lauricella, G., "Sulle Equazioni delle Deformazioni delle Piastre Elastiche Cilindriche," *Atti Accad. Nazl. Lincei, Rend., Serie 5*, Vol. 14, 1905, pp. 605–612.
32. ———, "Sur l'Integration de l'équation relative à l'équilibre des plaques élastiques encastrées," *Compt. Rend.*, 1909.
33. L'Hermite, R., "Résistance des Matériaux. Théorique et Expérimentale," T. 1, Paris, Dunod, France, 1954.
34. Levy, M., "Sur l'Equilibre d'une plaque rectangulaire," *Compt. Rend.*, Vol. 129, 1899, pp. 535–539.
35. Lheureux, P., "Calcul des plaques minces rectangulaires," Gauthier-Villars, Paris, France, 1949.

36. Lorenz, H., "Technische Elastizitätslehre," Oldenbourg, Munich, Germany, 1913.
37. Love, A. E. H., "A Treatise of the Mathematical Theory of Elasticity," Cambridge University Press, London, England, 1934.
38. Marcus, H., "Die vereinfachte Berechnung biegsamer Platten," Julius Springer, Berlin, Germany, 1929.
39. Mesnager, A., "Cours de beton armé," Dunod, Paris, France, 1921.
40. Nadai, A., "Elastische Platten," Julius Springer, Berlin, Germany, 1925.
41. Navier, L. M. H., "Recherches sur la flexion des plans élastiques," *Bul. Soc. Philomatique*, 1823, p. 92.
42. Olson, F. G. W., "Deflection of Uniformly Loaded Circular Plates," *J. Appl. Mech.*, 1943, p. 181.
43. Pigeaud, G., "Résistance des Matériaux," Paris, France, 1934.
44. Prescott, J., "Applied Elasticity," Longmans, Green & Company, Ltd., London, England, 1924.
45. Reissner, E., "On the Theory of Bending of Elastic Plates," *J. Math. Phys.*, **Vol. 23**, 1944, p. 184.
46. ———, "The Effects of Transverse Shear Deformation on the Bending of Elastic Plates," *J. Appl. Mech.* **Vol. 12**, 1945, p. A.68.
47. ———, "On Bending of Elastic Plates," *Quart. Appl. Mech.*, **Vol. 5**, 1947, p. 55.
48. ———, "Stresses in Elastic Plates over Flexible Subgrades," *Proc. Am. Soc. Civil Engrs.*, **Vol. 81**, Separate No. 690, 1955.
49. Ritz, W., "Oeuvres," Gauthier-Villars, Paris, France, 1911.
50. Salerno, V. L., and M. A. Goldberg, "Effect of Shear Deformations on the Bending of Rectangular Plates," *J. Appl. Mech.*, **Vol. 27**, 1960, pp. 54–58.
51. Salvati, M., "Il Calcolo della Lastra piana rettangolare," Bari, Accolti, 1936.
52. Schafer, M., "Über eine Verfeinerung der Klassischen Theorie dunner schwach gebogener Platten," *Z. Angew. Math. Mech.*, **B. 32, H. 6**, 1952, pp. 161–171.
53. Sen, B., "Note on the Bending of Thin Uniformly Loaded Plates Bounded by Cardioids, Lemniscates, and Certain Other Quartic Curves," *Phil. Mag.*, **Vol. 33**, 1942, p. 294.
54. Seth, B. R., "Bending of Clamped Rectilinear Plates," *Phil. Mag.*, **Vol. 38**, 1947, p. 292.
55. Stechkewicz, C., "Sul Calcolo della Deformazione della Piastra su Suolo Elastico," *Atti Accad. Nazl. Lincei, Rend.*, **Vol. V.**, 1948, pp. 339–344.
56. Tedone, O., "Sulla Deformazione delle Piastre di Grossezza Finita," *Atti Accad. Nazl. Lincei, Rend.*, Serie 5, **Vol. 10**, 1901, pp. 131–137.
57. ———, "Sull' Equilibrio di una Piastra Elastica, Isotropa, Indefinita," *Rend. Circolo Mat.*, **T. 18**, 1904, pp. 368–385.
58. Thomson, W., and P. G. Tait, "Treatise on Natural Philosophy," Cambridge University Press, London, England, 1879–1883.
59. Timoshenko, S. P., "History of Strength of Materials," McGraw-Hill Book Company, New York, N.Y., 1951.

60. Timoshenko, S. P., and S. Woinowsky-Krieger, "Theory of Plates and Shells," McGraw-Hill, Book Company, New York, N.Y., 1959.

61. Todhunter, I., and K. Pearson, "A History of the Theory of Elasticity and of Strength of Materials," Cambridge University Press, **Vol. 1**, 1886; **Vol. 2**, 1893. Reprinted by Dover Publications, Inc., New York, N.Y., 1960.

62. Trumpler, W. E., "Design Data for Flat Circular Plates with Central Hole," *J. Appl. Mech.*, 1943, p. 173.

63. Volterra, E., "Sul Problema Generale della Piastra Poggiata su Suolo Elastico," *Atti Accad. Nazl. Lincei, Rend.*, Serie 8, **Vol. II**, 1947, pp. 595–598.

64. ———, "Influenza del Taglio nella Dinamica e nella Statica delle Piastre Sottili," Nota 1 e 2, *Atti Accad. Nazl. Lincei, Rend.*, Serie 8, **Vol. 18, Fasc. 6**; also **Vol. 19, Fasc. 1, 2**, 1960.

65. ———, "Effect of Shear Deformations on the Bending of Rectangular Plates," *J. Appl. Mech.*, **Vol. 27**, 1960, pp. 594–596; discussion on paper of same title by V. L. Salerno and M. A. Goldberg. (See No. 50 of this bibliography.)

66. Weinstein, A., and D. H. Rock, "On the Bending of a Clamped Plate," *Quart. Appl. Math.*, **Vol. 2**, 1944, p. 262.

67. Westergaard, H. M., "New Formulas for Stresses in Concrete Pavements of Airfields," *Trans. Am. Soc. Civil Engrs.*, **Vol. 113**, 1948, pp. 425–444.

68. Winkler, E., "Die Lehre von der Elastizität und Festigkeit," Dominicus, Prague, Czechoslovakia, 1867.

69. Young, D., "Analysis of Clamped Rectangular Plates" *J. Appl. Mech.*, 1943, p. 139.

70. Zanaboni, O., "The Built-in Rectangular Plate," *Meccanica, J. Ital. Assoc. Theoret. Appl. Mech.*, **Vol. 1**, 1966, pp. 76–94.

PROBLEMS

8-1 to 8-12. For the circular plates shown in Figs. P.8.1, P.8.2, P.8.3, P.8.4, P.8.5, P.8.6, P.8.7, P.8.8, P.8.9, P.8.10, P.8.11, P.8.12, determine the equation of the elastic surface, the maximum deflection, the stresses, and the reactions on the supports. The plates are assumed to be symmetrically loaded with respect to their centers by uniformly distributed loads p_0 per unit area or by P per unit of length.

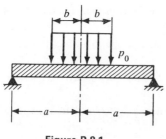

Figure P.8.1

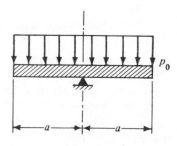

Figure P.8.2

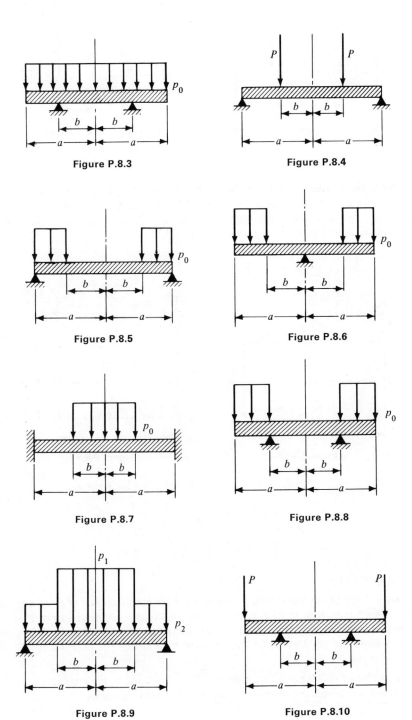

Figure P.8.3

Figure P.8.4

Figure P.8.5

Figure P.8.6

Figure P.8.7

Figure P.8.8

Figure P.8.9

Figure P.8.10

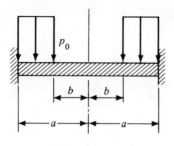

Figure P.8.11   Figure P.8.12

8-13. Determine, by applying Maurice Levy's method, the equation of the elastic surface, the maximum deflection, the stresses σ_x, σ_y, τ_{xy}, the shear forces, bending moments, and reactions at the supports for a rectangular plate, simply supported and subjected to a load $p = p_0(x/a)$ (see Fig. P.8.13).

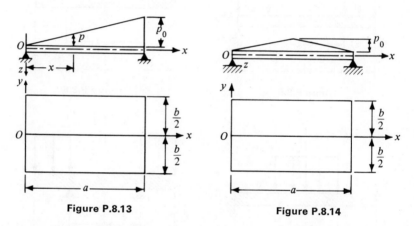

Figure P.8.13   Figure P.8.14

8-14. Solve Problem 8-13 for the case where the load is given by

$$P = \begin{cases} \dfrac{2p_0 x}{a} & \left(0 < x < \dfrac{a}{2}\right) \\ 2p_0\left(1 - \dfrac{x}{a}\right) & \left(\dfrac{a}{2} < x < a\right) \end{cases}$$

(See Fig. P.8.14.)

8-15. Establish an approximate equation for the elastic surface of a rectangular plate, which is built-in at its border and subjected to a concentrated force P at its center, by assuming that w is equal to the product of a constant and two polynomials conveniently chosen. Determine the constant by applying the principle of virtual displacements.

8-16. Solve Problem 8-15 by assuming that w is equal to the product of a constant and two simple trigonometric functions conveniently chosen.

8-17. Solve Problem 8-15 by assuming that w is equal to the sum of a limited number of terms each of which is the product of a constant and two trigonometric functions conveniently chosen. Compare the values for the maximum deflection of the plate given by solutions of Problems 8.15, 8.16, and 8.17.

8-18. Determine approximately the deflection of a rectangular plate of sides a and b ($b = 2a$) which is simply supported and subjected to a uniformly distributed load p_0. Solve by applying the method of finite differences. Apply the method assuming that the plate is divided respectively in sixteen, twenty-five, and thirty-six equal rectangles. Compare the results so obtained.

8-19. Solve Problem 8-18 for the case of a rectangular plate plate built-in at the edges.

8-20. Solve Problem 8-18 for the case of a rectangular plate with two opposite edges simply supported and the other two built-in.

AUTHOR INDEX

Page numbers followed by a single asterisk denote biographical material, followed by double asterisks denote in addition a picture of the author.

A

Abeles, P., 338
Adkins, A. W., 74
Airy, Sir George, 100, 101**, 120, 121, 134, 194, 196, 197
Albenga, G., 91
Alembert, J. B., Le Rond D', 19, 21**
Almansi, E., 33, 194, 252, 253, 499
Al-Rashid, N., x
Amaldi, U., 88
Andrews, E. S., 92
Anger, G., 336
Argyris, J. H., 87, 94
Arzelà, C., 252

B

Baker, Sir John, 89, 337, 338, 340
Barden, L., 424
Barrow, F. L. 377
Barsoum, F. R., 499
Bassali, W. A., 499
Beedle, L. S., 338
Beggs, G. E., 91
Bellavitis, G., 94
Belluzzi, O., 336
Beltrami, E., 5**, 6, 24, 25, 33, 34, 53, 54, 89
Benscoter, S. U. 336
Berio, A., 339
Bernoulli, Daniel, 143*, 145, 193, 215, 257, 269, 334, 431
Bernoulli, Jacob, 257, 335
Bernoulli, Johann, 143, 193
Bertrand, J. L. F., 95
Betti, E., 39, 37, 44, 62, 63, 64, 66 67**, 68, 70, 88, 91, 97
Biezeno, C. B., 34, 37, 88, 94, 194, 197, 353, 376, 499
Billevicz, V., 143, 197
Biot, M. A., 34, 424, 425
Bisplinghoff, R. L., 88
Bleich, F., 338
Bleich, H., 338
Block, V. I., 38
Boggio, T., 195, 252, 499
Boley, B. A., 36, 37, 38, 199
Bolle, L., 499
Bollinger, O. E., 336
Borchardt, C. W., 36
Boresi, A. P., 253

Borg, S. F., 336
Bouasse, H., 336
Boussart, R., 195
Boussinesq, M. J., 38, 424, 499
Bresler, B., 89
Bricas, M., 195
Bridgman, P. W., 51*, 90
Brown, E. H., 94
Brozzu, M., 339
Brungraber, R. S., 335
Budiansky, B., xi
Butty, E., 195

C

Campus, F., 339
Carnot, S. N. L., 71
Carother, S. D., 199
Carslaw, H. S., 36, 335, 340
Case, J., 90, 335
Castigliano, A. C., x, 44, 73, 75**, 76, 77, 79, 80, 81, 83, 84, 85, 86, 92, 94, 95
Cauchy, A. L., 4**, 5, 9, 10, 24, 34, 223
Ceradini, C., 88, 336
Cerruti, V., 38, 94, 252**, 499
Cesari, L., 336
Cesaro, E., 34, 38, 88, 252, 253
Chang, T. C., x
Charlton, T. M., 88, 93
Chung, R., 409
Clapeyron, B. P. E., 34, 38, 70*, 72, 73, 76, 77, 87, 88, 97, 250, 257, 304, 323
Clausius, R. J. E., 102
Clebsh, A., 34, 38, 103, 258
Coker, E. G., 195, 198
Colonnetti, G., 70*, 81, 88, 91, 93, 94, 253, 321, 338, 339
Conforto, F., 336
Cook, G., 90
Cornu, A., 216
Cotterill, J. H., 92, 94
Coulomb, C. A., 44, 47, 52, 205, 217*,

Coulomb, C. A. (*cont.*):
 223, 226, 229, 253, 254
Crandall, S. H., 335
Cremona, L., 252
Crotti, F., 92

D

Dahl, N. C., 335
Danusso, A., 338, 339
Den Hartog, J. P., 88, 93, 198, 254
De L'Hortet, 493, 499
Descartes, R., 101**
Diaz, J. B., 254
Dirichlet, P. G., 39, 327, 329
Domke, O., 94
Donati, L., 44, 80*, 93, 94
Donnell, L. H., xi, 499
Dorn, W. S., 5, 339
Dorna, A., 94
Drucker, D. C., 339
Duhamel, J. M., 5*, 6, 29, 36
Durelli, A. J., 195, 198
Dusterbehn F., 376

E

Eagle, A., 340
Eichinger, A., 91
Engesser, F., 74, 81, 93
Essenburg, F., 424
Eubanks, R. A., 38
Euler, L., 135, 192, 193**, 257, 269, 334, 335, 379, 424, 430, 476

F

Faraday, M., 67
Ferrari, C., x, 32
Filon, L. N. G., 195, 198
Finzi, B., 38
Finzi, L., 339
Flamant, A. A., 34, 38, 102*, 164, 166, 169, 170, 176, 179, 197, 253

AUTHOR INDEX

Flamard, E., 336
Flugge, W., 254
Föppl, A., 198, 254, 335, 500
Föppl, L., 88, 254, 500
Fourier, J., 257, 327, 328**, 329, 330, 331, 332, 333, 339
Franciosi, V., 340, 349, 351, 397, 495, 496
Fredholm, I., 38, 39, 196
Freudenthal, A. M., 339, 424
Freyssinet, H., 476
Fritsche, J., 339
Frocht, M. M., 195, 198
Fung, Y. C., 34, 36, 37, 38, 88, 252, 254, 424, 499

G

Gaj, A., x, 75
Galerkin, B. G., 38, 39, 499
Gatewood, B. E., 36
Gaulois, E., 67
Gauss, K. F., 55, 56**
Geiringer, H., 339
Gere, J. M., 383
Ghizzetti, A., ix
Giacchero, E., 339
Gibson, A. H., 376
Girtler, R., 88
Goldberg, M. A., 501
Golovin, H., 102, 140, 197, 198
Goodier, J. N., 36, 37, 38, 40, 89, 95, 197, 198, 199, 254, 255, 335
Goursat, E., 195
Graffi, D., 88
Grammel, R., 34, 37, 88, 94, 194, 197, 376, 499
Grandori, G., 339, 430
Grashof, F., 336, 488, 489, 490, 491, 499
Green, A. E., 34, 38, 88, 500
Green, G., 396
Greenberg, H. J., 81, 93, 339
Greenspan, M., 198, 205*, 248, 253, 255

Griffith, A. A., 90, 205*, 248
Grioli, G., 195
Griot, G., 336
Grossmann, G., 254
Guest, J. J., 44, 52, 90
Guidi, C., 88, 335
Guillemin, E. A., 336
Gunther, W., 255

H

Hadamard, J., 501
Haigh, B. P., 44, 53, 90
Hajnal-Konyi, K., 338
Helmholtz, H. Von, 102
Hencky, H., 44, 49, 51, 52, 53, 90, 500
Hendry, A. W., 430
Hertz, H., 102**, 177, 197, 198, 500
Hetenyi, M., 195, 254, 335, 425, 500
Heyman, J., 338, 339
Higgins, T. J., 254
Hill, R., 339
Hodge, P. G., Jr., 81, 93, 339, 340
Hoff, N. J., 36, 37, 40, 88, 94
Hoffman, O., 339
Hogan, B., 376
Holl, D. L., 500
Hooke, R., 3**, 5, 12, 34, 146, 155, 157, 263
Horne, M. R., 338
Horvay, G., 37
Huber, M. T., 44, 49, 51, 52, 53
Huth, J. H., 254

I

Iliouchine, A. A., 339
Inglis, Sir Charles, x, xi, 101**, 102, 159, 175, 177, 195, 198, 199, 337, 425
Irwin, G. R., 90

J

Jaeger, J. C., 36, 335
Jeffrey, G. B., 195

Jodi, C., 380, 381, 397, 400, 425, 497, 500
Jourawsky, D. J., 257, 258, 269, 334, 335

K

Kambo, L., 500
Kannenberg, B. G., 376
Karman, T. Von, 34, 90, 236, 425
Kelsey, S., 87, 94
Kelvin, Lord William, 36, 40, 95, 501
Kerr, A. D., 425
Ketter, R. L., 338
Kirchhoff, G., 32**, 33, 44, 60, 88, 431, 500
Kirsch, G., 102, 159, 197, 198
Klein, F., 195
Kleinlogel, A., 336
Koiter, W., 199
Kromm, A., 500
Krutkov, Y. A., 38
Krylov, A. N., 425

L

Lagrange, L., 429, 430**, 431, 435, 453, 461, 466, 492
Lamb, Sir Horace, 88
Lamé, G., 4**, 6, 13, 19, 20, 21, 34, 38, 70, 102, 145, 197, 198, 204, 205, 223, 251, 252, 335
Land, R., 70, 92
Langhaar, H. L., 88
Laplace, P. S., 21, 22**, 134, 191, 192, 194, 336
Lauricella, G., 35, 38, 39, 195, 252, 253, 500
Le Boiteux, H., 195
Lessells, J. M., 90
Levi, F., 339
Levi-Civita T., 54**, 88, 90, 195
Levy, M., 102*, 133, 197, 429, 460, 461, 464, 500
L'Hermite, R., 35, 36, 39, 88, 94, 195, 198, 254, 336, 500

Lheureux, P., 500
Ling, C. B., 254
Locatelli, P., 37
Lode, W., 90
Lorenz, H., 35, 39, 88, 94, 501
Lorsch, H. G., 424
Love, A. E. H., 35, 36, 39, 88, 196, 199**, 253, 376, 425, 501
Ludvik, P., 90

M

McGregor, C. W., 90
McLean, L., 89
Macaulay, W. H., 271, 335, 336
Macchi, G., 339
Maclaurin, C., 291, 292**, 295, 335, 479
Magnel, G., 335
Maier, A., 91
Malter, H., 425
Mar, J. W., 88
Marcolongo, R., 35, 39, 253, 254
Marcus, H., 353, 376, 377, 430, 476*, 478, 481, 501
Matheson, J. A. L., 92
Mathieu, E., 196
Mautner, K. W., 338
Maxwell, J. C., 44, 62, 67**, 68, 69, 70, 76, 77, 92, 102, 195, 196
Mayer, R., 377
Meier-Leibnitz, H., 339
Melan, E., 36, 316, 337, 339
Menabrea, L. F., 44, 75, 76**, 83, 84, 94, 95, 249
Mesnager, A., 197*, 501
Michell, A. G., 5
Michell, J. H., 5**, 6, 24, 25, 102, 163, 164, 196, 199, 203
Mikhlin, S. G., 196
Milne-Thomson, L. M., 196
Mindlin, R. D., 39, 254
Minelli, C., 336
Mises, R. von, 37, 44, 49, 51, 52, 53, 90

Mohr, O., 90, 100*, 104, 107, 108, 109, 110, 113, 117, 196, 200, 201, 202, 257, 300, 305, 321, 336, 483
Moorman, R. B., 377
Morera, G., 32**, 35, 39
Morkovin, D., 198
Morsch, E., 337
Muller, J., 197
Muller-Breslau, H., 89, 337
Murphy, G., 90
Murray, W. M., 198
Muskhelishvili, N. I., 196

N

Nadai, A., 89, 90, 91, 339, 429, 464*, 501
Naghdi, P. M., 37
Navier, L. M. H., 4*, 5, 19, 35, 143, 205, 215, 219, 223, 253, 254, 257, 258, 261, 263, 334, 335, 429, 431, 452, 464
Neal, B. G., 89, 95, 339
Neumann, F. E., 5*, 6, 29, 37, 225
Newton, Sir Isaac, 292, 430
Nowacki, W., 37, 199

O

Odquist, F. K. G., 339
Olson, F. C. W., 501
Olszak, W., 339
Oravas, G. A. E., 89
Osgood, W. R., 197

P

Padova, E., 254
Papkovitch, P. F., 39
Parkus, H., 36, 370
Pearson, K., 36, 40, 89, 198, 255, 502
Pérès, J., 35
Pestel, E., 254
Peterson, R. E., 91

Pfaff, J. F., 32
Phillips, E. A., 195, 198
Pian, T. H. H., 88
Picone, M., ix, 254
Pigeaud, G., 501
Piobert, G., 223
Pipes, L. A., 340
Pippard, A. J. S., 89, 95, 340, 353, 377
Pirlet, J., 89
Pister, K., 89
Pizzetti, G., 339
Poincaré, J. H., 35, 39
Poisson, S. D., 1, 2, 3**, 14, 35, 96, 216, 267, 431, 437, 438, 439, 443
Polya, G., 254
Poncelet, V. J., 223
Popov, E. P., 91, 335, 336
Poritsky, H., 196
Poschl, T., 95
Prager, W., 35, 81, 93, 337, 339, 340
Prandtl, L., 205, 235, 236**, 237, 239, 243, 244, 246, 253, 255
Prescott, J., 501

R

Ramsay, Sir William, 200
Rankine, W. J. M., 44, 52, 53**
Rayleigh, Lord John William, 200**
Raymondi, C., 425
Reissner, E., 93, 196, 501
Résal, J., 377
Ribière, M. C., 102, 140, 197, 198
Ricci, C. L., 92, 95
Ritchie, E. G., 376
Ritter, W., 337
Ritz, W., 205, 248, 249, 250, 253, 256, 429, 472, 473, 501
Rjanitsyn, A. R., 340
Roderick, J. W., 337, 338
Rodriguez, D. A., 425
Roš, M., 91
Routh, E. J., 67
Roy, M., 35

S

Sachs, G., 339
Sadowsky, M. A., 196, 198
Saint-Venant, Barré de, A. J. C., 4**, 6, 24, 31, 34, 37, 38, 44, 52, 102, 134, 205, 223, 225, 229, 235, 246, 253, 254, 255, 335, 353, 355, 358, 377, 401, 425
Salvadori, M. G., 254
Salvati, M., 501
Schaefer, H., 39
Schafer, M., 501
Schleicher, F., 377
Schmaus, F. T., xi
Scott, E. A., 338
Sechler, E. E., 35
Seely, F. B., 91
Seide, P., 425
Sen, B., 196, 501
Serret, J. A., 102
Seth, B. R., 501
Shanley, F. R., 89
Siebel, E., 91
Slobodiansky, M. G., 39
Smith, J. O., 91
Sneddon, I. N., 340
Sobrero, L., 196
Soderberg, C. R., 91
Sokolnikoff, E. S., 340
Sokolnikoff, I. S., 35, 40, 89, 93, 255, 340
Sokolowsky, W. W., 340
Somigliana, C., 32**, 35, 38, 40, 253
Sommerfeld, A., 89, 255
Sommerfeld, K. J., 35, 40, 338
Southwell, Sir Richard., 36, 37, 40, 89, 93, 95, 198, 253, 255
Squire, R. H., 338
Staack, J., 337
Stassi, D'Alia, F., 340
Steckiewicz, C., 495, 497, 501
Sternberg, E., 37, 38, 40, 198, 199
Stevenson, A. C., 196, 255
Strassner, A., 337
Straub, H., 335
Strutt, J. W. (see Rayleigh, Lord)
Stussi, E., 335, 339
Suter, E., 337

T

Tait, P. G., 36, 40, 501
Tate, M. B., 383
Taylor, B., 257, 291*, 295, 335, 344
Taylor, Sir Geoffrey, 91, 205, 248, 253, 255
Tedone, O., 36, 40, 197, 253, 501
Theodorescu, P. P., 197
Thomson, Sir William (see Kelvin, Lord)
Thomson, W. T., 336
Thurliman, B., 338
Timoshenko, S. P., 36, 37, 40, 81, 89, 95, 140, 197, 198, 199, 223, 255, 335, 337, 377, 493, 501, 502
Timpe, A., 197, 199
Todhunter, I., 36, 40, 89, 255, 502
Tresca, H., 91
Tricomi, F. G., x, 76
Trumpler, W. E., 502
Tsao, C. H., 195, 198

V

Vallette, R., 337
Van Den Brock, J. A., 340
Velutini, B., 377
Venkatraman, B., 339
Venske, O., 197
Voigt, W., 36
Volta, A., 32
Volterra, Vito, 3, 4, 5, 21, 22, 32, 54, 56, 67, 70, 92, 93, 101, 102, 193, 200, 236, 252, 292, 328, 340, 430

W

Wang, C. T., 36, 40, 89, 93, 95, 197, 255
Watson, G. N., 341
Weber, C., 40, 255
Weiner, J. H., 36, 199
Weingarten, J. L., 89
Weinstein, A., 254, 255, 502
Westergaard, H. M., 36, 40, 81, 89, 91, 94, 95, 197, 253, 337, 340, 502
Whittaker, Sir Edmund, 341
Wieghardt, K., 195
Williams, D., 95
Winkler, E., 143, 199, 379, 380, 381, 382, 384, 385, 386, 395, 396, 397, 399, 430, 492, 497, 502
Winter, G., 337, 340
Woinowsky-Krieger, S., 429, 466, 502

Y

Young, Dana, 502
Young, David M., ix
Young, D. H., 89, 95, 337
Young, T., 1, 3**, 431

Z

Zachmanoglou, E. C., 191, 292, 337, 340, 430
Zanaboni, O., 37, 426, 502
Zerna, W., 34, 38, 88
Zhudin, N. H., 340

SUBJECT INDEX

A

Airy stress function:
  cartesian coordinates, 120–134
  polar coordinates, 134–150
  use of polynomials, 121–134
Analogy, (see Prandtl's membrane analogy of torsion problem)
Angle of twist, 219–227, 235, 241, 250, 357
Application of finite difference equations to the bending of plates, 476–488
Application of trigonometric series to the study of:
  circular beams on elastic foundations, 417–424
  plates, 467–469, 492–495
  straight beams, 311–314
Arches bent out of the plane of initial curvature, 353–378
Average physical properties of common metals, 15

B

Bars (see Arches, Curved beams, Torsion)
Basic problems of theory of elasticity, 33

Beam deflections:
  by direct integration, 270–271
  Mohr's conjugate beam method, 300–303
  for straight beams on elastic foundations, 382–384
  by use of Maclaurin's and Taylor's series, 291–300
  by use of singularity functions, 271–291
  by use of trigonometric series, 311–314
Beams:
  bending of straight, 257–351
  curvature of, 261, 356
  curved, bent out of its initial plane, 353–378
  deflections of, 270–300
  normal stress in, 263
  shearing stress in, 264
  twist of, 219–227, 235, 241, 357
Beams, wedge-shaped (see Wedges)
Beams on elastic foundations, 379–427
  of finite length 390–396
  of infinite length, 384–390
Beltrami-Haigh strength theory, 53
Beltrami-Michell's compatibility equations, 25
Bending moment:
  diagram of, 303, 307, 309

Bending moment (*cont*):
  equation of, 258–261
  relation to curvature, 263
Bending of:
  beams, 257–324
    differential equations for deflections, 269–270
    differential equations of equilibrium, 258–261
    elastic behavior, 257–314
    elastic-plastic behavior, 314–324
    flexure formula, 261–263
    shear stress formula, 264–265
  circular arc bow girder, 363–376
  circular bar in pure bending, 140–145
  circular beam on an elastic foundation, 401–416
  curved beam out of its initial plane, 353–377
  plates, 429–505
  prismatic bars, 212–217
Bernoulli-Euler theory of deflection of beams, 269–270
Betti's reciprocity theorem, 61–66
Body forces, 2
Boundary conditions:
  in flexure problems, 271
  in plane strain problems, 112
  in plane stress problems, 104
  in three dimensional elasticity, 22–29
  in torsion problems, 222, 224–225, 230, 237–238, 241

C

Castigliano's first theorem, 73–80
  second theorem, 83–87
Cauchy's equations, 9
Circle of strain 113–120
  of stress, 104–110
Circular arc bent outside plane of initial curvature, 353–378
Circular bar, bending of, 140–145

Circular beam on elastic foundation, 417–424
Circular cylinder, torsion of, 223, 226
Circular disk, 153–157
Circular hole, stresses around a, 156–157, 159–163
Circular plates, deflection of, 435–449
Clapeyron's theorem, 70–73
Clapeyron's three moment equation, 304–311
Coefficient of thermal expansion, 29
Coefficients, elastic, 14, 15
Colonnetti reciprocity theorem, 70
Compatibility equations, 24–27, 103, 112
Complementary energy, 74, 81–82
Components of displacement, 9
Components of strain, 11, 136
Components of stress, 8, 135
Composite tube, 150
Compressibility of the fluid, 14, 103
Concentrated load acting on the free surface of a plate, 169–175
Concentrated load acting on the vertex of a wedge, 163–169
Concentration of stresses, 153–163
Conjugate beam method, 300–304
Constants, elastic (*see* Coefficients, elastic), 14
Constrained circular beam resting on an elastic foundation, 409–416
Continuous beams, 304
Coulomb-Guest's strength theory, 52
Coulomb's theory of torsion of circular shafts, 217–219
Curvature in bending of beams, 261, 317, 356
Curves, elastic, 269
Cylinder, hollow, 145–150, 158

D

D'Alembert's principle, 19
Dam, 111

SUBJECT INDEX

Deflections of:
  beams on elastic foundations, 382–424
  beams with circular axes, 355–376, 401–499
  plates, 430–499
  straight beams, 269–286, 291–304
  variable cross section beams, 286–291
Deflections of elastic beams:
  according to the Bernoulli-Euler theory, 269–270
  Macaulay's use of singularity functions, 271–291
  Mohr's conjugate beam method, 300–303
  Taylor's series, 291–300
  trigonometric series, 311–314
  solution by direct integration, 270–271
Deflection of plates due to shear, 439–440
Density, table of, 15
Diagrams:
  of bending moments, 303, 307, 309
  of shear forces, 303, 307, 309
  of tensile test, 2
Differential equations for deflections:
  of beams, 269–270
  of plates:
    in polar coordinates, 435–449
    in rectangular coordinates, 430–435
Differential equations of equilibrium:
  for curved beams, 258–259
  for plane strain, 111
  for plane stress, 103
  for straight beams, 258–261
  for three dimensions, 19–24
Disks:
  rotating, 153–157
  subjected to two opposite concentrated forces, 177–186
Dislocations, 324

Displacements, components of:
  in polar coordinates, 136–137
  in rectangular coordinates, 9
Donati's extension of Castigliano's theorem, 80
Duhamel-Neumann equations, 29

E

Elastic constants for isotropic materials, 14, 15
Elastic foundations, 379–427, 492–499
Elasticity, general equations of, 9–29
Elastic moduli (*see* Modulus)
Elastic-plastic bending of beams, 314–324
Energy:
  complementary, 74–75, 81–82
  distortion, 50–51
  methods, 44–99
  strain, 45–49
Engesser's theorem, 81–82
Equations:
  Beltrami-Michell's, 25–26
  Cauchy's 9, 24
  Compatibility, 24
  Duhamel-Neumann's 29–31
  Lame's 20–21
  Navier's, 19, 20
  Saint-Venant's, 24
Equations for deflection of elastic beams according to Bernoulli-Euler theory, 269–270
Equations of equilibrium, 19–24, 30, 103, 111
Equations of motion, 20, 21
Equilibrium equations, 19–24, 30, 103, 111
Equipollent load-systems (*see* Saint-Venant's principle)
Euler's equation, 135–136, 192–194
Expression of the Laplacian operator in polar coordinates, 191–192
External forces, 2

F

Failure of materials (*see* Strength theories)
Finite difference equations 476–488
Flamant's problem, 169–175
Flexure:
  approximate theory, 258–270
  Saint-Venant's theory, 355–359
Flexure formula for bent beam, 263
Forces, classification of:
  body forces, 2
  external forces, 2
  internal forces, 5
  surface forces, 2
Foundation, elastic, resting on:
  beam of finite length, 390–396
  beam of infinite length, 384–390
  circular beam, 401–424
  circular plate, 493–495
  constrained circular beam, 409–416
  rectangular plate, 492–493
Fourier's series, applied to:
  curved beams on elastic foundation, 417–424
  plates, 452–460, 467–469, 492–493
  straight beams, 311–314
Fourier's series expansion, 327–334

G

General equations of elasticity, 9–29
Generalized Hooke's law, 12–19
General problem of plate resting on an elastic foundation, 495–499
General problem of straight beam resting on an elastic foundation, 396–401
Golovin-Ribière problem, 140–145
Guns, 150
Green function, 402

H

Harmonic analysis of circular beams resting on elastic foundations, 417–424

Hertz's problem, 177–186
Hole:
  in plate, 159–163
  in rotating cylinder, 158
  in rotating disk, 157
Homogeneous, 1
Hooke's law:
  generalized, 12
  for shear, 16
  simple, 2
Huber-Von Mises-Hencky strength theory, 49–51

I

Indirect work, 63
Infinite plate with circular hole, 159–163
Influence function, 386
Inglis problem, 175–177
Integration of Euler's equation, 192–194
Internal forces, 5
Internal pressure, stresses due to:
  in thick shells, 145–150, 205–212
  in thick tubes 145–150
  in thin shells, 149, 211
  in thin tubes 149
Isotropic, 1
Isotropic materials, elastic constants of, 14, 15

J

Jourawsky's shearing stress formula, 264–268

K

Kirchhoff's theorem of uniqueness, 60–62
Kirsch's problem, 159–163

SUBJECT INDEX

L

Lagrange's equation of equilibrium for thin plates, 430–435
Lamé's:
  constant, 13
  equations of equilibrium, 21
  equations of motion, 21
  problem of thick hollow cylinder, 145–150
  problem of thick spherical shell, 205–212
Land-Colonnetti's reciprocity theorem, 70
Laplacian operator:
  in polar coordinates, 134
  in rectangular coordinates, 21
Lateral contraction, 2
Law of reciprocity:
  Betti's, 63–67
  Land-Colonnetti's, 70
  for shearing stresses, 8
  Volterra's, 70
Least work, theorem of, 75, 84
Levi-Civita's strength theory, 54
Levy's problems, 133–134
Limit design of beams, 324–327
Limit, proportional, 2
Linearly hardening materials, 316

M

Macaulay's use of singularity functions for studying deflections of beams, 271–286
Maxwell's reciprocity theorem, 67–70
Membrane analogy for pure torsion, 243–248
Menabrea's theorem, 83–87
Michell's problem, 163–169
Modulus:
  of compressibility, 13–14
  of rigidity, 13
  of Young, 1

Mohr's circle:
  for strain, 113–120
  for stress, 104–110
Mohr's conjugate beam method, 300–304
Moment acting on the vertex of a wedge, 175–177
Mutual work, 63

N

Navier's equations of equilibrium, 19
  equations of motion, 20
  flexure formula, 261–263
  theory of torsion, 219–223
Neutral axis, 261
Normal stress, 6

P

Perfectly plastic material, 316
Physical properties of common metals, 15
Planes:
  of maximum shearing stress, 108
  principal, 107, 117
Plane strain, 110–113
Plane stress, 102–104
Plastic curvature, 317
Plastic design of beams, 324–327
Plate, differential equations for deflection of, 430–435
Plates:
  application to bending of:
    finite difference equations, 476–488
    Grashof's method, 488–492
    Marcus's method, 476–492
    Ritz's method, 472–476
  circular, deflection of:
    built-in at edge and subjected to concentrated load at its center, 441–443
    built-in at edge and subjected to uniformly distributed load, 435–437

Plates (*cont.*):
  circular, deflection of (*cont.*):
    simply-supported at edge and subjected to concentrated load at its center, 443–444
    simply-supported at edge and subjected to uniformly distributed load, 437–439
    with a concentric circular hole bent by couples M_1 and M_2 uniformly distributed along edge, and supported at outer edge, 444–445
    with a concentric circular hole bent by uniformly distributed force applied at inner edge and simply-supported at outer edge, 446
  on elastic foundations:
    circular plates, 493–495
    general problem, 495–499
    rectangular plates, 492–493
  elliptic plate built-in at edge and subjected to uniformly distributed load, 449–452
  other cases of bent circular plates, 446–449
  rectangular plates, 452–464
    application of principle of virtual work:
      approximate solution, 469–472
      rigorous solution, 467–469
    Maurice Levy's solution, 460–464
    Navier's solution, 452–460
  triangular plate:
    Nadai's solution, 464–465
    Woinowsky-Krieger's solution 466
Plate with circular hole, 159–163
Poisson's number, 2
Poisson's ratio, 2
  experimental determination of, 216
  limits of, 14
Polar coordinates in two dimensional problems, 134–140
Polynomials, airy function, 121–133

Prandtl's membrane analogy of torsion problems, 243–248
Prandtl's theory of torsion, 235–242
Principal planes:
  for strain, 117
  for stress, 107
Principle:
  complementary energy, 81
  least work, 84
  Saint-Venant's, 31
  virtual work, 54–60
Prismatic bar in pure bending, 261
Problem of:
  Flamant, 169–175
  Golovin-Ribière, 140–145
  Hertz, 177–186
  Inglis, 175–177
  Kirsch, 159–163
  Lamé (thick cylinder), 145–150
  Lamé (thick spherical shell), 205–212
  Maurice Levy, 133–134
  Michell, 163–169
Proportional limit, 2

R

Rankine's strength theory, 52
Reciprocity, theorems of:
  Betti, 63
  Land-Colonnetti, 70
  Maxwell, 67
  Volterra, 70
Reciprocity laws for shearing stresses, 8
Rectangular plate, 452–464, 467–476, 481–493
Rectangular wall subjected to hydrostatic pressure, 133–134
Retaining wall, 133
Rigidity, modulus of, 133
Ritz's method:
  applied to plates, 472–476
  applied to torsion problems, 248–251

SUBJECT INDEX

Rotating disks and cylinders, 153–158
Rupture (*see* Strength theories)

S

Saint-Venant's:
  compatibility equations, 24
  equation for curved beams bent out of the plane of initial curvature, 355–359
  principle, 31–33
  semi-inverse method for solving torsion problems, 223–235
  strength theory, 52
Semi-inverse method, 223–235
Shear deflections of circular plates, 439–440
Shearing stress formula for bent beams, 264–268
Shear modulus, 13
Shear strain, 10
Shear stress, 6
Shell, spherical, 205–212
Shrink fits, 150–153
Simple Hooke's law, 2
Singularity functions used for study of deflection of beams, 271–291
Soap film (*see* Membrane analogy)
Solution of two-dimensional elasticity problems by use of polynomials of:
  second degree, 122
  third degree, 122
  fourth degree, 123
  fifth degree, 123
  sixth degree, 124
Spherical shell under uniform internal and external pressures, 205–212
Statical equivalence (*see* Saint-Venant's principle)
Strain, 9–11
  normal, 9
  shear, 10

Strain (*cont.*):
  sign conventions for, 11
Strain, components of:
  in polar coordinates, 136
  in rectangular coordinates, 11
Strain, Mohr's circle for, 113–118
Strain-displacement relations:
  in polar coordinates, 136–138
  in rectangular coordinates, 7–11
Strain energy, 45–49
Strain measurement, 118
Strain rosettes, 118–120
  equiangular or 60°, 119
  rectangular or 45°, 119
Strain-stress relations:
  in polar coordinates, 138
  in rectangular coordinates, 2, 12–19
Strain tensor:
  for plane strain, 110
  for three dimensions, 11
Strength theories:
  Beltrami-Haigh's, 53
  Coulomb-Guest's, 52
  Huber-Von Mises-Hencky's, 49
  Levi-Civita's, 54
  Rankine's, 52
  Saint-Venant's, 52
Stress, 6–9
  normal, 6
  plane, 102–104
  shear, 6
  sign convention, 7, 104
Stress, components:
  in polar coordinates, 135
  in rectangular coordinates, 8
Stress, Mohr's circle for, 105–110
Stress concentration due to a circular hole in a stressed plate, 159–163
Stress equations:
  of equilibrium, 19
  of motion, 20
Stresses, thermal, 29–31
Stress function, 120–121, 237
Stress-strain relations, 2, 12–19

Stress tensor:
    for plane stress, 103
    for three dimensions, 8
Surface forces, 2

T

Taylor and Griffith's applications of Prandtl's membrane analogy, 248
Taylor's series used for study of deflection of beams, 291–300
Temperature stresses (see Thermal stresses)
Tensile test diagram, 1
Theorem of:
    Castigliano, first, 73–80
    Castigliano, second, 83–87
    Clapeyron, 70–73
    Engesser, 81–82
    Kirchhoff, 60–62
    Menabrea, 83–87
    reciprocity (see Reciprocity theorems)
Theories of failure (see Strength theories)
Thermal expansion, coefficient of, 29
Thermal stresses, 29–31
Thick tube subjected to external and internal distributed pressures 145–150
Three basic problems in theory of elasticity, 33
Three dimensional elasticity, 204–256
Three moment equation, 304–311
Torsion, theories of:
    Coulomb, 217–219
    Navier, 219–223
    Prandtl, 235–242
    Saint-Venant, 223–235
Torsional rigidity, 225
Torsion of:
    circular shaft, 217–219, 226
    elliptic shaft, 226–229, 239–241
    rectangular shaft, 229–235, 246–247, 250–251

Torsion of (cont.):
    triangular shaft, 241–242
Triangular and rectangular walls subjected to hydrostatic pressures, 133–134
Triangular plate, 470–472
Triangular shaft, 241–242
Trigonometric series (see Fourier's series)
Tubes:
    thick walled, 145–150
    thin walled, 149
Tunnel, 111
Twist of a rod (see Torsion)
Two-dimensional elasticity, 100–194
Two-dimensional thermal stresses, 186–190

U

Ultimate strength, 314
Uniqueness of solution, 60–62
Unit doublet function, 278
Unit impulse function, 279
Unit ramp function, 273
Unit step function, 273

V

Variable cross section beams, 286–291
Virtual work, 54–60
Volterra's reciprocity theorems, 70
Volume expansion, 12

W

Water tanks, 387
Wedges, 163–169, 175–177
Winkler's hypothesis, 379, 382–384

Y

Yield stress, 314
Young's modulus, 1